Mathematical problems of classical nonlinear electromagnetic theory

Pitman Monographs and
Surveys in Pure and Applied Mathematics 63

Mathematical problems of classical nonlinear electromagnetic theory

Frederick Bloom
Northern Illinois University

Longman Scientific & Technical
Copublished in the United States with
John Wiley & Sons, Inc., New York

Longman Scientific & Technical
Longman Group UK Limited
Longman House, Burnt Mill, Harlow
Essex CM20 2JE, England
and Associated companies throughout the world.

Copublished in the United States with
John Wiley & Sons Inc., 605 Third Avenue, New York, NY 10158

First published 1993

AMS Subject Classification: 35, 78

ISSN 0269-3666

ISBN 0 582 21021 6

British Library Cataloguing in Publication Data

A catalogue record for this book is
available from the British Library

Library of Congress Cataloging-in-Publication Data

Bloom, Frederick, 1944–
Mathematical problems of classical nonlinear electromagnetic theory / F. Bloom.
p. cm. -- (Pitman monographs and surveys in pure and applied mathematics, ISSN 0269-3666)
ISBN 0-582-21021-6
1. Electromagnetic theory. 2. Nonlinear theories. I. Title.
II. Series.
QC670.B58 1993
530.1'41--dc20 92-31478
CIP

Printed and bound by Bookcraft (Bath) Ltd

This book is dedicated

to the

memory of my Brother

DONALD NORMAN MARTIN

née Bloom

(1933–1989)

and

to the

memory of my Mother

JEANETTE BLOOM

(1909–1991)

Contents

Preface

This book is a survey of a collection of problems arising in classical nonlinear electromagnetic theory with which the author has been involved for more than twelve years. The subject matter treated here has its origins in one of three basic subareas of nonlinear electromagnetic theory, namely, the propagation of electromagnetic waves in nonlinear dielectric materials, the evolution of the charge and current distributions in a nonlinear transmission line, and nonlocal problems in electromagnetic theory which depend, for their formulation, on the Biot–Savart Law. During the past several decades, it would seem that only two surveys have appeared in book form which touch on the broad range of subject matter covered here: the volume by I. G. Kataev on electromagnetic shock waves, which was published in 1966, and the monograph by A. C. Scott, whose focus was on wave propagation in transmission lines and which appeared in 1980; almost all of the results presented in the current volume have appeared since the publication of the two aforementioned surveys. It is hoped that this monograph will not only serve to introduce the reader to a beautiful (and somewhat neglected) area of applied mathematics, but that it may have the much-desired effect of stimulating further research on the many problems in this general area which are still unresolved, especially those which must be formulated in several space dimensions.

For the mathematician who has had little (or no) exposure to the subject matter of classical electromagnetic theory, as it is presented in most standard beginning graduate courses in American physics departments, we present, in Chapter 1, a brief survey of those parts of the subject which are deemed most useful to understanding what follows in the remainder of the book; the treatment in this first chapter, while concise, is strongly dependent on the presentation in several standard graduate level texts on electromagnetic theory.

In Chapter 2 we introduce the subject of wave propagation in nonlinear dielectrics, focusing on the problem of singularity formation in the wave-dielectric interaction problem, with much of the emphasis on problems which are dispersionless and can be formulated in one space dimension; the chapter begins with a survey of basic shock wave theory. Chapter 3 continues the discussion of the wave-dielectric interaction problem and is concerned with the derivation of growth estimates for solutions of the relevant initial-boundary value problems, as well as with the general issues of existence of smooth solutions for these problems, globally in time, and the asymptotic behavior of such solutions for large time.

Chapter 4 initiates our study of wave propagation in a distributed parameter nonlinear transmission line and covers both the issues of shock formation for "large" data, and the existence of smooth solutions for "small" data; in the discussion of both of these general problem areas we deal with a dispersionless line. In Chapter 5 we return to the transmission line problem to deal with the issue of existence of globally defined weak solutions in the presence of "large" initial data; the treatment in this chapter necessitates the introduction of the Young measure and the general concept of compensated compactness which have been instrumental, in recent years, in treating a variety of important nonlinear problems that arise in physical applications.

Chapter 6 offers a discussion of some nonlocal problems of electromagnetic theory with an emphasis on equilibrium problems that involve a self-interaction term arising from an application of the Biot–Savart Law; a particular focus is on problems for nonlinearly elastic, self-interacting current bearing wires in an ambient magnetic field.

This book was begun shortly after the tragic death of my brother during the summer of 1989 and was finished after the passing of my mother in the fall of 1991; it is dedicated, with much affection, to both of them. At those times in an individual's life when one must endure an emotional rollercoaster ride, it may be deemed a privilege to be a working scientist and to be involved with subject matter which appears to transcend, in a way that none of us yet seems to understand completely, our own mundane existence as human beings.

While the work of many colleagues has been referenced in this volume, several of them deserve to be singled out for special mention: Professor Alan Jeffrey, whose earlier work on some of the problems discussed in this book was instrumental, many years ago, in getting the author interested in these problems in the first place; Professors Constantine Dafermos and Marshall Slemrod, both friends of many years, whose mathematical work has inspired considerable portions of the material found in this book; Professor Stuart Antman, whose collaboration led to much of the work on nonlocal problems presented in the last chapter of this volume and, especially, my colleagues Professors Hamid Bellout and Jindřich Nečas, who have been friends as well as collaborators on many of the problems which are treated in this monograph. A special thanks goes to Mari–Anne Hartig, whose expert preparation of a long manuscript is very much appreciated by the author. Much of the work reported here, which is related to my own research in this area, was supported during the years 1984–1990 by a series of grants from the Applied Mathematics Program at NSF, for which the author is sincerely grateful. Finally, it is a pleasure to acknowledge the support of my family, my wife Leah, and my sons Daniel and Amir, whose constant encouragement during difficult times has made this book possible.

Frederick Bloom
Buffalo Grove, Illinois
September 1992

Chapter 1

ELEMENTS OF CLASSICAL ELECTROMAGNETIC THEORY

1.0 Introduction

Our goal in this first chapter is to provide the reader with a concise review of those elements of classical electromagnetic theory which will be particularly useful with regard to understanding the physical content of the remaining chapters. All of the material presented in this review is standard and may be gleaned from any one of the number of excellent graduate level texts available on the subject, e.g., [134], [73], [166]; however, even the reader who is more or less familiar with the structure and content of classical electromagnetic theory may find it helpful to quickly thumb through the material presented below in this introductory chapter. Our review of classical electromagnetic theory adheres to the following order: In § 1.1, we offer a quick synopsis of the classical theory of electrostatics emphasizing the theory associated with electrostatic fields in dielectric materials; the concepts of electric fields and potentials are introduced, as well as those of polarization and electric displacement, which will be so central to the material of Chapter 2. We continue in § 1.2 with a discussion of electric currents and electromagnetic induction, introducing, in the process, the current density vector (which also figures prominently in the discussion of wave-dielectric interactions in the next chapter) and the associated Ohm's laws (both the linear and nonlinear versions). Also introduced in § 1.2 are the concepts of resistance and magnetic induction, as well as the Biot-Savart law upon which much of the discussion in Chapter 6 concerning self-interacting current bearing wires revolves. The work in § 1.2 concludes with an analysis of the concepts of magnetic flux and electromotive force and leads us to the differential form of Faraday's law. Ampere's circuital law, which is yet one other consequence of the analysis in § 1.2, is generalized in § 1.3, in the manner first carried out by Maxwell, so as to take into account situations in which the charge density may change with time; this leads us, in § 1.3, to the full set of Maxwell's equations which are used in a classical fashion to investigate wave propagation in a linear medium. Our work in § 1.3 concludes, however, with the derivation of a coupled set of damped nonlinear wave equations for the components of the electric displacement field in a rigid nonlinear dielectric satisfying a simple nonlinear Ohm's law. Finally, in § 1.4, we present a discussion of the basic concepts associated with electrical circuits (or transmission lines); this includes an analysis of the concepts of capacitor, resistor, and inductor as components of a circuit and the delineation of Kirchhoff's laws in circuits excited by both constant and slowly varying voltages. In all the work in § 1.4 (unlike the analysis to be presented in Chapter 4), it is assumed that the resistance, capacitance, and self-inductance in the transmission line are constants and we exclude the presence in the circuit (until Chapter 4) of a leakage conductance per unit length of the line. The presentation in § 1.4 concludes with a standard analysis of the steady-state and transient behavior of the current in a series RLC transmission line.

1.1 Electrostatics and Dielectric Media

Electrostatic theory is based on Coulomb's law, which for a point charge $\bar{q}$, at the origin of some chosen coordinate system, and a point charge q at $\boldsymbol{r}$, gives the electrostatic force on q as

$$\boldsymbol{F}_e = \frac{1}{4\pi\epsilon_0}\frac{q\bar{q}}{r^2}\left(\frac{\boldsymbol{r}}{r}\right) \tag{1.1.1}$$

In (1.1.1), $\boldsymbol{r}$ is the position vector from the origin to the point occupied by the charge q, while $\epsilon_0 = 8.854 \times 10^{-12}$ $\mathrm{C^2/N{\cdot}m^2}$ in mks units[1]; the constant ϵ_0 is known as the permittivity of free space. With the assumption that we are working in a Cartesian coordinate system (x, y, z), $r = \|\boldsymbol{r}\| = \sqrt{x^2 + y^2 + z^2}$, i.e., the usual Euclidean norm. If we consider q to be a test charge, then we may use (1.1.1) to define the electric field $\boldsymbol{E}$ corresponding to the electric force $\boldsymbol{F}_e$ by $\boldsymbol{F}_e = q\boldsymbol{E}$, so that the electrostatic field at $\boldsymbol{r}$ which is due to a source charge $\bar{q}$ placed at $\boldsymbol{r} = \boldsymbol{0}$ is given by

$$\boldsymbol{E}(\boldsymbol{r}) = \frac{1}{4\pi\epsilon_0}\frac{\bar{q}}{r^2}\left(\frac{\boldsymbol{r}}{r}\right) \tag{1.1.2}$$

[1] In Gaussian units $\epsilon_0 = \dfrac{1}{4\pi}$.

For a system of n charges $q_i, i = 1, \ldots, n$, localized at positions $\boldsymbol{r}_i$, (1.1.1) generalizes in the expected fashion so as to produce, for the force on the i-th charge,

$$\boldsymbol{F}_i = q_i \sum_{k \neq i}^{n} \frac{q_k}{4\pi\epsilon_0} \frac{\boldsymbol{r}_{ik}}{r_{ik}^3} \tag{1.1.3}$$

where $\boldsymbol{r}_{ik} = \boldsymbol{r}_i - \boldsymbol{r}_k$ and $r_{ik} = \|\boldsymbol{r}_{ik}\|$. In view of the fact that $\nabla \times \left(\frac{\boldsymbol{r}}{r^3}\right) = \boldsymbol{0}$, while $\nabla \cdot \left(\frac{\boldsymbol{r}}{r^3}\right) = 4\pi\delta(\boldsymbol{r})$, δ being the Dirac delta function (distribution), we find that for a point charge (1.1.2) yields

$$\nabla \times \boldsymbol{E} = \boldsymbol{0}, \quad \nabla \cdot \boldsymbol{E} = \frac{1}{\epsilon_0}\bar{q}\delta(\boldsymbol{r}) \tag{1.1.4}$$

For a continuous distribution of charge density $\rho(\boldsymbol{r})$ defined over a region $\Omega \subseteq R^3$, so that the element of charge contained in a volume element dv is $dq = \rho(\boldsymbol{r})\,dv$, we have

$$\boldsymbol{E}(\boldsymbol{r}) = \frac{1}{4\pi\epsilon_0} \int_\Omega \frac{\boldsymbol{r} - \bar{\boldsymbol{r}}}{\|\boldsymbol{r} - \bar{\boldsymbol{r}}\|^3} \rho(\bar{\boldsymbol{r}})\,d\bar{v} \tag{1.1.5}$$

Clearly, $\|\boldsymbol{r} - \bar{\boldsymbol{r}}\| = \sqrt{(x - \bar{x})^2 + (y - \bar{y})^2 + (z - \bar{z})^2}$ in (1.1.5). Of course, $\rho(\boldsymbol{r}) = q_i\delta(\boldsymbol{r} - \boldsymbol{r}_i)$ for a point charge q_i located at $\boldsymbol{r}_i$. It follows directly, as a consequence of (1.1.5), that

$$\nabla \times \boldsymbol{E} = \boldsymbol{0}, \quad \nabla \cdot \boldsymbol{E} = \frac{1}{\epsilon_0}\rho(\boldsymbol{r}) \tag{1.1.6}$$

and these are the fundamental partial differential equations which are satisfied by all electrostatic fields.

If we integrate the second equation in (1.1.6) over $\Omega \subseteq R^3$, and apply the divergence theorem, we obtain Gauss's law, i.e.,

$$\int_{\partial\Omega} \boldsymbol{E} \cdot \boldsymbol{n}\,dS = \frac{1}{\epsilon_0}Q; \quad Q = \int_\Omega \rho(\boldsymbol{r})\,dv \tag{1.1.7}$$

where $\boldsymbol{n}$ is the (unit) exterior normal to $\partial\Omega$ and Q is the net charge inside Ω. From the first equation in (1.1.6), we infer immediately the existence (for a simply connected domain Ω) of an electrostatic potential function $\phi(\boldsymbol{r})$ satisfying $\boldsymbol{E} = -\nabla\phi$: the potential ϕ can be expressed either in terms of a given electric field $\boldsymbol{E}$ by

$$\phi(\boldsymbol{r}) = -\int_{\boldsymbol{r}_0}^{\boldsymbol{r}} \boldsymbol{E} \cdot d\boldsymbol{l} \tag{1.1.8}$$

or in terms of a given charge distribution $\rho(\boldsymbol{r})$ by

$$\phi(\boldsymbol{r}) = \frac{1}{4\pi\epsilon_0} \int_\Omega \frac{\rho(\boldsymbol{r})}{\|\boldsymbol{r} - \bar{\boldsymbol{r}}\|} d\bar{v} \tag{1.1.9}$$

Solutions of electrostatic problems in a domain Ω are obtained by combining the second equation in (1.1.6) with the consequence of the first equation, i.e., the existence of the potential function $\phi(\boldsymbol{r})$, so as to produce the Poisson equation $\nabla^2\phi = -\frac{\rho}{\epsilon_0}$, which is then solved subject to the specification of either ϕ or $\frac{\partial \phi}{\partial n}$ on $\partial\Omega$.

When we have two equal and opposite charges separated by a small distance, we say that an electric dipole exists. For a charge $-q$ located at $\bar{\boldsymbol{r}}$ and a charge q located at $\bar{\boldsymbol{r}} + \boldsymbol{l}$, the electric field is readily calculated to be

$$\boldsymbol{E}(\boldsymbol{r}) = \frac{q}{4\pi\epsilon_0}\left\{\frac{\boldsymbol{r}-\bar{\boldsymbol{r}}-\boldsymbol{l}}{\|\boldsymbol{r}-\bar{\boldsymbol{r}}-\boldsymbol{l}\|^3} - \frac{\boldsymbol{r}-\bar{\boldsymbol{r}}}{\|\boldsymbol{r}-\bar{\boldsymbol{r}}\|^3}\right\} \tag{1.1.10}$$

For a dipole field, the separation $\|\boldsymbol{l}\|$ is small compared with $\|\boldsymbol{r}-\bar{\boldsymbol{r}}\|$; expanding (1.1.10) by means of the binomial theorem, and retaining only those terms linear in $\boldsymbol{l}$, yields

$$\boldsymbol{E}(\boldsymbol{r}) = \frac{q}{4\pi\epsilon_0}\left\{\frac{3(\boldsymbol{r}-\bar{\boldsymbol{r}})\cdot\boldsymbol{p}}{\|\boldsymbol{r}-\bar{\boldsymbol{r}}\|^5}(\boldsymbol{r}-\bar{\boldsymbol{r}}) - \frac{\boldsymbol{p}}{\|\boldsymbol{r}-\bar{\boldsymbol{r}}\|^3}\right\} \tag{1.1.11}$$

where $\boldsymbol{p} = q\boldsymbol{l}$ is the electric dipole moment. For a point dipole it is assumed that $\|\boldsymbol{l}\| \to 0$, while $q \to \infty$, in such a way that $\|\boldsymbol{p}\|$ remains constant; in this case all terms in the expansion of (1.1.10) vanish in the limit, except those which are linear in $\boldsymbol{l}$, and (1.1.11) is exact for the point dipole. The potential corresponding to a point dipole may easily by shown to have the form

$$\phi(\boldsymbol{r}) = \frac{1}{4\pi\epsilon_0}\frac{\boldsymbol{p}\cdot(\boldsymbol{r}-\bar{\boldsymbol{r}})}{\|\boldsymbol{r}-\bar{\boldsymbol{r}}\|^3} \tag{1.1.12}$$

For a continuous distribution $\rho(\boldsymbol{r})$ of charge throughout a domain $\Omega \subseteq R^3$, the dipole moment of the distribution is defined to be

$$\boldsymbol{p} = \int_\Omega \bar{\boldsymbol{r}}\rho(\bar{\boldsymbol{r}})\,d\bar{v} \tag{1.1.13}$$

In this treatise considerable attention will be paid to the matter of electromagnetic wave-dielectric interactions (i.e., all of Chapters 2 and 3); as background material we now survey briefly the theory underlying electrostatic fields in dielectric media. Although an ideal dielectric material is one in which there exists no free charge, the term *bound charge*, when used in reference to dielectrics (as opposed to *free charge* for a conductor) is used to emphasize the fact that molecular charges are not free to travel significant distances or to be extracted from the dielectric; under the action of an electric field, the entire positive charge in a dielectric is viewed as being *displaced* relative to the negative charge and the dielectric is said to be *polarized.* A polarized dielectric, while electrically neutral (on the average) produces an electric field, both interior and exterior to the dielectric, which may, in turn, modify the free charge distribution on any conducting bodies in its vicinity and lead to changes in the external electric field which act back on the dielectric. If Δv denotes an infinitesimal volume element of a polarized dielectric, this volume element is characterized by an electric dipole moment which is given, according to (1.1.13), by

$$\Delta \boldsymbol{p} = \int_{\Delta v} \boldsymbol{r}\, dq \tag{1.1.14}$$

and gives rise to an *electric dipole moment* (per unit volume)

$$\boldsymbol{P} = \lim_{\Delta v \to 0} \frac{\Delta \boldsymbol{p}}{\Delta v} \tag{1.1.15}$$

When the limit in (1.1.15) exists, the point function $\boldsymbol{P}$ is termed the electric polarization of the dielectric medium and the electrostatic potential at a point $\boldsymbol{r}$ exterior to the dielectric, assumed to occupy a domain $\Omega \subseteq R^3$, is then given, according to (1.1.12), by

$$\phi(\boldsymbol{r}) = \frac{1}{4\pi\epsilon_0} \int_\Omega \frac{\boldsymbol{P}(\bar{\boldsymbol{r}}) \cdot (\boldsymbol{r} - \bar{\boldsymbol{r}}) d\bar{v}}{\|\boldsymbol{r} - \bar{\boldsymbol{r}}\|^3} \tag{1.1.16}$$

Inasmuch as $\bar{\nabla}\left(\frac{1}{\|\boldsymbol{r} - \bar{\boldsymbol{r}}\|}\right) = \frac{\boldsymbol{r} - \bar{\boldsymbol{r}}}{\|\boldsymbol{r} - \bar{\boldsymbol{r}}\|^3}$, where $\bar{\nabla}$ is the gradient operator with respect to the *barred* coordinates, (1.1.16) may be transformed into

$$\begin{aligned} \phi(\boldsymbol{r}) &= \frac{1}{4\pi\epsilon_0} \oint_{\partial\Omega} \frac{\boldsymbol{P} \cdot \boldsymbol{n}}{\|\boldsymbol{r} - \bar{\boldsymbol{r}}\|} d\bar{S} \\ &\quad + \frac{1}{4\pi\epsilon_0} \int_\Omega \frac{(-\bar{\nabla} \cdot \boldsymbol{P}(\bar{\boldsymbol{r}}))}{\|\boldsymbol{r} - \bar{\boldsymbol{r}}\|} d\bar{v} \end{aligned} \tag{1.1.17}$$

through use of the standard identity

$$\bar{\nabla} \cdot (f\boldsymbol{P}) = f\bar{\nabla} \cdot \boldsymbol{P} + \boldsymbol{P} \cdot \bar{\nabla} f$$

with $f = \frac{1}{\|\boldsymbol{r} - \bar{\boldsymbol{r}}\|}$. The quantities $\sigma_p = \boldsymbol{P} \cdot \boldsymbol{n}$ and $\rho_p = -\nabla \cdot \boldsymbol{P}$ are, respectively, the surface and volume densities of polarization charge. Inserting these quantities in (1.1.17), we see that we may express the potential due to the dielectric medium in such a way that it arises from appropriate distributions of polarization charge, i.e.,

$$\phi(\boldsymbol{r}) = \frac{1}{4\pi\epsilon_0} \left\{ \oint_{\partial\Omega} \frac{\sigma_p d\bar{S}}{\|\boldsymbol{r} - \bar{\boldsymbol{r}}\|} + \int_\Omega \frac{\rho_p d\bar{v}}{\|\boldsymbol{r} - \bar{\boldsymbol{r}}\|} \right\} \tag{1.1.18}$$

the definitions of σ_p, ρ_p being obvious upon comparison with (1.1.17).

We now recall the statement of Gauss's law, which states that the electric flux across an arbitrary closed surface is proportional to the total charge enclosed by the surface; in applying Gauss's law to a domain $\Omega \subseteq R^3$ occupied by a dielectric medium, the polarization charge, as well as the charge embedded in Ω, must be included. If $\partial\Omega'$ bounds a domain $\Omega' \subset \Omega$ and contains an amount of embedded charge $Q = \sum_{i=1}^{n} q_i$, with $q_i, i = 1, \ldots, n$ existing on the surfaces $\partial\Omega_i$ of n conductors, then by Gauss's law

$$\oint_{\partial\Omega'} \boldsymbol{E} \cdot \boldsymbol{n}\, dS = \frac{1}{\epsilon_0}(Q + Q_p) \tag{1.1.19}$$

where Q_p is the net polarization charge, i.e.

$$Q_p = \oint_{\cup_i(\partial\Omega_i)} \boldsymbol{P} \cdot \boldsymbol{n}\, dS + \int_{\Omega'} (-\nabla \cdot \boldsymbol{P})\, dv' \tag{1.1.20}$$

The surface integral in (1.1.20) does not contain a contribution from $\partial\Omega'$ inasmuch as there is no boundary of the dielectric there. Using the divergence theorem on the volume integral in (1.1.20), and noting that Ω' is bounded by $\partial\Omega' \bigcup (\cup_i \partial\Omega_i)$, we easily find that

$$Q_p = -\oint_{\partial\Omega'} \boldsymbol{P} \cdot \boldsymbol{n}\, dS \tag{1.1.21}$$

Combining (1.1.19) with (1.1.21) then yields

$$\oint_{\partial\Omega'} (\epsilon_0 \boldsymbol{E} + \boldsymbol{P}) \cdot \boldsymbol{n}\, dS = Q \tag{1.1.22}$$

Equation (1.1.22) naturally yields the definition of the field vector

$$\boldsymbol{D} = \epsilon_0 \boldsymbol{E} + \boldsymbol{P} \tag{1.1.23}$$

which is termed the *electric displacement*, in terms of which (1.1.22) becomes (we drop the prime)

$$\oint_{\partial\Omega} \boldsymbol{D} \cdot \boldsymbol{n}\, dS = Q \tag{1.1.24}$$

and (1.1.24) applies to any region Ω of $\mathbf{R}^3$ bounded by a closed surface $\partial\Omega$; applying (1.1.24) to an infinitesimal domain $\Omega \subseteq R^3$, with continuous distributed charge density ρ, dividing both sides of (1.1.24) by vol(Ω), and extracting the limit as vol(Ω) $\to 0$, we obtain the differential form of Gauss's law, namely,

$$\nabla \cdot \boldsymbol{D} = \rho \tag{1.1.25}$$

The degree of polarization in a dielectric depends not only on the impressed electric field, but also on the molecular properties of the material; macroscopically this behavior is specified by an experimentally determined relationship of the form $\boldsymbol{P} = \boldsymbol{P}(\boldsymbol{E})$.[2] Inasmuch as it is usually the case that $\boldsymbol{E} = \boldsymbol{0}$ implies that $\boldsymbol{P} = \boldsymbol{0}$, and that the dielectric is *isotropic*, so that the polarization vector points in the same direction as the impressed electric field, a common form of the $\boldsymbol{P}$-$\boldsymbol{E}$ relationship is $\boldsymbol{P} = \chi(\|\boldsymbol{E}\|)\boldsymbol{E}$, where $\chi(\cdot)$ is a scalar-valued function called the electric susceptibility; in this case it is clear that by virtue of (1.1.23) we have

$$\boldsymbol{D} = \epsilon(\|\boldsymbol{E}\|)\boldsymbol{E} \tag{1.1.26}$$

where the permittivity $\epsilon(\cdot)$ is defined to be

$$\epsilon(\|\boldsymbol{E}\|) = \epsilon_0 + \chi(\|\boldsymbol{E}\|) \tag{1.1.27}$$

For many dielectric materials, $\chi(\cdot)$, and hence $\epsilon(\cdot)$, are independent of the field strength $\|\boldsymbol{E}\|$, except for intense electric fields of the type common in laser beams; for such *linear dielectrics*, χ and ϵ are constants characteristic of the dielectric medium.

If we consider two dielectrics which are in contact along a common boundary S (a vacuum may be considered to be a dielectric with permittivity ϵ_0) which carries a surface density of external charge σ, then an application of Gauss's law to a pillbox-shaped surface S', which intersects the interface S, yields the relation

$$(\boldsymbol{D}_2 - \boldsymbol{D}_1) \cdot \boldsymbol{n} = \sigma \tag{1.1.28}$$

where $\boldsymbol{D}_i$ is the value of the displacement field in the i-th dielectric at a point $\boldsymbol{x} \in S$ and $\boldsymbol{n}$ is the exterior unit normal to S at $\boldsymbol{x}$; when $\sigma = 0$ on S, (1.1.28) states that the normal component of $\boldsymbol{D}$ is continuous across the interface between the two media. In an analogous fashion, as $\boldsymbol{E} = -\nabla\phi$, the line integral $\oint \boldsymbol{E} \cdot \boldsymbol{dl}$ around any closed path vanishes, and the application of this result to an arbitrary infinitesimal rectangular path which intersects the interface S yields the second fundamental boundary condition for dielectric media, namely

$$(\boldsymbol{E}_2)_t = (\boldsymbol{E}_1)_t \tag{1.1.29}$$

[2]Constitutive relations in which $\boldsymbol{P}$ depends both on the electric and magnetic fields may also occur.

where the subscript t denotes the tangential component of the indicated vector field. The boundary conditions (1.1.28), (1.1.29), in conjunction with (1.1.25), and the constitutive relation (1.1.26), yield a well-posed boundary value problem for the electrostatic field in any dielectric medium (or vacuum); if the first medium is both linear and isotropic, so that $\boldsymbol{D} = \epsilon \boldsymbol{E}$, as well as $\boldsymbol{E} = -\nabla \phi$, then (1.1.25) yields the Poisson equation

$$\nabla^2 \phi = -\frac{1}{\epsilon}\rho \tag{1.1.30}$$

for the electrostatic potential in the dielectric.

Before concluding this section on electrostatics and dielectric media, some remarks are in order about the electrostatic energy of a charge distribution and the corresponding concept of energy density in an electrostatic field. It is easily demonstrated that the electrostatic energy of an arbitrary charge distribution, which possesses volume density ρ and surface density σ, is given by

$$\mathsf{E} = \frac{1}{2}\int_{\Omega} \rho(\boldsymbol{r})\phi(\boldsymbol{r})\, dv + \frac{1}{2}\int_{\partial\Omega} \sigma(\boldsymbol{r})\phi(\boldsymbol{r})\, dS \tag{1.1.31}$$

provided that all dielectrics present are linear, the domain of integration Ω is large enough to include all of the charge density present, and the electrostatic potential ϕ is that due to the charge densities ρ and σ; for a domain $\Omega \subseteq R^3$ filled with a single, linear dielectric medium, of dielectric permittivity ϵ, the potential ϕ is given by the expression

$$\phi(\boldsymbol{r}) = \frac{1}{4\pi\epsilon}\int_{\Omega} \frac{\rho(\bar{\boldsymbol{r}})}{\|\boldsymbol{r} - \bar{\boldsymbol{r}}\|} d\bar{v} + \frac{1}{4\pi\epsilon}\int_{\partial\Omega} \frac{\sigma(\bar{\boldsymbol{r}})}{\|\boldsymbol{r} - \bar{\boldsymbol{r}}\|} d\bar{S} \tag{1.1.32}$$

The corresponding situation for point charges follows as a special case of (1.1.31), (1.1.32) if we set

$$\begin{cases} \rho(\boldsymbol{r}) &= \sum_{n=1}^{m} q_n \delta(\boldsymbol{r} - \boldsymbol{r}_n) \\ \rho(\bar{\boldsymbol{r}}) &= \sum_{\substack{k=1 \\ k \neq n}}^{m} q_k \delta(\bar{\boldsymbol{r}} - \boldsymbol{r}_k) \end{cases} \tag{1.1.33}$$

in which case

$$\mathsf{E} = \frac{1}{2}\sum_{j=1}^{m} q_j \phi_j \tag{1.1.34}$$

with

$$\phi_j = \sum_{\substack{k=1 \\ k \neq j}} \frac{q_k}{4\pi\epsilon_0 r_{jk}} \tag{1.1.35}$$

In a dielectric medium, we have, by virtue of (1.1.25) and (1.1.28), $\rho = \nabla \cdot \boldsymbol{D}$, $\sigma = \boldsymbol{D} \cdot \boldsymbol{n}$, provided Ω is constructed so that all surface densities of charge σ reside on conductor surfaces; then (1.1.31) becomes

$$\mathsf{E} = \frac{1}{2}\int_{\Omega} \phi \nabla \cdot \boldsymbol{D}\, dv + \frac{1}{2}\int_{\partial\Omega'} \phi \boldsymbol{D} \cdot \boldsymbol{n}\, dS' \tag{1.1.36}$$

In (1.1.36), Ω is the domain where $\nabla \cdot \boldsymbol{D} \neq 0$, while the surface integral is over the surface of the conductors. Employing the standard identity for $\nabla \cdot \phi \boldsymbol{D}$, and using the

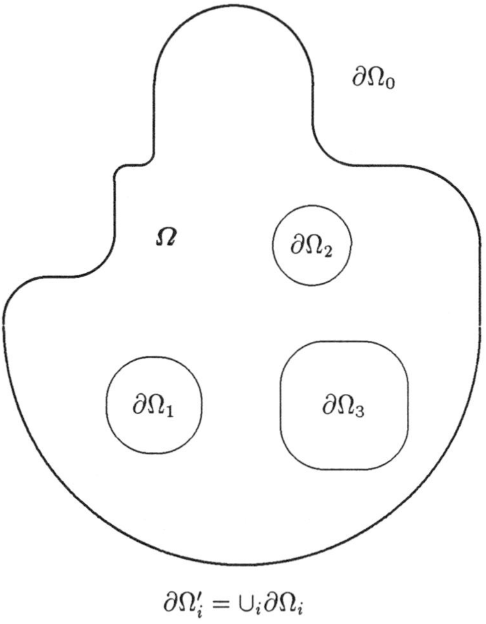

Figure 1.1

divergence theorem, as well as the fact that $\boldsymbol{E} = -\nabla\phi$, the expression (1.1.36) is easily seen to reduce to

$$\mathsf{E} = \frac{1}{2}\int_{\Omega} \boldsymbol{D} \cdot \boldsymbol{E}\, dv \tag{1.1.37}$$

Remark. If we refer to Figure 1.1 and set $\partial\Omega' = \bigcup_i \partial\Omega_i$, then we note that (1.1.36) first is reduced to

$$\mathsf{E} = \frac{1}{2}\int_{\partial\Omega_0 \cup \partial\Omega'} \phi \boldsymbol{D} \cdot \boldsymbol{n}\, dS \; + \; \frac{1}{2}\int_{\Omega} \boldsymbol{D} \cdot \boldsymbol{E}\, dv + \frac{1}{2}\int_{\partial\Omega'} \phi \boldsymbol{D} \cdot \boldsymbol{n}\, dS' \tag{1.1.38}$$

The two surface integrals over $\partial\Omega'$ mutually cancel each other; as $\phi \sim r^{-1}$, $\|\boldsymbol{D}\| \sim r^{-2}$, for large r, while $\partial\Omega_0 \sim r^2$ as we extend $\partial\Omega_0$ out to infinity, so $\int_{\partial\Omega_0} \phi\boldsymbol{D}\cdot\boldsymbol{n}\,dS \to 0$.

From (1.1.37) we infer that the energy density u_E in an electrostatic field is given by the expression

$$u_E = \frac{1}{2}\boldsymbol{D}\cdot\boldsymbol{E} \tag{1.1.39}$$

1.2 Electric Currents and Electromagnetic Induction

We now consider charges in uniform motion and thus will be dealing with conductors of electricity, which are materials in which the charge carriers are free to move under steady electric fields; this definition of conductor includes not only standard conductors such as metallic substances, but also imperfect dielectric media, and so the charge carriers may be either electrons or positive or negative ions.

The basic definition, of course, is that *current* is moving charge and a current I is then defined as *the rate at which charge is transported through a given surface*, i.e., $I = \frac{dQ}{dt}$; the unit of current in the mks system is the ampere, where 1 (ampere) = 1 (coulomb/second). Consider a conducting medium which possesses N charge carriers, each of charge q, per unit volume, with each carrier having the same drift velocity $\boldsymbol{v}$; the net charge δQ which crosses an element of surface in the conductor of area dS, in time δt, is easily seen to be

$$\delta Q = qN\boldsymbol{v}\cdot\boldsymbol{n}\,\delta t\,dS \tag{1.2.1}$$

where $\boldsymbol{n}$ is the external unit normal to the element of surface in the conductor. From (1.2.1) it is immediate that

$$dI \equiv \frac{\delta Q}{\delta t} = Nq\boldsymbol{v}\cdot\boldsymbol{n}\,dS \tag{1.2.2}$$

and if M different charge carriers are present, each carrying a charge q_i, and numbering N_i, $i = 1, \ldots, m$, then (1.2.1) clearly generalizes to

$$dI = \left\{\sum_{i=1}^{M} N_i q_i \boldsymbol{v}_i\right\}\cdot\boldsymbol{n}\,dS \tag{1.2.3}$$

The quantity in the brackets in (1.2.3) has the dimensions of current per unit area and is called the current density $\boldsymbol{J}$, i.e.,

$$\boldsymbol{J} = \sum_i N_i q_i \boldsymbol{v}_i \tag{1.2.4}$$

so that (1.2.3) may be expressed as $dI = \boldsymbol{J} \cdot \boldsymbol{n}\, dS$ and the current through an arbitrary surface $\partial\Omega$, bounding a domain $\Omega \subseteq R^3$ in the conductor, is then given by

$$I = \int_{\partial\Omega} \boldsymbol{J} \cdot \boldsymbol{n}\, dS \tag{1.2.5}$$

Inasmuch as the current entering Ω through $\partial\Omega$ is the negative of the quantity in (1.2.5), we have, by virtue of the divergence theorem,

$$I = -\int_{\Omega} \nabla \cdot \boldsymbol{J}\, dv \tag{1.2.6}$$

But, for a fixed, time-independent domain $\Omega \subseteq R^3$, it is also true that

$$I = \frac{dQ}{dt} = \frac{d}{dt}\int_{\Omega} \rho(\boldsymbol{x}, t)\, dv = \int_{\Omega} \frac{\partial \rho}{\partial t}\, dv \tag{1.2.7}$$

where we now denote the position of points in Ω by $\boldsymbol{x} = (x_1, x_2, x_3)$; combining (1.2.6), (1.2.7), assuming all indicated quantities are, at least, continuous functions of $\boldsymbol{x}$, and noting that the resulting integral relation must hold for each arbitrary bounded domain Ω in the conductor, we are led to the *equation of continuity*, namely,

$$\frac{\partial \rho}{\partial t} + \nabla \cdot \boldsymbol{J} = 0 \tag{1.2.8}$$

For many metallic substances at constant temperature, it has been noted experimentally that $\boldsymbol{J}$ is linearly proportional to $\boldsymbol{E}$ and the mathematical statement of this fact, i.e., $\boldsymbol{J} = \sigma \boldsymbol{E}$, is termed Ohm's law; the constant of proportionality σ (not to be confused with a measure of surface charge density) is called the *conductivity* of the material. In many instances, the conductivity may be demonstrated to be field dependent, e.g., conduction of electricity takes place in an imperfect dielectric, in which case the standard Ohm's law must be replaced by a constitutive relation of the form

$$\boldsymbol{J} = \sigma(\boldsymbol{E})\boldsymbol{E} \tag{1.2.9}$$

and, indeed, nonlinear Ohm's laws of this specific type will figure prominently in our considerations of wave-dielectric interactions in Chapter 2.

If we consider a homogeneous wire of uniform cross-section, which conducts electricity according to the linear Ohm's law $\boldsymbol{J} = \sigma \boldsymbol{E}$, and has its ends maintained at a constant potential difference $\delta\phi$, then an electric field will exist in the wire which satisfies the relation

$$\delta\phi = \int \boldsymbol{E} \cdot dl \tag{1.2.10}$$

and the electric field $\boldsymbol{E}$ is longitudinal with no component at a right angle to the axis of the wire; by virtue of the geometry of the situation described above, and the homogeneity of the wire, it is clear that (1.2.10) implies that $\delta\phi = El$, where l is the length of the wire and $E = \|\boldsymbol{E}\|$. However, inasmuch as $\boldsymbol{J} = \sigma \boldsymbol{E}$, with $\boldsymbol{E}$ longitudinal, and the cross-sectional area of the wire is constant, say A, the current through any cross-section of the wire is given by

$$I = \int_A \boldsymbol{J} \cdot \boldsymbol{n}\, dS = JA \tag{1.2.11}$$

From (1.2.11), the linear Ohm's law, and the relation $\delta\phi = El$, we easily obtain a linear relationship

$$I = \frac{\sigma A}{l}\delta\phi \equiv \frac{1}{R}\delta\phi \tag{1.2.12}$$

which serves to define the *resistance* R of the wire. It is not difficult to infer that the linear Ohm's law $\boldsymbol{J} = \sigma \boldsymbol{E}$ implies the relation (1.2.12) independently of the shape of the conductor.

For a homogeneous conducting medium in a steady-state conduction mode, which obeys a linear Ohm's law, $\frac{\partial \rho}{\partial t} = 0$, and the equation of continuity (1.2.8) reduces to the statement that $\nabla \cdot \sigma \boldsymbol{E} = 0$; with σ constant in a homogeneous medium, we infer that $\nabla \cdot \boldsymbol{E} = 0$ and if the field is static, so that $\nabla \times \boldsymbol{E} = 0$, we again infer the existence of a scalar potential ϕ satisfying Laplace's equation. The steady state conduction problem for any static system of homogeneous conductors, which conforms to a linear Ohm's law may, therefore, be solved in the same manner as electrostatic problems; to solve a boundary value problem for a conducting medium we must, of course, specify either ϕ or $\boldsymbol{J}$ at each point on the bounding surface of the medium.

Suppose we consider an isotropic homogeneous medium possessing conductivity σ and dielectric permittivity ϵ, i.e., for this medium, both $\boldsymbol{J} = \sigma \boldsymbol{E}$ and $\boldsymbol{D} = \epsilon \boldsymbol{E}$ apply. Suppose further that, at time $t = 0$, the medium is characterized by a prescribed volume density of charge $\rho_0(x, y, z)$; from the equation of continuity (1.2.8) and Ohm's law we have

$$\frac{\partial \rho}{\partial t} + \sigma \nabla \cdot \boldsymbol{E} = 0 \tag{1.2.13}$$

However, in view of (1.1.25) and the relation $\boldsymbol{D} = \epsilon \boldsymbol{E}$, we see that $\nabla \cdot \boldsymbol{E} = \dfrac{\rho}{\epsilon}$ so that (1.2.13) assumes the form

$$\frac{\partial \rho}{\partial t} + \left(\frac{\sigma}{\epsilon}\right) \rho = 0 \tag{1.2.14}$$

and integration of (1.2.14) subject to the initial condition $\rho(x, y, z, 0) = \rho_0(x, y, z)$ yields the elementary solution

$$\rho(x, y, z, t) = \rho_0(x, y, z) \exp(-\sigma t / \epsilon) \tag{1.2.15}$$

which shows that the conducting medium, if isolated from applied electric fields, will tend to an equilibrium state exponentially fast as $t \to \infty$. The quantity ϵ/σ in (1.2.15) obviously has the dimensions of time, and is termed the *relaxation time* of the medium.

We now turn our attention to a brief survey of those basic considerations related to a study of the magnetic field associated with steady currents in a conducting medium. We begin by recalling that the Coulomb force on a charge q located at position $\boldsymbol{r}$, due to a charge $\bar{q}$ at the origin of our chosen coordinate system, is given by (1.1.1), provided that both charges are at rest. When q and $\bar{q}$ are, instead, moving uniformly with velocities $\boldsymbol{v}$ and $\bar{\boldsymbol{v}}$, respectively, $\bar{q}$ exerts an additional *magnetic force* on q which is observed, experimentally, to have the form

$$\boldsymbol{F}_m = \left(\frac{\mu_0}{4\pi}\right) \frac{q\bar{q}}{r^2} \boldsymbol{v} \times \left(\bar{\boldsymbol{v}} \times \frac{\boldsymbol{r}}{r}\right) \tag{1.2.16}$$

The constant $\dfrac{\mu_0}{4\pi}$ in (1.2.16) plays the same role here that the constant $1/4\pi\epsilon_0$ did in the force law (1.1.1) and μ_0 is called the *magnetic permeability* of free space. If we now define the *magnetic induction* vector $\boldsymbol{B}$ by

$$\boldsymbol{B} = \left(\frac{\mu_0}{4\pi}\right) \frac{\bar{q}}{r^2} \bar{\boldsymbol{v}} \times \frac{\boldsymbol{r}}{r} \tag{1.2.17}$$

then the force law (1.2.16) assumes the form

$$\boldsymbol{F}_m = q\boldsymbol{v} \times \boldsymbol{B} \tag{1.2.18}$$

and when both electric and magnetic fields are present the net force on a moving charge q is given by the *Lorentz force*

$$\boldsymbol{F} = q(\boldsymbol{E} + \boldsymbol{v} \times \boldsymbol{B}) \tag{1.2.19}$$

We note that it is a direct consequence of (1.2.18) that, for any field $\boldsymbol{B}$, $\boldsymbol{v} \cdot \boldsymbol{F}_m = 0$ so that a magnetic force never does work on a charged particle. By multiplying both sides of (1.2.16) by ϵ_0/ϵ_0 we see, after comparison with the force law (1.1.1) that $\epsilon_0\mu_0$ has the dimensions of the square of an inverse velocity, say, $\epsilon_0\mu_0 = c^{-2}$, with c having the dimensions of velocity; the force law (1.2.16) may then be expressed as

$$\boldsymbol{F}_m = \left(\frac{1}{4\pi\epsilon_0}\right)\frac{q\bar{q}}{r^2}\frac{\boldsymbol{v}}{c} \times \left(\frac{\bar{\boldsymbol{v}}}{c} \times \frac{\boldsymbol{r}}{r}\right) \tag{1.2.20}$$

We note, in passing, that the experimentally determined values of μ_0 and ϵ_0 yield as a consequence that $c = (\epsilon_0\mu_0)^{-1/2}$ is in remarkable agreement with the experimentally measured velocity of light; this agreement is, in fact, well-known to be a direct consequence of Maxwell's equations and the assumption that light is a propagating electromagnetic wave (as we will observe in the next section). We also note here that it is a direct consequence of (1.1.1) and (1.2.20) that

$$\frac{\|\boldsymbol{F}_m\|}{\|\boldsymbol{F}_e\|} \leq \frac{\|\boldsymbol{v}\|\|\bar{\boldsymbol{v}}\|}{c^2} \tag{1.2.21}$$

so that if the particle velocities are small, in comparison to the velocity of light, then the magnetic force exerted is far smaller than the electric force.

From the magnetic force relation (1.2.18) one may compute an expression for the force exerted on an element $\boldsymbol{dl}$ of a current-carrying conductor. Assuming that $\boldsymbol{dl}$ is always in the direction of the current I in the conductor, so that $\boldsymbol{dl}$ is parallel to the velocity $\boldsymbol{v}$ of the charge carriers, we have, with K charge carriers per unit volume in the conductor, a force exerted on the element $\boldsymbol{dl}$ of the form

$$d\boldsymbol{F} = qKA\|\boldsymbol{dl}\|\boldsymbol{v} \times \boldsymbol{B} \tag{1.2.22}$$

with A being the constant cross-sectional area of the conductor and q the charge on each charge carrier. Because $\boldsymbol{v}$ and the element $\boldsymbol{dl}$ are parallel, we may rewrite (1.2.22) in the form

$$d\boldsymbol{F} = qKA\|\boldsymbol{v}\|\boldsymbol{dl} \times \boldsymbol{B} \tag{1.2.23}$$

and this expression is unchanged if more than one type of charge carrier is involved. If we examine (1.2.23), and note that the quantity $qKA\|\boldsymbol{v}\|$ is the current I associated with one type of charge carrier, then this equation for the force on an infinitesimal element of a charge-carrying conductor may be expressed as

$$d\boldsymbol{F} = Id\boldsymbol{l} \times \boldsymbol{B} \tag{1.2.24}$$

If $\mathcal{C}$ is a closed circuit in R^3, and we integrate both sides of (1.2.24) over $\mathcal{C}$, we obtain the force exerted on the current carrying circuit in the form

$$\boldsymbol{F} = I \oint_{\mathcal{C}} d\boldsymbol{l} \times \boldsymbol{B} \tag{1.2.25}$$

from which it is clear that $\boldsymbol{F} = \boldsymbol{0}$ when $\boldsymbol{B}$ is a uniform field (i.e., when $\boldsymbol{B}$ is independent of position). The infinitesimal torque on an element $d\boldsymbol{l}$ of the conductor may be expressed as $d\boldsymbol{\tau} = \boldsymbol{r} \times d\boldsymbol{F}$ so that, by virtue of (1.2.24), the torque $\boldsymbol{\tau}$ on a closed circuit $\mathcal{C}$ is given by

$$\boldsymbol{\tau} = I \oint_{\mathcal{C}} \boldsymbol{r} \times (d\boldsymbol{l} \times \boldsymbol{B}) \tag{1.2.26}$$

For a uniform field $\boldsymbol{B}$ an elementary analysis shows that (1.2.26) may be reduced to the relation $\boldsymbol{\tau} = I\boldsymbol{A}^* \times \boldsymbol{B}$ with $\boldsymbol{A}^*$ being the vector whose components are the areas of the projections of $\mathcal{C}$ on the yz, zx, and xy planes, respectively. It is customary to write $\boldsymbol{m} = I\boldsymbol{A}^*$ and to refer to $\boldsymbol{m}$ as the *magnetic moment* of the circuit; inasmuch as

$$\boldsymbol{A}^* = \frac{1}{2} \oint_{\mathcal{C}} \boldsymbol{r} \times d\boldsymbol{l} \tag{1.2.27}$$

we may express the magnetic moment as

$$\boldsymbol{m} = \frac{1}{2} I \oint_{\mathcal{C}} \boldsymbol{r} \times d\boldsymbol{l} \tag{1.2.28}$$

The discovery that currents produce magnetic effects was made by Oersted in 1820 and was followed, very closely, by Ampere's results on the magnetic interaction of two current-carrying circuits. For a circuit $\mathcal{C}_1$ (whose points are located by the position vector $\boldsymbol{r}_1$, which has line element $d\boldsymbol{l}_1$, and carries a current I_1) and a second circuit $\mathcal{C}_2$ (whose points are located by the position vector $\boldsymbol{r}_2$, which has line element $d\boldsymbol{l}_2$, and carries a current I_2), Ampere's experiments imply that the force $\boldsymbol{F}_2$ exerted on $\mathcal{C}_2$ due to the influence of $\mathcal{C}_1$ has the form

$$\boldsymbol{F}_2 = \left(\frac{\mu_0}{4\pi}\right) I_1 I_2 \oint_{\mathcal{C}_1} \oint_{\mathcal{C}_2} \frac{d\boldsymbol{l}_2 \times [d\boldsymbol{l}_1 \times (\boldsymbol{r}_2 - \boldsymbol{r}_1)]}{\|\boldsymbol{r}_2 - \boldsymbol{r}_1\|^3} \tag{1.2.29}$$

and, although not obvious from the form of the relation (1.2.29), it may be shown that $\boldsymbol{F}_2 = -\boldsymbol{F}_1$, where $\boldsymbol{F}_1$ is the force exerted on the circuit C_1 because of the influence of the magnetic field generated by the current flowing in C_2. In view of (1.2.25) we easily find that

$$\boldsymbol{B}(\boldsymbol{r}_2) = \frac{\mu_0}{4\pi} I_1 \oint_{C_1} \frac{d\boldsymbol{l}_1 \times (\boldsymbol{r}_2 - \boldsymbol{r}_1)}{\|\boldsymbol{r}_2 - \boldsymbol{r}_1\|^3} \tag{1.2.30}$$

and this is the *Biot-Savart* law which will form the basis for the work on nonlocal electromagnetic problems that is described in Chapter 6. For a continuous distribution of current, described by the current density $\boldsymbol{J}(\boldsymbol{r})$, we observe that the analog of (1.2.30) applies, namely,

$$B(\boldsymbol{r}_2) = \frac{\mu_0}{4\pi} \int_\Omega \frac{\boldsymbol{J}(\boldsymbol{r}_1) \times (\boldsymbol{r}_2 - \boldsymbol{r}_1)}{\|\boldsymbol{r}_2 - \boldsymbol{r}_1\|^3} \, dv \tag{1.2.31}$$

and it is, in fact, an experimental deduction that all magnetic induction fields can be described in terms of a current distribution, i.e., that $\boldsymbol{B}(\boldsymbol{r})$ has the form (1.2.31) for some $\boldsymbol{J}(\boldsymbol{r})$. This observation is the basis for the conclusion that isolated magnetic poles do not exist, for if we use the vector identity

$$\nabla \cdot (\boldsymbol{R} \times \boldsymbol{S}) = -\boldsymbol{R} \cdot (\nabla \times \boldsymbol{S}) + \boldsymbol{S} \cdot (\nabla \times \boldsymbol{R}), \tag{1.2.32}$$

in taking the divergence on both sides of (1.2.31), we find that

$$\nabla_2 \cdot \boldsymbol{B}(\boldsymbol{r}_2) = -\frac{\mu_0}{4\pi} \int_\Omega \boldsymbol{J}(\boldsymbol{r}_1) \cdot \nabla_2 \times \left[\frac{\boldsymbol{r}_2 - \boldsymbol{r}_1}{\|\boldsymbol{r}_2 - \boldsymbol{r}_1\|^3} \right] dv \tag{1.2.33}$$

But

$$\frac{\boldsymbol{r}_2 - \boldsymbol{r}_1}{\|\boldsymbol{r}_2 - \boldsymbol{r}_1\|^3} = \nabla_2 \left(\frac{-1}{\|\boldsymbol{r}_2 - \boldsymbol{r}_1\|} \right) \tag{1.2.34}$$

and as $\nabla \times (\nabla f) = \boldsymbol{0}$ for any scalar function f, it is immediate from (1.2.33) that

$$\nabla_2 \cdot \boldsymbol{B}(\boldsymbol{r}_2) = 0 \tag{1.2.35}$$

Equation (1.2.35) is the first of the two basic differential laws of magnetostatics, the second of which is Ampere's law. Suppose that the magnetic induction field which is described by (1.2.31) is generated by a steady current, i.e., that $\boldsymbol{J}$ satisfies $\nabla \cdot \boldsymbol{J} = 0$. If we compute the curl of the expression in (1.2.31), we have that

$$\begin{aligned} \nabla_2 \times \boldsymbol{B}(\boldsymbol{r}_2) &= \frac{\mu_0}{4\pi} \int_\Omega \left\{ \boldsymbol{J}(\boldsymbol{r}_1) \left(\nabla_2 \cdot \left[\frac{\boldsymbol{r}_2 - \boldsymbol{r}_1}{\|\boldsymbol{r}_2 - \boldsymbol{r}_1\|^3} \right] \right) \right. \\ &\qquad \left. - \boldsymbol{J}(\boldsymbol{r}_1) \cdot \nabla_2 \left[\frac{\boldsymbol{r}_2 - \boldsymbol{r}_1}{\|\boldsymbol{r}_2 - \boldsymbol{r}_1\|^3} \right] \right\} dv \end{aligned} \tag{1.2.36}$$

Changing from ∇_2 to ∇_1 in (1.2.36), and using the symmetry between $\boldsymbol{r}_2$ and $\boldsymbol{r}_1$, we compute that

$$\nabla_2 \times \boldsymbol{B}(\boldsymbol{r}_2) = \frac{\mu_0}{4\pi} \int_\Omega \Big\{ 4\pi \boldsymbol{J}(\boldsymbol{r}_1)\delta(\boldsymbol{r}_2 - \boldsymbol{r}_1) - \boldsymbol{J}(\boldsymbol{r}_1) \cdot \nabla_2 \left[\frac{\boldsymbol{r}_2 - \boldsymbol{r}_1}{\|\boldsymbol{r}_2 - \boldsymbol{r}_1\|^3} \right] \Big\} \, dv \tag{1.2.37}$$

where $\delta(\boldsymbol{r})$ is the Dirac delta function. An integration by parts applied to the second term in the integral in (1.2.37), when combined with the fact that $\nabla_1 \cdot \boldsymbol{J}(\boldsymbol{r}_1) = 0$, and followed by an application of the divergence theorem, shows that

$$\int_\Omega \boldsymbol{J}(\boldsymbol{r}_1) \cdot \nabla_1 \left[\frac{\boldsymbol{r}_2 - \boldsymbol{r}_1}{\|\boldsymbol{r}_2 - \boldsymbol{r}_1\|^3} \right] dv = 0 \tag{1.2.38}$$

provided $\partial\Omega$ contains the support of $\boldsymbol{J}(\boldsymbol{r})$; from what remains of the relation (1.2.37) it is now immediate that

$$\nabla_2 \times \boldsymbol{B}(\boldsymbol{r}_2) = \mu_0 \boldsymbol{J}(\boldsymbol{r}_2) \tag{1.2.39}$$

which is the differential form of *Ampere's Law*. By applying Stokes' theorem to (1.2.39), we obtain as additional information that

$$\oint_C \boldsymbol{B} \cdot d\boldsymbol{l} = \int_S (\nabla \times \boldsymbol{B}) \cdot \boldsymbol{n} \, da \equiv \mu_0 \int_S \boldsymbol{J} \cdot \boldsymbol{n} \, da \tag{1.2.40}$$

where C is any closed circuit in $\mathbb{R}^3$ that bounds the surface S having an area element da; thus the line integral of $\boldsymbol{B}$ around any closed circuit in $\mathbb{R}^3$ is equal to μ_0 times the total current passing through the surface S bounded by the circuit.

From the fact that $\boldsymbol{B}$ is divergence free, it follows that there exists a vector field $\boldsymbol{A}$, the *magnetic vector potential*, such that

$$\boldsymbol{B} = \nabla \times \boldsymbol{A} \tag{1.2.41}$$

In view of Ampere's law, (1.2.39), $\boldsymbol{A}$ must also satisfy $\nabla \times \nabla \times \boldsymbol{A} = \mu_0 \boldsymbol{J}$; if we make use of the vector identity $\nabla \times \nabla \times \boldsymbol{A} = \nabla(\nabla \cdot \boldsymbol{A}) - \nabla^2 \boldsymbol{A}$ and specify, without loss of generality, that $\nabla \cdot \boldsymbol{A} = 0$, we obtain for $\boldsymbol{A}$ the relation

$$\nabla^2 \boldsymbol{A} = -\mu_0 \boldsymbol{J} \tag{1.2.42}$$

which, it may be shown, easily leads to the representation

$$\boldsymbol{A}(\boldsymbol{r}) = \frac{\mu_0}{4\pi} \int_\Omega \frac{\boldsymbol{J}(\bar{\boldsymbol{r}})}{\|\boldsymbol{r} - \bar{\boldsymbol{r}}\|} \, d\bar{v} \tag{1.2.43}$$

An important quantity in magnetostatics and, especially, in discussions of electromagnetic induction, has yet to be introduced; this is the magnetic flux Φ defined by

$$\Phi = \int_S \boldsymbol{B} \cdot \boldsymbol{n} \, da \tag{1.2.44}$$

An application of the divergence theorem to (1.2.44), when coupled with (1.2.35), shows that $\Phi = 0$ for any closed surface S; a further consequence of this result is, of course, that the flux through a circuit C in $\mathbb{R}^3$ is independent of the particular surface (bounded by C) which is used to compute the flux Φ.

To this point, the basic physical laws, in differential form, for describing the magnetic effects of currents, are expressed by the equations (1.2.35) and (1.2.39); these equations, however, must be modified when the magnetic field includes a contribution from a magnetized material. If a current-carrying circuit consists of a closed loop of wire, then the magnetic field in the vacuum region which surrounds the wire can be computed using (1.2.30); but, if the region which surrounds the wire is filled with a material medium, then the magnetic induction is observed experimentally to be altered by the presence of the medium and this alteration is due to nothing more than the atomic currents which result from the motion of the electrons in the atoms comprising the material medium. Each atomic current, in turn, may be described as a magnetic dipole and, if the magnetic moment of the j-th atom in the medium is denoted by $\boldsymbol{m}_j$, then the macroscopic vector quantity $\boldsymbol{M}$, the *magnetization*, may be defined by the same procedure which led to the definition of the polarization vector $\boldsymbol{P}$, i.e. (1.1.15); we vectorially sum all the atomic dipole moments in an infinitesimal volume element Δv, divide by Δv, and extract the limit as $\Delta v \to 0$, namely,

$$\boldsymbol{M} = \lim_{\Delta v \to 0} \frac{1}{\Delta v} \sum_j \boldsymbol{m}_j \tag{1.2.45}$$

Whenever the process in (1.2.45) is well-defined, the magnetization $\boldsymbol{M}$ represents the magnetic dipole moment per unit volume of the medium. The important field vector

$$\boldsymbol{J}_M = \nabla \times \boldsymbol{M}, \tag{1.2.46}$$

the *magnetization current density*, can be shown to be the equivalent transport current density that would produce the same magnetic field as $\boldsymbol{M}$ itself and the magnetic induction due to a magnetized distribution of matter may be computed by taking the curl of the vector potential due to the magnetization, i.e., by taking the curl of

$$\boldsymbol{A}_M(\boldsymbol{r}) = \frac{\mu_0}{4\pi}\int_\Omega \frac{\boldsymbol{J}_M(\bar{\boldsymbol{r}})}{\|\boldsymbol{r}-\bar{\boldsymbol{r}}\|}\,d\bar{v} \tag{1.2.47}$$

We are now in a position to discuss the appropriate modification, to equations (1.2.35) and (1.2.39), which ensues whenever the magnetic field includes a contribution from a magnetized material. First of all, inasmuch as the field produced by a magnetized medium is derivable from atomic currents, $\boldsymbol{B}$ can always be written as the curl of a vector potential where, in fact, the vector potential due to the magnetization is given by (1.2.47); thus, equation (1.2.35) still applies. On the other hand, in (1.2.39), we must take care to include all the distinct types of currents that are capable of producing a magnetic field; the correct modification of (1.2.39), therefore, reads

$$\nabla \times \boldsymbol{B} = \mu_0(\boldsymbol{J} + \boldsymbol{J}_M) \tag{1.2.48}$$

with $\boldsymbol{J}$ the transport current density and $\boldsymbol{J}_M$ the magnetization current density. Inserting (1.2.46) into (1.2.48), we readily find that Ampere's law applies, namely,

$$\nabla \times \boldsymbol{H} = \boldsymbol{J} \tag{1.2.49}$$

where $\boldsymbol{H}$, the magnetic intensity field, is defined by

$$\boldsymbol{H} = \frac{1}{\mu_0}\boldsymbol{B} - \boldsymbol{M} \tag{1.2.50}$$

The basic magnetic field equations now assume the form (1.2.35) and (1.2.49) and must be supplemented by appropriate boundary conditions and an experimentally motivated relationship between $\boldsymbol{B}$ and $\boldsymbol{H}$; we note that, as an immediate consequence of Stokes' theorem, we obtain from (1.2.49) the fact that

$$\int_S \nabla \times \boldsymbol{H}\cdot\boldsymbol{n}\,da = \oint_C \boldsymbol{H}\cdot d\boldsymbol{l} = \int_S \boldsymbol{J}\cdot\boldsymbol{n}\,da \tag{1.2.51}$$

where $\mathcal{C}$ is any space curve bounding the surface S. Equation (1.2.51) implies that the line integral of the tangential component of the magnetic intensity around a closed path $\mathcal{C}$ in $\mathbb{R}^3$ is equal to the net transport current through the surface bounded by $\mathcal{C}$. As for the relationships which exist among the field vectors displayed in (1.2.50), there exist a large class of materials for which an approximate linear relationship holds between $\boldsymbol{M}$ and $\boldsymbol{H}$ and, if the material is isotropic as well as linear, then

$$\boldsymbol{M} = \chi_m \boldsymbol{H} \tag{1.2.52}$$

where χ_m is a dimensionless scalar called the *magnetic susceptibility*; for $\chi_m > 0$, the medium is termed *paramagnetic* and magnetic induction is strengthened by the presence of the material, while for $\chi_m < 0$ the medium is termed *diamagnetic* and the presence of the medium weakens magnetic induction. For both paramagnetic and diamagnetic materials, χ_m, which may be temperature dependent, is observed to be very small, i.e., $|\chi_m| \ll 1$. If we introduce the constitutive relation (1.2.52) into (1.2.50), and define the *permeability* μ by

$$\mu = \mu_0(1 + \chi_m) \tag{1.2.53}$$

then we clearly have a linear relationship between $\boldsymbol{B}$ and $\boldsymbol{H}$, namely,

$$\boldsymbol{B} = \mu \boldsymbol{H} \tag{1.2.54}$$

However, many materials, the *ferromagnets*, are characterized by a permanent magnetization and for such materials the magnetization, once established, does not disappear when $\boldsymbol{H}$ is brought to a value of zero; for studying the phenomenon of magnetization in ferromagnetic media, a nonlinear constitutive law for isotropic materials, of the form (1.2.54), but with $\mu = \mu(\boldsymbol{H})$, has been widely studied. Finally, we comment briefly on the problem of appropriate boundary conditions to be associated with (1.2.35) and (1.2.49). If we have an interface between two media with different magnetic properties, or between a magnetic material and a vacuum, then arguments analogous to those which led to (1.1.28) and (1.1.29), but based now on the magnetic field equations (1.2.35) and (1.2.49), lead to the results

$$(\boldsymbol{B}_1)_n = (\boldsymbol{B}_2)_n; \quad (\boldsymbol{H}_2 - \boldsymbol{H}_1)_t = \boldsymbol{j} \times \boldsymbol{n} \tag{1.2.55}$$

where $\boldsymbol{B}_i$, $\boldsymbol{H}_i$, $i = 1, 2$ are the magnetic induction and magnetic intensity fields, respectively, in the two different media, the n and t subscripts denote normal and tangential components of the indicated vector fields, $\boldsymbol{n}$ is the exterior unit normal to the interface, and $\boldsymbol{j}$ is the surface current density.

In the natural scheme of things, in the study of classical electromagnetic theory, we have now arrived at the point where we are able to address the phenomenon of electromagnetic induction. It was observed, principally by Faraday and Henry, that the equation which characterizes electrostatics, i.e. $\nabla \times \boldsymbol{E} = \mathbf{0}$ or, in integral form, $\oint_C \boldsymbol{E} \cdot d\boldsymbol{l} = 0$, remains valid provided that the only magnetic force present is that due to a steady current. However, $\nabla \times \boldsymbol{E} = \mathbf{0}$ does not hold in the presence of more general time-dependent fields. It turns out to be useful to define a new quantity, the EMF or *electromotive force*, around a closed circuit $\mathcal{C}$ in $\mathbb{R}^3$, by

$$\mathcal{E} = \oint_C \boldsymbol{E} \cdot d\boldsymbol{l} \tag{1.2.56}$$

so that $\mathcal{E} = 0$ for both static $\boldsymbol{E}$ and $\boldsymbol{B}$ fields. In the presence of time-dependent fields, however, the electric field $\boldsymbol{E}$ can no longer be derived from Coulomb's law but, rather, is now defined so that the Lorentz force, given by (1.2.19), is the electromagnetic force experienced by a test charge q. As a result of experiments, largely associated with Faraday, a relationship of the form

$$\mathcal{E} = -\frac{d\Phi}{dt}, \tag{1.2.57}$$

between the EMF and the change in magnetic flux through a circuit emerged and, indeed, (1.2.57) is known as Faraday's law of electromagnetic induction; it turns out that (1.2.57) is independent of the way in which Φ is varied, i.e., independent of the manner in which the values of $\boldsymbol{B}$ interior to the circuit are changed. If we employ the definitions of $\mathcal{E}$ and Φ, then (1.2.57) assumes the form

$$\oint_C \boldsymbol{E} \cdot d\boldsymbol{l} = -\frac{d}{dt}\int_S \boldsymbol{B} \cdot \boldsymbol{n}\, da \tag{1.2.58}$$

where $\mathcal{C}$ is any space curve in $\mathbb{R}^3$ bounding the surface S. If we now assume that $\mathcal{C}$ is a rigid, stationary circuit in $\mathbb{R}^3$, then by bringing the time derivative into the surface integral, transforming the line integral via Stokes' theorem, and noting that the resulting integral relation must hold for all fixed bounded surfaces embedded in $\mathbb{R}^3$, we are led to the differential form of Faraday's law, namely,

$$\nabla \times \boldsymbol{E} = -\frac{\partial \boldsymbol{B}}{\partial t} \tag{1.2.59}$$

The negative sign in (1.2.59) is indicative of the fact that the direction of the induced EMF is such as to oppose the change in the magnetic induction field which produces it; what this means in a practical sense is that if one tries to increase the flux through a circuit, then the induced EMF causes currents in a direction such as to decrease the flux, i.e., when we try to insert a pole of a magnet into a coil, the currents produced by the induced EMF set up a magnetic field which acts to repel the magnet. Having deduced (1.2.59) from experimental data, we are now in possession of the third of Maxwell's four equations and there remains only the task of generalizing the relation (1.2.49) so as to take into account the presence of time-varying fields; this generalization of (1.2.49) will be accomplished in the next section. Before proceeding to complete the full set of Maxwell's equations, however, we will conclude this section by looking at the concept of energy density in a magnetic field.

Suppose we consider a set of rigid current-carrying circuits, all of which are bounded (i.e. lie in a bounded domain of R^3) and embedded in a magnetic medium which conforms to the linear constitutive hypothesis (1.2.54); the energy of such a system is easily shown to have the form

$$\mathsf{E} = \frac{1}{2}\sum_{j=1}^{n} I_j \Phi_j \tag{1.2.60}$$

where I_j is the current in the j-th circuit, while Φ_j is the associated magnetic flux. For a single circuit C_j, we may write that

$$\Phi_j = \int_{S_j} \boldsymbol{B} \cdot \boldsymbol{n}\, da = \oint_{C_j} \boldsymbol{A} \cdot d\boldsymbol{l}_j \tag{1.2.61}$$

with $\boldsymbol{A}$ the associated vector potential, so that (1.2.60) assumes the form

$$\mathsf{E} = \frac{1}{2}\sum_{j=1}^{n} \oint_{C_j} I_j \boldsymbol{A} \cdot d\boldsymbol{l}_j \tag{1.2.62}$$

Now suppose that we are not dealing with currents in circuits defined by physical wires but, rather, that each circuit in our assumed conducting medium is a closed path of current density. Employing (1.2.62) for a large number of contiguous C_j, and identifying $I_j d\boldsymbol{l}_j$ with $\boldsymbol{J}\, dv$, a simple limiting argument leads us, for $\Omega \subseteq R^3$ a bounded domain, to

$$\mathsf{E} = \frac{1}{2}\int_{\Omega} \boldsymbol{J} \cdot \boldsymbol{A}\, dv \tag{1.2.63}$$

or, inasmuch as $\nabla \times \boldsymbol{H} = \boldsymbol{J}$, and

$$\nabla \cdot (\boldsymbol{A} \times \boldsymbol{H}) = \boldsymbol{H} \cdot (\nabla \times \boldsymbol{A}) - \boldsymbol{A} \cdot (\nabla \times \boldsymbol{H}),$$

to

$$\mathsf{E} = \frac{1}{2}\int_{\Omega} \boldsymbol{H}\cdot(\nabla\times\boldsymbol{A})\,dv - \frac{1}{2}\int_{\partial\Omega}(\boldsymbol{A}\times\boldsymbol{H})\cdot\boldsymbol{n}\,da \tag{1.2.64}$$

By allowing $\partial\Omega$ to move out to infinity, and noting that $\|\boldsymbol{A}\|$ decreases at least as fast as $\frac{1}{r}$, r being the distance from a point in Ω to the origin of whatever coordinate system we have chosen, we easily find that (1.2.64) yields

$$\mathsf{E} = \frac{1}{2}\int_{\mathbb{R}^3} \boldsymbol{H}\cdot\boldsymbol{B}\,dv \tag{1.2.65}$$

thus leading us to define, as the energy density in the magnetic field, the quantity

$$u_M = \frac{1}{2}\boldsymbol{H}\cdot\boldsymbol{B} \tag{1.2.66}$$

The definition (1.2.66) is the natural counterpart to (1.1.39), which serves to define the energy density in an electrostatic field. In the next section we will generalize these concepts and move on to a consideration of electromagnetic energy in the presence of time varying fields. First, however, we must consider Maxwell's generalization of Ampere's law, which is needed in all situations where $\frac{\partial\rho}{\partial t} \neq 0$.

1.3 Maxwell's Equations and the Propagation of Electromagnetic Waves

In the last section we saw that the magnetic field due to a current distribution satisfied Ampere's law (1.2.49), the integral form of which is given by (1.2.51); the aforementioned forms of Ampere's law assume, implicitly, that we are dealing with situations in which the charge density ρ does not change with time, for (1.2.49) implies that $\nabla\cdot\boldsymbol{J} = 0$ and this is compatible with the equation of continuity (1.2.8) only if $\frac{\partial\rho}{\partial t} = 0$.

For physical situations in which $\frac{\partial\rho}{\partial t} \neq 0$, Ampere's circuital law must be generalized; the appropriate generalization of Ampere's law, an achievement due to Maxwell, consists of adding a *displacement current* $\frac{\partial\boldsymbol{D}}{\partial t}$ to the right-hand side of Ampere's law (1.2.49). In fact, inasmuch as $\nabla\cdot\boldsymbol{D} = \rho$ we have, by virtue of the equation of continuity (1.2.8), that

$$\nabla\cdot\boldsymbol{J} + \frac{\partial}{\partial t}(\nabla\cdot\boldsymbol{D}) = \nabla\cdot(\boldsymbol{J} + \frac{\partial\boldsymbol{D}}{\partial t}) = 0 \tag{1.3.1}$$

In view of (1.3.1), it is clear that Ampere's law, for situations in which the fields are not "slowly varying" (so that the displacement current $\frac{\partial \boldsymbol{D}}{\partial t}$ cannot be ignored in comparison with the transport current $\boldsymbol{J}$) must be modified to read

$$\nabla \times \boldsymbol{H} = \boldsymbol{J} + \frac{\partial \boldsymbol{D}}{\partial t} \tag{1.3.2}$$

We are now in possession of the entire set of Maxwell's equations, namely, the extension of Ampere's law (1.3.2), Faraday's law of electromagnetic induction (1.2.59), Gauss's law (1.1.25), and (1.2.35), which remains valid for time-varying fields and is the mathematical expression of the fact that isolated magnetic poles have not been observed (at least not as of the time of preparation of this manuscript). For the sake of convenient future reference, we collect below the full set of Maxwell's equations and renumber them accordingly; it is important to state explicitly that these equations are the mathematical expressions of a vast collection of experimental results and observations and, as such, are not susceptible to mathematical proof. However, Maxwell's equations are the basic equations governing the classical electromagnetic fields produced by source charges and current densities and, as with basic field equations in mechanics, such as conservation of mass or balance of momentum, are thought to apply to all macroscopic electromagnetic phenomena. In summary, the fundamental equations are

$$\nabla \cdot \boldsymbol{D} = \rho \tag{1.3.3a}$$

$$\nabla \cdot \boldsymbol{B} = 0 \tag{1.3.3b}$$

$$\nabla \times \boldsymbol{E} = -\frac{\partial \boldsymbol{B}}{\partial t} \tag{1.3.3c}$$

$$\nabla \times \boldsymbol{H} = \boldsymbol{J} + \frac{\partial \boldsymbol{D}}{\partial t} \tag{1.3.3d}$$

and in the presence of material bodies we must also prescribe constitutive relations among the field vectors appearing above of the form, e.g., $\boldsymbol{D} = \boldsymbol{D}(\boldsymbol{E})$, $\boldsymbol{H} = \boldsymbol{H}(\boldsymbol{B})$, and $\boldsymbol{J} = \boldsymbol{J}(\boldsymbol{E})$. There is no need to add the equation of continuity to the set, as (1.2.8) follows as a direct consequence of (1.3.3d) and (1.3.3a). Once $\boldsymbol{E}$ and $\boldsymbol{B}$ have been determined (at least in principle) from the field equations (1.3.3a)-(1.3.3d), an appropriate set of constitutive relations, and whatever initial and boundary conditions are germane to a particular physical problem, and an associated geometry, then the

Lorentz force equation (1.2.19) is applicable and serves to describe the action of the electromagnetic field on charged particles.

We now consider the matter of electromagnetic energy in the presence of time-varying fields. In the two previous sections, we have noted that the electrostatic potential energy may be expressed in the form

$$\mathsf{E}_E = \frac{1}{2}\int_\Omega \boldsymbol{E}\cdot\boldsymbol{D}\,dv \tag{1.3.4}$$

with an analogous expression for the energy stored in a magnetic field, namely,

$$\mathsf{E}_M = \frac{1}{2}\int_\Omega \boldsymbol{H}\cdot\boldsymbol{B}\,dv \tag{1.3.5}$$

For time-varying fields, the relevant equations are those of Maxwell (1.3.3a)-(1.3.3d). If we take the dot product of (1.3.3d) with $\boldsymbol{E}$ and subtract the equation produced from the result of taking the dot product of (1.3.3c) with $\boldsymbol{H}$, we obtain

$$\boldsymbol{H}\cdot(\nabla\times\boldsymbol{E}) - \boldsymbol{E}\cdot(\nabla\times\boldsymbol{H}) = -\boldsymbol{H}\cdot\frac{\partial\boldsymbol{B}}{\partial t} - \boldsymbol{E}\cdot\frac{\partial\boldsymbol{D}}{\partial t} - \boldsymbol{E}\cdot\boldsymbol{J} \tag{1.3.6}$$

However, for sufficiently smooth vector fields $\boldsymbol{R}$, $\boldsymbol{S}$, we know that

$$\nabla\cdot(\boldsymbol{R}\times\boldsymbol{S}) = \boldsymbol{R}\cdot(\nabla\times\boldsymbol{S}) - \boldsymbol{S}\cdot(\nabla\times\boldsymbol{R}) \tag{1.3.7}$$

and an application of this vector identity to (1.3.6) yields

$$\nabla\cdot(\boldsymbol{E}\times\boldsymbol{H}) = -\boldsymbol{H}\cdot\frac{\partial\boldsymbol{B}}{\partial t} - \boldsymbol{E}\cdot\frac{\partial\boldsymbol{D}}{\partial t} - \boldsymbol{E}\cdot\boldsymbol{J} \tag{1.3.8}$$

When the substance to which we want to apply (1.3.8) is both linear and nondispersive, so that $\boldsymbol{D} = \epsilon\boldsymbol{E}$, $\boldsymbol{B} = \mu\boldsymbol{H}$, with ϵ, μ time-independent, as well as independent of the field vectors $\boldsymbol{E}$ and $\boldsymbol{H}$, respectively, then

$$\begin{aligned}
\boldsymbol{E}\cdot\frac{\partial\boldsymbol{D}}{\partial t} = \boldsymbol{E}\cdot\frac{\partial}{\partial t}(\epsilon\boldsymbol{E}) &= \frac{1}{2}\epsilon\frac{\partial}{\partial t}\|\boldsymbol{E}\|^2 = \frac{\partial}{\partial t}(\frac{1}{2}\boldsymbol{E}\cdot\boldsymbol{D}) \\
\boldsymbol{H}\cdot\frac{\partial\boldsymbol{B}}{\partial t} = \boldsymbol{H}\cdot\frac{\partial}{\partial t}(\mu\boldsymbol{H}) &= \frac{1}{2}\mu\frac{\partial}{\partial t}\|\boldsymbol{H}\|^2 = \frac{\partial}{\partial t}(\frac{1}{2}\boldsymbol{H}\cdot\boldsymbol{B})
\end{aligned}$$

and (1.3.8) assumes the form

$$\nabla\cdot(\boldsymbol{E}\times\boldsymbol{H}) = -\frac{\partial}{\partial t}\frac{1}{2}(\boldsymbol{E}\cdot\boldsymbol{D} + \boldsymbol{B}\cdot\boldsymbol{H}) - \boldsymbol{J}\cdot\boldsymbol{E} \tag{1.3.9}$$

Comparing the expression on the right-hand side of (1.3.9) with the integrands in (1.3.4) and (1.3.5), we see that it is still appropriate in this case to identify $\frac{1}{2}(\boldsymbol{E}\cdot\boldsymbol{D}+$

$\boldsymbol{H} \cdot \boldsymbol{B}$) as being the sum of the electric and magnetic energy densities; the physical interpretation of the term $\boldsymbol{J} \cdot \boldsymbol{E}$ is that it represents, in most instances (e.g., for $\boldsymbol{J} = \sigma \boldsymbol{E}$), the negative of the *Joule heating rate* per unit volume.

If we integrate (1.3.9) over a bounded domain $\Omega \subseteq R^3$, with sufficiently smooth boundary $\partial\Omega$, and then apply the divergence theorem to the left-hand side of the resulting identity, we obtain

$$\begin{aligned} -\int_\Omega \boldsymbol{J} \cdot \boldsymbol{E}\, dv &= \frac{d}{dt}\int_\Omega \frac{1}{2}[\boldsymbol{E} \cdot \boldsymbol{D} + \boldsymbol{B} \cdot \boldsymbol{H}]\, dv \\ &\quad + \oint_{\partial\Omega} \boldsymbol{E} \times \boldsymbol{H} \cdot \boldsymbol{n}\, da \end{aligned} \tag{1.3.10}$$

Equation (1.3.10) shows that $\int_\Omega \boldsymbol{J} \cdot \boldsymbol{E}\, dv$ is composed of two parts: the rate of change of the electromagnetic energy stored in Ω plus a surface integral representing the rate of energy flow across the bounding surface $\partial\Omega$; with this interpretation (1.3.10) expresses conservation of electromagnetic energy in a fixed volume $\Omega \subseteq R^3$. If in (1.3.9) we now set

$$\boldsymbol{S}^p = \boldsymbol{E} \times \boldsymbol{H} \tag{1.3.11}$$

and

$$u \equiv u_E + u_M = \frac{1}{2}(\boldsymbol{E} \cdot \boldsymbol{D} + \boldsymbol{B} \cdot \boldsymbol{H}) \tag{1.3.12}$$

then from (1.3.9) we obtain directly that

$$\nabla \cdot \boldsymbol{S}^p + \frac{\partial u}{\partial t} = -\boldsymbol{J} \cdot \boldsymbol{E} \tag{1.3.13}$$

If $\nabla \cdot \boldsymbol{S}^p = 0$ in (1.3.13), then the resulting relation expresses local conservation of energy, i.e., at any point $\boldsymbol{x} \in \Omega \subseteq R^3$, the rate of change of the electromagnetic field energy equals the power dissipation per unit volume. If, however, $\boldsymbol{J} \cdot \boldsymbol{E} = 0$ but $\nabla \cdot \boldsymbol{S}^p \neq 0$ (e.g. in a nonconducting medium), then (1.3.13) reduces to

$$\nabla \cdot \boldsymbol{S}^p + \frac{\partial u}{\partial t} = 0 \tag{1.3.14}$$

which is of the same form as the equation of continuity (1.2.8) except that $\boldsymbol{S}^p$ now assumes the role previously taken by $\boldsymbol{J}$, while the energy density of the electromagnetic field u assumes the role of the charge density ρ; in situations where (1.3.14) applies, $\nabla \cdot \boldsymbol{S}^p$ then represents the divergence of an energy current density or, equivalently, a rate of energy flow per unit area. The vector $\boldsymbol{S}^p$, defined by (1.3.11), is known in electromagnetic theory as the *Poynting vector.*

Probably the most important of all consequences of Maxwell's equations are the equations governing electromagnetic wave propagation; while the relevant wave equations are presented in most texts on the subject only for wave propagation in a linear medium, we will present here both the usual linear wave equation and a strongly coupled system of nonlinear wave equations for the components $D_i(\boldsymbol{x},t)$ of the electric displacement field, in a nonlinear dielectric medium which conforms to a constitutive hypothesis of the form

$$\begin{cases} \boldsymbol{D}(\boldsymbol{x},t) = \boldsymbol{D}(\boldsymbol{E}(\boldsymbol{x},t)) \\ \boldsymbol{B}(\boldsymbol{x},t) = \mu \boldsymbol{H}(\boldsymbol{x},t) \end{cases} \tag{1.3.15}$$

with $\boldsymbol{D}(\boldsymbol{0}) = \boldsymbol{0}$, $\mu > 0$ (constant), and

$$\det \left[\frac{\partial D_i}{\partial E_j}\right]\bigg|_{\boldsymbol{E}=\boldsymbol{0}} \neq 0 \tag{1.3.16}$$

In a linear medium, with $\boldsymbol{D} = \epsilon \boldsymbol{E}$, and $\boldsymbol{J} = \sigma \boldsymbol{E}$ ($\epsilon, \sigma > 0$ and constant) we first take the curl on both sides of (1.3.3d), and substitute from the constitutive relations, so as to obtain

$$\nabla \times \nabla \times \boldsymbol{H} = \sigma \nabla \times \boldsymbol{E} + \epsilon \frac{\partial}{\partial t}(\nabla \times \boldsymbol{E}) \tag{1.3.17}$$

Employing (1.3.3c) in (1.3.17), together with the constitutive hypothesis $\boldsymbol{B} = \mu \boldsymbol{H}$, $\mu > 0$ and constant, we find that

$$\nabla \times \nabla \times \boldsymbol{H} = -\sigma\mu \frac{\partial \boldsymbol{H}}{\partial t} - \epsilon\mu \frac{\partial^2 \boldsymbol{H}}{\partial t^2} \tag{1.3.18}$$

However, for any sufficiently smooth vector field $\boldsymbol{v}$, we have the well-known identity

$$\nabla \times \nabla \times \boldsymbol{v} = \nabla \nabla \cdot \boldsymbol{v} - \nabla^2 \boldsymbol{v}$$

whose use in (1.3.18) readily yields

$$\nabla \nabla \cdot \boldsymbol{H} - \nabla^2 \boldsymbol{H} = -\sigma\mu \frac{\partial \boldsymbol{H}}{\partial t} - \epsilon\mu \frac{\partial^2 \boldsymbol{H}}{\partial t^2} \tag{1.3.19}$$

or, as $\nabla \cdot \boldsymbol{H} = \dfrac{1}{\mu} \nabla \cdot \boldsymbol{B} = 0$, by virtue of (1.3.3b), the wave equation

$$\epsilon\mu \frac{\partial^2 \boldsymbol{H}}{\partial t^2} + \sigma\mu \frac{\partial \boldsymbol{H}}{\partial t} = \nabla^2 \boldsymbol{H} \tag{1.3.20}$$

for the evolution of the magnetic intensity field $\boldsymbol{H}$. If we begin a similar reduction by first taking the curl on both sides of the third of Maxwell's equations, (1.3.3c), we are easily led to the same wave equation (1.3.20) for the evolution of the electric field vector $\boldsymbol{E}$, namely,

$$\epsilon\mu\frac{\partial^2 \boldsymbol{E}}{\partial t^2} + \sigma\mu\frac{\partial \boldsymbol{E}}{\partial t} = \nabla^2 \boldsymbol{E} \tag{1.3.21}$$

provided $\nabla \cdot \boldsymbol{D} = 0$ in the linear medium, that is, provided the charge density is zero.

Inasmuch as (1.3.21) applies only to linear media, in which the charge density $\rho \equiv 0$, and the only current density $\boldsymbol{J}$ is that which arises from the passive response of the medium to the electric field of the wave, we now want to consider (still for a linear medium) the situation in which there are prescribed charge and current distributions $\rho(\boldsymbol{x}, t)$ and $\boldsymbol{J}(\boldsymbol{x}, t)$, respectively; the standard approach here consists of noting that, as the magnetic induction field $\boldsymbol{B}$ is divergence free, there exists a vector potential $\boldsymbol{A}(\boldsymbol{x}, t)$ such that (1.2.41) is satisfied at each point $\boldsymbol{x} \in \Omega \subseteq R^3$ and each $t \geq 0$. Employing (1.2.41) in Maxwell's third equation (1.3.3c), and interchanging the spatial and temporal differentiations, we are led to the equation

$$\nabla \times \left(\boldsymbol{E} + \frac{\partial \boldsymbol{A}}{\partial t}\right) = 0 \tag{1.3.22}$$

But (1.3.22) implies that there exists a scalar-valued function $\phi(\boldsymbol{x}, t)$ such that

$$\boldsymbol{E} = -\nabla\phi - \frac{\partial \boldsymbol{A}}{\partial t} \tag{1.3.23}$$

so that (1.2.41) and (1.3.23) now yield both the electric and magnetic fields in terms of a vector potential $\boldsymbol{A}$ and a scalar potential ϕ. By substituting (1.2.41) and (1.3.23) into the fourth of Maxwell's equations, (1.3.3d), after first setting $\boldsymbol{D} = \epsilon \boldsymbol{E}$ and $\boldsymbol{H} = \frac{1}{\mu}\boldsymbol{B}$, we are led to a wave equation for $\boldsymbol{A}$,

$$-\nabla^2 \boldsymbol{A} + \epsilon\mu\frac{\partial^2 \boldsymbol{A}}{\partial t^2} + \nabla\nabla \cdot \boldsymbol{A} + \epsilon\mu\nabla\left(\frac{\partial \phi}{\partial t}\right) = \mu \boldsymbol{J}, \tag{1.3.24}$$

where we have used the aforementioned vector identity for $\nabla \times \nabla \times \boldsymbol{A}$. Although $\nabla \times \boldsymbol{A} = \boldsymbol{B}$ is specified, the choice of $\nabla \cdot \boldsymbol{A}$ has, until this point, been arbitrary. If we now impose the *Lorentz condition*, namely,

$$\nabla \cdot \boldsymbol{A} = -\epsilon\mu\frac{\partial \phi}{\partial t} \tag{1.3.25}$$

then, clearly, (1.3.24) reduces to the wave equation

$$\epsilon\mu\frac{\partial^2 \boldsymbol{A}}{\partial t^2} - \nabla^2 \boldsymbol{A} = \mu \boldsymbol{J} \tag{1.3.26}$$

We next return to the representation of $\boldsymbol{E}$ given by (1.3.23) and substitute this relation in Maxwell's first equation (1.3.3a) after setting $\boldsymbol{D} = \epsilon \boldsymbol{E}$; this substitution yields as a consequence the equation

$$-\epsilon\left(\nabla\cdot\nabla\phi + \nabla\cdot\frac{\partial \boldsymbol{A}}{\partial t}\right) = \rho, \tag{1.3.27}$$

where $\rho = \rho(\boldsymbol{x}, t)$ is the prescribed charge density. Finally, if we write $\nabla \cdot \frac{\partial \boldsymbol{A}}{\partial t} = \frac{\partial}{\partial t}(\nabla \cdot \boldsymbol{A})$ in (1.3.27), and employ the Lorentz condition (1.3.25), we obtain for the scalar potential ϕ a wave equation entirely analogous to (1.3.26), namely,

$$\epsilon\mu\frac{\partial^2 \phi}{\partial t^2} - \nabla^2 \phi = \frac{1}{\epsilon}\rho \tag{1.3.28}$$

The solutions of the inhomogeneous wave equations (1.3.26), (1.3.28), for prescribed fields ρ, J, and associated initial and boundary data, consist, of course, of the general solution of the associated homogeneous linear wave equations plus particular solutions of the inhomogeneous equations; for the inhomogeneous wave equations (1.3.26), (1.3.28), particular solutions are well-known and are given by the *retarded scalar and vector potentials*

$$\phi(\boldsymbol{x}, t) = \frac{1}{4\pi\epsilon}\int_\Omega \frac{\rho(\bar{\boldsymbol{x}}, \bar{t})}{\|\boldsymbol{x} - \bar{\boldsymbol{x}}\|}\, d\bar{v} \tag{1.3.29a}$$

$$\boldsymbol{A}(\boldsymbol{x}, t) = \frac{\mu}{4\pi}\int_\Omega \frac{\boldsymbol{J}(\bar{\boldsymbol{x}}, \bar{t})}{\|\boldsymbol{x} - \bar{\boldsymbol{x}}\|}\, d\bar{v} \tag{1.3.29b}$$

where $\bar{t} = t - \sqrt{\epsilon\mu}\|\boldsymbol{x} - \bar{\boldsymbol{x}}\|$ is the retarded time. Once the scalar and vector potentials ϕ and $\boldsymbol{A}$ have been computed, the fields $\boldsymbol{B}$ and $\boldsymbol{E}$ may be determined from (1.2.41) and (1.3.23), respectively. Of course the fact that $\boldsymbol{A}$ and ϕ satisfy the inhomogeneous wave equations (1.3.26) and (1.3.28) is due to our imposition of the Lorentz condition (1.3.25); however, if a particular choice of the potentials $\boldsymbol{A}$ and ϕ yield, through (1.2.41) and (1.3.23), the appropriate electric and magnetic fields, then a *gauge transformation* of the form

$$\begin{cases} \hat{\boldsymbol{A}} = \boldsymbol{A} + \nabla\eta \\ \hat{\phi} = \phi - \dfrac{\partial \eta}{\partial t} \end{cases} \tag{1.3.30}$$

for arbitrary sufficiently smooth $\eta(\boldsymbol{x}, t)$, yields new potentials which produce the same $\boldsymbol{E}$ and $\boldsymbol{B}$ fields. If we substitute from (1.3.30) into (1.3.24), with $(\hat{\boldsymbol{A}}, \hat{\phi})$ replacing $(\boldsymbol{A}, \phi)$, we readily find that

$$\nabla^2\eta - \epsilon\mu\frac{\partial^2\eta}{\partial t^2} = -\left(\nabla \cdot \boldsymbol{A} + \epsilon\mu\frac{\partial\phi}{\partial t}\right) = 0 \tag{1.3.31}$$

so that $(\hat{\boldsymbol{A}}, \hat{\phi})$ satisfy the Lorentz condition, whenever $(\boldsymbol{A}, \phi)$ do, provided that η satisfies the usual scalar wave equation. Even if $(\boldsymbol{A}, \phi)$ did not satisfy the Lorentz condition, it is clear from (1.3.31) that $(\hat{\boldsymbol{A}}, \hat{\phi})$ will, provided the scalar-valued function η in (1.3.30) satisfies an inhomogeneous wave equation.

At this point it is worthwhile to comment on one very important aspect of the linear wave equations (1.3.26) and (1.3.28), which were deduced from Maxwell's equations (1.3.3a)-(1.3.3d) under the assumptions that the charge density ρ in the medium (either conducting or nonconducting) is zero and that the only current density $\boldsymbol{J}$ is the one which arises in passive response to the electric field in the wave: while (1.3.20) and (1.3.21) are consequences, in the aforementioned situation, of Maxwell's equations, the converse does not follow and Maxwell's equations (1.3.3a)-(1.3.3d) must still serve as a restriction on solutions of the wave equations (1.3.20) and (1.3.21).

The boundary conditions, at an interface between two media, are derived from Maxwell's equations, just as in the static case, using the familiar pillbox-shaped surfaces at the interface. From (1.3.3b) we readily deduce, just as in the static case, that $(\boldsymbol{B}_1 - \boldsymbol{B}_2)\cdot\boldsymbol{n} = 0$, where $\boldsymbol{B}_i$ is the limiting value of $\boldsymbol{B}$ in the i-th medium as we approach a point on the interface, while $\boldsymbol{n}$ is the exterior unit normal to the interface. Also, from (1.3.3c) it follows that $(\boldsymbol{E}_1 - \boldsymbol{E}_2)_t$ is proportional to

$$\oint_S \frac{\partial \boldsymbol{B}}{\partial t} \cdot \boldsymbol{n}\, da$$

where the subscript t denotes the tangential component of the indicated vector field on the interface, while S denotes the surface bounded by an infinitesimal rectangular path intersecting the interface. If $\dfrac{\partial \boldsymbol{B}}{\partial t}$ is bounded in a neighborhood of the interface, then the above integral vanishes as the sides of the rectangular path are shrunk down to zero length and we again find, as in the static case, that $(\boldsymbol{E}_1)_t = (\boldsymbol{E}_2)_t$. From the first of Maxwell's equations (1.3.3a), which we also integrate over a pillbox-shaped domain intersecting the interface, we find that $(\boldsymbol{D}_1 - \boldsymbol{D}_2)\cdot\boldsymbol{n} = \sigma_S$, where σ_S is the surface charge density on the interface; an analogous procedure, when applied to the equation of continuity (1.2.8), produces the fact that $(\boldsymbol{J}_1 - \boldsymbol{J}_2) \cdot \boldsymbol{n} = -\frac{\partial \sigma_S}{\partial t}$. Finally, if we integrate the last of Maxwell's equations (1.3.3d) over the same rectangular type of path that was employed to derive the boundary condition on $\boldsymbol{E}_t$ at the interface, we find that $(\boldsymbol{H}_1 - \boldsymbol{H}_2)_t = 0$, except in those cases where the conductivity σ, which appears in the constitutive relation (1.2.9), is infinite.

While we do not wish to dwell, in this monograph, on the matter of wave propagation in linear media, a brief discussion of special solutions of the wave equations (1.3.20), (1.3.21) should serve as a useful adjunct to the analysis of nonlinear wave propagation which will follow in the next two chapters. The class of solutions of the linear wave equations (1.3.20) and (1.3.21) which is most amenable to an elementary analysis consists of the so-called *monochromatic waves*, which are characterized by a single frequency. For monochromatic waves, one may obtain a solution of the wave equation (1.3.21) for $\boldsymbol{E}$ in order to compute $\nabla \times \boldsymbol{E}$ which, in turn, will yield $\frac{\partial \boldsymbol{B}}{\partial t}$; for a monochromatic wave the relationship between $\boldsymbol{B}$ and $\frac{\partial \boldsymbol{B}}{\partial t}$ is then sufficiently simple, so that a solution of (1.3.20) will follow directly. Thus, if we look for solutions of (1.3.21) of the form

$$\boldsymbol{E}(\boldsymbol{x}, t) = \boldsymbol{E}(\boldsymbol{x}) e^{-i\omega t}, \tag{1.3.32}$$

where the physically relevant electric field is obtained by taking the real part of (1.3.32), $\boldsymbol{E}(\boldsymbol{x})$, in general, also being complex-valued, then substitution in (1.3.21) yields, for the spatial part of $\boldsymbol{E}$, the equation

$$\nabla^2 \boldsymbol{E} + \omega^2 \epsilon \mu \boldsymbol{E} + i\omega\mu \boldsymbol{E} = 0 \tag{1.3.33}$$

Several simple cases of (1.3.33) are commonly encountered in practice: if the wave is propagating in empty space, then $\sigma = 0$, $\epsilon = \epsilon_0$, $\mu = \mu_0$. If, in addition, $\boldsymbol{E}(\boldsymbol{x})$ varies in only one space dimension, say the x-direction, then $\boldsymbol{E}(\boldsymbol{x}) = (E_1(x), E_2(x), E_3(x))$ and (1.3.33) reduces to

$$\frac{d^2 \boldsymbol{E}(x)}{dx^2} + \left(\frac{\omega}{c}\right)^2 \boldsymbol{E}(x) = 0 \tag{1.3.34}$$

with $c = \sqrt{\frac{1}{\epsilon_0 \mu_0}}$ the speed of light in a vacuum (indeed, this consequence of Maxwell's equations confirmed the electromagnetic nature of light). The Helmholtz equation

(1.3.34) has solutions of the form

$$\boldsymbol{E}(x) = \hat{\boldsymbol{E}} e^{\pm i\kappa x}, \tag{1.3.35}$$

with $\hat{\boldsymbol{E}}$ a constant vector, provided that the *wave number* $\kappa = \omega/c$. Combining (1.3.32) with (1.3.35), we have the following solution of (1.3.21) in this special case, namely,

$$\boldsymbol{E}(\boldsymbol{x}, t) = \hat{\boldsymbol{E}} e^{-i(\omega t \pm \kappa x)} \tag{1.3.36}$$

whose real part

$$\operatorname{Re} \boldsymbol{E}(\boldsymbol{x}, t) = \hat{\boldsymbol{E}} \cos(\omega t \pm \kappa x) \tag{1.3.37}$$

represents a sinusoidal wave traveling either to the right or left in the x-direction with speed $\frac{\omega}{\kappa} = c$. From (1.3.37) we gather that the wave has frequency $f = \frac{\omega}{2\pi}$ and wavelength $\lambda = \frac{2\pi}{\kappa}$, so that $\lambda f = c$.

If the medium is a nonconducting, linear, nonmagnetic dielectric then, again, $\sigma = 0$, $\mu = \mu_0$, but $\epsilon = K\epsilon_0$, where K is termed the *dielectric constant*; recall that, in general, $\boldsymbol{D} = \epsilon \boldsymbol{E}$, with the permittivity of the form (1.1.27), and for linear media the susceptibility χ is, most often, a constant so that

$$K = \frac{\epsilon}{\epsilon_0} = 1 + \frac{\chi}{\epsilon_0} \tag{1.3.38}$$

is constant. In this case the same results hold as in the vacuum situation, but with $\kappa = \dfrac{\sqrt{K}\omega}{c}$, and the velocity of the propagating wave is now c/n instead of c, where $n = \sqrt{\kappa}$ is termed the *index of refraction* of the linear dielectric medium. If, on the other hand, the medium is conducting, then $\sigma > 0$ and the last term in (1.3.33) cannot be dropped; two extreme subcases which arise here are those in which either $\sigma \ll \omega\epsilon$ or $\sigma \gg \omega\epsilon$. For $\sigma \gg \omega\epsilon$ it is common to ignore the second term on the left-hand side of (1.3.33), in which case, again for a one spatial-dimension dependence assumed for $\boldsymbol{E}$, we obtain

$$\frac{d^2 \boldsymbol{E}(x)}{dx^2} + i\omega\sigma\mu \boldsymbol{E}(x) = \boldsymbol{0} \tag{1.3.39}$$

If we assume a purely imaginary frequency ω, so that $\gamma = i\omega$ is real, and set $\kappa = \sqrt{\gamma\sigma\mu}$, then the same spatial dependence for $\boldsymbol{E}(x)$ as that obtained in (1.3.35) applies in this situation as well, but now, in place of (1.3.36), we obtain

$$\boldsymbol{E}(\boldsymbol{x}, t) = \hat{\boldsymbol{E}} e^{\pm i\kappa x} e^{-\gamma t} \tag{1.3.40}$$

so that the electric field in the wave, instead of oscillating, decays exponentially in time.

Finally, we want to consider general plane wave solutions of (1.3.33), beginning first with the case of plane monochromatic waves in nonconducting media and then proceeding to the case of conducting media. Recall that a plane wave propagating in the direction $\boldsymbol{l}$ is described by the function $\exp\{-i(\omega t - \kappa \boldsymbol{l} \cdot \boldsymbol{x})\}$ and if we define a propagation vector $\boldsymbol{\kappa} = \kappa \boldsymbol{l}$, then we may write this as $\exp\{-i(\omega t - \boldsymbol{\kappa} \cdot \boldsymbol{l})\}$. In the special cases considered above, we took $\boldsymbol{l} = \boldsymbol{i}$, the unit vector in the x direction, so that $\boldsymbol{l} \cdot \boldsymbol{x} = x$. By definition, the velocity of propagation of a plane monochromatic wave is just the velocity that the planes of constant phase move with, where by constant phase we mean that $\boldsymbol{\kappa} \cdot \boldsymbol{l} - \omega t =$ constant. As $\boldsymbol{\kappa} \cdot \boldsymbol{l} = \|\boldsymbol{\kappa}\| \|\boldsymbol{l}\| \cos\theta \equiv \kappa \tilde{l}$, $\kappa = \|\boldsymbol{\kappa}\|$ and $\tilde{l}$ is the projection of $\boldsymbol{l}$ on the direction of $\boldsymbol{\kappa}$, we have for the *phase velocity* v_p

$$v_p = \frac{d\tilde{l}}{dt} = \frac{\omega}{\kappa} \tag{1.3.41}$$

Now, in order to obtain the plane wave solutions $\boldsymbol{E}, \boldsymbol{B}$ of the wave equations (1.3.20) and (1.3.21), it actually turns out to be more convenient to go back to Maxwell's equations (1.3.3a)-(1.3.3d); in these equations, as we are first dealing with a nonconductor, and will assume that there are no prescribed charge or current densities, we have $\rho = 0$ in (1.3.3a) and $\boldsymbol{J} = \boldsymbol{0}$ in (1.3.3d). We begin by assuming that the electric field has the form

$$\boldsymbol{E}(\boldsymbol{x}, t) = \hat{E} \exp\{-i(\omega t - \boldsymbol{\kappa} \cdot \boldsymbol{l})\} \tag{1.3.42}$$

with analogous expressions for $\boldsymbol{D}$, $\boldsymbol{B}$, and $\boldsymbol{H}$; upon substituting $\boldsymbol{E}$ from (1.3.42), and the similar expressions for $\boldsymbol{D}$, $\boldsymbol{B}$, and $\boldsymbol{H}$, into (1.3.3a)-(1.3.3d), with $\rho = 0$, $\boldsymbol{J} = \boldsymbol{0}$, we obtain the following set of four equations to be satisfied by $\hat{\boldsymbol{E}}$, $\hat{\boldsymbol{D}}$, $\hat{\boldsymbol{B}}$, and $\hat{\boldsymbol{H}}$, the (constant) complex vector amplitudes associated with the plane wave:

$$\boldsymbol{\kappa} \times \hat{\boldsymbol{E}} = \omega \hat{\boldsymbol{B}} \quad ; \quad \boldsymbol{\kappa} \cdot \hat{\boldsymbol{D}} = 0 \tag{1.3.43}$$

$$\boldsymbol{\kappa} \cdot \hat{\boldsymbol{B}} = 0 \quad ; \quad \boldsymbol{\kappa} \times \hat{\boldsymbol{H}} = -\omega \hat{\boldsymbol{D}} \tag{1.3.44}$$

where, as a direct consequence of the linearity of the nonconducting medium,

$$\hat{\boldsymbol{D}} = \epsilon \hat{\boldsymbol{E}} \quad ; \quad \hat{\boldsymbol{H}} = \frac{1}{\mu} \hat{\boldsymbol{B}} \tag{1.3.45}$$

The ϵ and μ are assumed here to be constant scalars, so that the medium is also both isotropic and homogeneous and, if we also take the medium to be nonmagnetic, then we may as well set $\mu = \mu_0$. Noting again that $\epsilon = K\epsilon_0$, while $\epsilon_0\mu_0 = 1/c^2$, we easily find that as a consequence of (1.3.43)-(1.3.45) we have

$$K\boldsymbol{\kappa}\cdot\hat{\boldsymbol{E}} = 0 \quad ; \quad \boldsymbol{\kappa}\times\hat{\boldsymbol{E}} = \omega\hat{\boldsymbol{B}} \tag{1.3.46}$$

$$\boldsymbol{\kappa}\cdot\hat{\boldsymbol{B}} = 0 \quad ; \quad \boldsymbol{\kappa}\times\hat{\boldsymbol{B}} = -\frac{\omega}{c^2}K\hat{\boldsymbol{E}} \tag{1.3.47}$$

which for a given pair (ω, K) is a set of (vector) algebraic equations to be satisfied by $\boldsymbol{\kappa}, \hat{\boldsymbol{E}}$, and $\hat{\boldsymbol{B}}$. If $K \neq 0$, then from the first equation in each of the sets (1.3.46), (1.3.47) we see that both $\boldsymbol{E}$ and $\boldsymbol{B}$ must be orthogonal to $\boldsymbol{\kappa}$ and a plane wave with this property is termed *transverse.* But, from the second equation in the set (1.3.46), we have that $\hat{\boldsymbol{B}}$ is proportional to $\boldsymbol{\kappa}\times\hat{\boldsymbol{E}}$ so that $\hat{\boldsymbol{E}}\cdot\hat{\boldsymbol{B}} = 0$ and, thus, the electric and magnetic field vectors in any such propagating plane wave are also orthogonal to each other and $\boldsymbol{\kappa}$, $\boldsymbol{E}$, $\boldsymbol{B}$ form a (right-handed) orthogonal set. From the second equation in (1.3.46) it follows that $\|\hat{\boldsymbol{B}}\| = (\kappa/\omega)\|\hat{\boldsymbol{E}}\|$; to compute $\|\boldsymbol{\kappa}\|$ we employ the second equation in each of the sets (1.3.46) and (1.3.47) and find that

$$\boldsymbol{\kappa}\times(\boldsymbol{\kappa}\times\hat{\boldsymbol{E}}) = \omega\boldsymbol{\kappa}\times\hat{\boldsymbol{B}} = -K\left(\frac{\omega}{c}\right)^2\hat{\boldsymbol{E}} \tag{1.3.48}$$

which, when coupled with the vector identity $\boldsymbol{\kappa}\times(\boldsymbol{\kappa}\times\boldsymbol{v}) = (\boldsymbol{\kappa}\cdot\boldsymbol{v})\boldsymbol{\kappa} - \kappa^2\boldsymbol{v}$, and the fact that $\boldsymbol{\kappa}\cdot\hat{\boldsymbol{E}} = 0$, in a transverse wave, yields

$$-K\left(\frac{\omega}{c}\right)^2\hat{\boldsymbol{E}} = -\kappa^2\hat{\boldsymbol{E}} \tag{1.3.49}$$

From (1.3.49) it is immediate that

$$\kappa = \|\boldsymbol{\kappa}\| = \frac{\sqrt{K}\omega}{c} \tag{1.3.50}$$

which is often termed the *transverse dispersion relation.*

Now we turn our attention to the case of plane monochromatic waves in a conducting medium, again assuming that there are no prescribed charge or current distributions, but allowing for an induced current density of the form $\boldsymbol{J} = \sigma\boldsymbol{E}$, $\sigma > 0$ constant, which arises in response to the electric field in the wave. Our calculations are entirely analogous to those displayed above for a nonconducting medium, except that we now employ (1.3.3d) with $\boldsymbol{J} = \sigma\boldsymbol{E}$; we are led to

$$\boldsymbol{\kappa}\times\hat{\boldsymbol{H}} = -\omega\hat{\boldsymbol{D}} - i\sigma\hat{\boldsymbol{E}} \tag{1.3.51}$$

in lieu of the second equation in (1.3.44). Substituting, again, $\hat{\boldsymbol{H}} = \frac{1}{\mu}\hat{\boldsymbol{B}}$, $\hat{\boldsymbol{D}} = \epsilon\hat{\boldsymbol{E}}$, with $\epsilon = K\epsilon_0$, we obtain from (1.3.51) the relation

$$\boldsymbol{\kappa} \times \hat{\boldsymbol{B}} = -\frac{\omega}{c^2}\left(K + i\frac{\sigma}{\epsilon_0\omega}\right)\hat{\boldsymbol{E}} \tag{1.3.52}$$

so that if we define the complex dielectric constant K_c by

$$K_c = K + i\frac{\sigma}{\epsilon_0\omega} \tag{1.3.53}$$

then

$$\boldsymbol{\kappa} \times \hat{\boldsymbol{B}} = -\frac{\omega}{c^2}K_c\hat{\boldsymbol{E}} \tag{1.3.54}$$

which is analogous to the second equation in the set (1.3.47). With the assumptions $K_c \neq 0$, and $\boldsymbol{\kappa} \cdot \hat{\boldsymbol{E}} = 0$, a transverse dispersion relation similar to (1.3.40) results, namely, $\kappa = \frac{\sqrt{K_c}\omega}{c}$ which, in turn, leads to the definition of a complex refractive index $n_c = \sqrt{K_c}$. Now, in order to satisfy the relation $\kappa = n_c\omega/c$, clearly, either $\boldsymbol{\kappa}$ must be a complex-valued vector or ω must be a complex-valued scalar. Suppose that ω is real, while $\boldsymbol{\kappa} = \boldsymbol{\kappa}_R + i\boldsymbol{\kappa}_c$. Then, formally, we have plane wave solutions of Maxwell's equations (1.3.3a)-(1.3.3d) of the form

$$\begin{cases} \boldsymbol{E}(\boldsymbol{x},t) &= \hat{\boldsymbol{E}}e^{-\boldsymbol{\kappa}_c\cdot\boldsymbol{x}}e^{-i(\omega t-\boldsymbol{\kappa}_R\cdot\boldsymbol{x})} \\ \boldsymbol{B}(\boldsymbol{x},t) &= \hat{\boldsymbol{B}}e^{-\boldsymbol{\kappa}_c\cdot\boldsymbol{x}}e^{-i(\omega t-\boldsymbol{\kappa}_R\cdot\boldsymbol{x})} \end{cases} \tag{1.3.55}$$

The relations (1.3.55) represent a plane wave which propagates in the direction $\boldsymbol{\kappa}_R$ with wavelength $\lambda = \frac{2\pi}{\|\boldsymbol{\kappa}_R\|}$ and whose amplitude is exponentially decreasing, with the most pronounced decrease occurring if $\boldsymbol{x}$ is parallel to $\boldsymbol{\kappa}_c$. Surfaces of constant phase and constant amplitude may be defined, respectively, as planes orthogonal to the directions of $\boldsymbol{\kappa}_R$ and $\boldsymbol{\kappa}_c$. From the decomposition of $\boldsymbol{\kappa}$ into real and complex parts we have

$$\kappa = \left(\|\boldsymbol{\kappa}_R\|^2 - \|\boldsymbol{\kappa}_c\|^2 + 2i\boldsymbol{\kappa}_R \cdot \boldsymbol{\kappa}_c\right)^{1/2} \tag{1.3.56}$$

Since $\kappa = n_c\omega/c$, we may set $n_c = m + ip$ and proceed to investigate the phase velocity of the wave and the manner in which the amplitude is attenuated in space; however, we will do this in one particularly simple case only, i.e., for the case in which $\boldsymbol{\kappa}_R$ is parallel to $\boldsymbol{\kappa}_c$; in this case,

$$\boldsymbol{\kappa} = (\|\boldsymbol{\kappa}_R\| + i\|\boldsymbol{\kappa}_c\|)\,\boldsymbol{v} = \|\boldsymbol{\kappa}\|\boldsymbol{v}$$

with $\boldsymbol{v}$ a unit vector (real) in the direction of $\boldsymbol{\kappa}_R$ (and $\boldsymbol{\kappa}_c$). As $\hat{\boldsymbol{E}} \cdot \boldsymbol{v} = \hat{\boldsymbol{B}} \cdot \boldsymbol{v} = 0$, the electric and magnetic field vectors are perpendicular to the propagation direction $\boldsymbol{v}$ but $\hat{\boldsymbol{B}} = \frac{n_c}{c}\boldsymbol{v} \times \hat{\boldsymbol{E}}$, with n_c complex-valued, so that $\boldsymbol{E}$ and $\boldsymbol{B}$ will not be in phase with one another. With $n_c = m + ip$, we easily find that $\|\boldsymbol{\kappa}_R\| = m\omega/c$, and $\|\boldsymbol{\kappa}_c\| = p\omega/c$, and if we set $\boldsymbol{v} \cdot \boldsymbol{x} = \zeta$, then in this case we obtain

$$\boldsymbol{E}(\boldsymbol{x}, t) = \hat{\boldsymbol{E}} e^{-p\omega\zeta/c} e^{-i\omega(t - m\zeta/c)} \tag{1.3.57}$$

so that the plane wave propagates with phase velocity c/m and the attenuation constant is just $p\omega/c$. We note in passing that $\delta = c/p\omega$, the reciprocal of the attenuation constant, is called the skin depth; it is, by (1.3.57), the distance that the wave must propagate into the conducting medium in order for its magnitude to fall off to $1/e$ of the value that it had at the bounding surface where the wave entered the medium. For a nonconducting medium, $p = 0$ and the skin depth $\delta = \infty$. Except for the special case just discussed, i.e. when $\boldsymbol{\kappa}_R$ and $\boldsymbol{\kappa}_c$ are not parallel, $\boldsymbol{v}$ must, in general, be taken to be complex and both the phase velocity and attenuation constant will depend on m and p (the so-called optical constants) in a rather complicated fashion.

Although we will not begin, in earnest, the study of wave propagation in nonlinear electromagnetic media until we embark on our work in Chapter 2, we deem it feasible, at this juncture, to record here a general nonlinear wave equation for the evolution of the components of the electric displacement vector $\boldsymbol{D}$ in a nonlinear dielectric which conforms to the general constitutive hypothesis (1.2.26); if nothing else, making note of such a result here will give the reader a glimpse into the complexity which nonlinearity introduces into the relevant wave equations. The specific result we have in mind is the content of the following elementary

Lemma 1.1 [25] Let $\Omega \subseteq R^3$ be either a bounded or unbounded domain which is filled with a rigid nonlinear dielectric substance that conforms to the constitutive hypothesis

$$\boldsymbol{D} = \epsilon(\|\boldsymbol{E}\|)\boldsymbol{E}\ ; \quad \boldsymbol{H} = \mu^{-1}\boldsymbol{B}\ ; \quad \boldsymbol{J} = \sigma(\|\boldsymbol{E}\|)\boldsymbol{E} \tag{1.3.58}$$

where ϵ, σ are differentiable, $\epsilon, \sigma > 0$, with $(\epsilon(\zeta)\zeta)' > 0$, $\forall \zeta > 0$. Then $\exists \lambda, \eta$, both differentiable maps of $R^+ \to R^+$, such that for $i = 1, 2, 3$:

$$\mu \frac{\partial^2 D_i}{\partial t^2} + \frac{\partial}{\partial t}(\eta(\|\boldsymbol{D}\| D_i)) \;=$$

$$\nabla^2(\lambda(\|\boldsymbol{D}\|D_i) \quad - \quad \frac{\partial}{\partial x_i}(\nabla\lambda(\|\boldsymbol{D}\|)\cdot\boldsymbol{D}) \tag{1.3.59}$$

Proof: By (1.3.58), $\|\boldsymbol{D}\| = \epsilon(\|\boldsymbol{E}\|)\|\boldsymbol{E}\|$, so if $(\epsilon(\zeta)\zeta)' > 0$, $g(\|\boldsymbol{E}\|) \equiv \epsilon(\|\boldsymbol{E}\|)\|\boldsymbol{E}\|$ is invertible and $\exists g^{-1} : R^+ \to R^+$ such that $\|\boldsymbol{E}\| = g^{-1}(\|\boldsymbol{D}\|)$. Using (1.3.58), once again, we find that

$$\boldsymbol{E} = \frac{1}{\epsilon(\|\boldsymbol{E}\|)}\boldsymbol{D} = \frac{1}{\epsilon(g^{-1}(\|\boldsymbol{D}\|))}\boldsymbol{D} \equiv \lambda(\|\boldsymbol{D}\|)\boldsymbol{D} \tag{1.3.60}$$

Applying the identity

$$\Delta\boldsymbol{v} = \text{grad}(\nabla\cdot\boldsymbol{v}) - \text{curl}\,\text{curl}\,\boldsymbol{v}$$

to $\boldsymbol{v} = \boldsymbol{E}$ (assuming, of course, that $\boldsymbol{E}$ is sufficiently smooth) we have for $i = 1,2,3$

$$\nabla^2 E_i = \frac{\partial}{\partial x_i}(\nabla\cdot\boldsymbol{E}) - (\text{curl}\,\text{curl}\,\boldsymbol{E})_i \tag{1.3.61}$$

Using the third of Maxwell's equations (1.3.3c), then the constitutive hypothesis $(1.3.58)_2$, and finally the fourth Maxwell equation (1.3.3d), we find, in succession, that

$$\begin{aligned} \text{curl}\,\text{curl}\,\boldsymbol{E} &= -\text{curl}\,\frac{\partial\boldsymbol{B}}{\partial t} \\ &= -\mu\,\text{curl}\,\frac{\partial\boldsymbol{H}}{\partial t} \\ &= -\mu\frac{\partial}{\partial t}(\text{curl}\,\boldsymbol{H}) \\ &= -\mu\left[\frac{\partial^2\boldsymbol{D}}{\partial t^2} + \frac{\partial\boldsymbol{J}}{\partial t}\right] \end{aligned} \tag{1.3.62}$$

<u>But</u>

$$\begin{aligned} \frac{\partial}{\partial t}\boldsymbol{J}(\boldsymbol{E}) &= \frac{\partial}{\partial t}\{\sigma(\|\boldsymbol{E}\|)\boldsymbol{E}\} \\ &= \frac{\partial}{\partial t}\{\sigma(\lambda(\|\boldsymbol{D}\|)\|\boldsymbol{D}\|)\,\lambda(\|\boldsymbol{D}\|)\boldsymbol{D}\} \\ &= \frac{1}{\mu}\frac{\partial}{\partial t}\{\eta(\|\boldsymbol{D}\|)\boldsymbol{D}\} \end{aligned} \tag{1.3.63}$$

where we have set

$$\eta(\zeta) = \mu\sigma(\lambda(\zeta)\zeta)\lambda(\zeta), \quad \zeta \geq 0 \tag{1.3.64}$$

Combining (1.3.61)-(1.3.63) we have

$$\nabla^2 E_i = \frac{\partial}{\partial x_i}(\nabla \cdot \boldsymbol{E}) + \mu \frac{\partial^2 D_i}{\partial t^2} + \frac{\partial}{\partial t}\{\eta(\|\boldsymbol{D}\|)D_i\} \tag{1.3.65}$$

However, in the absence of a prescribed charge density, $\nabla \cdot \boldsymbol{D} = 0$ and, therefore,

$$\nabla \cdot \boldsymbol{E} = \nabla \cdot (\lambda(\|\boldsymbol{D}\|)\boldsymbol{D}) = \nabla\lambda(\|\boldsymbol{D}\|) \cdot \boldsymbol{D} \tag{1.3.66}$$

The relevant nonlinear wave equation for the components D_i of the electric displacement field, i.e., (1.3.59), now follows as an immediate consequence of (1.3.65), (1.3.66), and the constitutive relation $E_i = \lambda(\|\boldsymbol{D}\|)D_i$. □

Consequences of (1.3.59), as well as of the first-order systems that result by coupling Maxwell's equations to constitutive hypotheses of the form (1.3.58), will be explored, in detail, in the next chapter, as well as in Chapter 3.

1.4 Transmission Lines: Basic Concepts

An electrical circuit, or transmission line, is an assemblage of electrical conductors (usually in the form of wires) through which current from a power source, such as a battery or generator, flows. The components may be connected one after another (in series) or side by side (in parallel). In order to better understand the nature of a transmission line, we begin by examining, in detail, each of the constituent components of a circuit, starting with the concept of a capacitor.

A capacitor consists of two conductors that can store equal and opposite charges, say $\pm Q$, with a potential difference between them which is independent of whether other conductors in the system are charged. In such a situation, the potential which is contributed to each of the pair of conductors by other charges must be the same. It is easily shown that the potential difference between the conductors of a capacitor is directly proportional to the charge stored, so that

$$Q = C\Delta\phi, \tag{1.4.1}$$

$\Delta\phi = \phi_1 - \phi_2$ being the potential difference between the two conductors which form the capacitor. The quantity C is called the capacitance of the capacitor and for now, we will assume it to be a constant; in our later work (i.e., in Chapters 4 and 5), we will consider a more general situation in which $Q = Q(V)$, V being the voltage, or potential difference, and then $C = C(V)$ will be given by $C = \frac{dQ}{dV}$. The simplest configuration for a capacitor is that achieved in the parallel plate capacitor, idealized to consist of two oppositely charged parallel plates in which the plate separation λ is small in comparison with the dimensions of each of the plates. Assuming that the region between the plates is filled by a dielectric with constant permittivity ϵ, the electric field between the plates is given by $E = \frac{Q}{\epsilon A}$, A being the area of either of the two plates. As the potential difference between the plates is given by $\Delta\phi = E\lambda$, we obtain for the capacitance of this parallel plate capacitor $C = \epsilon A/\lambda$.

When a capacitor appears as a component in an electric circuit, it is commonly indicated by the symbol ⊣⊢ and capacitors in a circuit may be joined by either a series or parallel connection, each of which is depicted in Figure 1.2.

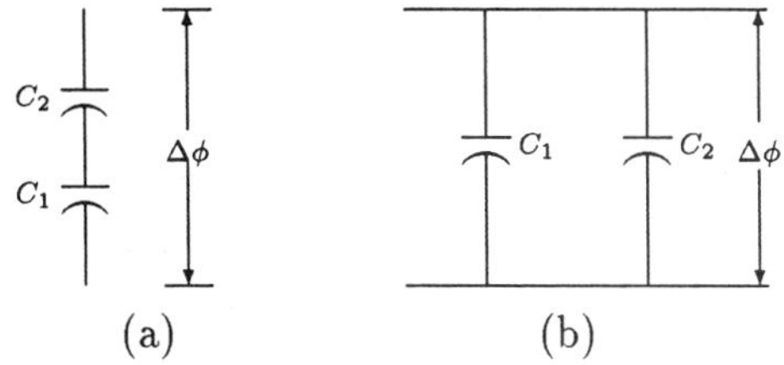

Fig. 1.2: (a) Series connection, (b) Parallel connection

Once two (or more) capacitors are connected, either in series or in parallel, we may speak of the combined capacitance of the combination: for two capacitors connected in series, the law of conservation of charge implies that each of the two capacitors shown in Figure 1.2(a) must acquire the same charge and this, in turn, gives an equivalent net capacitance C which satisfies $C^{-1} = C_1^{-1} + C_2^{-1}$; on the other hand, in the case of two capacitors connected in parallel, the voltage $\Delta\phi$, depicted in Figure 1.2(b), which acts across each of C_1 and C_2, must act across the combination so that $C = C_1 + C_2$.

Having considered, briefly, the concept of a capacitor, we now turn our attention to the equally important concept of a resistor. In § 1.2 we noted the experimental fact that in metals at constant temperature, the current density $\boldsymbol{J}$ is linearly proportional to the electric field $\boldsymbol{E}$, a fact which is embodied in Ohm's law $\boldsymbol{J} = \sigma \boldsymbol{E}$, σ being the conductivity. Furthermore, for a homogeneous wire of uniform cross section, which conducts electricity according to the linear Ohm's law, and has its ends maintained at a constant potential difference $\Delta\phi$, we showed (i.e., (1.2.12)) that the current I in the wire was given by $I = \Delta\phi/R$, where R, the resistance of the wire, is computed as $R = \sigma A/l$, A being the cross-sectional area of the wire and l its length.

We now suppose that our electric circuit or transmission line carries a steady current, i.e., we will consider, initially, direct current circuits. The basic problem of transmission line analysis is to find the current in each branch of an electric circuit given the resistance and applied voltage in each branch; the applied voltages are usually produced by a battery and although one can visualize an ideal source in the line which would provide for an applied voltage V, say, which is independent of the current drawn from the source, to some degree the voltage V provided by the source in the circuit always exhibits a dependence $V = V(I)$.

One may define a resistor as a conducting object that is characterized by its resistance R which was given, above, for a homogeneous wire of uniform cross-section, by $R = \sigma A/l$. When resistors appear in a circuit they are denoted by the symbol —\/\/— and such elements may, as was the case with capacitors, be connected either in series or parallel to form what is usually called a resistance network; these two possibilities are depicted in Figure 1.3. In the case of a series connection of two resistors the same current must pass through each resistor; applying the rule $V = IR$ to each resistor, we find that $V = IR_1 + IR_2 \equiv IR$, so that the equivalent resistance is just $R = R_1 + R_2$. In a parallel connection of the two resistors the potential drop across each resistor must be the same, while the net current through the combination is just $I = I_1 + I_2$; this implies that $I = \frac{V}{R_1} + \frac{V}{R_2} \equiv \frac{V}{R}$ so that the equivalent resistance of the combination follows as $\frac{1}{R} = \frac{1}{R_1} + \frac{1}{R_2}$.

Direct current electric circuits are subject to two basic rules which are known as Kirchhoff's laws, namely

(i) The algebraic sum of all currents which flow toward a branch point is zero, where a branch point is a point of the circuit joining three or more conductors, and

(ii) The algebraic sum of the voltage differences around any loop is zero, where a loop refers to any closed conducting path in the circuit; this second of Kirchhoff's laws is equivalent to the statement that the sum of the current-resistance products $\sum_j I_j R_j$ around any closed path must equal the total electromotive force in the loop.

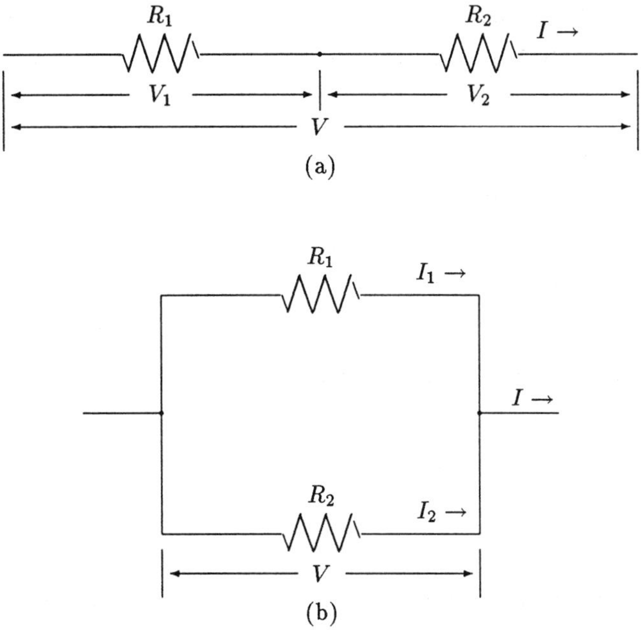

Fig. 1.3: (a) Series connection of two resistors; (b) Parallel connection of two resistors

In short order we will generalize the above-referenced Kirchhoff's laws so as to take into account slowly varying currents.

We now turn to the important circuit parameter known as the self-inductance, which arises from the relationship between the flux and current associated with an isolated circuit. The concept of magnetic flux was introduced in § 1.2 (i.e., (1.2.44)) as $\Phi = \int_S \boldsymbol{B} \cdot \boldsymbol{n}\, da$. By virtue of the Biot-Savart law (1.2.30), the magnetic flux linking an isolated circuit is linearly dependent on the current in the circuit and, therefore, in a rigid stationary circuit, the only changes in Φ which are possible are those which arise from changes in the current; we may write that

$$\frac{d\Phi}{dt} = \frac{d\Phi}{dI}\frac{dI}{dt} \equiv L\frac{dI}{dt} \tag{1.4.2}$$

where $L \equiv \frac{d\Phi}{dt}$, which is called the inductance, will be assumed, for our purposes in this section, to be a constant. In terms of the inductance L we may rewrite Faraday's law (1.2.56) in the form

$$\mathcal{E} = -L\frac{dI}{dt} \tag{1.4.3}$$

The mks unit of inductance is the henry (denoted by (H)) which is equal, as a consequence of (1.4.3), to one volt-second/ampere, since the unit of emf $\mathcal{E}$ in the mks system is the volt. The reader may easily locate in standard texts on electromagnetic theory [134], [73] the calculation of self-inductance for several important simple circuits, e.g., the self-inductance of a toroidal coil of constant cross-sectional area A, wound with N turns of wire of mean length l, and carrying a current I is given by

$$L = \frac{\mu_0 N^2 A}{l} \tag{1.4.4}$$

where (1.4.4) follows from (1.4.2) and the fact that the total flux linking the N turns is

$$\Phi = \frac{\mu_0 N^2 A}{l} I \tag{1.4.5}$$

Inasmuch as we will only consider isolated circuits in our work in Chapters 4 and 5, we shall not indulge here in a discussion of the interesting topic of mutual inductance, which must be considered when several circuits are involved and the emf induced in the i-th circuit results from current changes in all the circuits present. As is the case with resistors and capacitors, inductances are often connected in series (and in parallel) in a circuit and it is possible to compute an effective inductance in each case. In such situations, however, one must take into account the coupling which occurs between the inductors, i.e., for two inductors of strengths L_1 and L_2 in series, there is a mutual inductance of strength

$$M = \kappa\sqrt{L_1 L_2}, \quad -1 \leq \kappa \leq 1 \tag{1.4.6}$$

which must be taken into account and which leads to an effective inductance for such a series circuit of the form

$$L_{\text{eff}} = L_1 + L_2 + 2\kappa\sqrt{L_1 L_2} \tag{1.4.7}$$

For two inductors in parallel, again with strengths L_1 and L_2, the result is even somewhat more complicated than (1.4.7), namely,

$$L_{\text{eff}} = \frac{(L_1 L_2 - M^2)}{(L_1 + L_2 - 2M)} \tag{1.4.8}$$

As we shall find sufficient difficulties with the consequences of nonlinearity in the discussion of distributed parameter transmission lines in Chapters 4 and 5, we will bypass the derivation of the effective self-inductances (1.4.7) and (1.4.8) and confine our attention henceforth to circuits involving a single inductor.

To this point we have only considered circuits involving currents which are excited by a constant applied voltage. If instead of a constant applied voltage we have to deal with a slowly varying voltage, then a slowly varying current will arise in response, provided the line does not radiate away a considerable amount of electromagnetic energy. When such slowly varying applied voltages change periodically with time, it is found that an appreciable time after the application of such voltages, the currents in the line also vary periodically with time and the discussion of the behavior of such circuits is usually dependent on whether it is the periodic or the nonperiodic behavior that is of prime importance. The intrinsic behavior of an electrical circuit with a nonperiodic evolution of current is referred to as the steady-state behavior; the simplest possible instances of each type of behavior involve the transient analysis associated with excitation by a constant applied voltage and the steady-state analysis associated with excitation by a periodically (e.g., sinusoidal) varying applied voltage.

For slowly varying currents not only resistors, but also capacitors and inductors, must be included as circuit elements and each such element in the line also involves a potential difference which must be included in Kirchhoff's loop law; the terminology "counter voltage" has been applied most often in the literature to describe the potential difference between the terminals of passive elements such as these. Noting that each of Kirchhoff's laws has to apply to the instantaneous values of the currents, counter voltages, and applied voltages in a circuit, we may restate these laws, for slowly varying currents, as follows:

(i) The algebraic sum of the instantaneous currents which flow toward a branch point is zero.

(ii) The algebraic sum of the instantaneous applied voltages in a closed loop is equal to the algebraic sum of the instantaneous counter voltages in the loop.

As a simple example of the appropriate interpretation of the second of Kirchhoff's laws, above, for the case of slowly varying current in a circuit, we may consider the simple closed loop indicated in Figure 1.4 below:

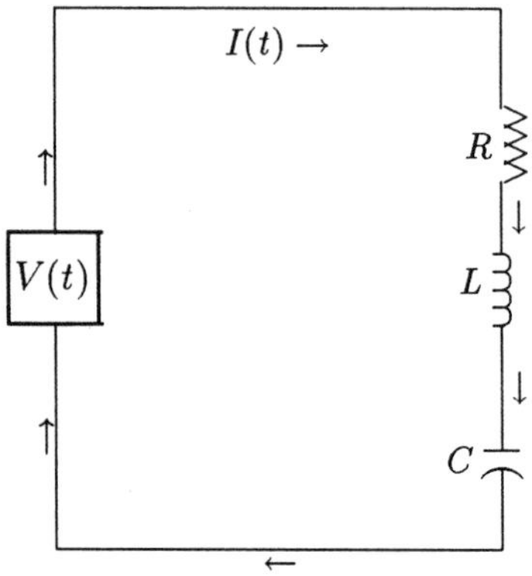

Figure 1.4.

The basic sign convention for the circuit depicted in Figure 1.4 is that the applied voltage $V(t)$, which is shown as being in series with a resistance R, an inductance L, and a capacitance C, is positive if it tends to induce the current to move in the indicated direction, i.e., clockwise in this case. The counter voltage due to the resistor is, of course, just IR, while that due to the inductor is $L\frac{dI}{dt}$; the capacitance counter voltage is the ratio $\frac{Q}{C}$, where $Q = \int_{t_0}^{t} I(\tau)\,d\tau$. Therefore, the second of Kirchhoff's laws for the elementary circuit shown in Figure 1.4 reads

$$V(t) = IR + L\frac{dI}{dt} + \frac{1}{C}\int_{t_0}^{t} I(\tau)\,d\tau \tag{1.4.9}$$

Now, consider the circuit of Figure 1.4 modified so as to include a battery which is capable of producing a constant applied voltage V_0 and a switch which we denote, in Figure 1.5 below, by S:

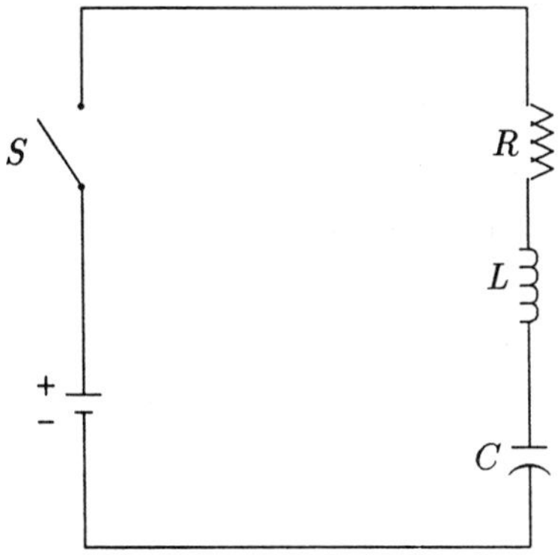

Figure 1.5.

For this series RLC circuit, the relation (1.4.9) applies, so that after the switch S is closed we have

$$V_0 = IR + L\frac{dI}{dt} + \frac{1}{C}\int_0^t I(\tau)\,d\tau \tag{1.4.10}$$

where, as t_0 may be interpreted as that time at which the charge on the capacitor is zero, we have simply taken $t_0 = 0$. Differentiation of (1.4.10) once with respect to time yields the familiar second-order linear ODE

$$L\frac{d^2I}{dt^2} + R\frac{dI}{dt} + \frac{1}{C}I = 0 \tag{1.4.11}$$

which has the solution

$$I(t) = (\alpha e^{i\omega_n t} + \beta e^{-i\omega_n t})e^{-\frac{R}{2L}t} \tag{1.4.12}$$

with

$$\omega_n = \left(\frac{1}{LC} - \frac{R^2}{4L^2}\right)^{1/2} \tag{1.4.13}$$

In view of the fact that the current I should be real-valued, the constant β in (1.4.12) must be the complex conjugate of α; also $I(0) = 0$ because the switch S is closed at $t = 0$. These considerations reduce (1.4.12) to

$$I(t) = \gamma \sin\omega_n t\left(e^{-\frac{R}{2L}t}\right) \tag{1.4.14}$$

Evaluating (1.4.10) at $t = 0$, we readily find that $V_0 = L\frac{dI}{dt}\,|_{t=0}$, in which case

$$\gamma = \frac{V_0}{\omega_n L} = \frac{V_0}{\left(\frac{L}{C} - \frac{R^2}{4}\right)^{1/2}} \tag{1.4.15}$$

Relations (1.4.14), (1.4.15) completely characterize the transient response of the basic series RLC circuit depicted in Figure 1.5; the current in this circuit possesses an amplitude $\gamma e^{-\frac{R}{2L}t}$, which is exponentially decreasing in time, and the current oscillates with the natural frequency given by (1.4.13).

Having characterized the transient response of an RLC circuit we now return to the situation depicted in Figure 1.4 and consider the steady-state behavior of such a circuit subject to an excitation of the form

$$V(t) = V_0 \cos \omega t \tag{1.4.16}$$

where ω is a given frequency. If we note that $V(t) = \operatorname{Re} V_0 e^{i\omega t}$, then we may consider a complex voltage of the form $V_1(t) + iV_2(t)$ applied to the circuit of Figure 1.4 which results in a complex valued current $I_1(t) + iI_2(t)$ that satisfies, by virtue of (1.4.9),

$$\begin{aligned}\frac{dV_1}{dt} + i\frac{dV_2}{dt} &= \left(L\frac{d^2 I_1}{dt^2} + R\frac{dI_1}{dt} + \frac{1}{C}I_1\right) \\ &+ i\left(L\frac{d^2 I_2}{dt^2} + R\frac{dI_2}{dt} + \frac{1}{C}I_2\right)\end{aligned} \tag{1.4.17}$$

Thus each of $I_1(t)$, $I_2(t)$ satisfies

$$\frac{dV}{dt} = L\frac{d^2 I}{dt^2} + R\frac{dI}{dt} + \frac{1}{C}I \tag{1.4.18}$$

with, respectively, $V_1(t)$, $V_2(t)$ appearing on the left-hand side of this equation. It is, therefore, sufficient in the present situation to solve (1.4.18), with $V(t) = V_0 e^{i\omega t}$, for $I(t) = I_0 e^{i\omega t}$, I_0 a complex constant; the physical current will then be $\operatorname{Re} I_0 e^{i\omega t}$. Substitution of $V(t) = V_0 e^{i\omega t}$, $I(t) = I_0 e^{i\omega t}$ into (1.4.18) produces the relation

$$i\omega V_0 e^{i\omega t} = \left[-\omega^2 L + i\omega R + \frac{1}{C}\right] I_0 e^{i\omega t} \tag{1.4.19}$$

or

$$V_0 e^{i\omega t} = Z I_0 e^{i\omega t} \tag{1.4.20}$$

where the quantity Z, the impedance of the circuit, is given by

$$Z = R + i\left(\omega L - \frac{1}{\omega C}\right) \tag{1.4.21}$$

We note that Z consists of a real part, the resistance R, and an imaginary part $X = \omega L - \dfrac{1}{\omega C}$ which is usually referred to as the reactance. Because Z is not, in general, real-valued, it follows, as a consequence, that the steady-state current in the RLC circuit will not be in phase with the applied voltage $V(t)$; in fact, if we define

$$\begin{cases} |Z| = \left(R^2 + \left[\omega L - \frac{1}{\omega C}\right]^2\right)^{1/2} \\ \theta = \tan^{-1}\left(\frac{\omega L - \frac{1}{\omega C}}{R}\right) \end{cases} \tag{1.4.22}$$

then $Z = |Z|e^{i\theta}$ and substitution of this form of the impedance into (1.4.20) produces

$$I(t) = \frac{V_0}{|Z|}e^{i(\omega t - \theta)} \tag{1.4.23}$$

so that the physical current in the circuit, which is now obtained by taking the real part of both sides of (1.4.23), is given by

$$I_P(t) = \frac{V_0}{|Z|}\cos(\omega t - \theta) \tag{1.4.24}$$

For $\theta > 0$ the current $I_P(t)$, in the circuit, will achieve a prescribed phase at a later time than the voltage, in which case we say that the current lags the voltage; the opposite situation applies, of course, if $\theta < 0$.

This concludes our discussion of the elementary transient and steady-state behavior of the simple RLC series circuit in which the inductance L and the capacitance C have been assumed to be constants. In our work in Chapters 4 and 5 on the behavior of nonlinear transmission lines, we will not only allow for the possibility that the charge in the line (and hence the capacitance) is voltage dependent, but will also introduce into the circuits under consideration a nonlinear, voltage dependent leakage conductance.

Chapter 2

WAVE-DIELECTRIC INTERACTIONS I: FORMATION OF SINGULARITIES

2.0 Introduction

In this chapter we initiate our study of the wave-dielectric interaction problem in one space dimension; we begin with the problem of trying to delineate a set of conditions under which solutions of the governing initial-boundary value problem, which are of class C^1 in space and time, break down in finite time. In order to provide some essential background for our discussion in this chapter, we begin in § 2.1 by presenting a concise outline of basic concepts associated with the dual topics of hyperbolic systems and conservation laws; this is followed by a rather thorough account of shock formation for a single hyperbolic conservation law in one-space dimension. In § 2.3 we characterize weak solutions of hyperbolic conservation laws; we discuss the Rankine-Hugoniot jump conditions and offer some simple examples. The concepts of Riemann invariant and genuine nonlinearity are then introduced and illustrated in § 2.4 within the context of a standard 2×2 system of conservation laws, as well as for the equations of gas dynamics in Eulerian form; for the sake of completeness we also introduce the idea of a k-rarefaction (or k-simple wave). In § 2.5 we present the original ideas of Lax relating to the formation of shock discontinuities in 2×2 systems of hyperbolic conservation laws.

Following the general discussion in § 2.1-§ 2.5 of shock formation in solutions of hyperbolic conservation laws, we offer, in § 2.6, a general formulation of the basic equations governing the evolution of linearly polarized waves in a nonlinear dielectric medium which is nondispersive; we assume the presence of a nontrivial conduction vector $\boldsymbol{J}$ which is governed by a nonlinear Ohm's law. The basic result of the analysis in § 2.6 is an inhomogeneous system of hyperbolic balance laws for the electric and magnetic induction fields; these equations then lead directly to a nonlinearly damped, nonlinear wave equation for the evolution of the electric induction (or displacement) field in the wave. The discussion in § 2.6 makes free use of the basic concepts from electromagnetic theory that were introduced in Chapter 1.

Shock formation for solutions of initial-value problems associated with the system formulated in § 2.6, in the special case where $\boldsymbol{J} = \mathbf{0}$, is shown, in § 2.7, to follow from earlier work of Lax [99] and Klainerman and Majda [92]; this is followed by a careful analysis of two special examples related to commonly studied forms of the constitutive relation between the electric displacement and electric fields in a nonlinear dielectric. Next, we turn in § 2.7 to an analysis of the case in which $\boldsymbol{J} \neq \mathbf{0}$ in the wave; we rewrite the basic governing system of equations in Riemann invariants form and study the system of nonlinear ordinary differential equations which results by restricting our attention to the relevant characteristic curves in the x, t plane. The highlight of § 2.7 (and, indeed, of this chapter) is Theorem 2.2, which establishes sufficient conditions for the breakdown of C^1 solutions $(B(x,t), D(x,t))$ to the wave-dielectric interaction problem in one-space dimension under specific conditions related directly to the size of the initial values of the electromagnetic field in the wave. Section 2.7 concludes with a survey of some of the more relevant earlier work on the problem of shock formation for wave-dielectric interactions in one-space dimension. Finally, in § 2.8, we offer a brief discussion of some ideas related to the problem of shock development in multidimensional wave-dielectric interactions.

2.1 Hyperbolic Systems and Conservation Laws

Let $\boldsymbol{u}(x,t)$ be a vector function of x and t and $\boldsymbol{A} = \boldsymbol{A}(\boldsymbol{u}, x, t)$ a matrix-valued function; then the system

$$\boldsymbol{u}_t + \boldsymbol{A}(x, t, \boldsymbol{u})\boldsymbol{u}_x = \mathbf{0} \tag{2.1.1}$$

is a first-order quasilinear system of partial differential equations in the two independent variables x (spatial) and t (time). The system (2.1.1) is said to be *strictly hyperbolic* if for each x, t and $\boldsymbol{u}$, the matrix $\boldsymbol{A}$ has real distinct eigenvalues, $\lambda_i = \lambda_i(x, t, \boldsymbol{u})$, $i = 1, \ldots, n$. In an analogous manner, a quasilinear system in the $n+1$ variables $x_1, \ldots, x_n, t$ of the form

$$\boldsymbol{u}_t + \sum_{j=1}^{n} \boldsymbol{A}_j(x_1, \ldots, x_n, t, \boldsymbol{u})\boldsymbol{u}_{x_j} = \boldsymbol{0} \tag{2.1.2}$$

is *strictly hyperbolic* if for each $\boldsymbol{x} = (x_1, \ldots, x_n)$, t, and $\boldsymbol{u}$, and each unit vector $\boldsymbol{\nu}$, the matrix $\sum_{j=1}^{n} \boldsymbol{A}_j \nu_j$ has real, distinct eigenvalues $\lambda_j(\boldsymbol{x}, t, \boldsymbol{u}, \boldsymbol{\nu})$, $j = 1, \ldots, n$. These familiar definitions of hyperbolicity for quasilinear systems of partial differential equations may be generalized slightly, i.e., if $\boldsymbol{B}_0, \boldsymbol{B}_1, \ldots, \boldsymbol{B}_n$ are $k \times k$ matrix-valued functions of $\boldsymbol{x}$, t, $\boldsymbol{u}$, and $\boldsymbol{c}$ is a k-vector field of $\boldsymbol{x}$, t, $\boldsymbol{u}$, then the quasilinear system

$$\boldsymbol{B}_0 \cdot \boldsymbol{u}_t + \sum_{j=1}^{n} \boldsymbol{B}_j \cdot \boldsymbol{u}_{x_j} + \boldsymbol{c} = \boldsymbol{0} \tag{2.1.3}$$

is *hyperbolic* if for every fixed $\boldsymbol{x}$, t, $\boldsymbol{u}$, and any unit vector $\boldsymbol{\nu}$, the eigenvalue problem

$$\left(\sum_{j=1}^{n} \boldsymbol{B}_j \nu_j - \lambda \boldsymbol{B}_0 \right) \cdot \boldsymbol{v} = \boldsymbol{0} \tag{2.1.4}$$

has k real eigenvalues, not necessarily distinct, and a corresponding set of k linearly independent eigenvectors; if the eigenvalues are distinct, then the system (2.1.3) is again termed strictly hyperbolic. Very often, in classical physical theories, hyperbolic quasilinear systems of partial differential equations of the first order arise from *conservation laws.* With $\boldsymbol{x} = (x_1, \ldots, x_n)$, again, and t the time variable, a conservation law is an integral relation of the form

$$\begin{aligned} \int_\Omega \boldsymbol{g}(\boldsymbol{x}, t)\Big|_{t_1}^{t_2} d\boldsymbol{x} \quad &+ \quad \int_{t_1}^{t_2} \oint_{\partial\Omega} \boldsymbol{A}(\boldsymbol{x}, t) \cdot \boldsymbol{\nu} \, dS \, d\tau \\ &+ \quad \int_{t_1}^{t_2} \int_\Omega \boldsymbol{f}(\boldsymbol{x}, t) \, d\boldsymbol{x} \, dt = \boldsymbol{0} \end{aligned} \tag{2.1.5}$$

which holds on every time interval (t_1, t_2) and every domain $\Omega \subseteq R^n$ with sufficiently smooth boundary $\partial\Omega$. In (2.1.5), $\boldsymbol{g}$ and $\boldsymbol{f}$ are k-vector fields, while $\boldsymbol{A}$ is a $k \times n$ matrix-valued field and $\boldsymbol{\nu}$ is the exterior unit normal on $\partial\Omega$; the physical interpretation of (2.1.5) is that the change in the function

$$G(\Omega, t) = \int_\Omega \boldsymbol{g}(\boldsymbol{x}, t) \, d\boldsymbol{x}$$

between times t_1 and t_2 is balanced by an influx through $\partial\Omega$ and a production in Ω during (t_1, t_2). When a conservation law of the type (2.1.5) is augmented by constitutive hypotheses of the form

$$\boldsymbol{g} = \boldsymbol{g}^*(\boldsymbol{x}, t, \boldsymbol{u}), \quad \boldsymbol{A} = \boldsymbol{A}^*(\boldsymbol{x}, t, \boldsymbol{u}), \quad \boldsymbol{f} = \boldsymbol{f}^*(\boldsymbol{x}, t, \boldsymbol{u}), \tag{2.1.6}$$

with $\boldsymbol{u} = \boldsymbol{u}(x_1, \ldots, x_n, t)$ a k-vector describing the state of the physical system, then, provided the $\boldsymbol{g}^*, \boldsymbol{A}^*, \boldsymbol{f}^*$ are $\boldsymbol{u}$ are differentiable, (2.1.5) can, with the help of the divergence theorem, be rewritten in the form

$$\int_{t_1}^{t_2} \int_{\Omega} \{\boldsymbol{g}_t + \nabla \cdot \boldsymbol{A} + \boldsymbol{f}\}\, d\boldsymbol{x}\, dt = \boldsymbol{0} \tag{2.1.7}$$

The integral relation (2.1.7) then holds on the arbitrary domain $\Omega \times (t_1, t_2)$, in $R^n \times [0, \infty)$, if and only if

$$\boldsymbol{g}_t + \nabla \cdot \boldsymbol{A} + \boldsymbol{f} = \boldsymbol{0} \tag{2.1.8}$$

and (2.1.8) is a quasilinear first-order system of partial differential equations of the form (2.1.3). As a simple example of a system of conservation laws, we note the equations of gas dynamics for an inviscid, non-heat conducting, gas in Lagrangian coordinates, viz.,

$$v_t - u_x = 0 \text{ (conservation of mass)} \tag{2.1.9a}$$

$$u_t - p_x = 0 \text{ (conservation of momentum)} \tag{2.1.9b}$$

$$E_t + (up)_x = 0 \text{ (conservation of energy)} \tag{2.1.9c}$$

where $v = 1/\rho$ is the specific volume of the gas, with ρ the density, u is the velocity, $E = e + \frac{1}{2}u^2$ the specific energy, with e the internal energy density, and $p = p(E, v)$ the equation of state which yields the pressure. Another important system, generalizations of which will appear frequently in our work, both on wave-dielectric interactions, as well as in the study of signal propagation in nonlinear transmission lines, is the so-called p-system

$$v_t - u_x = 0 \tag{2.1.10a}$$

$$u_t + p(v)_x = 0 \tag{2.1.10b}$$

For p of the form $p(v) = kv^{-\gamma}$, $\gamma \geq 1$, $k > 0$, the system (2.1.10) serves as a model for isentropic gas dynamics. We note that as a consequence of (2.1.10), $v(x,t)$ satisfies the nonlinear wave equation

$$v_{tt} + p(v)_{xx} = 0 \tag{2.1.11}$$

Also, in a simply connected domain, the relation $v_t = u_x$ implies the existence of $\phi(x,t)$ such that $v = \phi_x$ and $u = \phi_t$ so that

$$\phi_{tt} + p'(\phi_x)\phi_{xx} = 0 \tag{2.1.12}$$

which, with $p' < 0$, is a nonlinear wave equation with speed of propagation $\sqrt{-p'}$ depending on ϕ_x. Various generalizations of the nonlinear wave equations (2.1.11), (2.1.12) will also appear later on in this chapter, as well as in Chapter 3.

Before proceeding to our discussion of shock formation, and generalized solutions of conservation laws, it is worth noting two other standard interpretations of hyperbolicity. First of all, suppose that Γ is a smooth surface in $\boldsymbol{x}, t$ space, described by $h(\boldsymbol{x}, t) = 0$, and that $\boldsymbol{u}(\boldsymbol{x}, t)$ is a continuous k-vector field such that $\boldsymbol{u}$ is of class C^1, at every point $(\boldsymbol{x}, t) \notin \Gamma$, and satisfies (2.1.3); we also suppose that the limits of $\nabla_{(\boldsymbol{x},t)}\boldsymbol{u}$ exist as we approach Γ but are unequal, i.e., there is a finite jump in the first derivatives of $\boldsymbol{u}$ across Γ. By visualizing Γ as a one-parameter family of surfaces in space, Γ will appear to an observer of the physical phenomena governed by (2.1.3) to be a propagating wave and to distinguish such waves from the class of shock waves, described in the next section, we will call them weak waves. Now, at fixed t, the

$$\nu_j = \frac{h_{x_j}}{\left(\sum_{l=1}^{n} h_{x_l}^2\right)^{1/2}}, \quad j = 1, \ldots, n \tag{2.1.13}$$

are the components of the exterior unit normal to this weak wave and if v is the speed of propagation of Γ in the direction of $\boldsymbol{\nu}$, then

$$\frac{dh}{dt} = h_t + \sum_{l=1}^{n} h_{x_l} \frac{dx_l}{dt} = 0 \tag{2.1.14}$$

along the trajectory $\dfrac{dx_l}{dt} = v\nu_l$. Combining (2.1.13) and (2.1.14) we readily find that

$$v = -\frac{h_t}{\left(\sum_{l=1}^{n} h_{x_l}^2\right)^{1/2}} \tag{2.1.15}$$

Now, suppose that we take the jump of (2.1.3) across Γ, i.e., take the jump of each quantity in (2.1.3) across Γ and sum (by $[\boldsymbol{f}(\boldsymbol{x}, t)]_\Gamma \equiv [\boldsymbol{f}] \equiv \boldsymbol{f}^+ - \boldsymbol{f}^-$ we mean the difference of the limits of $\boldsymbol{f}$ as we approach Γ from either side). As $\boldsymbol{B}_0, \boldsymbol{B}_1, \ldots, \boldsymbol{B}_n$ depend at most on $\boldsymbol{x}, t$, and $\boldsymbol{u}$ and are, therefore, continuous across Γ, we obtain:

$$\boldsymbol{B}_0 \cdot [\boldsymbol{u}_t] + \sum_{j=1}^{n} \boldsymbol{B}_j \cdot [\boldsymbol{u}_{x_j}] = \boldsymbol{0} \tag{2.1.16}$$

Because $\boldsymbol{u}$ is continuous across Γ, the tangential derivative of $\boldsymbol{u}$ is also continuous across Γ; thus, only the normal derivative of $\boldsymbol{u}$ can suffer a jump discontinuity across

Γ. However, by virtue of the relations (2.1.13) and (2.1.15), the vector $(\nu_1, \ldots, \nu_n, -v)$ is normal to Γ. If we let $\boldsymbol{u}^{\#}$ denote the jump of the normal derivative of $\boldsymbol{u}$ across Γ, then

$$[\boldsymbol{u}_{x_j}] = \nu_j \boldsymbol{u}^{\#}, \qquad [\boldsymbol{u}_t] = -v\boldsymbol{u}^{\#} \tag{2.1.17}$$

and substitution of the relations (2.1.17) into (2.1.16) now yields

$$\left(\sum_{j=1}^{n} \boldsymbol{B}_j \nu_j - v\boldsymbol{B}_0\right) \cdot \boldsymbol{u}^{\#} = \boldsymbol{0} \tag{2.1.18}$$

By comparing (2.1.18) with (2.1.4) we are led to conclude that the system (2.1.3) is hyperbolic if in each direction in space k independent weak waves can propagate; the speeds of propagation v are determined by the eigenvalues of (2.1.4) while the amplitudes $\boldsymbol{u}^{\#}$ of the propagating weak waves are determined, up to a constant multiplicative factor, by the eigenvectors of (2.1.14).

The other common mode of interpreting hyperbolicity for a system such as (2.1.3) involves the concept of characteristic surface; if we seek a solution of (2.1.3) with values assigned on some smooth surface Γ in space-time (the Cauchy problem), then a basic consideration is whether or not the restriction of a solution $\boldsymbol{u}(\boldsymbol{x}, t)$ of (2.1.3) to the surface Γ determines the restriction of $\nabla_{(\boldsymbol{x},t)}\boldsymbol{u}$ on Γ. If we again let $h(\boldsymbol{x}, t) = 0$ be the equation defining Γ, and take $(y_1, \ldots, y_n, h)$ to be a curvilinear coordinate system which has h as a coordinate surface, we may proceed as follows: Through a standard application of the chain rule, we rewrite (2.1.3) in the coordinate system $(y_1, \ldots, y_n, h)$ as

$$\begin{aligned} &\left(h_t \boldsymbol{B}_0 + \sum_{j=1}^{n} h_{x_j} \boldsymbol{B}_j\right) \cdot \boldsymbol{u}_h \\ &\qquad + \sum_{l=1}^{n} \left(y_{l,t} \boldsymbol{B}_0 + \sum_{j=1}^{n} y_{l,x_j} \boldsymbol{B}_j\right) \cdot \boldsymbol{u}_{y_l} + \boldsymbol{c} = \boldsymbol{0} \end{aligned} \tag{2.1.19}$$

As the derivatives $\boldsymbol{u}_{y_l}$ are tangential derivatives, they are determined by the restriction of $\boldsymbol{u}$ on Γ; but this means that $\nabla_{(\boldsymbol{y},h)}\boldsymbol{u}$ will be completely determined on Γ if we can calculate $\boldsymbol{u}_h$ from (2.1.19). In order to compute $\boldsymbol{u}_h$ through (2.1.19) it is clear that we must have that the matrix

$$h_t \boldsymbol{B}_0 + \sum_{j=1}^{n} h_{x_j} \boldsymbol{B}_j \tag{2.1.20}$$

is nonsingular, which leads to the definition of Γ as a *characteristic surface* for the system (2.1.3) if, at each point on Γ, $h(\boldsymbol{x}, t)$ satisfies the equation

$$\det[h_t \boldsymbol{B}_0 + \sum_{j=1}^{n} h_{x_j} \boldsymbol{B}_j] = 0 \tag{2.1.21}$$

By contrasting equation (2.1.21) with (2.1.4) and (2.1.18), we readily conclude that as a consequence of (2.1.13), (2.1.15), each weak wave must propagate along a characteristic; for this reason, the eigenvalues of (2.1.4) are also called the *characteristic speeds* of (2.1.3). The concept of characteristic speeds (values) and the associated characteristic curves in the x, t plane will play a key role in the sequel for problems of wave-dielectric interactions posed in one-space dimension. The material summarized in this section on quasilinear systems, hyperbolic conservation laws, weak waves, and characteristic surfaces is, by now, quite standard. Among the references which cover all of the aforementioned material that has been treated here we may single out the expository works of Lax [103] and Dafermos [41] and the text by Smoller [165].

2.2 Shock Formation: Single Equations in One Space Dimension

The most often cited example used to illustrate the formation of shock discontinuities in an initial-value (or Cauchy) problem, for a single hyperbolic conservation law in one-space dimension, is that of the Burger's equation, i.e.,

$$u_t(x,t) + u(x,t)u_x(x,t) = 0 \tag{2.2.1}$$

Suppose that $u(x, t)$ is a solution of (2.2.1) for $x \in R^1$, $t > 0$ which is of class C^1 in x and t. For any point (x_0, t_0), with $t_0 > 0$, we consider the uniquely defined solution curve $x = x(t)$ of the initial-value problem

$$\frac{dx}{dt} = u(x(t), t); \qquad x(t_0) = x_0 \tag{2.2.2}$$

A curve defined by (2.2.2) is a *characteristic curve* for the Burger's equation (2.2.1) and along this curve

$$\frac{d}{dt}u(x(t), t) = u_x \frac{dx}{dt} + u_t \equiv uu_x + u_t = 0 \tag{2.2.3}$$

In other words, $u(x,t)$ is constant along the characteristics, so that by (2.2.2), the characteristics have constant slope $dt/dx = 1/u(x(t),t) \equiv 1/u(x_0,t_0)$, i.e., they are straight lines; this clearly implies that for each $t \geq 0$, we must have $u(x_1,t) \leq u(x_2,t)$ if $x_1 < x_2$, for otherwise the characteristic straight lines would meet at some $t > 0$ and, at such a point of intersection, $u(x,t)$ would be multivalued. In particular, if $u(x,0) = u_0(x)$ and $u_0'(\bar{x}) < 0$ at some $\bar{x}$, then the solution of the Cauchy problem for (2.2.1) cannot be of class C^1 for all $t > 0$. In order to illustrate this situation somewhat more graphically, suppose that $u_0 \in C^\infty(R^1)$ with

$$u_0(x) = \begin{cases} 1, & x \leq 0 \\ 0, & x \geq 1 \end{cases} \quad \text{and} \quad u_0'(x) \leq 0, \; 0 \leq x \leq 1$$

Such an initial data function is sketched below.

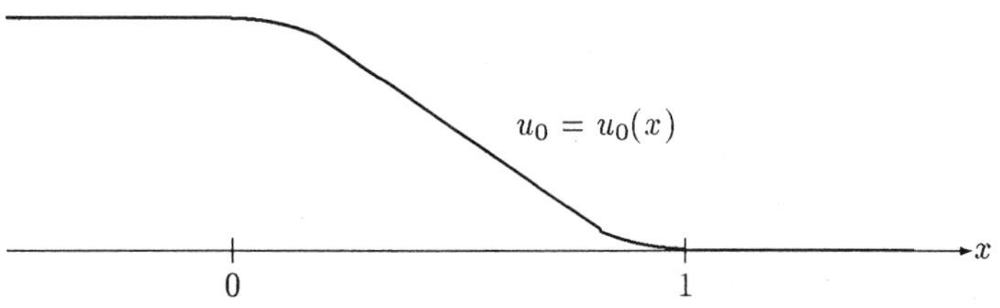

Figure 2.1

The characteristics for this situation are drawn below in the strip $\{(x,t) | -\infty < x < \infty, 0 \leq t \leq 1\}$:

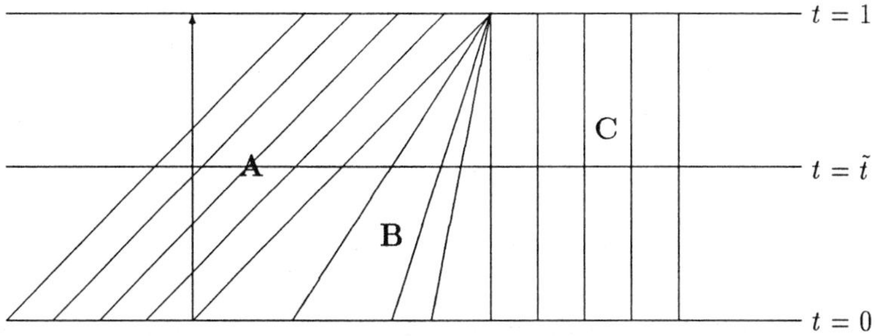

Figure 2.2

By virtue of (2.2.2) and (2.2.3) we have

$$u(x,t) = \begin{cases} 1, & \text{in domain } \mathbf{A} \\ \in [0,1], & \text{in domain } \mathbf{B} \\ 0, & \text{in domain } \mathbf{C} \end{cases}$$

and as t increases from $t = 0$ to $t = 1$ the decrease of $u(x,t)$ from the value 1 to the value 0 takes place over successively shorter spatial intervals until, at $t = 1$, this decrease is represented by a sudden jump in the value of u. This progressive decrease in the value of u is illustrated by the figure below:

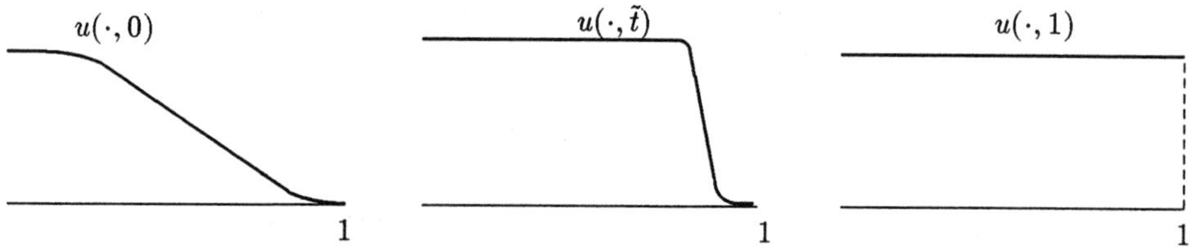

Figure 2.3

The phenomenon illustrated above by the Cauchy problem for the Burger's equation (2.2.1) carries over directly to the general initial-value problem for the single conservation law in one-space dimension

$$\left.\begin{aligned} u_t + f(u)_x = 0, \quad & x \in R^1, t > 0 \\ u(x,0) = u_0(x), \quad & x \in R^1 \end{aligned}\right\} \tag{2.2.4}$$

Indeed, along the characteristic curves defined by

$$\frac{dx}{dt} = f(u(x(t),t)) \tag{2.2.5}$$

we again have $\frac{d}{dt}u(x(t),t) = 0$, so that $u(x,t)$ is constant along the characteristics which have slope

$$\frac{dt}{dx} = \frac{1}{f'(u)} \equiv \frac{1}{f'(u_0(x))} \tag{2.2.6}$$

If, therefore,

$$m_1 = \frac{1}{f'(u_0(x_1))} < m_2 = \frac{1}{f'(u_0(x_2))}, \quad \text{for } x_1 < x_2 \tag{2.2.7}$$

then the characteristics, which emanate at $t = 0$ from x_1, x_2, must intersect at some $t > 0$ and, at such a t, any C^1 solution must become discontinuous. For f convex, i.e., $f'' > 0$, and $u_0'(\bar{x}) < 0$, it is clear that (2.2.7) must obtain; in order to explore this issue slightly further we refer to Figure 2.4 below:

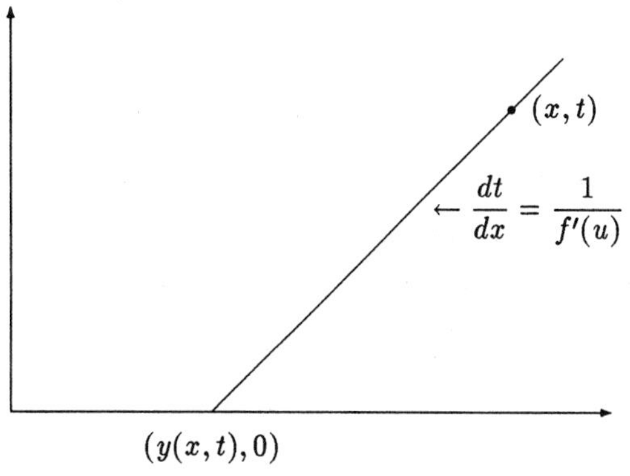

Figure 2.4

We let (x,t) be any point in the plane with $t > 0$ and $y(x,t)$ the unique point on the x-axis which lies on the straight line characteristic through (x,t). Then $t/(x-y(x,t)) = 1/f'(u)$, so

$$u(x,t) = u_0(y(x,t)) = u_0(x - tf'(u(x,t)) \tag{2.2.8a}$$

For $u_0(\cdot)$ differentiable we may now invoke the implicit function theorem to solve (2.2.8a) for u and then a direct calculation yields

$$u_t = \frac{-f'(u)u_0'}{1+u_0'f''(u)t}, \qquad u_x = \frac{u_0'}{1+u_0'f''(u)t} \tag{2.2.8b}$$

In view of (2.2.8b), it is clear that if $u_0'(x) \geq 0$, $\forall\, x \in R^1$, then $\nabla_{(x,t)}u$ is bounded for all $t > 0$ and, therefore, the solution of the initial-value problem (2.2.4) exists $\forall\, t \geq 0$. However, if $u_0'(\bar{x}) < 0$, at some $\bar{x} \in R^1$, then at such a point both u_t and u_x must become unbounded as $1 + u_0'(\bar{x})f''(u)t \to 0$, i.e., for t sufficiently large.

As an alternative means of exhibiting the breakdown of classical solutions $u(x,t)$ for the initial-value problem (2.2.4) we may argue as follows: we set $a(u) = f'(u)$ and rewrite (2.2.4) in the form

$$u_t + a(u)u_x = 0 \tag{2.2.9}$$

a differentiation of which, with respect to x, leads to

$$u_{tx} + a(u)u_{xx} + a'(u)u_x^2 = 0 \tag{2.2.10}$$

If we set $v = u_x$ then (2.2.10) can be rewritten as

$$v' + a'(u)v^2 = 0 \tag{2.2.11}$$

where v' is the directional derivative of $v(x,t)$ along the characteristic curve defined by $\frac{dx}{dt} = a(u)$, i.e.,

$$v' = v_t + a(u)v_x \tag{2.2.12}$$

As (2.2.11) is just an ordinary differential equation for v along the characteristic, it may be integrated so as to yield

$$v(t) = v(x(t),t) = \frac{v(0)}{1+v(0)ct} \tag{2.2.13}$$

where $c = a'(u)$ is constant along the characteristic. Thus, if $cv(0) > 0$, $v(t)$ is bounded for all $t > 0$ while if $cv(0) < 0$, then $v(t)$ will blow up at time $t = -1/cv(0)$.

The arguments put forth above, both the analytic and geometric ones, clearly indicate that, with respect to the initial-value problem (2.2.4), there cannot exist a function $u(x,t)$, for all $t > 0$, which solves this problem in the classical sense if $a(u_0(x))$ is not an increasing function of x; we turn our attention, therefore, to more general solutions of such initial-value problems, i.e., to generalized (or weak) solutions containing shocks.

2.3 Weak Solutions of Hyperbolic Conservation Laws

The considerations in § 2.2 suggest that we should look for weak (discontinuous) solutions of the conservation law (2.2.4). Inasmuch as the passage from the integral form of the general conservation law (2.1.7) to the differential form (2.1.8) depends on an *a priori* smoothness assumption, a reasonable thing to do, with regard to (2.2.4), would be to look for solutions of

$$\frac{d}{dt}\int_a^b u(x,t)\,dx = f(u(a)) - f(u(b)) \tag{2.3.1}$$

which is, of course, for arbitrary $a, b \in R^1$, the integral form of (2.2.4). Suppose that u is a solution of $u_t + f(u)_x = 0$ in the ordinary sense on each side of a smooth curve $x = X(t)$ across which u is discontinuous (the locus of points $x = X(t)$ will be the *shock wave*). We let u_l, u_r be, respectively, the values of u to the (immediate) left and right of $x = X(t)$ and we choose a, b so that $X(t)$ intersects the interval $[a,b]$ at time t; the situation is depicted in Figure 2.5. If we set $U(t) = \int_a^b u(x,t)\,dx$ and

$s = \dfrac{dX}{dt}$, then writing

$$U(t) = \int_a^{X(t)} u(x,t)\,dx + \int_{X(t)}^b u(x,t)\,dx$$

we have

$$\frac{dU}{dt} = \int_a^X u_t(x,t)\,dx + u_l s + \int_X^b u_t(x,t)\,dx - u_r s \tag{2.3.2}$$

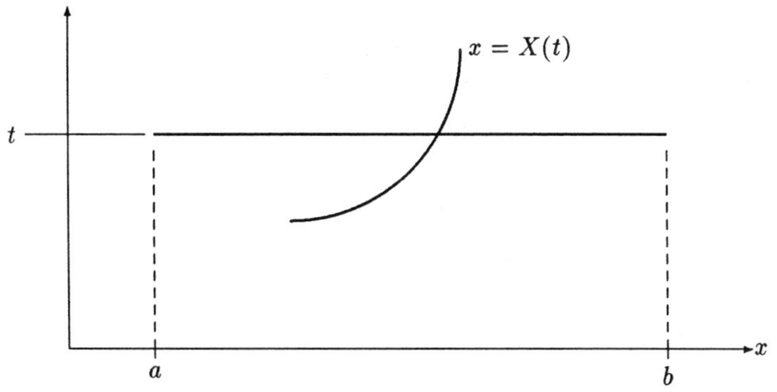

Figure 2.5

However, for both $x < X$ and $x > X$, $u_t = -f_x$ so, incorporating this fact into (2.3.2), and combining the result with (2.3.1), we are led to

$$\begin{aligned}\frac{dU}{dt} &= f(u(a)) - f(u_l) + u_l s - f(u(b)) + f(u_r) - u_r s \\ &= f(u(a)) - f(u(b))\end{aligned}$$

which implies, in turn, that

$$s[u] = [f(u)] \tag{2.3.3}$$

where $[u] = u_r - u_l$ and $[f(u)] = f(u_r) - f(u_l)$. The relation (2.3.3) is the *Rankine-Hugoniot* condition and must be satisfied by weak solutions of (2.2.4), i.e., solutions of the integral relation (2.3.1) which exhibit jump discontinuities; unfortunately, such solutions are not uniquely determined without the imposition of additional (entropy) conditions, as the following simple example for the Burger's equation (2.2.1) shows: We consider (2.2.4) with $f(u) = \frac{1}{2}u^2$ and $u_0(x) = \begin{cases} 0, x < 1 \\ 1, x > 0 \end{cases}$. Then, as depicted below, the method of characteristics determines the solution $u(x,t)$ everywhere for $t > 0$ except in the sector $0 < x < t$.

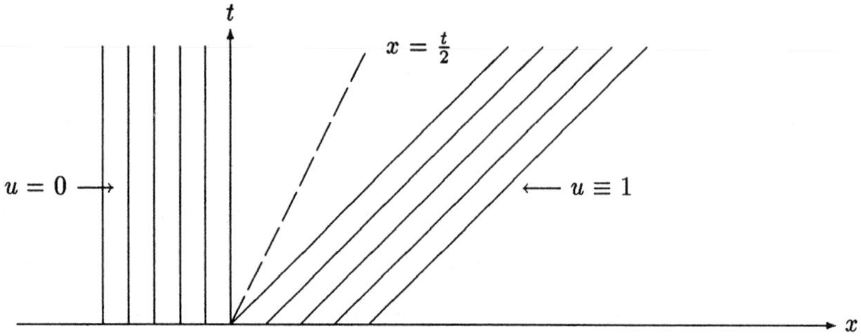

Figure 2.6

The following two functions now yield solutions to the initial-value problem at hand, and differ precisely in the sector $0 < x < t$:

$$u_1(x,t) = \begin{cases} 0, & x < t/2 \\ 1, & x > t/2 \end{cases} \qquad u_2(x,t) = \begin{cases} 0, & x < 0 \\ x/t, & 0 < x < t \\ 1, & x \geq t \end{cases} \tag{2.3.4}$$

We note that $u_1(x,t)$ satisfies the Rankine-Hugoniot condition across the shock at $x = t/2$ while $u_2(x,t)$ is continuous across this line; the solution $u_2(x,t)$ of (2.3.4) is referred to as a centered simple wave. The solution $u_1(x,t)$, which is illustrated below, may be ruled inadmissible in the class of weak (or generalized) solutions of the initial-value problem by the following elementary conditions: For the single

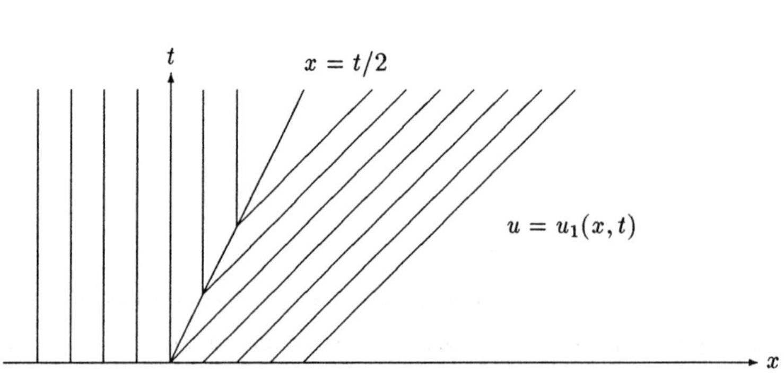

Figure 2.7

conservation law $u_t + a(u)u_x = 0$, let $a'(u) = f''(u) > 0$ and suppose that $u_r > u_l$. Then

$$s = \frac{f(u_r) - f(u_l)}{u_r - u_l} = f'(\xi), \quad u_l < \xi < u_r \tag{2.3.5}$$

Now, as f' is increasing, $f'(u_l) < s < f'(u_r)$, so that the shock speed is intermediate to the characteristic speeds on both sides of the shock; for the weak solution depicted in Figure 2.7, $s = \frac{1}{2}$, $f(u) = \frac{1}{2}u^2$, $u_l = 0$, and $u_r = 1$ and the above shock admissibility criteria, which was introduced by Lax [100], is clearly violated. Beyond the fact that uniqueness is lost for weak solutions of conservation laws, and must be recovered through the imposition of admissibility criteria, such as the elementary one described above, we should also note that different conservation laws in integral form may lead to the same conservation law in differential form; the following example has been noted in Dafermos [41]: The (integral) conservation laws

$$\begin{cases} \int_{x_1}^{x_2} u\Big|_{t_1}^{t_2} dx + \frac{1}{2}\int_{t_1}^{t_2} u^2\Big|_{x_1}^{x_2} dt & = 0 \\ \int_{x_1}^{x_2} u^2\Big|_{t_1}^{t_2} dx + \frac{2}{3}\int_{t_1}^{t_2} u^3\Big|_{x_1}^{x_2} dt & = 0 \end{cases} \tag{2.3.6}$$

both yield the Burger's equation (2.2.1), but the (different) Rankine-Hugoniot jump conditions

$$\begin{cases} -s[u] + \frac{1}{2}[u^2] & = 0 \\ -s[u^2] + \frac{2}{3}[u^3] & = 0 \end{cases} \tag{2.3.7}$$

and, therefore, the corresponding classes of generalized solutions do not coincide.

As a simple example of how the notion of weak solution, as described above, can be used to give a complete description of the solution to the Cauchy problem (2.2.4), after shock discontinuities develop, consider the Burger's equation (2.2.1) with initial data function

$$u_0(x) = \begin{cases} 1, & x < 0 \\ 1 - x, & 0 \leq x \leq 1 \\ 0, & x > 1 \end{cases} \tag{2.3.8}$$

The solution, up until time $t = 1$, when the first shock discontinuity forms, is again given by the method of characteristics and the basic idea is described the sketch in Figure 2.8 below:

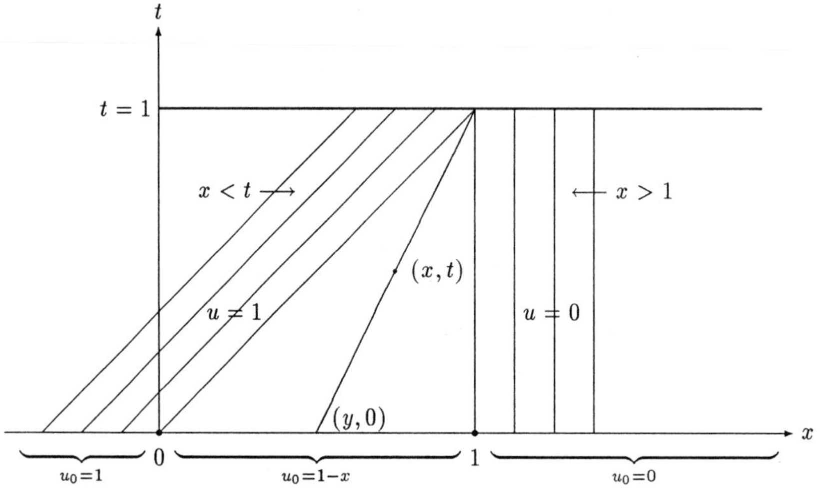

Figure 2.8

As the characteristics are described by the equations $\dfrac{dt}{dx} = \dfrac{1}{u(x,t)} = \dfrac{1}{u_0(x)}$, we have, for (x,t) on a characteristic through $(y,0)$, with $0 \leq y \leq 1$:

$$\frac{t}{x-y} = \frac{1}{u_0(y)} = \frac{1}{1-y} \tag{2.3.9}$$

Therefore,

$$t = \frac{x-y}{1-y} = \frac{x-1}{1-y} + 1 \Rightarrow \frac{1}{1-y} = \frac{t-1}{x-1} \tag{2.3.10}$$

and

$$u(x,t) \equiv u_0(y) = 1 - y = \frac{x-1}{t-1}, \qquad t \leq x \leq 1 \tag{2.3.11}$$

For $t < 1$ we then have a classical solution, namely,

$$u(x,t) = \begin{cases} 1, & x < t \\ \dfrac{1-x}{1-t}, & t \leq x \leq 1 \\ 0, & x > 1 \end{cases} \tag{2.3.12}$$

For $t \geq 1$ we use the Rankine-Hugoniot relation $s(u_l - u_r) = \frac{1}{2}(u_l^2 - u_r^2)$, with $u_l = 1$, $u_r = 0$, so that $s = \frac{1}{2}$; thus, for $t \geq 1$ the curve of discontinuity is the straight line with (reciprocal) slope $\dfrac{dx}{dt} = \dfrac{1}{2}$ emanating from $(1,1)$, i.e., $x = 1 + \frac{1}{2}(t-1)$. In the region $t \geq 1$ we may now define u by

$$u(x,t) = \begin{cases} 1, & x < 1 + \frac{1}{2}(t-1) \\ 0, & x > 1 + \frac{1}{2}(t-1) \end{cases}$$

The complete solution of the initial-value problem is depicted in Figure 2.9.

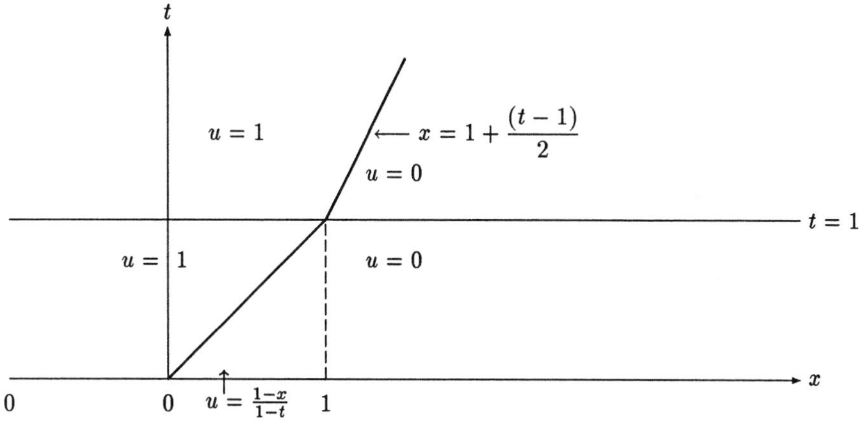

Figure 2.9

Although we illustrated, above, the genesis of the Rankine-Hugoniot jump conditions for a single conservation law, involving a single spatial variable, we will need, in the sequel, the analogous results for systems, albeit still in terms of a single spatial variable. To this end we consider the system of conservation laws

$$\boldsymbol{u}_t + \boldsymbol{f}(\boldsymbol{u})_x = 0, \qquad (x,t) \in R^1 \times R^+ \tag{2.3.13}$$

where $\boldsymbol{u} = (u_1, \dots, u_n) \in R^n$, $n \geq 1$, and $\boldsymbol{f} : R^n \to R^n$. Associated with (2.3.13) will be the initial values

$$\boldsymbol{u}(x,0) = \boldsymbol{u}_0(x), \qquad x \in R^1 \tag{2.3.14}$$

Suppose now that $\boldsymbol{u}(x,t)$ is a classical solution of the initial-value problem (2.3.13), (2.3.14) and let ψ be a C^1 function which is identically zero outside a compact subset Ω in the domain $t \geq 0$. Suppose also that $\mathcal{R}$ is a rectangle of the form $a \leq x \leq b, 0 \leq t \leq T$, such that $\Omega \subseteq \mathcal{R}$ and $\psi \equiv 0$ on the sides $x = a$, $x = b$, and $t = T$ of $\mathcal{R}$. The situation is illustrated in Figure 2.10 below.

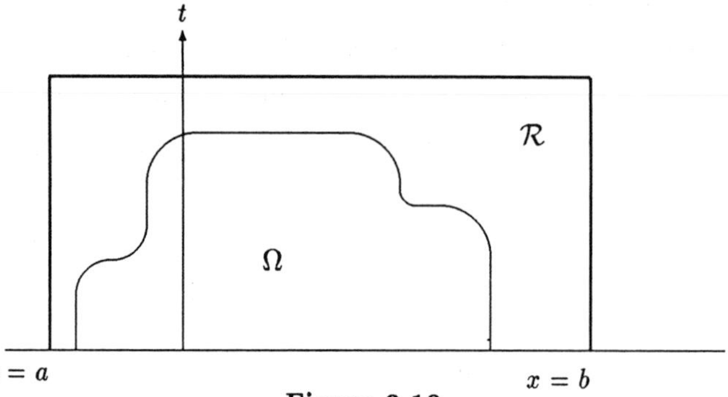

Figure 2.10

We multiply (2.3.13) by ψ and integrate over $t > 0$ so as to obtain

$$\iint_{t>0} (\boldsymbol{u}_t + \boldsymbol{f}_x)\psi \, dx \, dt = \iint_{\Omega} (\boldsymbol{u}_t + \boldsymbol{f}_x)\psi \, dx \, dt = \int_a^b \int_0^T (\boldsymbol{u}_t + \boldsymbol{f}_x)\psi \, dt \, dx = 0 \tag{2.3.15}$$

The following integrations by parts can now be executed:

$$\int_a^b \int_0^T \boldsymbol{u}_t \psi \, dt \, dx = \int_a^b \int_0^T \left[\frac{\partial}{\partial t}(\boldsymbol{u}\psi) - \boldsymbol{u}\psi_t\right] dt \, dx = \int_a^b \boldsymbol{u}\psi \Big|_0^T dx - \int_0^T \int_a^b \boldsymbol{u}\psi_t \, dx \, dt = -\int_a^b \boldsymbol{u}_0(x)\psi(x,0)\, dx - \int_0^T \int_a^b \boldsymbol{u}\psi_t \, dx \, dt$$

and

$$\int_0^T \int_a^b \boldsymbol{f}_x \psi \, dx \, dt = \int_0^T \int_a^b \left[\frac{\partial}{\partial x}(\boldsymbol{f}\psi) - \boldsymbol{f}\psi_x\right] dx \, dt = -\int_0^T \int_a^b \boldsymbol{f}\psi_x \, dx \, dt$$

Employing the above two results, in (2.3.15), we are led to

$$\iint_{t \geq 0} (\boldsymbol{u}\psi_t + \boldsymbol{f}(\boldsymbol{u})\psi_x)\, dx \, dt + \int_{-\infty}^{\infty} \boldsymbol{u}_0(x)\psi(x,0)\, dx = 0 \tag{2.3.16}$$

Therefore, for $\boldsymbol{u}$ a classical solution of the initial-value problem (2.3.13), (2.3.14), it follows that (2.3.16) holds for all C^1 functions ψ with compact support in $t \geq 0$.

We note, however, that (2.3.16) still has a sensible mathematical interpretation if $\boldsymbol{u}(x,t)$ and $\boldsymbol{u}_0(x)$ are only bounded, measurable functions. Following Smoller [165] we term a bounded measurable function $\boldsymbol{u}(x,t)$ a *weak solution* of the initial-value problem (2.3.13), (2.3.14), with bounded, measurable initial data $\boldsymbol{u}_0(x)$, if (2.3.16) holds $\forall\,\psi \in C^1$ with compact support in $t \geq 0$. If, on the other hand, (2.3.16) holds $\forall\,\psi \in C^1$ with compact support in $t \geq 0$, and $\boldsymbol{u} \in C^1$, then $\boldsymbol{u}(x,t)$ must be a classical solution of the initial-value problem (2.3.13), (2.3.14); to see this we choose $\tilde{\psi} \in C^1$ with compact support in $t > 0$ and integrate (2.3.16), with $\psi \to \tilde{\psi}$, by parts to obtain

$$\iint_{t>0} (\boldsymbol{u}_t + \boldsymbol{f}(\boldsymbol{u})_x)\tilde{\psi}\,dx\,dt = 0 \tag{2.3.17}$$

and, thus, (2.3.13). Multiplying (2.3.13) again by ψ and integrating by parts we find

$$\iint_{t\geq 0} (\boldsymbol{u}\psi_t + \boldsymbol{f}(\boldsymbol{u})\psi_x)\,dx\,dt + \int_{-\infty}^{\infty} \boldsymbol{u}(x,0)\psi(x,0)\,dx = 0 \tag{2.3.18}$$

Then from (2.3.16), (2.3.18) it follows that

$$\int_{-\infty}^{\infty} (\boldsymbol{u}(x,0) - \boldsymbol{u}_0(x))\psi(x,0)\,dx = 0 \tag{2.3.19}$$

and, as $\boldsymbol{u}_0(x)$ is continuous, we obtain (2.3.14).

The requirement that (2.3.16) holds places restrictions on the type of discontinuities that weak solutions may experience and brings us back, for the case of systems of conservation laws in one-space variable, to the natural extension of the Rankine-Hugoniot condition (2.3.3). Suppose that $\Gamma : x = X(t)$ is a smooth curve across which $\boldsymbol{u}$ experiences a jump discontinuity, but such that $\boldsymbol{u}$ has well-defined limits on both sides of Γ. Let $p \in \Gamma$ and let B be a ball centered at p; also, let B_1 and B_2 be the components of B determined by Γ. The situation is depicted in Figure 2.11.

We take $\psi \in C_0^1(B)$ so that $\psi\big|_{t=0} = 0$ for B as depicted in Figure 2.11. By virtue of (2.3.16)

$$\begin{aligned}\iint_B (\boldsymbol{u}\psi_t + \boldsymbol{f}(\boldsymbol{u})\psi_x)\,dx\,dt &= \iint_{B_1} (\boldsymbol{u}\psi_t + \boldsymbol{f}(\boldsymbol{u})\psi_x)\,dx\,dt \\ &+ \iint_{B_2} (\boldsymbol{u}\psi_t + \boldsymbol{f}(\boldsymbol{u})\psi_x)\,dx\,dt = 0\end{aligned} \tag{2.3.20}$$

Now, by hypothesis, $\boldsymbol{u} \in C^1(B_i)$, $i = 1,2$, so applying the divergence theorem we have

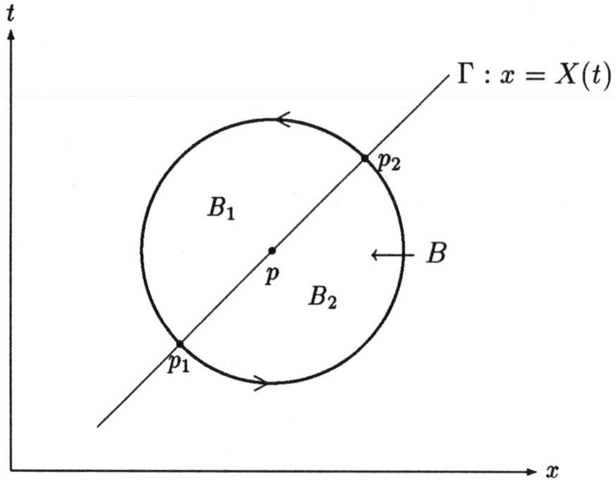

Figure 2.11

$$\iint_{B_i} (\boldsymbol{u}\psi_t + \boldsymbol{f}\psi_x)\,dx\,dt = \iint_{B_i} [(\boldsymbol{u}\psi)_t + (\boldsymbol{f}\psi)_x]\,dx\,dt = \int_{\partial B_i} \psi(-\boldsymbol{u}\,dx + \boldsymbol{f}\,dt) \tag{2.3.21}$$

where we have used the fact that (2.3.13) holds both in B_1 and B_2. However, $\psi\big|_{\partial B} = 0$, by hypothesis, so the line integrals $\int_{\partial B_i} \psi(-\boldsymbol{u}\,dx + \boldsymbol{f}\,dt)$ are nonvanishing only along Γ. We set

$$\boldsymbol{u}_l = \boldsymbol{u}(X(t)^-, t), \qquad \boldsymbol{u}_r = \boldsymbol{u}(X(t)^+, t) \tag{2.3.22}$$

and then

$$\begin{cases} \int_{\partial B_1} \psi(-\boldsymbol{u}\,dx + \boldsymbol{f}\,dt) = \int_{p_1}^{p_2} \psi(-\boldsymbol{u}_l\,dx + \boldsymbol{f}(\boldsymbol{u}_l)\,dt) \\ \int_{\partial B_2} \psi(-\boldsymbol{u}\,dx + \boldsymbol{f}\,dt) = -\int_{p_1}^{p_2} \psi(-\boldsymbol{u}_r\,dx + \boldsymbol{f}(\boldsymbol{u}_r)\,dt) \end{cases} \tag{2.3.23}$$

so that with $[\boldsymbol{u}] = \boldsymbol{u}_r - \boldsymbol{u}_l$, $[\boldsymbol{f}(\boldsymbol{u})] = \boldsymbol{f}(\boldsymbol{u}_r) - \boldsymbol{f}(\boldsymbol{u}_l)$,

$$\int_\Gamma \psi\,([\boldsymbol{u}]\,dx - [\boldsymbol{f}(\boldsymbol{u})]\,dt) = \mathbf{0} \tag{2.3.24}$$

Since $\psi \in C_0^1(B)$ has been arbitrarily chosen, we are led to the natural generalization of the condition (2.3.3), namely, the set of n conditions

$$s[\boldsymbol{u}] = [\boldsymbol{f}(\boldsymbol{u})], \qquad s = \frac{dX}{dt} \tag{2.3.25}$$

The vector equation (2.3.24) is again termed the Rankine-Hugoniot condition.

2.4 Riemann Invariants and Genuine Nonlinearity

The existence of Riemann invariants, for the hyperbolic system of quasilinear equations associated with wave-dielectric interactions in one-space dimension, is central to our work in this chapter, where we study shock formation in dielectrics; similar considerations will arise in our work in the next chapter on distributed parameter nonlinear transmission lines.

We begin by considering a system of two conservation laws

$$\begin{cases} u_t + f(u,v)_x &= 0 \\ v_t + g(u,v)_x &= 0 \end{cases} \tag{2.4.1}$$

or

$$\begin{pmatrix} u \\ v \end{pmatrix}_{,t} + \boldsymbol{A} \begin{pmatrix} u \\ v \end{pmatrix}_{,x} = \boldsymbol{0} \tag{2.4.2}$$

where

$$\boldsymbol{A} = \begin{pmatrix} f_u & f_v \\ g_u & g_v \end{pmatrix} \tag{2.4.3}$$

Inasmuch as we assume hyperbolicity, $\boldsymbol{A}$ has real distinct eigenvalues, say, λ, ρ ($\lambda < \rho$) with $\lambda = \lambda(u,v)$, $\rho = \rho(u,v)$; the corresponding left and right eigenvectors of $\boldsymbol{A}$ then satisfy

$$\begin{cases} \boldsymbol{l}_\lambda \boldsymbol{A} = \lambda \boldsymbol{l}_\lambda, \quad \boldsymbol{A}\boldsymbol{r}_\lambda = \lambda \boldsymbol{r}_\lambda \\ \boldsymbol{l}_\rho \boldsymbol{A} = \rho \boldsymbol{l}_\rho, \quad \boldsymbol{A}\boldsymbol{r}_\rho = \rho \boldsymbol{r}_\rho \end{cases} \tag{2.4.4}$$

If we suppose that $w = w(u,v)$ and $z = z(u,v)$ satisfy

$$\operatorname{grad} w \cdot \boldsymbol{r}_\lambda = \operatorname{grad} z \cdot \boldsymbol{r}_\rho = 0, \tag{2.4.5}$$

then w is constant along trajectories of the vector field $\boldsymbol{r}_\lambda$, while z is constant along trajectories of the vector field $\boldsymbol{r}_\rho$, for

$$\frac{dw}{dt} = \operatorname{grad} w \cdot \begin{pmatrix} u \\ v \end{pmatrix}_{,t} = \operatorname{grad} w \cdot \boldsymbol{r}_\lambda(u,v) = 0 \tag{2.4.6}$$

if $(u_t, v_t) = \boldsymbol{r}_\lambda(u(t), v(t))$, with a similar result for z. It can be shown that the mapping $(u,v) \mapsto (w,z)$ is one-to-one on a simply connected domain. Because the left and right eigenvectors of a matrix with distinct eigenvalues are biorthogonal, i.e.,

$\boldsymbol{l}_\rho \cdot \boldsymbol{r}_\lambda = 0$, etc., it follows from (2.4.5) that $\operatorname{grad} w$ must be a left eigenvector of $\boldsymbol{A}$ corresponding to the eigenvalue ρ ; thus,

$$\operatorname{grad} w \cdot \boldsymbol{A} = \rho \operatorname{grad} w \tag{2.4.7}$$

From (2.4.2) we then have

$$\operatorname{grad} w \cdot \begin{pmatrix} u \\ v \end{pmatrix}_{,t} + \operatorname{grad} w \cdot \boldsymbol{A} \begin{pmatrix} u \\ v \end{pmatrix}_{,x} = 0$$

from which it is immediate that

$$w_t + \rho \operatorname{grad} w \cdot \begin{pmatrix} u \\ v \end{pmatrix}_{,x} = 0 \tag{2.4.8}$$

or

$$w' = w_t + \rho w_x = 0 \tag{2.4.9}$$

where $'$ denotes the directional derivative of $w(u,v)$ along the (characteristic) curves defined by $\frac{dx}{dt} = \rho$. As a direct consequence of (2.4.9), we infer that $w(u,v)$ is, in fact, constant along the trajectories defined by $\frac{dx}{dt} = \rho$, and in an entirely analogous fashion,

$$\grave{z} = z_t + \lambda z_x = 0, \tag{2.4.10}$$

with $\grave{}$ denoting the directional derivative along the (characteristic) curves defined by $\frac{dx}{dt} = \lambda$, so that $z(u,v)$ is constant along such curves. The functions $w(u,v), z(u,v)$, satisfying (2.4.5), are termed *Riemann invariants.*

As a simple, but important, example to illustrate the above considerations for a 2×2 system of conservation laws, consider the system

$$\begin{cases} u_t - v_x &= 0 \\ v_t - \sigma(u)_x &= 0 \end{cases} \tag{2.4.11}$$

which arises both in fluid mechanics and nonlinear elasticity; inhomogeneous versions of (2.4.11) will be seen to arise in our work on wave-dielectric interactions in one-space dimension as well as in the propagation of signals in nonlinear transmission lines. With $u = y_x$, $v = y_t$ the second equation in (2.4.11) yields the nonlinear wave equation $y_{tt} = \sigma(y_x)_x$. From (2.4.11) we infer immediately that

$$\boldsymbol{A} = \begin{pmatrix} 0 & -1 \\ -\sigma'(u) & 0 \end{pmatrix} \tag{2.4.12}$$

which has eigenvalues $\rho = \sqrt{\sigma'(u)}$, $\lambda = -\sqrt{\sigma'(u)}$; we assume, of course, that $\sigma' > 0$ so that (2.4.11) is a hyperbolic system. The eigenvectors corresponding to ρ, λ are, respectively, $(1, -\sqrt{\sigma'(u)})$ and $(1, \sqrt{\sigma'(u)})$ and the two families of characteristic curves are defined by the solutions of $\frac{dx}{dt} = \pm\sqrt{\sigma'(u)}$. The characteristic curves are no longer straight lines, as they were, for example, for the single conservation law (2.4.4), nor are u or v constant along the characteristics; however, the Riemann invariants defined by

$$\begin{cases} w(u,v) = & v - \int_0^u \sqrt{\sigma'(\xi)}\, d\xi \\ z(u,v) = & v + \int_0^u \sqrt{\sigma'(\xi)}\, d\xi \end{cases} \tag{2.4.13}$$

do possess this important property of constancy along the characteristic curves, i.e., they satisfy (2.4.9) and (2.4.10), respectively. That the functions (2.4.13) satisfy (2.4.9), (2.4.10) may be seen either by computing directly that

$$\operatorname{grad} w = (-\sqrt{\sigma'(u)}, 1), \qquad \operatorname{grad} z = (\sqrt{\sigma'(u)}, 1) \tag{2.4.14}$$

and then noting that (2.4.5) is satisfied with $\boldsymbol{r}_\lambda = (1, \sqrt{\sigma'(u)})$, $\boldsymbol{r}_\rho = (1, -\sqrt{\sigma'(u)})$, or by observing that, by virtue of (2.4.11), (2.4.13),

$$\begin{aligned} \frac{w_t}{w_x} &= \frac{v_t - \sqrt{\sigma'(u)}u_t}{v_x - \sqrt{\sigma'(u)}u_x} \qquad (2.4.15) \\ &= \frac{\sigma'(u)u_x - \sqrt{\sigma'(u)}v_x}{v_x - \sqrt{\sigma'(u)}u_x} = -\sqrt{\sigma'(u)} \end{aligned}$$

so that

$$w_t + \sqrt{\sigma'(u)}w_x = 0, \tag{2.4.16}$$

which is (2.4.9), and (2.4.10) may be verified in a similar fashion.

While we will not, in this exposition, deal with quasilinear systems involving more than two equations in two unknowns (in a single spatial variable), the above considerations apply to systems of n equations in n unknowns $(u_1, \ldots, u_n)$ of the form (2.3.13) as long as there is only a single space variable; for the sake of completeness on this particular point, we will render the appropriate definitions and a single illustrative example from gas dynamics. Thus, consider (2.3.13) with $\boldsymbol{u} = (u_1, \ldots, u_n)$ and $\boldsymbol{f}(\boldsymbol{u}) = (f_1(\boldsymbol{u}), \ldots, f_n(\boldsymbol{u}))$ with $\boldsymbol{f}$ at least C^1 in some open domain $\mathcal{D} \subset R^n$. We assume that the Jacobian matrix $\nabla \boldsymbol{f}(\boldsymbol{u})$ possesses n real, distinct eigenvalues $\lambda_1(\boldsymbol{u}) <$

$\cdots < \lambda_n(\boldsymbol{u})$ in $\mathcal{D}$. For each $\lambda_i(\boldsymbol{u})$, $i = 1, \ldots, n$ there is a right eigenvector $\boldsymbol{r}_i(\boldsymbol{u})$ and a left eigenvector $\boldsymbol{l}_i(\boldsymbol{u})$; the eigenvalues and eigenvectors of $\nabla \boldsymbol{f}(\boldsymbol{u})$ are assumed to be at least of class C^1 for $\boldsymbol{u} \in \mathcal{D}$. Then a *k-Riemann invariant* is a smooth function $w : \mathcal{D} \to R$ such that $\nabla w(\boldsymbol{u}) \cdot r_k(\boldsymbol{u}) = 0$; it may be shown (e.g. Smoller [165]) that $\exists (n-1)$ k-Riemann invariants whose gradients are linearly independent in $\mathcal{D}$. The following example arises in the Eulerian description of the equations of gas dynamics: we consider the system

$$\begin{cases} \rho_t + (\rho u)_x = 0 \\ u_t + u u_x + \dfrac{1}{\rho} p_x = 0 \\ s_t + u s_x = 0 \end{cases} \tag{2.4.17}$$

where ρ is density, u velocity, and s entropy and where $p = p(\rho, s)$, p being the pressure in the gas. We assume that $p_\rho > 0$ so that the sound speed $c = \sqrt{p_\rho}$ is well-defined. A direct calculation produces

$$\nabla \boldsymbol{f}(\boldsymbol{u}) = \begin{pmatrix} u & \rho & 0 \\ p_\rho/\rho & u & p_s/\rho \\ 0 & 0 & u \end{pmatrix} \tag{2.4.18}$$

and

$$\lambda_1(\boldsymbol{u}) = u - c, \qquad \lambda_2(\boldsymbol{u}) = u, \qquad \lambda_3(\boldsymbol{u}) = u + c \tag{2.4.19}$$

as well as

$$\boldsymbol{r}_1(\boldsymbol{u}) = (\rho, -c, 0)^t, \qquad \boldsymbol{r}_2(\boldsymbol{u}) = (p_s, 0, -p_\rho)^t, \qquad \boldsymbol{r}_3(\boldsymbol{u}) = (\rho, c, 0)^t \tag{2.4.20}$$

with the t superscript denoting transpose. For each $\boldsymbol{r}_k$, $k = 1, 2, 3$, there are $n - 1 = 2$ associated independent k-Riemann invariants, namely,

$$\begin{cases} w_1 = s \quad \text{or} \quad w_1 = (u + h) \\ w_2 = u \quad \text{or} \quad w_2 = p \\ w_3 = s \quad \text{or} \quad w_3 = (u - h) \end{cases} \tag{2.4.21}$$

where the *enthalphy* $h = h(\rho, s)$ satisfies $h_\rho = c/\rho$. For example, with $k = 3$ and $\boldsymbol{r}_3(\boldsymbol{u}) = (\rho, c, 0)^t$, any Riemann invariant $w(\rho, u, s)$ must satisfy

$$\rho w_\rho + c w_u + 0 \cdot w_s = 0 \tag{2.4.22}$$

so that we may, trivially, take $w = s$. With $w = u - h$, (2.4.22) becomes $\rho w_\rho + c w_u = -\rho h_\rho + c = 0$, by virtue of the definition of $h(\rho, s)$.

It is worth noting that if $\boldsymbol{u}$ is a C^1 solution of (2.3.13), in some domain $\mathcal{D} \subseteq R^n$, and all the associated k-Riemann invariants are constant in $\mathcal{D}$, then $\boldsymbol{u}$ is called a *k-rarefaction wave* (or, sometimes, a *k-simple wave*). As a generalization of (2.4.9) and (2.4.10) we also note the following result: if $\boldsymbol{u}(\eta)$ is an integral curve of $r_k(\boldsymbol{u})$ parametrized by η, i.e., $\frac{d}{d\eta}\boldsymbol{u}(\eta) = r_k(\boldsymbol{u}(\eta))$, and w is a k-Riemann invariant, then

$$\frac{d}{d\eta}w(\boldsymbol{u}(\eta)) = \nabla w(\boldsymbol{u}(\eta)) \cdot \frac{d\boldsymbol{u}}{d\eta} = 0 \tag{2.4.23}$$

by virtue of the definition of $w(\boldsymbol{u})$. Thus any k-Riemann invariant w must be constant along any integral curve of $r_k(\boldsymbol{u})$.

We close this section with a definition and illustrative example which will be useful to us in the sequel. Consider the system (2.3.13); then the k-th characteristic family is said to be *genuinely nonlinear* in a region $\mathcal{D} \subseteq R^n$ if for all $\boldsymbol{u} \in \mathcal{D}$, $\nabla\lambda_k(\boldsymbol{u})\cdot r_k(\boldsymbol{u}) \neq 0$. For the scalar equation (2.2.4) we have $\lambda(u) = f'(u)$, and $r(u) \equiv 1$, so that $\nabla\lambda \cdot r = f''(u)$ and genuine nonlinearity is equivalent to the convexity or concavity of $f(u)$. For the 2×2 system (2.4.11) we have $\lambda_1(u, v) = \sqrt{\sigma'(u)}$, $\lambda_2(u, v) = -\sqrt{\sigma'(u)}$ and $\boldsymbol{r}_1(u, v) = (1, -\sqrt{\sigma'(u)})$, $\boldsymbol{r}_2(u, v) = (1, \sqrt{\sigma'(u)})$. Therefore,

$$\begin{aligned} \nabla\lambda_1(u, v) \cdot \boldsymbol{r}(u, v) &= \left(\frac{\sigma''(u)}{2\sqrt{\sigma'(u)}}, 0\right) \cdot \left(1, -\sqrt{\sigma'(u)}\right) \\ &= \frac{\sigma''(u)}{2\sqrt{\sigma'(u)}} \end{aligned} \tag{2.4.24}$$

and the first characteristic family is genuinely nonlinear if, e.g., $\sigma'' > 0$.

2.5 Shock Formation in 2×2 Systems

In this section we exhibit, within the framework of the 2×2 system (2.4.1), the methodology that will be used in the sequel in order to establish the formation of discontinuities both in the wave-dielectric interaction problem, in a single space dimension, as well as in the problem of signal propagation in a nonlinear transmission line. We choose Riemann invariants $w = w(u, v)$ and $z = z(u, v)$ satisfying (2.4.5) and, therefore, (2.4.9) and (2.4.10), respectively. Next, we differentiate (2.4.9) through with respect to x, thereby obtaining

$$w_{tx} + \rho w_{xx} + \rho_w w_x^2 + \rho_z w_x z_x = 0 \tag{2.5.1}$$

where we have used the inverse transformation $u = u(w, z), v = v(w, z)$ to express ρ as a function of the Riemann invariants w and z. We now set $y(x, t) = w_x(x, t)$ and note that we may rewrite (2.5.1) in the form

$$y' + \rho_w y^2 + \rho_z y z_x = 0 \tag{2.5.2}$$

$y' = y_t + \rho y_x$ being the directional derivative of $y(x, t)$ in the direction of the tangent to the characteristic curve defined by $\frac{dx}{dt} = \rho$. By virtue of (2.4.10), and the fact that $z' = z_t + \rho z_x$, we have

$$z_x = \frac{z'}{\rho - \lambda} \tag{2.5.3}$$

Combining (2.5.2) and (2.5.3) we are led to the equation

$$y' + \rho_w y^2 + \frac{\rho_z}{\rho - \lambda} z' y = 0 \tag{2.5.4}$$

Now choose $g(w, z)$ so as to satisfy

$$g_z = \frac{\rho_z}{\rho - \lambda} \tag{2.5.5}$$

so that

$$g' = g_w w' + g_z z' = \frac{\rho_z}{\rho - \lambda} z' \tag{2.5.6}$$

because $w' = 0$. Employing (2.5.6), we rewrite (2.5.4) as

$$y' + \rho_w y^2 + g' y = 0 \tag{2.5.7}$$

If we multiply (2.5.7) through by e^g, and set $Y = e^g y$ and $h = e^{-g} \rho_w$, then it is easily seen that (2.5.7) becomes

$$Y' + hY^2 = 0 \tag{2.5.8}$$

which is an ordinary differential equation for $Y = e^g w_x(x, t)$ along the characteristic defined by $\frac{dx}{dt} = \rho$; integration of (2.5.8) along this characteristic produces

$$Y(x(t), t) = \frac{Y(x_0, 0)}{1 + Y(x_0, 0) H(t)} \tag{2.5.9}$$

where $H(t) = \int_0^t h(x(\tau), \tau)\, d\tau$. Now, suppose that $\rho_w > 0$ and that the initial values of w and z are bounded, i.e., $|w_0|_\infty, |z_0|_\infty < M$ for some $M > 0$. Then $|w|_\infty, |z|_\infty < M$ for all t because the value of w at any point (x, t) must be equal, by virtue of (2.4.9), to the value of w at that point on the line $t = 0$ to which it can be connected by a characteristic defined by $\frac{dx}{dt} = \rho$, with an analogous result for z. Because (w, z) remains in a bounded set of $\mathbf{R}^2$, for all $t > 0$, when (w_0, z_0) is in a bounded set, $\exists h_0 > 0$ such that $h(x,t) \geq h_0$ for all $(x,t) \in R^2$ (recall that $h = e^{-g}\rho_w$). From the definition of $H(t)$ we now have $H(t) \geq h_0 t$, $\forall t \geq 0$. Therefore, by (2.5.9), if

$$Y(x_0, 0) = e^{g(x_0,0)} w_x(x_0, 0) > 0 \tag{2.5.10}$$

$$\Leftrightarrow \; w_x(x_0, 0) > 0$$

then $Y = e^g w_x(x,t)$ remains bounded for all $t > 0$; however, if $w_x(x_0, 0) < 0$, then again, by (2.5.9) and the fact that $H(t) \geq h_0 t$, it follows that $w_x(x,t)$ must become unbounded in finite time. We now note the following: Suppose we consider as a special case of (2.4.1) the system (2.4.11) for which $\rho = \sqrt{\sigma'(u)}$; then

$$\rho_w = \rho_u u_w + \rho_v v_w \equiv \rho_u u_w = \frac{\sigma''(u)}{2\sqrt{\sigma'(u)}} u_w \tag{2.5.11}$$

However u_w may be computed as follows: from $w = w(u(w,z), v(w,z))$ we have

$$1 = w_u \cdot u_w + w_v \cdot v_w \tag{2.5.12}$$

By (2.4.13), $w_u = -\sqrt{\sigma'(u)}$, and $w_v = 1$, so

$$1 = -\sqrt{\sigma'(u)} u_w + v_w \tag{2.5.13}$$

Also, $z = z(u(w,t), v(w,z))$ so

$$0 = z_u u_w + z_v v_w$$

in which case, using (2.4.13) again,

$$0 = \sqrt{\sigma'(u)} u_w + v_w \tag{2.5.14}$$

Combining (2.5.13) and (2.5.14) we easily find that

$$u_w = \frac{-1}{2\sqrt{\sigma'(u)}} \tag{2.5.15}$$

so that (2.5.11) becomes

$$\rho_w = \frac{-\sigma''(u)}{4\sigma'(u)} \tag{2.5.16}$$

Therefore, the critical assumption that $\rho_w > 0$ is seen, by virtue of (2.4.24), to be equivalent to the requirement that the first characteristic family for (2.4.11), i.e., $\lambda_1(u,v) \equiv \rho(u,v)$, be genuinely nonlinear. Our results may be summarized in the following:

Theorem 2.1 Let (u,v) be a solution pair for the 2×2 system of conservation laws (2.4.1); let $w(u,v)$, $z(u,v)$ be a corresponding pair of Riemann invariants and suppose that $\rho_w > 0$ where $\rho(w,z)$ is the eigenvalue of $\boldsymbol{A}$, as given by (2.4.3), which defines the first characteristic family. Then if $w_x(x,0) < 0$ at any x, the derivatives u_x, v_x of the solution become unbounded in a finite time span.

Variants of the methodology which led to Theorem 2.1 have been used to establish finite-time breakdown of smooth solutions to initial-value problems for quasilinear hyperbolic systems arising in a wide variety of applications, e.g., the work of Nishida [125], Slemrod [163], Nohel [126], Hattori [67], Klainerman and Majda [92], Lax [99], J. Wang and C. Li [178], and Bloom [23], [22].

2.6 Basic Equations for Wave-Dielectric Interactions

In this section we begin our study of the propagation of an electromagnetic wave into a domain in R^3 occupied by a nonlinear dielectric substance. For now, $\Omega \subseteq R^3$ can be any bounded, or unbounded, open domain which is filled with a nonlinear, isotropic dielectric substance that conforms to the constitutive relations

$$\boldsymbol{D}(\boldsymbol{x},t) = \epsilon(\boldsymbol{E}(\boldsymbol{x},t))\boldsymbol{E}(\boldsymbol{x},t) \tag{2.6.1}$$

$$\boldsymbol{B}(\boldsymbol{x},t) = \mu(\boldsymbol{H}(\boldsymbol{x},t))\boldsymbol{H}(\boldsymbol{x},t) \tag{2.6.2}$$

for $\boldsymbol{x} \in \Omega$, where $\boldsymbol{B}$, $\boldsymbol{H}$, $\boldsymbol{E}$, and $\boldsymbol{D}$ are, respectively, the magnetic field, magnetic intensity, electric field, and electric induction field, as introduced in Chapter 1 and $\boldsymbol{D} = \epsilon_0\boldsymbol{E} + \boldsymbol{P}(\boldsymbol{E})$, $\boldsymbol{P}$ begin the polarization vector and ϵ_0 the permittivity of free space. Throughout Ω, for all $t > 0$, we assume that Maxwell's equations, in the form (1.3.3a)-(1.3.3d) apply, where $\rho = \rho(\boldsymbol{x},t)$ is the free-charge density and $\boldsymbol{J}$ is the conduction vector. To simplify matters somewhat, we will assume here that $\mu(\boldsymbol{H}) = \mu_0 =$ const. (the permeability of free space) and while we do not, at this point, assume a specific form for the scalar-valued constitutive function ϵ, a typical assumption, which appears quite frequently in the nonlinear optics [18] literature, is that

$$\epsilon(\boldsymbol{E}) = \epsilon_0 + \epsilon_2\|\boldsymbol{E}\|^2, \quad \epsilon_0 > 0,\ \epsilon_2 > 0 \tag{2.6.3}$$

Relations of the form (2.6.3), but with $\epsilon_2 < 0$, have also been studied extensively.

While $\Omega \subseteq R^3$ may be any bounded or unbounded open domain, the situation of primary interest to us here will be the one in which $\Omega \equiv \Omega_c$, an infinite cylindrical domain of the form $\{(x,y,z)| -\infty < x < \infty,\ f(y,z) = c > 0\}$. If we assume that a linearly polarized wave is propagating in Ω_c, then

$$\boldsymbol{E} = (0, E(x,t), 0) \quad , \quad \boldsymbol{D} = (0, D(x,t), 0) \tag{2.6.4}$$

$$\boldsymbol{B} = (0, 0, B(x,t)) \quad , \quad \boldsymbol{H} = (0, 0, H(x,t)) \tag{2.6.5}$$

In this particular case, it is clear that the constitutive relations (2.6.1), (2.6.2) reduce to the set of scalar relations

$$D(x,t) = \tilde{\epsilon}(E(x,t))E(x,t) \equiv \mathcal{D}(E(x,t)) \tag{2.6.6}$$

$$B(x,t) = \mu_0 H(x,t) \tag{2.6.7}$$

provided we define

$$\tilde{\epsilon}(\zeta) \equiv \epsilon((0,\zeta,0)), \quad \forall \zeta \in R^1 \tag{2.6.8}$$

If we assume that

$$a(\zeta) \equiv (\zeta \tilde{\epsilon}(\zeta))' > 0, \quad \forall \zeta \in R^1 \tag{2.6.9}$$

then (2.6.6) clearly can be rewritten in the form

$$D(x,t) = \int_0^{E(x,t)} a(\zeta)\, d\zeta \tag{2.6.10}$$

and by virtue of the monotonicity of the integral, as a function of E, $\exists \mathsf{E}$ such that $\forall x \in R^1$ and $t > 0$

$$E(x,t) = \mathsf{E}(D(x,t)) \tag{2.6.11}$$

Alternatively, we may assume that (2.6.11) is prescribed, *a priori*, as e.g., in Grindlay [63]. Clearly, E in (2.6.11) satisfies $\mathsf{E}'(D) = \dfrac{1}{a(\mathsf{E}(D))} > 0$.

Finally, we require a constitutive relation for the conduction vector $\boldsymbol{J}$. In all of our work in this chapter we will assume that $\boldsymbol{J}$ is given by a nonlinear Ohm's law of a type formally similar to that of (1.3.57), i.e., that $\boldsymbol{J} = \sigma(\boldsymbol{E})\boldsymbol{E}$ with $\sigma(\cdot)$ of class C^1. Analogous to (2.6.8) we set

$$\tilde{\sigma}(\zeta) = \sigma((0,\zeta,0)), \quad \forall \zeta \in R^1 \tag{2.6.12}$$

so that

$$\boldsymbol{J} = (0, \tilde{\sigma}(E(x,t))E(x,t), 0) \tag{2.6.13}$$

If we assume, as we will here, that the free-charge density $\rho(\boldsymbol{x}, t)$ is negligible, then with $\boldsymbol{J}$ given by (2.6.13), the equation expressing conservation of charge, namely, (1.2.8), is trivially satisfied.

In all of the work done on the problem of plane wave-nonlinear dielectric interaction prior to the author's paper [23], e.g. [145], [46], [29], [79], [87], [50], and [75], it was assumed *a priori* that the conduction vector $\boldsymbol{J} = \boldsymbol{0}$; for the case where $\boldsymbol{J} = \boldsymbol{0}$, it had been shown by the authors of these papers, and others as well, that singularities form in the solutions of the initial-value problems associated with the homogeneous quasi-linear hyperbolic systems and equations that result when Maxwell's equations are combined with constitutive hypotheses of the type delineated above, and that this is true even for initial values of the electromagnetic field that are arbitrarily small, smooth, and compactly supported. One such example of the kind of results that are obtainable in the case $\boldsymbol{J} = \boldsymbol{0}$ will be expounded upon below in § 2.7. However, no nonlinear dielectric is a perfect nonconductor and thus an *a priori* assumption to the effect that $\boldsymbol{J} = \boldsymbol{0}$ is not entirely reasonable with respect to the problem at hand. In referring to the problem of the plane wave-nonlinear dielectric interaction, Broer [29] expressed the view that "in reality the solution of the physical problem may be univalent throughout. This means that there must be an effect ...which becomes operative whenever waves steepen. This effect must put a bound on the steepness of the wave fronts. An analogous situation exists in the theory of non-linear acoustics (where) the neglected effects are viscosity and heat conduction. In the optical case the wave propagation will certainly be rendered univalent by some dissipative process not included in the field equations, which admit an energy equation without dissipation. There is at present no theory of this effect." In this treatise we will effect, at least, a partial answer to the hypothesis of Broer, even for the case of wave propagation in nondispersive nonlinear dielectrics. We will show in § 2.7 that the inclusion of nonlinear conduction in the model of wave-dielectric interaction leads to the conclusion that singularities form provided that initial gradients of the electromagnetic fields in the wave are (pointwise) sufficiently large; no such restriction on these initial gradients is needed for singularities to form when $\boldsymbol{J} = \boldsymbol{0}$. Then, in § 2.7, it will be shown that with $\boldsymbol{J} \neq \boldsymbol{0}$, the assumption of initial data which is both smooth and sufficiently small, in a sense which will be made precise, assures the existence of a globally smooth electromagnetic field in the wave and, thus, precludes the formation of singularities.

If, at this point, we combine our assumption of plane wave propagation, with the above delineated constitutive hypotheses, we find that Maxwell's equations (1.3.3a)-(1.3.3d) reduce to the scalar quasilinear system

$$\begin{cases} \dfrac{\partial \mathcal{D}(E)}{\partial t} + \tilde{\sigma}(E)E = -\dfrac{\partial H}{\partial x} \\ \dfrac{\partial B}{\partial t} = -\dfrac{\partial E}{\partial x} \end{cases} \tag{2.6.14}$$

However, $\dfrac{\partial \mathcal{D}}{\partial E} = a(E)$, and $B = \mu_0 H$, so we may rewrite (2.6.14) in the form

$$\begin{pmatrix} E \\ H \end{pmatrix}_{,t} + \begin{pmatrix} 0 & 1/a(E) \\ 1/\mu_0 & 0 \end{pmatrix} \begin{pmatrix} E \\ H \end{pmatrix}_{,x} = \begin{pmatrix} -\tilde{\sigma}(E)E/a(E) \\ 0 \end{pmatrix} \tag{2.6.15}$$

The system (2.6.15) is a strictly hyperbolic inhomogeneous quasilinear system whose characteristics are defined by the nonlinear ordinary differential equations

$$\frac{dx}{dt} = \pm 1/\sqrt{\mu_0 a(E(x,t))} \tag{2.6.16}$$

It is, however, not very convenient to deal with the wave-dielectric interaction problem in the form given by (2.6.15), inasmuch as the corresponding homogeneous problem which results from setting $\tilde{\sigma} = 0$ is not in conservation form; instead, we use the fact that (2.6.11) applies to rewrite (2.6.14) in the form

$$\begin{cases} \dfrac{\partial D}{\partial t} + \dfrac{1}{\mu_0}\dfrac{\partial B}{\partial x} = -\Sigma(D) \\ \dfrac{\partial B}{\partial t} + \mathsf{E}'(D)\dfrac{\partial D}{\partial x} = 0 \end{cases} \tag{2.6.17}$$

where

$$\Sigma(\zeta) = \tilde{\sigma}(\mathsf{E}(\zeta))\mathsf{E}(\zeta), \quad \forall \zeta \in R^1 \tag{2.6.18}$$

The system (2.6.17) is an (inhomogeneous) hyperbolic conservation law of the form

$$\frac{\partial u_i}{\partial t} + \frac{\partial f_i}{\partial x}(\boldsymbol{u}) = \boldsymbol{g}(\boldsymbol{u}) \tag{2.6.19}$$

with

$$\boldsymbol{u} = \begin{pmatrix} D \\ B \end{pmatrix}, \quad \boldsymbol{f}(\boldsymbol{u}) = \begin{pmatrix} B/\mu_0 \\ \mathsf{E}(D) \end{pmatrix}, \quad \boldsymbol{g}(\boldsymbol{u}) = \begin{pmatrix} -\Sigma(D) \\ 0 \end{pmatrix} \tag{2.6.20}$$

and associated characteristics defined by the nonlinear ordinary differential equations

$$\frac{dx}{dt} = \pm\sqrt{\frac{\mathsf{E}'(D(x,t))}{\mu_0}} \tag{2.6.21}$$

For $\tilde{\sigma} \not\equiv 0$ it is a simple matter to show that (2.6.17) implies that $D(x,t)$ satisfies the nonlinear wave equation

$$\frac{\partial^2 D}{\partial t^2} + \Sigma'(D)\frac{\partial D}{\partial t} = \frac{1}{\mu_0}\frac{\partial^2 \mathsf{E}(D)}{\partial x^2} \tag{2.6.22}$$

Our study of the (one-dimensional) wave-nonlinear dielectric interaction problem in the cylinder Ω_c will involve both the system (2.6.17) as well as the nonlinearly damped nonlinear wave equation (2.6.21).

2.7 Shock Formation in Dielectrics

We will begin with the case in which $\sigma \equiv 0$ so that $\boldsymbol{J} = \boldsymbol{0}$; in this case (2.6.17) reduces to the standard hyperbolic conservation law

$$\begin{cases} \dfrac{\partial D}{\partial t} + \dfrac{1}{\mu_0}\dfrac{\partial B}{\partial x} = 0 \\ \dfrac{\partial B}{\partial t} + \mathsf{E}'(D)\dfrac{\partial D}{\partial x} = 0 \end{cases} \tag{2.7.1}$$

whose characteristics are defined by (2.6.21) and with which we associated initial data of the form

$$D(x,0) = D_0(x), \quad B(x,0) = B_0(x), \quad x \in R^1 \tag{2.7.2}$$

System (2.7.1) is, clearly, of the form (2.4.1) and, therefore, Theorem 2.1 on the finite time breakdown of class C^1 solutions applies with the Riemann invariants $w(D,B)$ and $z(D,B)$ given, respectively, by

$$\begin{cases} w(D,B) &= \dfrac{-B}{\mu_0} + \dfrac{1}{\sqrt{\mu_0}}\displaystyle\int_0^D \sqrt{E'(\zeta)}\,d\zeta \\ z(D,B) &= \dfrac{-B}{\mu_0} - \dfrac{1}{\sqrt{\mu_0}}\displaystyle\int_0^D \sqrt{E'(\zeta)}\,d\zeta \end{cases} \tag{2.7.3}$$

For $w(D,B)$, $z(D,B)$ to be well-defined we must have either $\mathsf{E}'(\zeta) > 0$, $\forall\,\zeta > 0$ (which is equivalent to the strict hyperbolicity of the system (2.7.1)) or $\mathsf{E}'(D(x,t)) > 0$, for as long as a C^1 solution of the initial-value problem (2.7.1), (2.7.2) exists. By virtue of our discussion in § 2.5 (actually, the proof of Theorem 2.1) if $\sup_{x\in R^1} |D_0(x)|$ and $\sup_{x\in R^1} |B_0(x)|$ are sufficiently small, and $\mathsf{E}'(0) > 0$, then, in fact, we will have $\mathsf{E}'(D(x,t)) > 0$ for as long as a C^1 solution $(D(x,t), B(x,t))$ of (2.7.1), (2.7.2) exists. For sufficiently small initial data, therefore, the (real) characteristics (2.7.3) are well-defined on the maximal interval of existence of a C^1 solution even if only local hyperbolicity of (2.7.1) obtains, i.e., even if we have only $\mathsf{E}'(0) > 0$. Note also that in view of our discussion in § 2.4, $w(D,B)$ and $z(D,B)$ are constant along their respective characteristics, i.e.,

$$\begin{cases} w'(x,t) &= \dfrac{\partial w}{\partial t} - \sqrt{\dfrac{\mathsf{E}'(D(x,t))}{\mu_0}}\dfrac{\partial w}{\partial x} &= 0 \\ \grave{z}\,(x,t) &= \dfrac{\partial z}{\partial t} + \sqrt{\dfrac{\mathsf{E}'(D(x,t))}{\mu_0}}\dfrac{\partial z}{\partial x} &= 0 \end{cases} \tag{2.7.4}$$

Two fundamental blow-up results may be associated with the solutions of (2.7.1), (2.7.2), based solely on the work of Lax [99] and Klainerman and Majda [92], namely:

(i) If $D_0(x)$ is periodic on R^1, while $B_0(x) \equiv 0$, and $\mathsf{E}''(0) \neq 0$ (so that the problem exhibits genuine nonlinearity) then finite-time blow-up must occur for

$$w_x(x,t) = -\mu_0 D_t(x,t) + \sqrt{\frac{\mathsf{E}'(D(x,t))}{\mu_0}}D_x(x,t) \tag{2.7.5}$$

i.e., $\nabla_{(x,t)}D \equiv (D_t, D_x)$ must blow up in finite time and a shock develops. Furthermore, it is a consequence of the work of Lax [99] that if $t_{\max}$ denotes the maximal time of existence of a C^1 solution $(D(x,t), B(x,t))$, then

$$t_{\max} \cong \frac{\mu_0}{\max_{R^1} |D_0'(x)|} \cdot \left(\sqrt{\mathsf{E}'(0)}/|\mathsf{E}''(0)|\right) \tag{2.7.6}$$

(ii) Suppose that $D_0(x)$ and $B_0(x)$ both have compact support in R^1; then so will

$$\begin{cases} w(x,0) &= B_0(x) + \dfrac{1}{\sqrt{\mu_0}}\displaystyle\int_0^{D_0(x)} \sqrt{\mathsf{E}'(\zeta)}\,d\zeta \\ z(x,0) &= B_0(x) - \dfrac{1}{\sqrt{\mu_0}}\displaystyle\int_0^{D_0(x)} \sqrt{\mathsf{E}'(\zeta)}\,d\zeta \end{cases} \tag{2.7.7}$$

It then follows from the analysis in [92] that if $w(\cdot,0)$, $z(\cdot,0)$ are of class C^1 on R^1, any C^1 solution of the initial-value problem for the diagonalized system (2.7.4) must develop singularities in finite time in the first derivatives w_x, z_x if $\mathsf{E}'(\zeta)$ is not constant on any open interval. For example, if $\mathsf{E}(\zeta) = \lambda_0\zeta + \lambda_2\zeta^3$ with $\lambda_0 > 0$, $\lambda_2 \neq 0$ (such constitutive relations are considered in Grindlay [63]) then $\mathsf{E}'(0) = \lambda_0 > 0$ (local hyperbolicity) and $\mathsf{E}''(0) = 0$ (loss of genuine nonlinearity) but $\mathsf{E}''(\zeta) = 6\lambda_2\zeta \neq 0$ provided $\zeta \neq 0$ and the result cited above applies to the initial value problem (2.7.1), (2.7.2) in the case of C^1, compactly supported initial data.

Once a shock forms in the solution of the initial-value problem (2.7.1), (2.7.2), the Rankine-Hugoniot conditions require that the speed s of the shock satisfy $s[u_k] = [f_k]$ where $\boldsymbol{u}$ and $\boldsymbol{f}$ are given by (2.6.20). We obtain, therefore, the conditions

$$\begin{cases} s[D] &= \dfrac{1}{\mu_0}[B] = [H] \\ s[B] &= [\mathsf{E}(D)] \end{cases} \tag{2.7.8}$$

from which it follows that

$$\begin{cases} [D][E] &= [H][B] = \dfrac{1}{\mu_0}[B]^2 \\ s^2[D] &= \dfrac{s}{\mu_0}[B] = \dfrac{1}{\mu_0}[E] \end{cases} \tag{2.7.9}$$

The shock speed is, therefore, given by

$$s = \pm(1/\sqrt{\mu_0})\sqrt{[\mathsf{E}(D)]/[D]} \tag{2.7.10}$$

so that two possible shocks may occur, one moving to the right and one moving to the left. One possible formulation of an admissibility criterion (entropy condition), for the shocks whose speeds are determined by (2.7.10), is represented by the k-shock conditions of Lax [103], [100] as applied to the system (2.6.19) with $\boldsymbol{g} \equiv \boldsymbol{0}$ and $i = 1, 2$, i.e., we require that for either $k = 1$ or $k = 2$

$$\lambda_k(\boldsymbol{u}_-) > s > \lambda_k(\boldsymbol{u}_+) \tag{2.7.11}$$

when $\mathsf{E}''(0) \neq 0$, where $\boldsymbol{u} = \begin{pmatrix} D \\ B \end{pmatrix}$ and the λ_k, $k = 1, 2$ are the distinct real eigenvalues of the matrix $\nabla_{\boldsymbol{u}}\boldsymbol{f}$ with $\boldsymbol{f} = \begin{pmatrix} B/\mu_0 \\ \mathsf{E}(D) \end{pmatrix}$. In view of (2.6.21) we clearly have

$$\lambda_1 = \frac{1}{\sqrt{\mu_0}}\sqrt{\mathsf{E}'(D)}, \qquad \lambda_2 = -\frac{1}{\sqrt{\mu_0}}\sqrt{\mathsf{E}'(D)} \tag{2.7.12}$$

Of course, in (2.7.11), $\boldsymbol{u}_- = \begin{pmatrix} D_- \\ B_- \end{pmatrix}$ and $\boldsymbol{u}_+ = \begin{pmatrix} D_+ \\ B_+ \end{pmatrix}$ denote, respectively, the values of $\boldsymbol{u}$ immediately behind and in front of the shock. Using (2.7.12), the relations (2.7.11) become either

$$\sqrt{\mathsf{E}'(D_-)} > \sqrt{\mu_0} s > \sqrt{\mathsf{E}'(D_+)} \tag{2.7.13a}$$

or

$$-\sqrt{\mathsf{E}'(D_-)} > \sqrt{\mu_0} s > -\sqrt{\mathsf{E}'(D_+)} \tag{2.7.13b}$$

However, $E = \mathsf{E}(D) = \mathsf{E}(\mathcal{D}(E))$ where $\mathcal{D}(E) = \tilde{\epsilon}(E)E$ so that (2.7.13a), (2.7.13b) are equivalent to

$$1/\sqrt{\mathcal{D}'(E_-)} > \sqrt{\mu_0} s > 1/\sqrt{\mathcal{D}'(E_+)} \tag{2.7.14a}$$

or

$$-1/\sqrt{\mathcal{D}'(E_-)} > \sqrt{\mu_0} s > -1/\sqrt{\mathcal{D}'(E_+)} \tag{2.7.14b}$$

For the shock moving to the right with speed s_r, $s_r = \dfrac{1}{\sqrt{\mu_0}} \cdot \sqrt{[E]/[\mathcal{D}(E)]}$, only (2.7.14a) makes sense in this case and we must have

$$\sqrt{\mathcal{D}'(E_-)} < \sqrt{[\mathcal{D}(E)]/[E]} < \sqrt{\mathcal{D}'(E_+)} \tag{2.7.15}$$

while for the shock moving to the left with speed $s_l = -\dfrac{1}{\sqrt{\mu_0}} \cdot \sqrt{[E]/[\mathcal{D}(E)]}$, only the restriction represented by (2.7.14b) is applicable, and for this shock the relation

$$\sqrt{\mathcal{D}'(E_-)} > \sqrt{[\mathcal{D}(E)]/[E]} > \sqrt{\mathcal{D}'(E_+)} \tag{2.7.16}$$

obtains. The Lax k-shock conditions, therefore, restrict the realizable shocks (moving to the right and left) to satisfy, respectively, the inequalities (2.7.15) and (2.7.16).

We now offer, below, two examples so as to illustrate the utility of the results on shock formation, previously stated, which are based on the general results found in Klainerman and Majda [92] and Lax [99]; in these examples, the basic constitutive relation is taken in the form $\mathcal{D}(E) = \tilde{\epsilon}(E)E$ instead of (*a priori*) in the form $E = \mathsf{E}(D)$.

Example 1 Suppose that $\tilde{\epsilon}(E) = \epsilon_0 + \epsilon_2 E^2$ with $\epsilon_0 > 0, \epsilon_2 > 0$. By virtue of the definitions of E and $\mathcal{D}$, $\forall \zeta \in R^1$, $\zeta = \mathsf{E}(\mathcal{D}(\zeta))$ with $\mathcal{D}(\zeta) = \epsilon_0\zeta + \epsilon_2\zeta^3$. Thus $\mathcal{D}(\zeta) = 0$ if and only if $\zeta = 0$. A direct computation produces

$$\mathsf{E}''(\mathcal{D}(\zeta)) = -\frac{\mathsf{E}'(\mathcal{D}(\zeta))\mathcal{D}''(\zeta)}{\mathcal{D}'^2(\zeta)} \tag{2.7.17}$$

where

$$\mathsf{E}'(\mathcal{D}(\zeta)) = \left.\frac{d\mathsf{E}}{d\mathcal{D}}\right|_{\zeta}, \quad \mathcal{D}'(\zeta) = \frac{d\mathcal{D}}{d\zeta}$$

For $|\zeta|$ sufficiently small, it is easy to see that $\mathsf{E}'(\mathcal{D}(\zeta)) > 0$ and

$$\mathsf{E}''(\mathcal{D}(\zeta)) = -\mathsf{E}'(\mathcal{D}(\zeta)) \cdot \left(\frac{6\epsilon_2\zeta}{(\epsilon_0 + 3\epsilon_2\zeta^2)^2}\right) \tag{2.7.18}$$

so that $\mathsf{E}''(\mathcal{D}(0)) = \mathsf{E}''(0) = 0$, but $\mathsf{E}''(\mathcal{D}(\zeta)) \neq 0, \quad \forall \zeta \neq 0$. Thus, for C^1 initial data which is compactly supported, the results of [92] will apply to the initial-value problem (2.7.1), (2.7.2), with $D = \epsilon_0 E + \epsilon_2 E^3$, and shocks will develop in finite time.

Example 2 We assume that the dielectric substance in Ω_c conforms to the constitutive hypothesis $D = \epsilon_0 E + P(E)$ with the polarization given by $P(E) = \chi_0 E + \chi_1 E^2$, $\chi_0 > 0$, $\chi_1 > 0$; the quantities χ_0, χ_1 are, respectively, the linear and (first) nonlinear susceptibilities of the dielectric. Therefore, $\mathcal{D}(E) = \tilde{\epsilon}(E)E$, $\tilde{\epsilon}(E) = \bar{\epsilon}_0 + \bar{\epsilon}_2 E$, $\bar{\epsilon}_0 = \epsilon_0 + \chi_0$, $\bar{\epsilon}_2 = \chi_1$. From the relation

$$D = (\epsilon_0 + \chi_0)E + \chi_1 E^2 \tag{2.7.19}$$

we find that

$$E(x,t) = -\frac{(\epsilon_0 + \chi_0)}{2\chi_1} \pm \frac{1}{2\chi_1}\sqrt{(\epsilon_0 + \chi_0)^2 + 4\chi_1 D(x,t)} \tag{2.7.20}$$

By choosing the positive sign on the radical, and expanding in a power series, we obtain a relation of the form $E = \lambda(D)D + \mathcal{O}(D^3)$ where $\lambda(D) = \lambda_0 + \lambda_1 D$ and

$$\lambda_0 = \frac{\chi_1}{(\epsilon_0 + \chi_0)} > 0, \qquad \lambda_1 = \frac{-\chi_1^2}{(\epsilon_0 + \chi_0)^{3/2}} < 0 \tag{2.7.21}$$

Thus, up to terms of order $\mathcal{O}(D^3)$ we have

$$\mathsf{E}(D) = \frac{\chi_1}{(\epsilon_0 + \chi_0)} D - \frac{\chi_1^2}{(\epsilon_0 + \chi_0)^{3/2}} D^2 \tag{2.7.22}$$

so that for $|D| < (\epsilon_0 + \chi_0)^{1/2}/2\chi_1$ we have $\mathsf{E}'(D) > 0$. For the initial data we assume that $B_0(x) \equiv 0$, $D_0(x)$ is periodic, $x \in R^1$, and sufficiently small, so that the standard *a priori* estimates for the Riemann invariants associated with (2.7.1) imply that $\mathsf{E}'(D) > 0$ for as long as a C^1 field $D(x,t)$ exists; the aforementioned *a priori* estimates for the Riemann invariants (2.7.3) will be described below. Now

$$D_0(x) = (\epsilon_0 + \chi_0)E_0(x) + \chi_1 E_0^2(x), \qquad x \in R^1 \tag{2.7.23}$$

so that

$$D_0'(x) = [(\epsilon_0 + \chi_0) + 2\chi_1 E_0(x)]\, E_0'(x) \qquad (2.7.24)$$
$$\cong 2\chi_1 E_0(x) E_0'(x)$$

if, for example, one assumes that the initial electric field strength is similar to that found in a high intensity laser beam; in such a situation, $\|\boldsymbol{E}_0\|$ is of the order of magnitude of 10^9 (volts/meter) while ϵ_0, the permittivity of free space, is of the order 10^{-13} (C/N·m^2), χ_0 is of the order 10^{-13} (C/N·m^2) and χ_1 of the order 10^{-13} (C/V^2). For example, $\chi_1 = 4 \times 10^{-13}$ for index matched KDP [6]. The various units (in the mks system) are related by the identification N/C $\equiv$ V/m, which are the dimensions of the electric field E (that of D, the electric induction field, as well as of P, the polarization, are then C/m^2). Therefore, in this scenario, the quantity $(\epsilon_0 + \chi_0)$ in (2.7.24) is of the order 10^{-13}, while $\chi_1 E_0$ is of the order 10^{-4}.

From (2.7.22) it follows that $\mathsf{E}''(0) = -2\chi_1^2/(\epsilon_0+\chi_0)^{3/2} \neq 0$ so our sample problem exhibits genuine nonlinearity and the results of Lax [99] apply. Using, in particular, (2.7.6) we find that

$$t_{\max} \cong \frac{\mu_0(\epsilon_0 + \chi_0)}{\chi_1^{5/2}} \left(\max|E_0 E_0'|\right)^{-1} \qquad (2.7.25)$$

From (2.6.16) and (2.6.9), (2.7.19), we have for the characteristic speed of the beam in the dielectric

$$s_b = 1/\sqrt{\mu_0(\tilde{\epsilon}(E)E)'} \qquad (2.7.26)$$
$$= 1/\sqrt{\mu_0((\epsilon_0 + \chi_0) + 2\chi_1 E)}$$

However, $\mu_0(\epsilon_0 + \chi_0)$ is of the order of magnitude 10^{-19} (μ_0, the permeability of free space, being $4\pi \cdot 10^{-7}$ N·sec^2/C) while $\mu_0\chi_1 E$ is of the order of magnitude 10^{-10}. Therefore

$$s_b \cong \frac{1}{\sqrt{2\mu_0\chi_1|E|}} \geq \frac{1}{\sqrt{2\mu_0\chi_1}}(\max|E|)^{-1/2} \qquad (2.7.27)$$

As an upper bound for the distance traveled by the beam in the dielectric, until a shock develops, we now have the estimate

$$s_{\max} \cong s_b \cdot t_{\max} \qquad (2.7.28)$$
$$\leq \frac{1}{\sqrt{2}} \frac{\mu_0^{1/2}(\epsilon_0 + \chi_0)}{\chi_1^3}(\max|E|)^{-1/2}(\max|E_0 E_0'|)^{-1}$$

If, in addition, we suppose that over the distance traveled by the beam until shock development, $\max|E| \cong \max|E_0|$, then we have the (crude but, nonetheless, informative) estimate

$$s_{\max} \leq C_0(\max|E_0|)^{-3/2}(\max|E_0'|)^{-1} \tag{2.7.29}$$

where

$$C_0 \equiv \frac{1}{\sqrt{2}}\left[\mu_0^{1/2}(\epsilon_0+\chi_0)/\chi_1^3\right] \tag{2.7.30}$$

can be considered to be a characteristic material coefficient associated with a particular nonlinear dielectric substance, e.g., for one of the most common such materials, index matched KDP, C_0 is a very large number of the order of magnitude 10^{21}. What (2.7.29) indicates, therefore, is that even for an initial high intensity beam of the order of magnitude 10^9 V/m, a steep gradient on the initial beam will be required if shock development is to occur within distances that are realistic in a laboratory setting. Results somewhat similar to that embodied in (2.7.29), emphasizing the importance of the magnitude of the initial gradient of the electric field in the mechanism of shock formation in the wave-dielectric interaction problem, may be found, e.g., in [46], [29], [79], [87], [50], and [75].

Before turning to a discussion of the system (2.6.17), (2.6.18), in which we do not necessarily have $\tilde{\sigma} \equiv 0$ (i.e., $\boldsymbol{J} \equiv \boldsymbol{0}$), we remark on the development of shocks in the travelling wave solutions of the initial value problem associated with the system (2.6.15) with $\tilde{\sigma}(E) \equiv 0$. In fact, we may consider the slightly more general system

$$\begin{cases} E_t + \tilde{a}(E)H_x &= 0 \\ H_t + \tilde{a}(H)E_x &= 0 \end{cases} \tag{2.7.31}$$

where $\tilde{a}(E) = a(E)^{-1} = 1/(\tilde{\epsilon}(E)E)'$ and $\tilde{b}(H) = 1/(\tilde{\mu}(H)H)'$; system (2.7.31) results from the (typical) constitutive assumption (2.6.2), coupled with (2.6.5), and the obvious definition of $\tilde{\mu}$, i.e., $\tilde{\mu}(\zeta) = \mu((0,0,\zeta))$, $\forall \zeta \in R^1$. We assume, of course, that $\forall \zeta \in R^1$ we have $(\tilde{\mu}(\zeta)\zeta)' > 0$. The analysis is entirely analogous to that which led us from (2.2.8a) to (2.2.8b) and has been developed in (among other places) [29] and [50].

For the system (2.7.31) it is a simple matter to show that there exist solutions of the form $E(x,t) = E_0(x-\lambda t)$, $H(x,t) = H_0(x-\lambda t)$ where $E_0(x) = E(x,0)$, $H_0(x) = H(x,0)$, $x \in R^1$, and $\lambda = \pm\sqrt{\tilde{a}(E)\tilde{b}(H)}$. Therefore, we have travelling wave solutions with the (implicit) form

$$\begin{cases} E = E_0(x \pm \sqrt{\tilde{a}(E)\tilde{b}(H)}t) \\ H = H_0(x \pm \sqrt{\tilde{a}(E)\tilde{b}(H)}t) \end{cases} \tag{2.7.32}$$

which are well-defined, at least for small values of t. For the common situation in which $\tilde{\mu}(H) = \mu_0$, so that $\tilde{b}(H) = \mu_0^{-1}$, (2.7.32) reduces to

$$\begin{cases} E = E_0(x \pm \mu_0^{-1/2}\sqrt{\tilde{a}(E)}t) \\ H = H_0(x \pm \mu_0^{-1/2}\sqrt{\tilde{a}(E)}t) \end{cases} \tag{2.7.33}$$

We now define:

$$\mathcal{F}(x,t,E) \equiv E - E_0(x \pm \mu_0^{-1/2}\sqrt{\tilde{a}(E)}t) \tag{2.7.34}$$

For any value of $x_0 \in R^1$ such that $E_0(x_0) = 0$, we have $\mathcal{F}(x_0,0,0) = 0$, and

$$\begin{aligned} \mathcal{F}_E(x,t,E) = 1 \; \pm \; & \frac{1}{2}\mu_0^{-1/2}E_0'(x \pm \mu_0^{-1/2}\sqrt{\tilde{a}(E)}t) \\ \times \; & (\tilde{a}(E))^{-1/2}\tilde{a}'(E) \end{aligned} \tag{2.7.35}$$

In a typical situation, e.g., $\tilde{\epsilon}(E) = \epsilon_0 + \epsilon_2 E^2, \epsilon_0 > 0, \epsilon_2 > 0$, we have $\tilde{a}(0) = \epsilon_0$ and $\tilde{a}'(0) = 0$ so that $\mathcal{F}_E(x_0,0,0) = 1$; in such a situation, the implicit function theorem guarantees us the existence of a solution $E = E(x,t)$ of $\mathcal{F}(x,t,E) = 0$, for $|x - x_0|$ and $|t|$ sufficiently small. Differentiating the first relation in (2.7.33) through with respect to x, and then solving for $\mathcal{F}$, we obtain

$$E_x = E_0'(x \pm \mu_0^{-1/2}\sqrt{\tilde{a}(E)}t)/(1 \pm \frac{1}{2}\sqrt{\mu_0^{-1}/\tilde{a}(E)}\tilde{a}'(E)t) \tag{2.7.36}$$

Therefore, if $\exists\, \epsilon^* > 0$ such that

$$\sqrt{\tilde{a}(\zeta)}/\tilde{a}'(\zeta) < \frac{1}{\epsilon^*}, \quad \forall \zeta \in R^1 \text{ such that } |\zeta| \text{ is sufficiently small,} \tag{2.7.37}$$

then for the wave moving to the left, with velocity $\mu_0^{-1/2}\sqrt{\tilde{a}(E)}$, we have

$$1 - \frac{1}{2}\mu_0^{-1/2}\left(\tilde{a}'(E)/\sqrt{\tilde{a}(E)}\right)t < 1 - \frac{1}{2}\mu_0^{-1/2}\epsilon^* t \to 0 \tag{2.7.38}$$

as $t \to t^* = 2/\mu_0^{-1/2}\epsilon^*$. From (2.7.36) we now have

$$|E_x(x,t)| \to +\infty, \quad \text{as } t \to t^* \tag{2.7.39}$$

and a shock can be expected to develop provided the constitutive relation for the dielectric conforms to the hypothesis (2.7.37). We note, however, that (2.7.37) cannot be satisfied for the situation in which $\tilde{\epsilon}(\zeta) = \epsilon_0 + \epsilon_2\zeta^2$, $\epsilon_0 > 0$, $\epsilon_2 > 0$, inasmuch as it is equivalent to the requirement that

$$\sqrt{\epsilon_0 + 3\epsilon_2\zeta^2} < \frac{6\epsilon_2}{\epsilon^*}\zeta \tag{2.7.40}$$

which is not satisfied $\forall\,\zeta \in R^1$, such that $|\zeta|$ is sufficiently small, no matter how small $\epsilon^* > 0$ is chosen; this is not surprising, as in this case genuine nonlinearity fails and shock development cannot be shown to follow directly from (2.7.36). As an example in which the line of thought developed above can be applied, we could consider $\tilde{\epsilon}(E) = \epsilon_0 + \epsilon_1 E$ with $\epsilon_0 > 0$, $\epsilon_1 > 0$; then $\tilde{a}(0) = \epsilon_0$, $\tilde{a}'(0) = 2\epsilon_1$ and

$$\mathcal{F}_E(x,t,E) = 1 \pm (\epsilon_1\mu_0^{-1/2}/\epsilon_0^{1/2})E_0'(x_0) \tag{2.7.41}$$

Thus, if x_0 is such that not only do we have $E_0(x_0) = 0$ but, also, $\epsilon_0^{1/2} \pm \epsilon_1\mu_0^{-1/2}E_0'(x_0) \neq 0$, then a solution $E = E(x,t)$ of $\mathcal{F}(x,t,E) = 0$ again will exist, in this case, for $|x - x_0|$ and $|t|$ sufficiently small, while (2.7.37) assumes the form

$$\sqrt{\epsilon_0 + 2\epsilon_1\zeta} < 2\epsilon_1/\epsilon^* \tag{2.7.42}$$

which is certainly satisfied for $|\zeta|$ sufficiently small provided $\epsilon^* > 0$ is sufficiently small. For the constitutive relation $\boldsymbol{D} = \epsilon(\boldsymbol{E})\boldsymbol{E}$ with (2.6.4) applicable and $\tilde{\epsilon}(\zeta) = \epsilon_0 + \epsilon_1\zeta$, $\zeta \in R^1$, $\tilde{\epsilon}(\zeta) = \epsilon((0,\zeta,0))$, (2.7.39) can be expected to follow for some $t^* > 0$ finite; in fact, $t^* = 2\mu_0^{1/2}/\epsilon^*$ with ϵ^* chosen so as to satisfy (2.7.42) $\forall\,\zeta \in R^1$ such that $|\zeta|$ is sufficiently small, i.e., with $\epsilon^* < 2\epsilon_1/\sqrt{\epsilon_0}$.

We now want to turn our attention to the case in which $\tilde{\sigma}(\cdot) \not\equiv 0$, i.e., to the system (2.6.17) or, equivalently, the nonlinearly damped nonlinear wave equation (2.6.22); in both formulations, $\Sigma(\cdot)$ is given by (2.6.18). Both Nishida [125] and Slemrod [163] studied the damped quasilinear system

$$\begin{cases} \dfrac{\partial v}{\partial t} - \Gamma'(u)\dfrac{\partial u}{\partial x} = -\alpha v \ (\alpha > 0) \\ \dfrac{\partial u}{\partial t} - \dfrac{\partial v}{\partial x} = 0 \end{cases} \tag{2.7.43}$$

which leads to the nonlinear wave equation with linear damping

$$\frac{\partial^2 u}{\partial t^2} + \alpha\frac{\partial u}{\partial t} = \frac{\partial^2 \Gamma(u)}{\partial x^2} \tag{2.7.44}$$

Clearly (2.6.22) and (2.7.44) have the same form iff $\Sigma(\cdot)$ is a linear function. By virtue of (2.6.18) it is clear that $\Sigma(\cdot)$ will not be linear in a nonlinear dielectric, even if one assumes a linear Ohm's law for $\boldsymbol{J}$ as an approximation, i.e., even if $\tilde{\sigma}$ is constant.

Nishida [125] considered the system (2.7.43), with associated periodic initial data $u(x,0) = u_0(x)$, $v(x,0) = v_0(x)$, assumed local hyperbolicity, i.e., $\Gamma'(0) > 0$, and proved an *a priori* estimate which shows that, for as long as a C^1 solution (u,v) in (x,t) exists, $\sup_x |u|$ will be small, when the L^∞ norms of the initial data $u_0(\cdot)$, $v_0(\cdot)$ are sufficiently small; thus $\Gamma' > 0$ for as long as a smooth solution exists. Working with the Riemann invariants

$$\left\{\begin{array}{rcl} r(u,v) &=& v + \int_0^u \sqrt{\Gamma'(\zeta)}\,d\zeta \\ s(u,v) &=& v - \int_0^u \sqrt{\Gamma'(\zeta)}\,d\zeta \end{array}\right. \tag{2.7.45}$$

that are naturally associated to (2.7.43), and studying the behavior of $\frac{\partial r}{\partial x}, \frac{\partial s}{\partial x}$ along the characteristic curves in the x,t plane, Nishida was able to derive *a priori* estimates for $|r_x(x,t)|$ and $|s_x(x,t)|$ along the characteristics which, in turn, were used to prove that for $\Gamma''(0) > 0$, with the C^1 norms of $u_0'(\cdot)$ and $v_0'(\cdot)$ sufficiently small, a C^1 (in (x,t)) solution (u,v) of the initial-value problem associated with (2.7.43) exists for all $t > 0$; an analog of Nishida's result for the nonlinearly damped nonlinear wave equation (2.6.22) will be presented in § 3.5. In the remainder of this section, however, we will provide for the system (2.6.17) an analog of the nonexistence result proven by Slemrod [163] for the system (2.7.43), namely, that for the Riemann invariants r, s, as defined by (2.7.45), $\exists\, t_\infty < \infty$ such that $|r_x(x,t)| \to \infty$, as $t \to t_\infty$, when $\Gamma''(0) > 0$ and $r_0'(x)$ is sufficiently large and positive at some x.

In terms of the Riemann invariants r, s introduced in (2.7.45), the system (2.7.43) may be written in the form

$$r' = -\frac{\alpha}{2}(r+s), \qquad \grave{s} = -\frac{\alpha}{2}(r+s) \tag{2.7.46}$$

where

$$' = \frac{\partial}{\partial t} - \sqrt{\tilde{\Gamma}'(r-s)}\frac{\partial}{\partial x}, \qquad ` = \frac{\partial}{\partial t} + \sqrt{\tilde{\Gamma}'(r-s)}\frac{\partial}{\partial x} \tag{2.7.47}$$

and $\tilde{\Gamma}(r-s) = \Gamma(u(r-s))$. Now, we will be showing, below, that the system (2.6.17) can be written, in terms of the Riemann invariants $w(D,B)$, $z(D,B)$ given by (2.7.3), in the form

$$w' = \Phi(w-z), \qquad \grave{z} = -\Phi(w-z) \tag{2.7.48}$$

for an appropriate choice of $\Phi(\cdot)$, where

$$' = \frac{\partial}{\partial t} + \lambda(x,t)\frac{\partial}{\partial x}, \qquad \grave{} = \frac{\partial}{\partial t} - \lambda(x,t)\frac{\partial}{\partial x} \tag{2.7.49}$$

and

$$\lambda = -\sqrt{\mathsf{E}'(D(x,t))/\mu_0} \tag{2.7.50}$$

In [67], Hattori established finite-time breakdown of C^1 solutions for a system (in Riemann invariants form) of the type

$$\begin{cases} r' &= -\frac{\alpha}{2}(r+s) + \bar{\Phi}(x,t) \\ \grave{s} &= -\frac{\alpha}{2}(r+s) + \bar{\Phi}(x,t) \end{cases} \tag{2.7.51}$$

by using an argument, due to Rozhdestvenskii [143], which involves showing that characteristics of the same family must cross in finite time. Central to the analysis in [67], however, is the derivation and application of the *a priori* estimate

$$|r(x,t)| + |s(x,t)| \le \sup_x |r_0(x)| \tag{2.7.52}$$

$$+ \sup_x |s_0(x)| + \left(\frac{4}{\alpha}\right) \sup_x |\bar{\Phi}(x,t)|$$

where $r_0(x) = r(x,0)$, $s_0(x) = s(x,0)$. An estimate of the type (2.7.52) has, however, no relevance for a system of the form (2.7.48) in which $\alpha = 0$ and, thus, the finite-time breakdown results in [67] cannot be carried over directly to the system of interest to us here, namely, (2.6.17).

We begin our analysis of finite-time breakdown of C^1 solutions for the initial-value problem associated with (2.6.17) with the following:

Lemma 2.1 ([23]) For the wave-dielectric interaction system (2.6.17), the associated Riemann invariants $w(D,B)$, $z(D,B)$, as given by (2.7.3), satisfy (2.7.48)-(2.7.50), where $\forall\, \kappa \in R^1$,

$$\Phi(\kappa) = -\sqrt{\frac{\mathsf{E}'(\hat{\eta}^{-1}(\kappa))}{\mu_0}}\,\Sigma(\hat{\eta}^{-1}(\kappa)) \tag{2.7.53}$$

with

$$\hat{\eta}(\kappa) = 2\int_0^{\kappa} \sqrt{\frac{\mathsf{E}'(\zeta)}{\mu_0}}\, d\zeta \tag{2.7.54}$$

Proof: By virtue of the definitions (2.7.3) and (2.7.50) for w, z, and λ, respectively, we have

$$\left\{ \begin{aligned} \frac{\partial w}{\partial t}(x,t) &= -\frac{1}{\mu_0}\frac{\partial B(x,t)}{\partial t} - \lambda(x,t)\frac{\partial D(x,t)}{\partial t} \\ \frac{\partial w}{\partial x}(x,t) &= -\frac{1}{\mu_0}\frac{\partial B(x,t)}{\partial x} - \lambda(x,t)\frac{\partial D(x,t)}{\partial x} \end{aligned} \right. \tag{2.7.55}$$

with analogous results for $\dfrac{\partial z(x,t)}{\partial t}$ and $\dfrac{\partial z(x,t)}{\partial x}$. Thus,

$$\begin{aligned} \frac{\partial w}{\partial t} + \lambda\frac{\partial w}{\partial x} &= -\lambda\left(\frac{\partial D}{\partial t} + \frac{1}{\mu_0}\frac{\partial B}{\partial x}\right) - \left(\frac{1}{\mu_0}\frac{\partial B}{\partial t} + \lambda^2\frac{\partial D}{\partial x}\right) \\ &= \lambda\Sigma(D) - \left(\frac{1}{\mu_0}\frac{\partial B}{\partial t} + \frac{\mathsf{E}'(D)}{\mu_0}\frac{\partial D}{\partial x}\right) \\ &= \lambda\Sigma(D), \end{aligned}$$

in view of the definition of $\lambda(x,t)$ and (2.6.17). Therefore, by virtue of (2.7.49), we have

$$w'(x,t) = -\sqrt{\frac{\mathsf{E}'(D(x,t))}{\mu_0}}\Sigma(D(x,t)) \tag{2.7.56}$$

Directly from (2.7.3) we obtain

$$w(x,t) - z(x,t) = 2\int_0^{D(x,t)} \sqrt{\frac{\mathsf{E}'(\zeta)}{\mu_0}}\, d\zeta \equiv \eta(x,t) \tag{2.7.57}$$

with $\eta = \hat{\eta}(D(x,t))$ satisfying $\dfrac{d\hat{\eta}}{dD} = 2\sqrt{\dfrac{\mathsf{E}'(D)}{\mu_0}} > 0$. Thus, $\hat{\eta}$ is monotonically increasing in D so we may define for $\hat{\eta}$ an inverse $\hat{\eta}^{-1}$ such that

$$D(x,t) = \hat{\eta}^{-1}(w(x,t) - z(x,t)) \tag{2.7.58}$$

in which case

$$\begin{aligned} \lambda(x,t) &= -\sqrt{\mathsf{E}'\left(\hat{\eta}^{-1}(w(x,t) - z(x,t))\right)/\mu_0} \\ &\equiv \hat{\lambda}(w(x,t) - z(x,t)) \end{aligned} \tag{2.7.59}$$

The desired result now follows from (2.7.56), (2.7.58) and the analogous result for $\grave{z}(x,t)$. □

As a direct consequence of Lemma 2.1 we see that along the characteristic curves $x_1(t;\beta_1)$, $x_2(t;\beta_2)$ defined by the solutions of the initial-value problems

$$\begin{cases} \dfrac{dx_1}{dt} &= \hat{\lambda}(w(x_1,t) - z(x_1,t));\ x_1(0,\beta_1) = \beta_1 \\ \dfrac{dx_2}{dt} &= -\hat{\lambda}(w(x_2,t) - z(x_2,t));\ x_2(0,\beta_2) = \beta_2 \end{cases} \tag{2.7.60}$$

the system of partial differential equations (2.7.48) reduces to the pair of ordinary differential equations

$$\begin{cases} \dfrac{d}{dt}w(x_1(t,\beta_1),t) &= \Phi(w(x_1(t,\beta_1),t) - z(x_1(t,\beta_1),t)) \\ \dfrac{d}{dt}z(x_2(t,\beta_2),t) &= -\Phi(w(x_2(t,\beta_2),t) - z(x_2(t,\beta_2),t)) \end{cases} \tag{2.7.61}$$

Associated with the system (2.7.48) we also have initial data of the form

$$w(x,0) = w_0(x), \quad z(x,0) = z_0(x), \qquad x \in R^1 \tag{2.7.62}$$

and, at this juncture, we will assume that both $w_0(\cdot)$ and $z_0(\cdot)$ are periodic and of class C^1 on R^1.

Up to this point we have required of our constitutive theory only that $\forall\, \zeta \in R^1$, $(\zeta\tilde{\epsilon}(\zeta))' > 0$, which implies, of course, that $\mathsf{E}'(\zeta) > 0$, $\forall\, \zeta \in R^1$. To make further progress, some additional assumptions must be made, namely, we will assume that $\mathsf{E}''(0) > 0$ and that $\mathsf{E}(\cdot)$ and $\tilde{\sigma}(\cdot)$ satisfy, jointly, the condition

$$\sup_{\zeta} \left| \frac{d}{d\zeta} \left\{ \sqrt{\mathsf{E}'(\hat{\eta}^{-1}(\zeta))}\Sigma(\hat{\eta}^{-1}(\zeta)) \right\} \right| < \infty \tag{2.7.63}$$

which, of course, is equivalent, by virtue of (2.7.53), to the assumption that $\sup_{\zeta} |\Phi'(\zeta)| < \infty$. With hypothesis (2.7.63) in hand we are prepared to state the following:

Lemma 2.2 ([23]) Let (w,z) be a C^1 solution of the initial-value problem (2.7.48), (2.7.62) for $0 \le t \le T$, $T < \infty$, where Φ is defined by (2.7.53). Then if (2.7.63) holds, $\exists\, M$, $0 < M < \infty$ such that for all $t \le T$, and $x \in R^1$,

$$|w(x,t)| + |z(x,t)| \le (|w_0| + |z_0|)\, e^{MT} \tag{2.7.64}$$

where $|w_0| = \sup_x |w_0(x)|$, $|z_0| = \sup_x |z_0(x)|$.

Proof: As $\hat{\eta}(0) = 0$ we note that

$$\Phi(0) = -\sqrt{\frac{\mathsf{E}'(0)}{\mu_0}}\Sigma(0) = -\sqrt{\frac{\mathsf{E}'(0)}{\mu_0}}\tilde{\sigma}(\mathsf{E}(0))\mathsf{E}(0) = 0$$

and, therefore,

$$\Phi(w-z)=\int_0^{w-z}\Phi'(\zeta)\,d\zeta.$$

By virtue of the hypothesis (2.7.63), then,

$$\begin{aligned}|\Phi(w-z)| &\le \sup_\zeta |\Phi'(\zeta)|\,(|w|+|z|)\\ &\equiv \frac{M}{2}(|w|+|z|)\end{aligned}\tag{2.7.65}$$

Now, consider the characteristics $x_1(t,\beta_1)$, $x_2(t,\beta_2)$ defined by the initial-value problems (2.7.60); as already noted, along these curves the system (2.7.48) reduces to the set of nonlinear ordinary differential equations (2.7.61). If we integrate (2.7.61) along their respective characteristics, we obtain

$$\left\{\begin{aligned} w(x_1,t) &= w_0(\beta_1)+\int_0^t \Phi(w(x_1,\tau)-z(x_1,\tau))\,d\tau\\ z(x_2,t) &= z_0(\beta_2)+\int_0^t \Phi(w(x_2,\tau)-z(x_2,\tau))\,d\tau\end{aligned}\right.\tag{2.7.66}$$

so that for $t\le T$,

$$\left\{\begin{aligned} |w(x_1,t)| &= |w_0(\beta_1)|+\int_0^t |\Phi(w(x_1,\tau)-z(x_1,\tau))|\,d\tau\\ |z(x_2,t)| &= |z_0(\beta_2)|+\int_0^t |\Phi(w(x_2,\tau)-z(x_2,\tau))|\,d\tau\end{aligned}\right.\tag{2.7.67}$$

By appealing to the estimate (2.7.65), we may now infer from (2.7.67) that for $t\le T$

$$\begin{aligned}|w(x_1,t)|\le|w_0(\beta_1)| &+ \frac{M}{2}\int_0^t |w(x_1,\tau)|\,d\tau\\ &+ \frac{M}{2}\int_0^t |z(x_1,\tau)|\,d\tau\end{aligned}\tag{2.7.68a}$$

and

$$\begin{aligned}|z(x_2,t)|\le|z_0(\beta_2)| &+ \frac{M}{2}\int_0^t |z(x_2,\tau)|\,d\tau\\ &+ \frac{M}{2}\int_0^t |w(x_2,\tau)|\,d\tau\end{aligned}\tag{2.7.68b}$$

Setting $W(t)=\sup_x|w(x,t)|$ and $Z(t)=\sup_x|z(x,t)|$, we now find that

$$|w(x_1,t)| \le |w_0|+\frac{M}{2}\int_0^t (W(\tau)+Z(\tau))\,d\tau \tag{2.7.69a}$$

$$|z(x_2,t)| \le |z_0|+\frac{M}{2}\int_0^t (W(\tau)+Z(\tau))\,d\tau \tag{2.7.69b}$$

Inasmuch as the initial data $w_0(\cdot)$, $z_0(\cdot)$ were assumed to be periodic in x, for each $t\le T$, we also have that $w(\cdot,t)$ and $z(\cdot,t)$ are periodic in x. Therefore, for $t=\tilde{t}$,

$\exists\, \tilde{x}_1, \tilde{x}_2$ such that $W(\tilde{t}) = |w(\tilde{x}_1, \tilde{t})|$ and $Z(\tilde{t}) = |z(\tilde{x}_2, \tilde{t})|$. We now choose $\alpha_1 = \alpha_1(t)$, $\alpha_2 = \alpha_2(t)$ such that $\tilde{x}_1 = x_1(\alpha_1(\tilde{t}), \tilde{t})$, $\tilde{x}_2 = x_2(\alpha_2(\tilde{t}), \tilde{t})$. Then for each $t = \tilde{t}$

$$\left\{ \begin{array}{rcl} W(\tilde{t}) & = & \left|w(x_1(\alpha_1(\tilde{t}), \tilde{t}), \tilde{t})\right| \\ Z(\tilde{t}) & = & \left|z(x_2(\alpha_2(\tilde{t}), \tilde{t}), \tilde{t})\right| \end{array} \right. \tag{2.7.70}$$

We now choose, for each $t \leq T$, $x_1 = x_1(\alpha_1(t), t)$ and $x_2 = x_2(\alpha_2(t), t)$ on the left-hand side, respectively, of (2.7.69a)-(2.7.69b), in which case we find that

$$\left\{ \begin{array}{rcl} W(t) & \leq & |w_0| + \dfrac{M}{2} \displaystyle\int_0^t (W(\tau) + Z(\tau))\, d\tau \\ Z(t) & \leq & |z_0| + \dfrac{M}{2} \displaystyle\int_0^t (W(\tau) + Z(\tau))\, d\tau \end{array} \right. \tag{2.7.71}$$

By defining $U(t) = W(t) + Z(t)$ and adding the estimates in (2.7.71), we now obtain

$$U(t) \leq |w_0| + |z_0| + M \int_0^t U(\tau)\, d\tau \tag{2.7.72}$$

Our desired result, i.e. (2.7.64), is now a direct consequence of the standard Gronwall inequality ([176] or [66], for example) applied to (2.7.72) and the definitions of W, Z and U. □

With Lemma 2.2 in hand, we are now in a position to state and prove the main result of this section; this result will imply, of course, the development of singularities in smooth (i.e. C^1) solutions of the initial-value problem (2.6.17) associated with the plane wave-nonlinear dielectric interaction problem in one-space dimension, with an associated nonlinear Ohm's law governing the evolution of the conduction current vector $\boldsymbol{J}$.

Theorem 2.2 ([23]) Consider the inhomogeneous quasilinear system (2.6.17) with periodic initial data $B(x,0) = B_0(x)$, $D(x,0) = D_0(x)$, $x \in R^1$. Suppose that $\mathsf{E}'(\zeta) > 0$, $\forall \zeta \in R^1$, $\mathsf{E}''(0) > 0$, and that the constitutive functions $\tilde{\epsilon}$, $\tilde{\sigma}$ are such that (2.7.63) is satisfied. Then a C^1 solution $(B(x,t), D(x,t))$ cannot exist for all $t > 0$ if $\sup_x |B_0(x)|$ and $\sup_x |D_0(x)|$ are chosen sufficiently small while

$$F(x) \equiv \sqrt{\frac{\mathsf{E}'(D_0(x))}{\mu_0}} D_0'(x) - \frac{1}{\mu_0} B_0'(x), \qquad x \in R^1 \tag{2.7.73}$$

is chosen so as to be positive and sufficiently large at some $x \in R^1$. More precisely, given $T > 0$, $\exists\, \rho_1(T), \rho_2(T) > 0$ such that for

$$\max\left(\sup_x |B_0(x)|, \sup_x |D_0(x)|\right) \leq \rho_1(T) \tag{2.7.74a}$$

and

$$F(x) \geq \rho_2(T) \tag{2.7.74b}$$

there exists $t_\infty < T$ such that

$$\lim_{t \to t_\infty} \left(\left[\frac{\partial B(x,t)}{\partial x} \right]^2 + \left[\frac{\partial D(x,t)}{\partial x} \right]^2 \right) = \infty. \tag{2.7.75}$$

Remark: It is clear that the conditions relative to $F(x)$, as defined by (2.7.73), are satisfied if, for example, $D_0'(x) > 0$, $\forall\, x \in R^1$, while $B_0'(x) < 0$, $\forall\, x \in R^1$ with $|B_0'(x)|$ sufficiently large at some $x \in R^1$.

Proof: (Theorem 2.2) We begin by assuming that there exists a C^1 solution $(w(x,t), z(x,t))$ of the initial-value problem (2.7.48), (2.7.62) for $0 \leq t \leq T$, with $T > 0$ arbitrary; we will derive a contradiction to this assumption by showing that for $|w_0|$, $|z_0|$ sufficiently small, and $w'(x,0)$ sufficiently large, at some $x \in R^1$, $\left|\frac{\partial w}{\partial x}(x,t)\right| \to \infty$ as $t \to t_\infty$, for some $t_\infty < T$. We initiate the argument that will take us along the aforementioned path by first rewriting the first equation in (2.7.48) in the form

$$\frac{\partial w}{\partial t} - \hat{\nu}(w-z)\frac{\partial w}{\partial x} = \Phi(w-z) \tag{2.7.76}$$

with

$$\hat{\nu}(w-z) \equiv \sqrt{\frac{\mathsf{E}^{|}(\hat{\eta}^{-1}(w-z))}{\mu_0}} \tag{2.7.77}$$

so that

$$\Phi(w-z) = -\hat{\nu}(w-z)\Sigma(\hat{\eta}^{-1}(w-z)) \tag{2.7.78}$$

and $\hat{\nu} = -\hat{\lambda}$, as given by (2.7.59). Next, we differentiate (2.7.76) through with respect to x so as to obtain the partial differential equation

$$\frac{\partial^2 w}{\partial t \partial x} - \hat{\nu}\frac{\partial^2 w}{\partial x^2} - \frac{\partial \hat{\nu}}{\partial x}\frac{\partial w}{\partial x} = \Phi^{|}(w-z)\left(\frac{\partial w}{\partial x} - \frac{\partial z}{\partial x}\right) \tag{2.7.79}$$

with

$$\Phi^{|}(w-z) = \left.\frac{d\Phi(\zeta)}{d\zeta}\right|_{\zeta = w-z}$$

In view of the definition of the derivative $/$, namely, $/ = \frac{\partial}{\partial t} - \hat{\nu}\frac{\partial}{\partial x}$, (2.7.79) may be put in the form

$$\left(\frac{\partial w}{\partial x}\right)' - \frac{\partial \hat{\nu}}{\partial x}\frac{\partial w}{\partial x} = \Phi^{|}(w-z)\left(\frac{\partial w}{\partial x} - \frac{\partial z}{\partial x}\right) \tag{2.7.80}$$

However,

$$\frac{\partial\hat{\nu}}{\partial x}=\frac{\partial\hat{\nu}}{\partial w}\frac{\partial w}{\partial x}+\frac{\partial\hat{\nu}}{\partial z}\frac{\partial z}{\partial x}=\frac{\partial\hat{\nu}}{\partial w}\left(\frac{\partial w}{\partial x}-\frac{\partial z}{\partial x}\right)$$

so, if we set $\Psi(w-z)=\Phi^{\mathsf{I}}(w-z)$, we may rewrite (2.7.80) as

$$\left(\frac{\partial w}{\partial x}\right)'+\frac{\partial\hat{\nu}}{\partial w}\frac{\partial w}{\partial x}\frac{\partial z}{\partial x}=\frac{\partial\hat{\nu}}{\partial w}\left(\frac{\partial w}{\partial x}\right)^2+\Psi(w-z)\left(\frac{\partial w}{\partial x}-\frac{\partial z}{\partial x}\right) \tag{2.7.81}$$

Now, combining equations (2.7.48) we have

$$w'-\grave{z}=2\Phi(w-z) \tag{2.7.82}$$

However, $z'=\dfrac{\partial z}{\partial t}-\hat{\nu}\dfrac{\partial z}{\partial x}$, so

$$\frac{\partial z}{\partial x}=-\frac{\partial\Phi}{\partial\hat{\nu}}+\frac{(w'-z')}{2\hat{\nu}}, \tag{2.7.83}$$

as a consequence of (2.7.82) and the definition of $\grave{z}$. We now substitute from (2.7.83) into (2.7.81) and simplify so as to obtain

$$\left(\frac{\partial w}{\partial x}\right)'+\frac{\partial\hat{\nu}}{\partial w}\cdot\frac{(w'-z')}{2\hat{\nu}}\cdot\frac{\partial w}{\partial x}-\frac{\Phi}{\hat{\nu}}\frac{\partial\hat{\nu}}{\partial w}\frac{\partial w}{\partial x} \tag{2.7.84}$$

$$=\frac{\partial\hat{\nu}}{\partial w}\left(\frac{\partial w}{\partial x}\right)^2+\Psi\frac{\partial w}{\partial x}+\frac{\Phi\Psi}{\hat{\nu}}-\frac{\Psi(w'-z')}{2\hat{\nu}}$$

An elementary calculation shows that

$$(\log\hat{\nu})'=\frac{\partial\hat{\nu}/\partial w}{\hat{\nu}}(w'-z') \tag{2.7.85}$$

and, using this equivalence, we may further rewrite (2.7.84) as

$$\left(\frac{\partial w}{\partial x}\right)'\frac{1}{2}(\log\hat{\nu})'\frac{\partial w}{\partial x}=\frac{\partial\hat{\nu}}{\partial w}\left(\frac{\partial w}{\partial x}\right)^2 \tag{2.7.86}$$

$$+\left(\Psi+\frac{\Phi}{\hat{\nu}}\frac{\partial\hat{\nu}}{\partial w}\right)\frac{\partial w}{\partial x}+\frac{\Phi}{\hat{\nu}}\Psi-\Psi\left(\frac{w'-z'}{2\hat{\nu}}\right)$$

We now multiply (2.7.86) through by $\hat{\nu}^{1/2}$; using the fact that $\hat{\nu}^{1/2}(\log\hat{\nu})'=\hat{\nu}^{-1/2}\hat{\nu}'$, we then find that

$$\left(\hat{\nu}^{1/2}\frac{\partial w}{\partial x}\right)'=\hat{\nu}^{1/2}\frac{\partial\hat{\nu}}{\partial w}\left(\frac{\partial w}{\partial x}\right)^2+\left(\Psi+\frac{\Phi}{\hat{\nu}}\frac{\partial\hat{\nu}}{\partial w}\right)\hat{\nu}^{1/2}\frac{\partial w}{\partial x} \tag{2.7.87}$$

$$-\frac{1}{2}\Psi\hat{\nu}^{-1/2}(w'-z')+\hat{\nu}^{-1/2}\Phi\Psi.$$

If we set $\chi \equiv \hat{\nu}^{1/2}\dfrac{\partial w}{\partial x}$, we may further reduce (2.7.87) to

$$\begin{aligned}\chi' &= \hat{\nu}^{-1/2}\frac{\partial \hat{\nu}}{\partial w}\chi^2 + \left(\Psi + \frac{\Phi}{\hat{\nu}}\frac{\partial \hat{\nu}}{\partial w}\right)\chi \\ &\quad -\frac{1}{2}\Psi\hat{\nu}^{-1/2}(w' - z') + \hat{\nu}^{-1/2}\Phi\Psi\end{aligned} \tag{2.7.88}$$

If, in addition, we set

$$\mathcal{F}(w-z) = -\frac{1}{2}\int_0^{w-z} \Psi(\zeta)\hat{\nu}^{-1/2}(\zeta)\, d\zeta \tag{2.7.89}$$

then we obtain from (2.7.88) the equation

$$\chi' = \hat{\nu}^{-1/2}\frac{\partial \hat{\nu}}{\partial w}\chi^2 + \left(\Psi + \frac{\Phi}{\hat{\nu}}\frac{\partial \hat{\nu}}{\partial w}\right)\chi + \hat{\nu}^{-1/2}\Phi\Psi + \mathcal{F}' \tag{2.7.90}$$

Our (next to) last modification of the notation involves the introduction of the following functions:

$$\left\{\begin{aligned}\phi &= \Psi + \frac{\Phi}{\hat{\nu}}\frac{\partial \hat{\nu}}{\partial w} \\ \psi &= \hat{\nu}^{-1/2}\Phi\Psi \\ \theta &= \chi - \mathcal{F}\end{aligned}\right. \tag{2.7.91}$$

In terms of ϕ, ψ, θ, as defined above, (2.7.90) now becomes

$$\theta' = \hat{\nu}^{-1/2}\frac{\partial \hat{\nu}}{\partial w}\theta^2 + \mathcal{G}\theta + \mathcal{H} \tag{2.7.92}$$

where

$$\left\{\begin{aligned}\mathcal{G}(w-z) &\equiv 2\hat{\nu}^{-1/2}\frac{\partial \hat{\nu}}{\partial w}\mathcal{F} + \phi \\ \mathcal{H}(w-z) &\equiv \hat{\nu}^{-1/2}\frac{\partial \hat{\nu}}{\partial w}\mathcal{F}^2 + \phi\mathcal{F} + \psi\end{aligned}\right. \tag{2.7.93}$$

At this junction in the proof we pause to look carefully at the coefficient of the quadratic term in (2.7.92), i.e., at $\hat{\nu}^{-1/2}\dfrac{\partial \hat{\nu}}{\partial w}$. Directly from the definitions (2.7.54) and (2.7.77) of $\hat{\eta}$ and $\hat{\nu}$, respectively, we find that

$$\frac{\partial D}{\partial w} = \frac{1}{2}\mu_0^{1/2}(\mathsf{E}^{\mathsf{I}}(\hat{\eta}^{-1}(w-z))^{-1/2} \tag{2.7.94}$$

and

$$\frac{\partial \hat{\nu}}{\partial w} = \frac{1}{4}(\mathsf{E}^{\mathsf{I}}(\hat{\eta}^{-1}(w-z))^{-1}\mathsf{E}^{\mathsf{II}}(\hat{\eta}^{-1}(w-z)) \tag{2.7.95}$$

where $D = \hat{\eta}^{-1}(w - z)$. In arriving at (2.7.95), we have used the fact that $\frac{\partial \hat{\nu}}{\partial w} = \frac{\partial \hat{\nu}}{\partial D} \cdot \frac{\partial D}{\partial w}$. It is now a simple matter to see that the crucial coefficient in (2.7.92) is given by

$$\hat{\nu}^{-1/2}\frac{\partial \hat{\nu}}{\partial w} = \frac{1}{4}\mu_0^{1/4}\frac{\mathsf{E}''(D)}{(\mathsf{E}'(D))^{5/4}} \tag{2.7.96}$$

However, by virtue of our assumption of genuine nonlinearity, i.e., $\mathsf{E}''(0) > 0$, it follows that $\exists\, \Lambda > 0$ such that $\mathsf{E}''(\zeta) > 0$ for $|\zeta| \leq \Lambda$. Using Lemma 2.2 we find that for $\Lambda^* > 0$

$$|w(x,t) - z(x,t)| \leq \Lambda^*, \qquad t \leq T \tag{2.7.97}$$

provided that we choose $w_0(\cdot)$, $z_0(\cdot)$ so that $|w_0| + |z_0| \leq \Lambda^* \exp(-MT)$; with such a choice of the initial data it then follows that for $-\infty < x < \infty$, and $t \leq T$

$$\mathsf{E}''(\hat{\eta}^{-1}(w(x,t) - z(x,t))) > 0 \tag{2.7.98}$$

if Λ^* is sufficiently small. Therefore, with $|w_0| + |z_0| \leq \Lambda^* \exp(-MT)$, and Λ^* sufficiently small, we find, as a consequence of (2.7.96), that $\hat{\nu}^{-1/2}\frac{\partial \hat{\nu}}{\partial w} > 0$, $0 \leq t \leq T$.

We now introduce our last set of symbols. We define

$$\Gamma = \inf \hat{\nu}^{-1/2}\frac{\partial \hat{\nu}}{\partial w}; \quad g = \sup|\mathcal{G}|; \quad h = \inf \mathcal{H} \tag{2.7.99}$$

where the inf and sup above are taken over the bounded set of arguments $\{\eta = w - z |\, |\eta| \leq \Lambda^*,\ 0 \leq t \leq T\}$. From the definitions (2.7.99), and the nonlinear ordinary differential equation (2.7.29), along the backward characteristic curve, we obtain now the ordinary differential inequality

$$\theta' \geq \Gamma\theta^2 + \mathcal{G}\theta + h \tag{2.7.100}$$

However,

$$\mathcal{G}\theta \geq -\frac{\Gamma}{2}\theta^2 - \frac{1}{2\Gamma}\mathcal{G}^2 \geq -\frac{\Gamma}{2}\theta^2 - \frac{1}{2\Gamma}g^2 \tag{2.7.101}$$

so for $0 \leq t \leq T$

$$\theta' \geq \frac{\Gamma}{2}\theta^2 - \frac{1}{2\Gamma}g^2 + h \tag{2.7.102}$$

Without loss of generality we may assume that $h \leq 0$, in which case we may rewrite (2.7.102) in the form

$$\theta' \geq \frac{\Gamma}{2}(\theta^2 - j^2), \qquad 0 \leq t \leq T \tag{2.7.103}$$

with $j = \left[\left(\frac{g}{\Gamma}\right)^2 - \frac{2}{\Gamma}g\right]^{1/2}$. Associated with (2.7.103) we have, as a consequence of the definitions of θ, χ, and $\mathcal{F}$, the initial data

$$\theta(x,0) = \hat{\nu}^{1/2}(w_0(x) - z_0(x))w_0^{|}(x) + \frac{1}{2}\int_0^{w_0(x)-z_0(x)} \Psi(\zeta)\hat{\nu}^{-1/2}(\zeta)\,d\zeta \tag{2.7.104}$$

We now compare the solution $\theta(x,t)$ of (2.7.103), (2.7.104) with the solution of the initial-value problem

$$\begin{cases} \tilde{\theta}' = \frac{\Gamma}{2}(\tilde{\theta}^2 - j^2), & 0 \le t \le T \\ \tilde{\theta}(x,0) = \theta(x,0), & x \in R^1 \end{cases} \tag{2.7.105}$$

In fact, by virtue of a standard comparison theorem (e.g., [95]) we have $\theta(x,t) \ge \tilde{\theta}(x,t)$, for $-\infty < x < \infty$, and $0 \le t \le T$. But (2.7.105) may be solved in closed form as

$$\frac{1}{\tilde{\theta}(x,t)+j} = \frac{e^{j\Gamma t}}{\theta(x,0)+j} + \frac{1}{2j}(1 - e^{j\Gamma t}) \tag{2.7.106}$$

A simple inspection of the relation (2.7.106) now reveals that $\exists\, t_\infty$, $0 < t_\infty < \infty$, such that $\tilde{\theta}(x,t) \to \infty$, as $t \to \infty$, provided

$$\lim_{t\to t_\infty}\left[\frac{e^{j\Gamma t}}{\theta(x,0)+j} + \frac{1}{2j}(1 - e^{j\Gamma t})\right] = 0$$

which, in turn, is satisfied iff for some $t_\infty < \infty$

$$\frac{2j}{\theta(x,0)+j} = 1 - e^{-j\Gamma t_\infty} \tag{2.7.107}$$

Inasmuch as $0 < 1 - e^{-j\Gamma t_\infty}$, for all $t > 0$, it follows that (2.7.107) is satisfied iff $\theta(x,0)$ can be chosen so as to satisfy the restriction

$$0 < \frac{2j}{\theta(x,0)+j} < 1 \tag{2.7.108}$$

For $\theta(x,0)$ satisfying (2.7.108) we then have

$$t_\infty = -\frac{1}{j\Gamma}\ln\left[1 - \frac{2j}{\theta(x,0)+j}\right] > 0 \tag{2.7.109}$$

An inspection of (2.7.104) clearly indicates that $\theta(x,0)$ will satisfy the requirement stipulated by (2.7.108) if, at some $x \in R^1$, $w_0^{|}(x)$ is both positive and sufficiently large. Therefore, for $w_0^{|}(x)$ both positive and sufficiently large, at some $x \in R^1$, $\tilde{\theta}(x,t)$, and also $\theta(x,t)$, by the comparison theorem cited above, will become unbounded as $t \to t_\infty$, with t_∞ given by (2.7.109). Now

$$\theta(x,t) = \hat{\nu}^{1/2}(w(x,t) - z(x,t))\frac{\partial w}{\partial x}(x,t) \qquad (2.7.110)$$
$$+\frac{1}{2}\int_0^{w(x,t)-z(x,t)} \Psi(\zeta)\hat{\nu}^{1/2}(\zeta)\,d\zeta$$

and, moreover, for any fixed $T > 0$, t_∞ as given by (2.7.109) will satisfy $t_\infty < T$, provided $w_0^{|}(x) > 0$ is chosen so large that

$$\theta(x,0) > \frac{2j}{1-e^{-j\Gamma t}} - j \qquad (2.7.111)$$

Therefore, for $w_0^{|}(x)$ both positive and sufficiently large, at some $x \in R^1$, (2.7.97) and (2.7.110) combine to yield the conclusion that $\left|\frac{\partial w}{\partial x}(x,t)\right| \to \infty$ as $t \to t_\infty < T$; this contradicts, of course, our assumption that $w(x,t) \in C^1(x,t)$ for all t, $0 \le t \le T$. The theorem now follows directly from the finite-time breakdown of the assumed C^1 solution to the initial-value problem (2.7.48), (2.7.62) coupled with (2.7.55). □

Before moving on in the next chapter to a discussion of growth estimates for both electric fields and electric displacement fields in plane wave-nonlinear dielectric interaction problems, as well as the problem of existence of globally defined smooth solutions, we offer, in the remainder of this section, a brief survey of some of the literature on shock formation in nonlinear dielectrics. One of the earliest pieces of relevant work would appear to be that of Broer [29], in which the system (2.6.14) is considered, with $\tilde{\sigma}(\cdot) \equiv 0$, and travelling wave solutions of the type (2.7.33) are studied; Broer [29] also considers, in this context, the case where the electromagnetic wave is incident upon the half-space $\Omega_+ = \left\{(x,y,z)\middle| -\infty < y < \infty,\ -\infty < z < \infty,\ x > 0\right\}$ in which case $E = E(x,t)$ must be prescribed as a function of t for $x = 0$. Also mentioned in [29] is the more complicated problem of the reflection of a normally incident wave on the boundary at $x = 0$. Beyond the problem of a normally incident plane electromagnetic wave on the boundary of a half-space Ω_+, occupied by a nonlinear dielectric substance, the author treats, in [29], the situation in which the dielectric occupies the half-space and the incident ray is in the x-y plane and polarized in the z-direction. In this case both the electric and polarization fields in the medium, if the medium is uniaxial with the axis oriented in the z-direction, are in the z-direction with $P_z = P(E_z)$, the z subscript here denoting vector component. Using the notation of [29], the relevant field equations are then

$$\begin{cases} \dfrac{\partial E_z}{\partial x} - \dfrac{\partial B_y}{\partial t} &= 0 \\ \dfrac{\partial E_z}{\partial y} + \dfrac{\partial B_x}{\partial t} &= 0 \\ \dfrac{\partial B_y}{\partial x} - \dfrac{\partial B_y}{\partial y} &= \dfrac{\partial E_z}{\partial t} \end{cases} \tag{2.7.112}$$

provided the conduction field $\boldsymbol{J}$ is taken to be identically zero. Travelling wave solutions of (2.7.112) are examined by Broer [29] and are obtained in the form, e.g.,

$$E_z = f(t - xp_x - yp_y) \tag{2.7.113}$$

where the propagation vector $\boldsymbol{p} = (p_x, p_y, 0)$ satisfies

$$\frac{\partial E_z}{\partial x} = -p_x(E_z)\frac{\partial E_z}{\partial t}, \qquad \frac{\partial E_z}{\partial y} = -p_y(E_z)\frac{\partial E_z}{\partial t} \tag{2.7.114}$$

with $p_x^2 + p_y^2 = \dfrac{1}{c^2(E_z)}$, $c > 0$ being the speed of propagation of the wave.

In [75], [79] A. Jeffrey and Jeffrey and Korobeinikov, respectively, continued the studies of Broer on electromagnetic wave propagation through nonlinear non-dispersive media. Beginning with the system (2.6.14), with $\tilde{\sigma}(\cdot) \equiv 0$, and introducing the associated Riemann invariants, Jeffrey [75] makes use of the results in [77] to prove that solutions of the initial value problem remain unique and differentiable only up to some time t_c and obtains bounds of the form $t_{\text{inf}} < t_c < t_{\text{sup}}$ as well as sharp expressions for the quantities t_{inf} and t_{sup}. These bounds, in the case where the relevant Riemann invariants r, s are always close to constant values, say, $\tilde{r}_0$, $\tilde{s}_0$, assume the form

$$\begin{cases} t_{\text{inf}} &= 4\left(\mu_0 \max\left[c^4(d^2P/dE^2)\right]_0 \beta\right)^{-1} \\ t_{\text{sup}} &= 4\left(\mu_0 \min\left[c^4(d^2P/dE^2)\right]_0 \beta\right)^{-1} \end{cases} \tag{2.7.115}$$

In (2.7.115), $D = \epsilon_0 E + P(E)$, $c^2 = (\epsilon_0\mu_0 + \mu_0 \dfrac{dP}{dE})^{-1}$, β is the larger of $\max(\partial r/\partial x)_0$, $\max(\partial s/\partial x)_0$, and the zero subscript refers to evaluation in the constant state that gives rise to $\tilde{r}_0$, $\tilde{s}_0$. In [79] the authors consider (again with $\boldsymbol{J} \equiv \boldsymbol{0}$) the slightly more complex situation of plane waves which depend only on the z-coordinate, but allow for a situation in which $H_x \neq 0$, $E_y \neq 0$ so that the governing system of equations is

$$\begin{cases} -\dfrac{\partial H_y}{\partial z} = \dfrac{1}{c}\dfrac{\partial D_x}{\partial t}, & \dfrac{\partial H_x}{\partial z} = \dfrac{1}{c}\dfrac{\partial D_y}{\partial t} \\ \dfrac{\partial E_y}{\partial z} = \dfrac{1}{c}\dfrac{\partial B_x}{\partial t}, & \dfrac{\partial E_x}{\partial z} = -\dfrac{1}{c}\dfrac{\partial B_y}{\partial t} \end{cases} \tag{2.7.116}$$

with $H_z =$ const. , $D_z =$ const., the subscripts once again denoting vector components. Employing results of Jeffrey and Taniuti [78] the authors in [79] demonstrate that discontinuities can develop in finite time for solutions of the system (2.7.116) that evolve from continuous initial data, specifically, that a strong discontinuity will first form on a wavefront which comprises a propagating weak (or Lipschitz) discontinuity. An estimate for t_c, the critical time lapse to the formation of a strong discontinuity, is given under the assumption that the wave is propagating into a constant state. Also considered in [79] is the decay of plane electromagnetic shock waves of small amplitude propagating along the z-axis in a nonlinear dielectric body, with particular emphasis placed on the derivation of an asymptotic growth law for the shock wave coordinate z_s.

The fundamental paper [46] discusses the onset of shock formation (self-steepening of wavefronts) in nonlinear dielectric media, obtaining estimates for the shock coordinate (steepening distance) z_s for the case of an initial Gaussian pulse; the work in [46] is presented in terms of the evolution of the energy density defined by

$$\mathcal{E} = \frac{1}{16\pi}\left[\langle ED\rangle + \langle H^2\rangle\right]$$

where the angular brackets denote averaging over an optical period of the wave (again assumed to be a linearly polarized plane wave propagating in the z-direction). Unlike the aforementioned references, the work in [46] also considers memory effects (relaxation) in the nonlinear dielectric medium, as well as the combined effects of both relaxation and dispersion; as in the previous references, *a priori*, nonlinear conduction currents are ignored.

We mention briefly here three other related works, the paper of Rogers, Cekirge, and Askar [136], the study by Kazakia and Venkataraman [90], and the paper of Donato and Fusco [50]. In the first work the system (2.6.14) with $\tilde{\sigma} \equiv 0$ is again considered, but the authors refer the equations to the appropriate hodograph plane taking E and H ($E = E_x, H = H_y$) as independent variables and x, t as dependent variables with the resulting hodograph system having the form

$$\begin{bmatrix} x \\ t \end{bmatrix}_{,H} = \begin{bmatrix} 0 & -1/D'(E) \\ -B'(H) & 0 \end{bmatrix} \begin{bmatrix} x \\ t \end{bmatrix}_{,E} \tag{2.7.117}$$

Introducing the definitions

$$h = \int_{H_0}^{H} [B'(\zeta)]^{1/2}\, d\zeta, \qquad e = \int_{E_0}^{E} [D'(\zeta)]^{1/2}\, d\zeta \tag{2.7.118}$$

as well as

$$A(e,h) = [B'(H)D'(E)]^{-1/2} \tag{2.7.119}$$

and then setting

$$t^* = A^{1/2}t, \qquad x^* = A^{-1/2}x \tag{2.7.120}$$

transforms (2.7.117) into the system

$$\begin{cases} \Box t^* - \{\Box A^{1/2}/A^{1/2}\}t^* &= 0 \\ \Box x^* - \{A^{1/2}\Box A^{-1/2}\}x^* &= 0 \end{cases} \tag{2.7.121}$$

where $\Box \equiv \dfrac{\partial^2}{\partial e^2} - \dfrac{\partial^2}{\partial h^2}$. The authors [136] note that both equations in the set (2.7.121) are of the form

$$\Box f - \lambda(e,h)f = 0,$$

making possible an application of the Bergman linear integral operator method [17]. Special classes of constitutive relations are considered for which $\Box A^{\pm 1/2} = 0$ and for these classes of constitutive relations the hodograph equations are then integrated exactly. We note here that the relations (2.7.121) for the modified hodograph variables x^* and t^* were also obtained by Donato and Fusco [50] in a slightly different fashion. In [50] the authors first transform the system (2.6.14) to Riemann invariants form and then employ the hodograph transformation in the form $x = x(r,s)$, $t = t(r,s)$, r and s being the two Riemann invariants. There results, for $t(r,s)$, the second-order equation

$$t_{rs} - \frac{\nu_s}{\lambda - \nu}t_r + \frac{\lambda_r}{\lambda - \nu}t_s = 0 \tag{2.7.122}$$

where λ, ν are the characteristic values given by $\lambda = (B'(H)D'(E))^{-1/2}$, $\nu = -(B'(H)D'(E))^{-1/2}$, and the subscripts denote differentiation with respect to the indicated variable. Riemann's method of integration may be applied (in a manner similar to that followed by Ludford [108] for the one-dimensional gas dynamics equations) so as to write the solution of (2.7.122) in the form

$$t(\xi,\eta) = \int_{x_1}^{x_2} \frac{R(\varphi,\psi,\xi,\eta)}{\lambda(\varphi,\psi) - \nu(\varphi,\psi)}\,dx \tag{2.7.123}$$

where $r = \varphi(x)$, $s = \psi(x)$ describe an initial curve C in the hodograph plane, x_1 and x_2 are the values of x at the points P_1, P_2 in Figure 2.12 below, and $R(r,s,\xi,\eta)$ is the Riemann function (see, e.g., Courant and Hilbert [37]). The relation (2.7.122) is subsequently transformed in [50] into the form

$$t^*_{rs} - \left(\left(\frac{\partial^2 \lambda^{1/2}}{\partial r \partial s}\right) \Big/ \lambda^{1/2}\right) t^* = 0 \tag{2.7.124}$$

which is, essentially, equivalent to the first equation in the set (2.7.121). In (2.7.124), $t^* = e^g t$, with

$$g(r,s) = \int \frac{\nu_s}{\nu - \lambda}\, ds + \int \frac{\lambda_r}{\lambda - \nu}\, dr \tag{2.7.125}$$

e.g., for the model nonlinear dielectric given in [50] by the constitutive relations

$$D = qE + \frac{1}{2}\eta E^2, \qquad B = \mu_0 H \tag{2.7.126}$$

where q and η are constant,

$$g = -\frac{1}{4} \log\left(\mu_0(q + \eta E)\right) \tag{2.7.127}$$

For the constitutive theory given by (2.7.126), the Riemann method of integration is applied, in [50], to (2.7.124) so as to yield the following interesting expression for t_c, the critical time until shock formation: $t_c = \min\left(t_c^{(r)}, t_c^{(s)}\right)$ where

$$\left\{\begin{array}{rcl} t_c^{(r)} & = & \displaystyle\min_{x \in C} \frac{4\sqrt{\mu_0}(q + \eta E_0(x))^2}{\eta\left(\sqrt{\mu_0}H_0'(x) + \sqrt{q + \eta E_0(x)}E_0'(x)\right)} \\ t_c^{(s)} & = & \displaystyle\min_{x \in C} \frac{4\sqrt{\mu_0}(q + \eta E_0(x))^2}{\eta\left(\sqrt{\mu_0}H_0'(x) - \sqrt{q + \eta E_0(x)}E_0'(x)\right)} \end{array}\right. \tag{2.7.128}$$

with C again being the contour depicted, in Figure 2.12 below, in the computation of the Riemann function $R(r,s,\xi,\eta)$. We note, in passing, that the Riemann invariants relative to the choice (2.7.126) are

$$\left\{\begin{array}{rcl} r & = & \sqrt{\mu_0}H + \frac{2}{3}(q + \eta E)^{3/2} \\ s & = & \sqrt{\mu_0}H - \frac{2}{3}(q + \eta E)^{3/2} \end{array}\right. \tag{2.7.129}$$

In [90] the authors study the evolution of a large amplitude centered fan inside a plane parallel nonlinear dielectric slab which is embedded between two linear dielectrics; such a disturbance is generated, for example, by the arrival of an electromagnetic shock wave at one of the interfaces of the slab with the surrounding medium. When the initial disturbance traverses the slab, it interacts in a complicated fashion

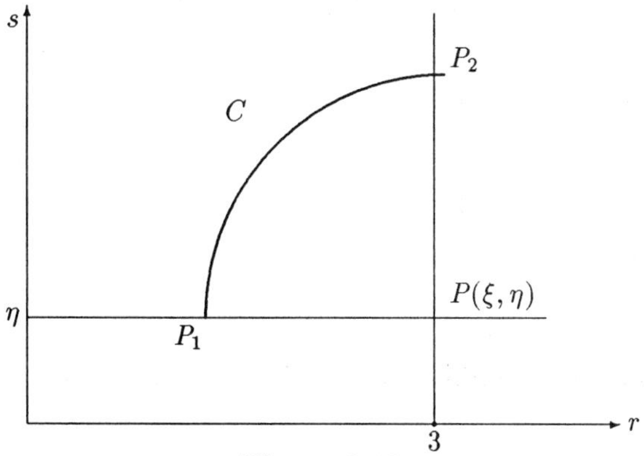

Figure 2.12

with the reflected signal from the other boundary. Using the method of characteristics, the authors [90] provide a detailed description of the nonlinear interaction process. Various model constitutive relations for the behavior of the nonlinear dielectric in the slab are introduced so as to allow for the construction of analytical solutions in [90]; it is worth noting here that these model relations in [90] have the form

$$\begin{cases} D = D_0 + (ab)^{-1}\left[1 - \left\{1 + \frac{3a}{b^3}(E - E_0)\right\}^{-1/3}\right] \\ B = B_0 + (cd)^{-1}\left[1 - \left\{1 + \frac{3c}{d^3}(H - H_0)\right\}^{-1/3}\right] \end{cases} \tag{2.7.130}$$

with a, b, c, d being material constants and, as D and B arise from the integration of certain differential constitutive relations, E_0, H_0, D_0, and B_0 are integration constants. What is interesting to observe here is the fact that the model equations (2.7.130) can be used to approximate a range of both experimentally and theoretically determined material dielectric responses, e.g., if we consider, for small field intensities, the isotropic constitutive relations

$$\begin{cases} D = \epsilon_1 E + \epsilon_2 |E| E \\ B = \mu_1 H + \mu_2 |H| H \end{cases} \tag{2.7.131}$$

then with the material constants in (2.7.130) chosen so that

$$\begin{cases} a = \mp\frac{1}{2}\epsilon_2\epsilon_1^{-7/4}, \quad b = \epsilon_1^{-1/4}\left[1 \mp \frac{3}{2}\frac{\epsilon_2}{\epsilon_1}E_0\right]^{1/3} \\ c = \mp\frac{1}{2}\mu_2\mu_1^{-7/4}, \quad d = \mu_1^{-1/4}\left[1 \mp \frac{3}{2}\frac{\mu_2}{\mu_1}H_0\right]^{1/3} \end{cases} \tag{2.7.132}$$

the model laws locally approximate the constitutive relations (2.7.131) to within terms of order $\mathcal{O}(E^3)$ and $\mathcal{O}(H^3)$, the $\mp$ signs corresponding, respectively, to negative and positive values of the field variables E and H. The authors are careful to point out in [90] that the interaction analysis they carry out inside the plane parallel nonlinear dielectric slab, idealized so as to occupy, in R^3, the domain between parallel planes through $x = 0$ and $x = 1$, is valid only until the disturbance reflected from the boundary at $x = 0$ forms a shock.

2.8 Multidimensional Problems: Shock Formation

In Lemma 1.1 we have shown that, if $\Omega \subseteq R^3$ is either a bounded or unbounded domain which is filled with a (rigid) nonlinear dielectric material, satisfying the constitutive hypotheses (1.3.57), then the components $D_i(\boldsymbol{x}, t)$ of the electric displacement field satisfy the nonlinear wave equations (1.3.58). In (1.3.58), $\boldsymbol{E} = \lambda(\|\boldsymbol{D}\|)\boldsymbol{D}$, while

$$\eta(\|\boldsymbol{D}\|) = \sigma(\lambda(\|\boldsymbol{D}\|)\|\boldsymbol{D}\|)\lambda(\|\boldsymbol{D}\|) \tag{2.8.1}$$

Without loss of generality we will set $\mu = 1$ throughout the course of our considerations in this section. Two frequently considered examples of multidimensional wave-dielectric interaction problems, which fit into the present framework, would be the following:

A. Unidirectional Propagation In this case $\boldsymbol{D} = (0, 0, D_z(x, y, t))$, so that $\nabla \cdot \boldsymbol{D} = 0$ is automatically satisfied, where D_z simply denotes the third component of the vector $\boldsymbol{D}$. Taking $\lambda(\|\boldsymbol{D}\|) = \lambda_0 + \lambda_2\|\boldsymbol{D}\|^2$ and extracting the third component of the vector equation (1.3.58) we find that

$$\frac{\partial^2 D_z}{\partial t^2} + \frac{\partial}{\partial t}(\eta(|D_z|)D_z) = \nabla^2(\gamma(|D_z|)D_z) \tag{2.8.2}$$

where

$$\gamma(|D_z|) = \lambda_0 + \lambda_2 D_z^2 \tag{2.8.3}$$

Equation (2.8.2), with $\eta = 0$, is implicit in the work of Broer [29] which we have referenced in § 2.7, i.e., simply substitute $\boldsymbol{E} = \lambda(\|\boldsymbol{D}\|)\boldsymbol{D}$, with $\boldsymbol{D} = (0, 0, D_z(x, y, t))$, into the field equations (2.7.112), and then eliminate among the equations obtained.

B. Bidirectional Propagation We begin with

$$\begin{cases} \boldsymbol{D} &= (D_x(x,y,z,t), D_y(x,y,z,t,0) \\ \boldsymbol{H} &= (0,0,H_z(x,y,z,t)) \end{cases} \tag{2.8.4}$$

where the subscripts denote, as in the system (2.7.112), components of the relevant vector field and not partial differentiation. Substituting into Maxwell's equations we have

$$\begin{cases} \partial_t D_x = \partial_y H_z = \partial_y B_z \\ \partial_t D_y = -\partial_x H_z = -\partial_x B_z \\ \partial_z E_y = \partial_z E_x = 0 \\ \partial_t B_z = \partial_y E_x - \partial_x E_y \end{cases} \tag{2.8.5}$$

while $\nabla \cdot \boldsymbol{D} = 0$ may be satisfied by taking $D_x = D_x(y,z,t)$ and $D_y = D_y(x,z,t)$. Then (2.8.5) and the constitutive relation $\boldsymbol{E} = \lambda(\|\boldsymbol{D}\|)\boldsymbol{D}$ imply that, in fact, $D_x = D_x(y,t)$ while $D_y = D_y(x,t)$. By eliminating among the remaining equations in (2.8.5) we are led to the system

$$\begin{cases} \partial_t^2 D_x &= \partial_y^2(\lambda D_x) - \partial_x\partial_y(\lambda D_y) \\ \partial_t^2 D_y &= \partial_x^2(\lambda D_x) - \partial_x\partial_y(\lambda D_x) \end{cases} \tag{2.8.6}$$

with

$$\lambda(x,y,t) = \lambda\left(\sqrt{D_x^2(y,t) + D_y^2(x,t)}\right) \tag{2.8.7}$$

For the two-dimensional problem considered by Broer [29], i.e. case A above, we may choose to work, in the situation where $\eta \equiv 0$, with the system (2.7.112) in lieu of the wave equation (2.8.2). If we set $E_z = \lambda(|D_z|)D_z$ in (2.7.112), then we obtain, as in [29], travelling wave solutions of the form (2.7.113), where $p_x = p_x(E_z)$, $p_y = p_y(E_z)$ are the components of the *propagation vector*, which satisfies

$$\frac{\partial E_z}{\partial x} = -p_x(E_z)\frac{\partial E_z}{\partial t}, \qquad \frac{\partial E_z}{\partial y} = -p_y(E_z)\frac{\partial E_z}{\partial t} \tag{2.8.8}$$

with

$$p_x^2(E_z) + p_y^2(E_z) = 1/c^2(E_z), \tag{2.8.9}$$

$c > 0$ being the speed of the propagating wave.

It is established, in [29], that, in general, $\exists\, t_\infty < \infty$ such that, as a consequence of (2.7.113), $|\partial_x E_z| \to \infty$, $|\partial_y E_z| \to \infty$, as $t \to t_\infty^-$. Thus, finite-time breakdown of smooth solutions of the nonlinear wave equation (2.8.2), or the equivalent system (2.7.112), with $E_z = \lambda(|D_z|)D_z$, is to be expected. If we now modify the last equation in (2.7.112) so as to include the influence of nonlinear conduction currents, i.e.,

$$\partial_x B_y - \partial_y B_x = \partial_t E_z + J(E_z) \tag{2.8.10}$$

with $E_z = \lambda(|D_z|D_z)$, and $J = \sigma(|E_z|)E_z$, we are, of course, back to the situation governed by (2.8.2), as we have already indicated above.

Suppose now that we set

$$\begin{cases} u_1 = D_z, \quad u_2 = B_x, \quad u_3 = B_y \\ x_1 = x, \qquad x_2 = y \end{cases} \tag{2.8.11}$$

Then it is an easy matter to verify that the system (2.7.112), with the last equation in this system replaced by (2.8.10), assumes the form

$$\frac{\partial \boldsymbol{u}}{\partial t} + \sum_{k=1}^{2} \boldsymbol{A}^{(k)} \frac{\partial \boldsymbol{u}}{\partial x_k} + \boldsymbol{g}(\boldsymbol{u}) = \boldsymbol{0} \tag{2.8.12}$$

with

$$\boldsymbol{u} = \begin{bmatrix} u_1 \\ u_2 \\ u_3 \end{bmatrix}, \qquad \boldsymbol{g}(\boldsymbol{u}) = \begin{bmatrix} J(E_z(u_1)) \\ 0 \\ 0 \end{bmatrix} \tag{2.8.13a}$$

and

$$\begin{cases} \boldsymbol{A}^{(1)} = \begin{bmatrix} 0 & 0 & -1 \\ 0 & 0 & 0 \\ -E_z'(u_1) & 0 & 0 \end{bmatrix} \\ \boldsymbol{A}^{(2)} = \begin{bmatrix} 0 & 1 & 0 \\ E_z'(u_1) & 0 & 0 \\ 0 & 0 & 0 \end{bmatrix} \end{cases} \tag{2.8.13b}$$

If, in (2.8.12), we had $A_{ij}^{(k)} = A^{(k)}\delta_{ij}$, then the system would reduce to one of the form

$$\frac{\partial u_i}{\partial t} + \sum_{k=1}^{2} A^{(k)} \frac{\partial u_i}{\partial x_k} + g_i(\boldsymbol{u}) = 0, \qquad i = 1, 2, 3 \tag{2.8.14}$$

which has been treated, with regard to breakdown of smooth solutions, by Levine and Protter [106]; it may also be possible to treat the system (2.8.12) by employing the underlying concepts established in [106] relating finite-time breakdown of solutions to quasilinear systems with oscillations in solutions of a related system of ordinary differential equations. Other promising approaches to establishing finite-time breakdown of smooth solutions, e.g., for the problem of (two-space dimension) unidirectional propagation in a nonlinear dielectric, would be more closely tied to the nonlinearly damped, nonlinear wave equation (2.8.2); in this regard the reader may be interested in consulting the work of John [26], [81], Glassey [61], [62], Levine [104], [105], Knops, Levine, and Payne [93], and Sideris [157].

When $\boldsymbol{J} = \boldsymbol{0}$, (2.8.12) reduces to a system of conservation laws on R^2, i.e.

$$\frac{\partial \boldsymbol{u}}{\partial t} + \sum_{j=1}^{2} \frac{\partial \boldsymbol{F}^{(j)}(\boldsymbol{u})}{x_j} = \boldsymbol{0} \tag{2.8.15}$$

with

$$\boldsymbol{F}^{(1)} = \begin{bmatrix} -u_3 \\ 0 \\ -1/E_z(u_1) \end{bmatrix}, \qquad \boldsymbol{F}^{(2)} = \begin{bmatrix} u_2 \\ 1/E_z(u_1) \\ 0 \end{bmatrix} \tag{2.8.16}$$

For (2.8.15), (2.8.16), with Cauchy data $\boldsymbol{u}(\boldsymbol{x},0) = \boldsymbol{u}^0(\boldsymbol{x}), \boldsymbol{x} \in R^2$, the classical shock front problem is the following: Let $\mathcal{M}$ be a smooth hypersurface, parametrized by α, with unit normal $\boldsymbol{N}(\alpha)$. Suppose that $\Omega_\pm$ have $\mathcal{M}$ as a common boundary, that the data is piecewise smooth, i.e.

$$\boldsymbol{u}^0(\boldsymbol{x}) = \begin{cases} \boldsymbol{u}^0_+(\boldsymbol{x}), & \boldsymbol{x} \in \Omega_+ \\ \boldsymbol{u}^0_-(\boldsymbol{x}), & \boldsymbol{x} \in \Omega_- \end{cases} \tag{2.8.17}$$

with

$$-\,s(\alpha)(\boldsymbol{u}^0_+(\alpha) - \boldsymbol{u}^0_-(\alpha)) + \sum_{j=1}^{2} \left(\boldsymbol{F}^{(j)}(\boldsymbol{u}^0_+(\alpha)) - \boldsymbol{F}^{(j)}(\boldsymbol{u}^0_-(\alpha))\right) \boldsymbol{N}_j(\alpha) = \boldsymbol{0}, \tag{2.8.18}$$

for some smooth scalar function $s(\alpha)$ which does not define a characteristic direction, i.e., $\min_{1\le j_\alpha \le 2} \left|s(\alpha) - \lambda_j(\boldsymbol{u}^0_\pm(\alpha))\right| > 0$, with λ_j the j-th eigenvalue of $\boldsymbol{A}^{(j)} = \nabla_u \boldsymbol{F}^{(j)}$. We seek a unique C^2 space-time normal (N_t, N_1, N_2), together with two unique C^1 vector-valued functions $\boldsymbol{u}_-(x,t), \boldsymbol{u}_+(x,t)$ defined on domains $G^\pm$, on either side of $S(t)$, satisfying

$$\begin{cases} \dfrac{\partial \boldsymbol{u}_\pm}{\partial t} + \displaystyle\sum_{j=1}^{2} \boldsymbol{A}^{(j)}(\boldsymbol{u}_\pm) \frac{\partial \boldsymbol{u}_\pm}{\partial x_j} = \boldsymbol{0}, \quad \text{in } G^\pm \\ \boldsymbol{u}_\pm(\boldsymbol{x},0) = \boldsymbol{u}^0_\pm(\boldsymbol{x}), \quad \boldsymbol{x} \in \Omega_\pm \end{cases} \tag{2.8.19}$$

and the Rankine-Hugoniot condition

$$\boldsymbol{N}_t(\boldsymbol{u}_+ - \boldsymbol{u}_-)|_S = \sum_{j=1}^{2} \boldsymbol{N}_j \left(\boldsymbol{F}^{(j)}(\boldsymbol{u}_+) - \boldsymbol{F}^{(j)}(\boldsymbol{u}_-)\right)|_S \tag{2.8.20}$$

In recent work [111], [110], [109] Majda has proposed a constructive (iterative) method for computing a solution of (2.8.19), (2.8.20), for small t, of the form

$$S(t) = \{(x,y,t)|x = \phi(y,t), \phi(y,0) = 0\} \tag{2.8.21}$$

for the case where the data $\boldsymbol{u}^0(x,y)$ is given by

$$\boldsymbol{u}^0(x,y) = \begin{cases} \boldsymbol{u}_+^0 + \boldsymbol{v}_+^0(x,y), & x \geq 0 \\ \boldsymbol{u}_-^0, & x < 0 \end{cases} \tag{2.8.22}$$

with $\boldsymbol{u}_-^0, \boldsymbol{u}_+^0$ defining a steady planar shock along the x axis (i.e., (2.8.18) is satisfied with $s = 0$) and $\boldsymbol{v}_+^0(x,y) \in (C_0^\infty(R^2))^3$ having compact support in $\{(x,y)|x \geq 0\}$. Central to the construction is the iteration scheme

$$\begin{cases} L(\boldsymbol{u}_+^N, d\phi^N)\boldsymbol{v}_+^{N+1} = \boldsymbol{0}, & x > 0, t > 0 \\ \boldsymbol{v}_+^{N+1}(x,y,0) = \boldsymbol{v}_+^0(x,y), & x > 0 \end{cases} \tag{2.8.23}$$

$\boldsymbol{u}_+^N = \boldsymbol{u}_+^0 + \boldsymbol{v}_+^N$, $\boldsymbol{v}_+^0(x,y,t) \equiv \boldsymbol{v}_+^0(x,y)$, $\phi^0 = 0$, where the linearized operator L is defined by

$$\boldsymbol{L}(\boldsymbol{u}_+, d\phi)\boldsymbol{v} \equiv \boldsymbol{v}_t + \boldsymbol{A}^{(2)}(\boldsymbol{u}_+)\boldsymbol{v}_y + \left(\boldsymbol{A}^{(1)}(\boldsymbol{u}_+) - \phi_y \boldsymbol{A}^{(2)}(\boldsymbol{u}_+) - \phi_t \boldsymbol{I}\right)\boldsymbol{v}_x \tag{2.8.24}$$

In our problem the $\boldsymbol{A}^{(j)}$ are given by (2.8.13b) and elementary calculations yield the system of equations

$$\begin{cases} \partial_t v_1^{N+1} + \partial_y v_2^{N+1} - \partial_t\phi^N \partial_x v_1^{N+1} - \partial_y \phi^N \partial_x v_2^{N+1} - \partial_x v_3^{N+1} = 0 \\ \partial_t v_2^{N+1} + \dfrac{1}{\mu} E'(u_{1+}^N)\partial_y v_1^{N+1} - \dfrac{1}{\mu}\partial_y\phi^N E'(u_{1+}^N)\partial_x v_1^{N+1} - \partial_t \phi^N \partial_x v_2^{N+1} = 0 \\ \partial_t v_3^{N+1} - \dfrac{1}{\mu} E'(u_{1+}^N)\partial_x v_1^{N+1} - \partial_t \phi^N \partial_x v_3^{N+1} = 0 \end{cases} \tag{2.8.25}$$

with the iterates ϕ^N satisfying (2.8.20) at each step.

It would be interest to try to effect, as completely as possible, a concrete solution of the shock front problem, for the two-dimensional wave-dielectric interaction, by implementing the iterative procedure outlined above and to delineate conditions on

$J(E_z(\cdot))$ which might yield convergence for the extension of this iterative scheme to the Cauchy problem for the quasilinear system of balance laws (2.8.12). For other relevant work on shock formation and propagation in multidimensional situations which can, in all likelihood, be tied in to considerations germane to the system (2.8.12), we refer the interested reader to the monograph of Majda [109], as well as to the recent papers [68], [107], and [182] and the references cited therein.

Chapter 3

WAVE-DIELECTRIC INTERACTIONS II: GROWTH ESTIMATES AND EXISTENCE OF SMOOTH SOLUTIONS

3.0 Introduction

The natural counterpart to the problem of shock formation in the solutions of wave-dielectric interaction problems in one-space dimension, that was studied in the last chapter, is the problem of globally defined smooth solutions; this problem will be the focus of the present chapter, although we will also present several kinds of growth estimates that are valid on the maximal (time) interval of existence for a sufficiently smooth solution, i.e., on $[0, t_{\max})$, with the understanding that without appropriate restrictions on the size of the initial data (as given, e.g., in § 3.5), we may have $t_{\max} < \infty$. We begin in § 3.1 by presenting a concise discussion of those differential inequality techniques that are usually known as logarithmic convexity and concavity arguments; the basic ideas are illustrated through an application to the equations of (isothermal) dynamic nonlinear hyperelasticity. The arguments developed in § 3.1 are then applied, in § 3.2, to derive some growth estimates for the L^2 norm of the electric displacement vector, in the one-dimensional wave-dielectric interaction problem, under the assumption that $\boldsymbol{J} = \boldsymbol{0}$. In § 3.3 we return to the Riemann invariants form of the inhomogeneous system of hyperbolic balance laws governing the evolution of the fields $B(x,t)$ and $D(x,t)$ in the one-space dimension wave-dielectric interaction problem; with $\boldsymbol{J} \neq \boldsymbol{0}$, the analysis in Chapter 2 served to point to the possibility that nonlinear conduction currents, in a nonlinear dielectric, might serve as a natural dissipative mechanism. In § 3.3 we use a Riemann invariants argument in order to derive a lower bound for the life span of a classical solution to the wave-dielectric interaction problem for the specific case in which the initial electromagnetic fields in the wave are assumed to be periodic.

In § 3.4 we begin the discussion of global existence of smooth solutions for the one-space dimension wave-dielectric interaction problem under the assumption of small initial data. The specific aim of § 3.4 is to illustrate the essential nature of the energy argument that is applied in § 3.5 to the nonlinearly damped, nonlinear wave equation which governs the evolution of the electric displacement field in the wave; we do this by employing the basic technique to establish the global existence of a C^1 solution to the initial-value problem for the Burger's equation with frictional damping and sufficiently small initial data. The energy argument technique of § 3.4 is then used, in § 3.5, to establish the global (in time) existence of smooth solutions for the wave-dielectric interaction problem with associated small initial data where the precise nature of what we mean by small data is delineated in Theorem 3.5; as a consequence of this theorem we also obtain asymptotic results on the decay of solutions, both in $L^\infty(R^1)$, as well as in $L^2(R^1)$, as the time $t \to \infty$. The globally defined smooth solutions $D(x,t)$ obtained in Theorem 3.5 belong to the class of functions that are $C^2(R^1 \times [0,\infty))$. It is also possible to prove, for the wave-dielectric interaction problem, the global existence of weak solutions, i.e., solutions in $L^\infty(R^1 \times [0,\infty))$, for arbitrarily large initial data; however, such a result for the wave-dielectric interaction problem may be achieved through an analysis which is formally similar to the one presented in Chapter 5 in order to establish the global existence of a weak solution to the system which governs the evolution of the current and voltage in a distributed parameter nonlinear transmission line. Also, the proof of existence of weak solutions, for both the wave-dielectric system and the nonlinear transmission line system, requires, first, a preliminary discussion of basic concepts related to the notions of weak convergence, compensated compactness, and the Young measure; these will be presented in Chapter 4 in connection with our discussion of the distributed parameter nonlinear transmission line problem. Finally, in § 3.6 we study, again, through the use of an energy argument, the nonlinearly damped, nonlinear wave equation for $D(x,t)$, in the wave-dielectric interaction problem in one-space dimension, but this time we assume that the constitutive relation between J and E is of the form: $J = \sigma_\mu E$

with the conductivity $\sigma_\mu \to +\infty$ as the magnetic permeability $\mu \to 0^+$; the basic result obtained for the model in question is that a wave which propagates with a finite speed of propagation for fixed permeability $\mu > 0$ approaches, in the limit as $\mu \to 0^+$, one which propagates with infinite speed. The aforementioned result lends some credence to our characterization of the model presented in § 3.6 as one which may be associated, albeit in a somewhat naive fashion, with wave propagation in a superconducting nonlinear dielectric. Finally, in § 3.7, we offer a few comments on the difficulties involved with proving the existence of global (in time) solutions for the nonlinearly damped, nonlinear wave equation governing a wave-dielectric interaction problem in two space dimensions; we also discuss, briefly, some aspects of wave-dielectric interactions in one space dimension which are not governed by strictly hyperbolic systems of equations.

3.1 Logarithmic Convexity and Concavity

We will review in this section certain basic types of differential inequality arguments that are known in the literature as logarithmic convexity and concavity inequalities; such arguments have been applied with success to study the behavior of solutions to broad classes of non-well-posed problems for partial differential, as well as partial-integrodifferential, equations and are useful for obtaining estimates for the growth (in time) of the solution (to associated initial and initial-boundary value problems) as measured by a suitable norm. For a full discussion of the techniques discussed below, the reader may, for example, consult either the monograph [131] of Payne or that of the author [27], both of which contain an extensive list of references and cover a wide range of applications to problems arising in both linear and nonlinear elasticity, viscoelasticity, and electromagnetic theory; the arguments promulgated here will be applied to a particular class of wave-dielectric interaction problems in § 3.2. We begin with a discussion of the ideas underlying the basic types of logarithmic convexity and concavity arguments that will be most useful to us in the next section and will include an illustrative application to a problem in hyperelasticity.

Recall that a function $f(\cdot) \in C^2(0,T)$ is convex provided $f''(t) > 0, 0 < t < T$; geometrically, the curve representing a convex function must be (i) below the straight line segment joining any two points on the curve and (ii) above the tangent drawn at any point to the curve. In the application of the technique of logarithmic convexity to problems arising in mathematical physics, we must choose a real-valued, twice continuously differentiable function, say, $F(t)$ which is defined on solutions $\boldsymbol{u}(\cdot,t)$ of the problem, and which satisfies

$$\begin{cases} F(t) \geq 0, & 0 < t < T \\ F(t) = 0 \Leftrightarrow \boldsymbol{u}(\cdot,t) = 0, & 0 < t < T \end{cases} \tag{3.1.1}$$

The function $\boldsymbol{u}(\cdot,t), 0 < t < T$, will usually be defined on some Hilbert or Banach space of functions with values in a related space, the two spaces in question being defined by the problem under investigation. Once $F(\cdot)$ has been chosen (it will usually be a norm on the space in which $\boldsymbol{u}(\cdot,t)$ takes its values) then one must establish that it satisfies a differential inequality of the form

$$F\ddot{F} - (\dot{F})^2 \geq -a_1 F\dot{F} - a_2 F^2 \tag{3.1.2}$$

for some constants a_1 and a_2. It is easy to see that (3.1.2) is but a generalization of $\ddot{F} \geq 0$, for if $F(t) > 0$, for $0 < t_1 < t < t_2 < t$, then (3.1.2) may be rewritten as

$$\overset{\cdot\cdot}{\overline{(\ln F(t))}} + a_1 \overset{\cdot}{\overline{(\ln F)}} + a_2 \geq 0, \quad t_1 < t < t_2 \tag{3.1.3}$$

We now make the change of variables $\lambda = \exp(-a_1 t)$ in which case a straightforward calculation shows that (3.1.3) becomes

$$\frac{d^2}{d\lambda^2} \ln\left\{F(\lambda)\lambda^{-a_2/a_1^2}\right\} \geq 0, \qquad \lambda_1 < \lambda < \lambda_2 \tag{3.1.4}$$

where $\lambda_j = \exp(-a_1 t_j)$, $j = 1,2$. From (3.1.4) it follows that $G(\lambda) = F(\lambda)\lambda^{-a_2/a_1^2}$ is a convex function of λ on (λ_1, λ_2). The differential inequality (3.1.2) is often a basis for uniqueness and continuous data dependence arguments in situations where it is applicable, i.e., if we integrate (3.1.4) by means of Jensen's inequality (which is equivalent to the first geometric property of convex functions referred to above) we obtain the estimate

$$F(t) \leq e^{-\frac{a_2}{a_1}t} \left[F(t_1)e^{\frac{a_2}{a_1}t_1}\right]^{\mu} \left[F(t_2)e^{\frac{a_2}{a_1}t_2}\right]^{1-\mu} \tag{3.1.5}$$

for $0 \leq t_1 < t < t_2 < T$, where $\mu = \dfrac{e^{-a_1 t} - e^{-a_1 t_1}}{e^{-a_1 t_1} - e^{-a_1 t_2}}$. Uniqueness is then established by employing (3.1.5) to show that $F(0) = 0$ implies that $F(t) = 0$, $0 \leq t \leq T$. On the other hand, (3.1.5) may be easily used to establish continuous dependence of the solution $\boldsymbol{u}(\cdot, t)$, to the problem at hand, upon its initial data, say, if we restrict ourselves to that class of solutions which satisfy $F(T) \leq M \exp(-a_2 T/a_1)$ for some $M > 0$; in fact, for such a class of solutions it follows from (3.1.5) that $\exists\, k(T) > 0$ such that

$$F(t) \leq k(T) M^{1-\mu(t)} (F(0))^{\mu(t)}, \quad 0 < \mu(t) < 1 \tag{3.1.6}$$

showing that the solution $\boldsymbol{u}(\cdot, t)$ depends Hölder continuously on the initial data, in the measure defined by F, on compact sub-intervals of $[0, T]$. For well-posed problems, or problems in which the solution is Lyapunov stable in the usual sense, there is no need to resort to estimates of the form (3.1.6), such estimates being well-suited however to large classes of non-well-posed problems. Having made the somewhat obligatory statements concerning the utility of estimates like (3.1.5) with regard to the issues of uniqueness and continuous dependence of solutions, we now turn to our primary interest in differential inequalities of the form (3.1.3), namely, their use in understanding how solutions evolve in time.

In the simplest scenario we might have an estimate of the type (3.1.3) with both $a_1 = a_2 = 0$; in such a circumstance it is easily seen that (3.1.3) yields, for $F(t)$, the lower bound,

$$F(t) \geq F(0) \exp\left\{t \frac{F'(0)}{F(0)}\right\}, \qquad t \geq 0, \tag{3.1.7}$$

whenever $F'(0) \neq 0$, so that for $F'(0) > 0$ we have $F(t)$ bounded from below by an exponentially increasing function of t. For $a_1 \neq 0$, the differential inequality (3.1.3) may be integrated over $(0, T)$ using the second geometric property of convex functions alluded to above; we easily obtain the lower bound

$$F(t) \geq F(0) \exp\left[\left\{\frac{F'(0) + \frac{a_2}{a_1} F(0)}{a_1 F(0)}\right\}\left(1 - e^{-a_1 t}\right) - \frac{a_2}{a_1} t\right] \tag{3.1.8}$$

provided $F(0) \neq 0$. Various results on the growth of $\boldsymbol{u}(\cdot, t)$ in the measure defined by F, follow directly from (3.1.8), e.g., $F(t)$ is bounded from below by an exponentially increasing function of t if $a_1 < 0$ while $a_1 F'(0) + a_2 F(0) < 0$. Among the wide range of problems in mathematical physics where differential inequalities have proven useful we may, for example, mention linear thermoelasticity in which the measure F is given by

$$F(t) = \int_0^t \int_{B(\tau)} \rho u_i u_i \, d\boldsymbol{x} \, d\tau + (T - t) \int_{B(0)} \rho u_i u_i \, d\boldsymbol{x} + \kappa \tag{3.1.9}$$

(with u_i the i-th component of the displacement vector, κ a computable constant, ρ the density, and $B(t)$ the region of $\mathbf{R}^3$ occupied by the body at time $t \leq T$) of problems involving a mixture of two linear elastic materials, for which it has been found useful to take

$$F(t) = \int_0^t \int_{B(\tau)} (\rho_1 u_i u_i + \rho_2 v_i v_i) \, d\boldsymbol{x} \, d\tau \tag{3.1.10}$$

(where u_i, v_i are the i-th components of the displacements of the constituent materials and ρ_1, ρ_2 are their densities). In general, inequalities of the form (3.1.3) have proven to be quite useful for studying the growth behavior of weak solutions to equations of the type

$$\boldsymbol{M}\ddot{\boldsymbol{u}} + \boldsymbol{D}\dot{\boldsymbol{u}} = \boldsymbol{N}\boldsymbol{u} \tag{3.1.11}$$

in which $\boldsymbol{M}, \boldsymbol{N}$, and $\boldsymbol{D}$ are linear, time-independent symmetric operators with values in a Hilbert space H and $\boldsymbol{M}, \boldsymbol{N}$ are also positive-definite; an appropriate choice for the measure in many of these situations has turned out to be

$$F(t) = \| \cdot \|_H^2 + \beta(t + t_0)^2 \tag{3.1.12}$$

for suitably chosen values of $\beta, t_0 \geq 0$.

A method closely associated with the logarithmic convexity technique is that of the concavity argument in which a differential inequality of the form $\ddot{\overline{F^{-\gamma}}}(t) \leq 0$ is generated for a suitably chosen measure $F(\cdot)$ and a $\gamma > 0$. In order to illustrate the type of problem which is amenable to a concavity inequality of the type just alluded to, or some simple generalization thereof, we present below a simple example that arises in hyperelasticity.

Let $u_i(x_j, t)$ be the displacement at time t of a point in the hyperelastic body, with strain-energy function per unit volume W, whose position vector in the reference configuration $B \subseteq R^3$ is $\boldsymbol{x} = (x_1, x_2, x_3)$. The equations of motion, with zero external forcing function, are given by

$$\rho_0 \ddot{u}_i = \frac{\partial}{\partial x_j} \sigma_{ij}, \qquad i = 1, 2, 3 \tag{3.1.13}$$

where $\rho_0 > 0$ is the mass density in the reference configuration and

$$\sigma_{ij} = \frac{\partial W}{\partial u_{i,j}}, \qquad W = W(u_{i,j}) \tag{3.1.14}$$

We assign homogeneous boundary data of the form

$$\begin{cases} u_i = 0, & \text{on } \overline{\partial B_1} \times [0,T) \\ \sigma_{ij} n_j = 0, & \text{on } \overline{\partial B_2} \times [0,T) \end{cases} \tag{3.1.15}$$

for $T > 0$, $i = 1,2,3$, where $\overline{\partial B_1} \cup \partial B_2 = \partial B$, with $\partial B_2 \neq \emptyset$, and $\boldsymbol{n}$ is the unit normal on ∂B_2. If we define the energy

$$\mathcal{E}(t) = \frac{1}{2}\int_B \rho_0 \dot{u}_i \dot{u}_i \, d\boldsymbol{x} + \int_B W \, d\boldsymbol{x} \tag{3.1.16}$$

then it follows directly from (3.1.13), (3.1.14), and (3.1.15) that $\mathcal{E}(t) = \mathcal{E}(0)$, at least for all t, $0 \le t < T$, such that a classical solution of the initial-boundary value problem for (3.1.13) exists (associated with (3.1.13) are, of course, initial conditions of the form $u_i(\boldsymbol{x},0) = u_{i0}(\boldsymbol{x})$, $\frac{\partial u_i}{\partial t}(\boldsymbol{x},0) = u^*_{i0}(\boldsymbol{x})$, $\boldsymbol{x} \in B$, in addition to (3.1.15)). Suppose now that $\exists\, \lambda > 2$ such that

$$\int_B \left\{ \lambda W - \frac{\partial W}{\partial u_{i,j}} u_{i,j} \right\} d\boldsymbol{x} \ge 0 \tag{3.1.17}$$

and consider the measure (compare with (3.1.12)):

$$F(t) = \int_B \rho_0 u_i u_i \, d\boldsymbol{x} + \beta(t+t_0)^2, \qquad 0 \le t \le T \tag{3.1.18}$$

Then straightforward calculations yield

$$\dot{F}(t) = 2\int_B \rho_0 u_i \dot{u}_i \, d\boldsymbol{x} + 2\beta(t+t_0) \tag{3.1.19}$$

and

$$\ddot{F}(t) \ge (2+\lambda)\int_B \rho_0 \dot{u}_i \dot{u}_i \, d\boldsymbol{x} - 2\lambda\mathcal{E}(0) + 2\beta \tag{3.1.20}$$

if we employ the hypothesis (3.1.17). An application of Schwarz's inequality now produces the differential inequality

$$F\ddot{F} - \left(\frac{2+\lambda}{4}\right)\dot{F}^2 \ge -\lambda(2\mathcal{E}(0)+\beta)F, \tag{3.1.21}$$

for $0 \le t < T$, which, as $\lambda > 2$, by hypothesis, is easily seen to be equivalent to

$$\overline{\ddot{(F^{-\gamma})}} \le 2\gamma(1+2\gamma)F^{-(\gamma+1)}(2\mathcal{E}(0)+\beta) \tag{3.1.22}$$

with $\gamma = (\lambda - 2)/4 > 0$. How one integrates the differential inequality (3.1.22) depends now on the assumptions relative to $\mathcal{E}(0)$, e.g., if $\mathcal{E}(0) \leq 0$ and we choose β such that $2\mathcal{E}(0) + \beta = 0$, then $F^{-\gamma}(t)$ is concave on $[0, T]$ and by virtue of Jensen's inequality we obtain the estimate

$$F^{\gamma}(t) \leq \frac{F^{\gamma}(0)F^{\gamma}(T)}{(1 - t/T)F^{\gamma}(T) + (t/T)F^{\gamma}(0)}, \qquad 0 \leq t \leq T \tag{3.1.23}$$

from which we may infer information about the uniqueness of solutions to the initial-boundary value problem associated with (3.1.13), given the hypothesis (3.1.17) relative to the strain-energy W, as well as show that the solution depends Hölder continuously on its initial data. We may also integrate the inequality $\overset{\cdot\cdot}{\overline{F^{-\gamma}(t)}} \leq 0$, $0 \leq t \leq T$, so as to obtain the lower bound

$$F^{\gamma}(t) \geq F^{\gamma}(0)/[1 - \gamma t \frac{\dot{F}(0)}{F(0)}], \qquad 0 \leq t \leq T \tag{3.1.24}$$

No matter how the initial data $\boldsymbol{u}_0(\cdot), \boldsymbol{u}_0^*(\cdot)$ are chosen, it is clear, by virtue of (3.1.19), that $t_0 \geq 0$ can be chosen so that $\dot{F}(0) > 0$ and, thus, (3.1.24) implies that $F(t)$ must become unbounded in finite time $t_\infty < T$, for $T > 0$ sufficiently large; this, in turn, allows us to conclude that the norm $\|\boldsymbol{u}\|_{(L^2(B))^3}$ must become unbounded in finite time $t_\infty < T$, for $T > 0$ sufficiently large (unless, of course, the solution breaks down and ceases, for other reasons, to be a classical solution of (3.1.13) prior to the time t_∞). The precise nature of the type of conclusions that one can deduce from differential inequality arguments of the type surveyed in this section has been carefully examined by Ball [9],[10]. It is also possible to derive interesting estimates from (3.1.22) when the initial data and form of W are such that $\mathcal{E}(0) > 0$; such estimates will be derived within the context of the wave-dielectric interaction problem, which we return to in the next section.

3.2 Growth Estimates for Wave-Dielectric Interactions: Differential Inequalities

We return in this section to the nonlinear wave equation (2.6.21) governing the plane-wave nonlinear dielectric interaction in the cylinder Ω_c; we will examine, further, the behavior of solutions to (2.6.21), for $-\infty < x < \infty, t \le t_{\max}$, $[0, t_{\max})$ being the maximal interval of existence of a classical solution, under the following assumptions:

(i) $\tilde{\sigma}(\cdot) \equiv 0$, so the nonlinear damping term is absent from (2.6.21) (3.2.1)

(ii) $\mathsf{E}(\zeta) = \zeta\lambda(\zeta),\ -\infty < \zeta < \infty$, with $\lambda(\cdot) \in C^1(R;[0,\infty))$, and (3.2.2)

(iii) $0 < \zeta\lambda'(\zeta) + \lambda(\zeta) < \infty$, $\forall\, \zeta \in R^1$, such that $|\zeta|$ is sufficiently small, (3.2.3)

(iv) for all $\zeta \in R^1$ and some $\alpha > 2$, $\alpha \int_0^\zeta \rho\lambda(\rho)\,d\rho \ge \rho^2\lambda(\rho)$ (3.2.4)

(v) we have associated initial data $D(x,0) = D_0(x)$, $D_t(x,0) = D_1(x)$, $x \in R^1$, with $D_0(\cdot), D_1(\cdot)$ having compact support on R^1 and of class $C^2(R^1)$.

By virtue of the hypotheses stated above our initial-value problem is

$$\mu_0 \frac{\partial^2 D}{\partial t^2} = \frac{\partial}{\partial x^2}(D\lambda(D)), \quad -\infty < x < \infty, t > 0 \tag{3.2.5}$$

$$D(x,0) = D_0(x), \quad D_t(x,0) = D_1(x), \quad -\infty < x < \infty \tag{3.2.6}$$

subject to (3.2.3), (3.2.4) and with $D_0(\cdot), D_1(\cdot)$ in the space $C_0^2(R^1)$; the *a priori* estimates of Chapter 2 can be used to show that for $|D_0(\cdot)|$, $|D_1(\cdot)|$ sufficiently small, in the sup norm on R^1, it follows from (3.2.3) that $(D\lambda(D))' > 0$ for as long as a classical solution of (3.2.5), (3.2.6) exists. Standard results on hyperbolic equations [37],[85] imply that for each t, $0 \le t < t_{\max}$, the support of $D(\cdot,t)$, supp $D(\cdot,t) \subset (-\delta(t), \delta(t))$, where $\delta(t)$ is finite for each $t < \infty$. If $t_{\max} < \infty$ (and, as indicated in Chapter 2, this is usually the case) then, clearly, $\sup_{[0,t_{\max})} \delta(t) = \delta_{\max} < \infty$. However, if $t_{\max} = \infty$ then $\sup_{[0,\infty)} \delta(t)$ may be either finite or infinite. We are going to show below that under an appropriate set of assumptions on the initial data (3.2.6)

$$\sup_{[0,t_{\max})} \delta(t) = \delta < \infty \Rightarrow t_{\max} < \infty \tag{3.2.7}$$

i.e., if supp $D(\cdot,t)$ is bounded on $[0, t_{\max})$ then $t_{\max} < \infty$ so that a globally defined classical solution cannot exist for (3.2.5), (3.2.6); this will be accomplished by showing that with the assumption $\delta < \infty$, the $L^2(-\infty,\delta)$ norm of D must be bounded from below by a real-valued nonnegative function of t which tends to infinity as $t \to t_\infty < \infty$. In the course of the analysis we will also obtain several growth estimates for solutions of the initial-value problem (3.2.5), (3.2.6) which are valid on the maximal time-interval of existence $[0, t_{\max})$. Before proceeding, we note the following: if we set $\gamma(\zeta) = \zeta\lambda(\zeta)$, $\zeta \in R^1$, and $\Gamma(\zeta) = \int_0^\zeta \gamma(\rho)\,d\rho$, then $\Gamma'(\zeta) = \zeta\lambda(\zeta)$, $\zeta \in R^1$, and the hypothesis (3.2.4) is equivalent to

$$\begin{cases} \alpha\Gamma(\zeta) \geq \zeta\Gamma'(\zeta),\ \forall\,\zeta \in R^1, \text{ and} \\ \text{some } \alpha > 2 \end{cases} \tag{3.2.8}$$

As we proceed, it should be clear that the assumption (3.2.8) fulfills the same role in the analysis presented below as that of the hypothesis (3.1.17) in the analysis of the dynamic hyperelasticity problem (3.1.13), (3.1.14), (3.1.15) of the previous section; much of what follows appeared originally in the author's paper [24].

We begin by associating an energy functional with the initial-value problem (3.2.5), (3.2.6), namely, for $0 \leq t < t_{\max}$, we define

$$\begin{aligned} \mathcal{E}(t) &= \frac{\mu_0}{2}\int_{-\infty}^{\delta}\left(\int_{-\infty}^{x} D_t(y,t)\,dy\right)^2 dx \\ &\quad + \int_{-\infty}^{\delta}\left(\int_0^{D(x,t)} \rho\lambda(\rho)\,d\rho\right) dx \end{aligned} \tag{3.2.9}$$

where, as above, $\delta = \sup\limits_{[0,t_{\max})} \delta(t)$ and we will assume that $\delta < \infty$.

Lemma 3.1 ([24]) With $\mathcal{E}(t)$ defined as in (3.2.9) we have, for $0 \leq t < t_{\max}$,

$$\begin{aligned} \mathcal{E}(t) &= \frac{\mu_0}{2}\int_{-\infty}^{\delta}\left(\int_{-\infty}^{x} D_1(y)\,dy\right)^2 dx \\ &\quad + \int_{-\infty}^{\delta}\left(\int_0^{D_0(x)} \rho\lambda(\rho)\,d\rho\right) dx \end{aligned} \tag{3.2.10}$$

Proof: By virtue of the definitions of $\gamma(\zeta)$ and $\Gamma(\zeta)$

$$\mathcal{E}(t) = \frac{\mu_0}{2}\int_{-\infty}^{\delta}\left(\int_{-\infty}^{x} D_t(y,t)\,dy\right)^2 dx + \int_{-\infty}^{\delta} \Gamma(D(x,t))\,dx \tag{3.2.11}$$

and, therefore,

$$\begin{aligned} \dot{\mathcal{E}}(t) &= \frac{\mu_0}{2}\int_{-\infty}^{\delta}\left(\int_{-\infty}^{x} D_t(y,t)\,dy\right)\left(\int_{-\infty}^{x} D_{tt}(y,t)\,dy\right) dx \\ &\quad + \int_{-\infty}^{\delta} \Gamma'(D(x,t))D_t(x,t)\,dx \\ &= \int_{-\infty}^{\delta}\left(\int_{-\infty}^{x} D_t(y,t)\,dy\right)\left(\int_{-\infty}^{x} \frac{\partial^2}{\partial y^2}\gamma(D(y,t))\,dy\right) dx \\ &\quad + \int_{-\infty}^{\delta} \Gamma'(D(x,t))D_t(x,t)\,dx \end{aligned}$$

However, by virtue of the compact support of $D(\cdot,t)$ on R^1, and the fact that $\gamma(0)=0$,

$$\int_{-\infty}^{x} \frac{\partial^2}{\partial y^2}\gamma(D(y,t))\,dy = \gamma(D(x,t))_{,x} \tag{3.2.12}$$

so that

$$\begin{aligned}\dot{\mathcal{E}}(t) &= \int_{-\infty}^{\delta}\left(\int_{-\infty}^{x} D_t(y,t)\,dy\right)\gamma(D(x,t))_{,x}\,dx \\ &\quad + \int_{-\infty}^{\delta}\Gamma'(D(x,t))D_t(x,t)\,dx \end{aligned}\tag{3.2.13}$$

Therefore,

$$\begin{aligned}\dot{\mathcal{E}}(t) &= \int_{-\infty}^{\delta}\frac{\partial}{\partial x}\left[\gamma(D(x,t))\int_{-\infty}^{x} D_t(y,t)\,dy\right]dx \\ &- \int_{-\infty}^{\delta}\gamma(D(x,t))D_t(x,t)\,dx \\ &+ \int_{-\infty}^{\delta}\Gamma'(D(x,t))D_t(x,t)\,dx = 0\end{aligned}\tag{3.2.14}$$

as $\gamma(\zeta)=\Gamma'(\zeta)$, $\forall\,\zeta\in R^1$, and supp $D(\cdot,t)\subset(-\delta,\delta)$ for $0\le t<t_{\max}$. The required result, i.e. (3.2.11), now follows by integration of (3.2.14) over $[0,t]$, $0\le t<t_{\max}$. □

The next lemma is geared toward establishing a particular differential inequality (of the type (3.1.21)) for a real-valued, nonnegative functional F defined on classical solutions $D(x,t)$ of the initial-value problem (3.2.5), (3.2.6).

Lemma 3.2 Let $u\in C^2(R^1\times[0,T))$, $T\le t_{\max}$ be a classical solution of (3.2.5), (3.2.6) and set

$$F(t)=\mu_0\int_{-\infty}^{\delta}\left(\int_{-\infty}^{x} D(y,t)\,dy\right)^2 dx+\beta(t+t_0)^2 \tag{3.2.15}$$

with $\beta, t_0\ge 0$. Then for $0\le t<T$

$$F\ddot{F}-(\eta+1)\dot{F}^2\ge -2(2\eta+1)(\beta+2\mathcal{E}(0))F \tag{3.2.16}$$

where $\eta=\dfrac{1}{4}(\alpha-2)>0$ with α as in the constitutive hypothesis (3.2.8).

Proof: For $T\le t_{\max}$ we may differentiate (3.2.15) directly so as to obtain

$$\dot{F}(t)=2\mu_0\int_{-\infty}^{\delta}\left(\int_{-\infty}^{x} D(y,t)\,dy\right)\left(\int_{-\infty}^{x} D_t(y,t)\,dy\right)dx+2\beta(t+t_0) \tag{3.2.17}$$

and

$$\begin{aligned}\ddot{F}(t) &= 2\mu_0 \int_{-\infty}^{\delta} \left(\int_{-\infty}^{x} D_t(y,t)\, dy \right)^2 dx \\ &+2 \int_{-\infty}^{\delta} \left(\int_{-\infty}^{x} D(y,t)\, dy \right) \gamma(D(x,t))_{,x}\, dx + 2\beta\end{aligned} \tag{3.2.18}$$

where we have employed (3.2.5), the definition of $\gamma(\zeta), \zeta \in R^1$, and the compact support of D as in (3.2.12). Integrating, by parts, the second term on the right-hand side of (3.2.18) now yields

$$\begin{aligned}\ddot{F}(t) &= 2\mu_0 \int_{-\infty}^{\delta} \left(\int_{-\infty}^{x} D_t(y,t)\, dy \right)^2 dx \\ &-2 \int_{-\infty}^{\delta} D(x,t)\Gamma'(D(x,t))\, dx + 2\beta\end{aligned} \tag{3.2.19}$$

We now add and subtract the expression $2\alpha \int_{-\infty}^{\delta} \Gamma(D(x,t))\, dx$ on the right-hand side of (3.2.19) and then employ the hypothesis (3.2.8) so as to obtain

$$\ddot{F}(t) = 2\mu_0 \int_{-\infty}^{\delta} \left(\int_{-\infty}^{x} D_t(y,t)\, dy \right)^2 dx - 2\alpha \int_{-\infty}^{\delta} \Gamma(D(x,t))\, dx \tag{3.2.20}$$

However, in view of the definitions of $\mathcal{E}(t)$, and $\Gamma(\zeta), \zeta \in R^1$, the lower bound (3.2.20) may be replaced by

$$\begin{aligned}\ddot{F}(t) \;&\geq\; \int_{-\infty}^{\delta} \left(\int_{-\infty}^{x} D_t(y,t)\, dy \right)^2 dx \\ &-\; 2\alpha \left[\mathcal{E}(t) - \frac{\mu_0}{2} \int_{-\infty}^{\delta} \left(\int_{-\infty}^{x} D_t(y,t)\, dy \right)^2 dx \right] + 2\beta\end{aligned}$$

or, with the use of Lemma 3.1,

$$\begin{aligned}\ddot{F}(t) \;&\geq\; (2+\alpha) \left[\mu_0 \int_{-\infty}^{\delta} \left(\int_{-\infty}^{x} D_t(y,t)\, dy \right)^2 dx + \beta \right] \\ &-\alpha[\beta + 2\mathcal{E}(0)].\end{aligned} \tag{3.2.21}$$

We now combine (3.2.15), (3.2.17), and (3.2.21) and we find that

$$\begin{aligned}&F\ddot{F} - \frac{1}{4}(\alpha+2)\dot{F}^2 \geq -\alpha(\beta + 2\mathcal{E}(0)) \\ &+(2+\alpha)\Bigg\{ \Bigg(\left[\mu_0 \int_{-\infty}^{\delta} \left(\int_{-\infty}^{x} D(y,t)\, dy \right)^2 dx + \beta(t+t_0)^2 \right] \\ &\times \left[\mu_0 \int_{-\infty}^{\delta} \left(\int_{-\infty}^{x} D_t(y,t)\, dy \right)^2 dx + \beta \right] \Bigg) \\ &-\mu_0 \left(\int_{-\infty}^{\delta} \left(\int_{-\infty}^{x} D(y,t)\, dy \right) \left(\int_{-\infty}^{x} D_t(y,t)\, dy \right) dx + \beta(t+t_0) \right)^2 \Bigg\}\end{aligned} \tag{3.2.22}$$

The Cauchy-Schwartz inequality can now be applied directly to the expression in the brackets $\{\cdot\}$ on the right-hand side of (3.2.22) so as to demonstrate that this term is, in fact, nonnegative for $0 \le t < T$; therefore

$$F\ddot{F} - \frac{1}{4}(\alpha + 2)\dot{F}^2 \ge -\alpha(\beta + 2\mathcal{E}(0))F \tag{3.2.23}$$

for $0 \le t < T$, with $T \le t_{\max}$, and the indicated result, i.e., (3.2.16), then follows directly by setting $\eta = \frac{1}{4}(\alpha - 2)$. □

With the differential inequality (3.2.16) in hand, we are now in a position to make a definitive statement relative to the thesis that (3.2.7) is applicable to solutions of the initial-value problem; to facilitate matters, we define the following two functions of the initial datum:

$$I(D_0) = \mu_0 \int_{-\infty}^{\delta} \left(\int_{-\infty}^{x} D_0(y)\, dy \right)^2 dx \tag{3.2.24}$$

$$J(D_0, D_1) = 2\mu_0 \int_{-\infty}^{\delta} \left(\int_{-\infty}^{x} D_0(y)\, dy \right) \left(\int_{-\infty}^{x} D_1(y)\, dy \right) dx \tag{3.2.25}$$

Our first result is relevant to situations in which $\mathcal{E}(0)$, as given by (3.2.10), satisfies $\mathcal{E}(0) \le 0$, namely, we have the following:

Theorem 3.1 ([24]) Under the assumptions which prevail in Lemma 3.2, as well as (3.2.2)-(3.2.4) relative to $\mathsf{E}(\cdot)$, if $J(D_0, D_1) > 0$ and

$$\int_{-\infty}^{\delta} \left(\int_0^{D_0(x)} \gamma(\rho)\, d\rho \right) dx \le -\frac{\mu_0}{2} \int_{-\infty}^{\delta} \left(\int_{-\infty}^{x} D_1(y)\, dy \right)^2 dx \tag{3.2.26}$$

then $\exists\, \kappa = \kappa(\mu_0, \delta) > 0$, and $t_\infty < \infty$, such that

$$\|D(\cdot, t)\|^2_{L^2(-\infty,\delta)} \ge \kappa G(t), \qquad 0 \le t \le t_\infty, \tag{3.2.27}$$

with $t_{\max} \le t_\infty$, and $\lim_{t \to t_\infty} G(t) = +\infty$. Thus either $t_{\max} < \infty$ or, if $t_{\max} = \infty$, then $\sup_{[0,\infty)} \delta(t) = +\infty$.

Remark: So as to emphasize the point, once again, what Theorem 3.1 states is that, under the assumed hypotheses, we cannot have for the wave-dielectric interaction in the cylinder Ω_c both $t_{\max} = \infty$ and $\delta < \infty$, thus yielding the implication (3.2.7).

Proof: (Theorem 3.1): By virtue of (3.2.6), and the definition of the energy functional (3.2.9), we have $\mathcal{E}(0) \leq 0$. Defining $F(t)$ as in (3.2.15), and setting $\beta = 0$, we then find that (3.2.16) reduces to the statement that

$$F_0(t)\ddot{F}_0(t) - (\eta + 1)\dot{F}_0^2(t) \geq 0, \qquad 0 \leq t \leq t_{\max} \tag{3.2.28}$$

where

$$F_0(t) = \mu_0 \int_{-\infty}^{\delta} \left(\int_{-\infty}^{x} D(y,t)\, dy \right)^2 dx \tag{3.2.29}$$

However, (3.2.28) is equivalent to

$$\overline{(F_0^{-\eta})}^{\,\cdot\cdot}(t) \leq 0, \qquad 0 \leq t \leq t_{\max} \tag{3.2.30}$$

where $\eta = \frac{1}{4}(\alpha - 2)$, $\alpha > 2$ being the constitutive parameter defined by the hypothesis (3.2.4). If we now integrate (3.2.30) twice in succession, we easily obtain the estimate

$$F_0^{-\eta}(t) \leq -\eta F_0^{-\eta-1}(0)\dot{F}_0(0)t + F_0^{-\eta}(0), \qquad 0 \leq t \leq t_{\max} \tag{3.2.31}$$

and as $F_0(t) > 0$, $\eta > 0$, we obtain from (3.2.31) the lower bound

$$F_0(t) \geq \left[\frac{F_0^{\eta}(0)}{1 - \eta(\dot{F}_0(0)/F_0(0))t} \right]^{1/\eta} \equiv G(t), \qquad 0 \leq t \leq t_{\max} \tag{3.2.32}$$

From the definition of $G(t)$ it is clear that $\lim_{t \to t_\infty} G(t) = +\infty$ where

$$0 < t_\infty = \frac{1}{\eta} \left(\frac{F_0(0)}{\dot{F}_0(0)} \right) \equiv \frac{1}{\eta} \frac{I(D_0)}{J(D_0, D_1)} < \infty \tag{3.2.33}$$

Now, as supp $D(\cdot, t) \subset (-\delta, \delta)$, $0 \leq t \leq t_{\max}$, we have the following series of estimates:

$$\begin{aligned}
\int_{-\infty}^{\delta} \left(\int_{-\infty}^{x} D(y,t)\, dy \right)^2 dx &= \int_{-\infty}^{x} \left(\int_{-\delta}^{x} D(y,t)\, dy \right)^2 dx \\
&\leq \int_{-\delta}^{\delta} (x + \delta) \left(\int_{-\delta}^{x} D^2(y,t)\, dy \right) dx \\
&\leq \left(\int_{-\delta}^{\delta} (x+\delta)^2\, dx \right)^{1/2} \left(\int_{-\delta}^{\delta} \left[\int_{-\delta}^{x} D^2(y,t)\, dy \right]^2 dx \right)^{1/2} \\
&\leq (2\delta)^{3/2} \left(\int_{-\delta}^{\delta} \left[\int_{-\delta}^{\delta} D^2(y,t)\, dy \right]^2 dx \right)^{1/2} \\
&= 4\delta^2 \int_{-\delta}^{\delta} D^2(y,t)\, dy
\end{aligned}$$

so that

$$\|D(\cdot,t)\|^2_{L^2(-\infty,\delta)} \geq \frac{1}{4\delta^2}\int_{-\infty}^{\delta}\left(\int_{-\infty}^{x} D(y,t)\,dy\right)^2 dx \tag{3.2.34}$$

Combining (3.2.34) with (3.2.32), and the definition of $F_0(t)$, i.e., (3.2.29), we are led to (3.2.27) with $\kappa(\mu_0,\delta) \equiv 1/4\mu_0\delta^2$. □

Several other sets of hypotheses are possible which lead to conclusions that are similar in nature to those presented in Theorem 3.1. For example, suppose that $\mathcal{E}(0) < 0$ and $D_1(x) \equiv 0$, $x \in R^1$; in such a situation, we may choose $\beta = \beta_0 > 0$ such that $\beta_0 + 2\mathcal{E}(0) = 0$. We now see that, once again, the differential inequality (3.2.16) reduces to (3.2.30) but this time with $F_0(t)$ replaced by

$$F(t;\beta_0,t_0) = \mu_0\int_{-\infty}^{\delta}\left(\int_{-\infty}^{x} D(y,t)\,dy\right)^2 dx + \beta_0(t+t_0)^2 \tag{3.2.35}$$

so that, for $0 \leq t \leq t_{\max}$, $F(t;\beta_0,t_0)$ satisfies

$$\begin{aligned} F(t;\beta_0,t_0) &\geq \left[\frac{F^\eta(0;\beta_0,t_0)}{1-\eta(\dot F(0;\beta_0,t_0)/F(0;\beta_0,t_0))t}\right]^{1/\eta} \\ &\equiv H(t) \end{aligned} \tag{3.2.36}$$

Clearly, $\lim_{t\to t_\infty(t_0)} H(t) = +\infty$ now, where

$$\begin{aligned} t_\infty(t_0) &= \frac{1}{\eta}\frac{F(0;\beta_0,t_0)}{\dot F(0;\beta_0,t_0)} \\ &= \left[\frac{I(D_0)+\beta_0 t_0^2}{2\beta_0 t_0}\right] \end{aligned} \tag{3.2.37}$$

Now, by virtue of our assumption that $\mathcal{E}(0) < 0$, $D_1(x) \equiv 0$, $x \in R^1$, as well as our choice of $\beta_0 = -2\mathcal{E}(0)$, we have

$$\beta_0 = -2\int_{-\infty}^{\delta}\left(\int_0^{D_0(x)}\gamma(\rho)\,d\rho\right) dx > 0 \tag{3.2.38}$$

A straightforward calculation shows that the minimum value of $t_\infty(t_0)$, as given by (3.2.37), is obtained when

$$t_0 = t_0^* = \sqrt{\frac{I(D_0)}{\beta_0}} \tag{3.2.39}$$

in which case $t_\infty(t_0^*) = \frac{1}{\eta}t_0^*$. Therefore, choosing $t_0 = t_0^*$ in the lower bound (3.2.36) for $F(t;\beta_0,t_0)$ we find the estimate

$$\begin{aligned} &\mu_0\int_{-\infty}^{\delta}\left(\int_{-\infty}^{x} D(y,t)\,dy\right)^2 dx + \beta_0(t+t_0^*)^2 \\ &\geq I(D_0)/[1-\sqrt{\beta_0/I(D_0)}t]^{\frac{1}{\eta}} \end{aligned} \tag{3.2.40}$$

for $0 \leq t \leq t_{\max}$. By availing ourselves of the lower bound represented in (3.2.34) we thus achieve the growth estimate

$$\frac{1}{\kappa}\|D(\cdot,t)\|^2_{L^2(-\infty,\delta)} + \beta_0\left(t+\sqrt{\frac{I(D_0)}{\beta_0}}\right)^2 \geq I(D_0)\Big/[1-\sqrt{\beta_0/I(D_0)}t]^{\frac{1}{\eta}}, \tag{3.2.41}$$

for $0 \leq t \leq t_{\max} \leq \sqrt{I(D_0)/\beta_0}$, where β_0 is given by (3.2.38); it should be clear that given (3.2.41), the same conclusions as those obtained in Theorem 3.1 now hold forth in the case where $\mathcal{E}(0) < 0$ with $D_1(x) \equiv 0$, $x \in R^1$ (in this case, of course, $J(D_0, D_1) = 0$ where $J(D_0, D_1)$ is the functional defined by (3.2.25)).

Having looked at two situations in which, respectively, we had $\mathcal{E}(0) \leq 0$ with $J(D_0, D_1) > 0$, and $\mathcal{E}(0) < 0$ with $J(D_0, D_1) = 0$, we now want to examine the case in which $\mathcal{E}(0) < 0$ but $J(D_0, D_1) < 0$; in such a situation we may again choose $\beta = \beta_0$ such that $2\mathcal{E}(0) + \beta_0 = 0$ and we find that $F(t; \beta_0, t_0)$ satisfies (3.2.36). Now, however, in place of (3.2.37) we find that

$$t_\infty(t_0) = \frac{1}{\eta}\left[\frac{I(D_0)+\beta_0 t_0^2}{2\beta_0 t_0 - |J(D_0,D_1)|}\right] \tag{3.2.42}$$

where, by hypothesis,

$$\beta_0 = -\mu_0\int_{-\infty}^{\delta}\left(\int_{-\infty}^{x} D_1(y)\,dy\right)^2 dx - 2\int_{-\infty}^{\delta}\left(\int_0^{D_0(x)}\gamma(\rho)\,d\rho\right) dx > 0 \tag{3.2.43}$$

Our choice of t_0 is, therefore, restricted so as to satisfy $t_0 > |J(D_0, D_1)|/2\beta_0$; an elementary calculation now shows that $t_\infty(t_0)$, as given by (3.2.42), achieves a minimum when

$$t_0 = \hat{t}_0 = \frac{1}{\beta_0}\,[|J(D_0,D_1)| + 2\beta_0 I(D_0)] \tag{3.2.44}$$

in which case, for $0 \leq t \leq t_{\max} \leq t_\infty(\hat{t}_0)$,

$$\frac{1}{\kappa}\|D(\cdot,t)\|^2_{L^2(-\infty,\delta)} + \beta_0(t+\hat{t}_0^2) \geq I(D_0)\big/[1-t_\infty^{-1}(\hat{t}_0)t]^{1/\eta} \tag{3.2.45}$$

and the conclusions of Theorem 3.1 again apply.

Remark: For the three situations examined to this point, it should be evident that up to the maximal time of existence, $t_{\max}$, of a classical solution to the initial value problem (3.2.5), (3.2.6), where in each case $t_{\max}$ is less than or equal to the relevant time t_∞, the respective growth estimates (3.2.27), with $G(t)$ given by (3.2.32), (3.2.41), and (3.2.45), are valid.

Our final results in this section concern themselves with the situation in which the initial energy $\mathcal{E}(0) > 0$; one such result is embodied in the statement of

Theorem 3.2 ([24]): Under the assumptions which prevail in Lemma 3.2, as well as (3.2.2)-(3.2.4) relative to $\mathsf{E}(\cdot)$, if $\mathcal{E}(0) > 0$, $J(D_0, D_1) > 0$, with

$$J^2(D_0, D_1) > 8\mathcal{E}(0)I(D_0) \tag{3.2.46}$$

we cannot have, for the initial-value problem (3.2.5), (3.2.6), $t_{\max} = \infty$ with $\delta < \infty$, i.e., the implication expressed by (3.2.7) is valid with respect to classical solutions of the initial-value problem.

Proof: We assume that $t_{\max} = \infty$ with $\delta < \infty$ and will derive a contradiction. As $\mathcal{E}(0) > 0$, we begin by setting $\beta = 0$ in (3.2.16) so as to obtain the differential inequality

$$F\ddot{F} - (\eta + 1)\dot{F}^2 \geq -2\nu^2(2\eta + 1)F \tag{3.2.47}$$

where we have set $\nu^2 = 2\mathcal{E}(0)$. Now, inasmuch as $\dot{F}(0) = J(D_0, D_1) > 0$, when $\beta = 0$, we clearly have

$$\dot{\overline{(F^{-\eta})}}(0) = -\eta F^{-(\eta+1)}(0)\dot{F}(0) < 0 \tag{3.2.48}$$

and thus, by continuity, $\dot{\overline{F^{-\eta}}}(t) < 0$, provided that t is sufficiently small. Suppose now that $\exists\, t = t^*$ such that

$$\dot{\overline{F^{-\eta}}}(t) < 0, t < t^*, \text{ but } \dot{\overline{F^{-\eta}}}(t^*) = 0 \tag{3.2.49}$$

We will show that (3.2.49) cannot be true. As $F(t) > 0$, $t \in [0, t^*]$, we can rewrite (3.2.47) in the form

$$\ddot{\overline{F^{-\eta}}}(t) \leq 2\eta\nu^2(2\eta + 1)F^{-(\eta+1)}, \quad 0 \leq t < t^* \tag{3.2.50}$$

Using the hypothesis (3.2.49) we now multiply (3.2.50) through by $2(\dot{\overline{F^{-\eta}}})(t)$, $0 \leq t < t^*$, so as to obtain the following estimate:

$$\frac{d}{dt}\left[\dot{\overline{(F^{-\eta})}}\right]^2 \geq 4\eta^2\nu^2\{-2\eta+1\}F^{-2(\eta+1)}\dot{F} = 4\eta^2\nu^2\frac{d}{dt}F^{-(2\eta+1)} \tag{3.2.51}$$

Integrating (3.2.51) over $[0,t]$, $t < t^*$, we find that

$$\left[\dot{\overline{(F^{-\eta})}}(t)\right]^2 - 4\eta^2\nu^2 F^{-(2\eta+1)}(t) \geq \mathcal{C}_0 \tag{3.2.52}$$

where

$$\mathcal{C}_0 = \eta^2 F^{-(2\eta+1)}(0)\left[F^{-1}(0)\dot{F}^2(0) - 4\nu^2\right] > 0 \tag{3.2.53}$$

by virtue of our hypotheses relative to the choice of the initial data (recall that $\nu^2 = 2\mathcal{E}(0)$). At this point we factor the left-hand side of the estimate (3.2.52) so as to rewrite this inequality in the form

$$\left\{\dot{\overline{(F^{-\eta})}}(t) - 2\eta\nu\left[F^{-(2\eta+1)}(t)\right]^{1/2}\right\} \times \left\{\dot{\overline{(F^{-\eta})}}(t) + 2\eta\nu\left[F^{-(2\eta+1)}(t)\right]^{1/2}\right\} \geq \mathcal{C}_0 > 0 \tag{3.2.54}$$

As $\dot{\overline{(F^{-\eta})}}(t) < 0$, for $t \in [0,t^*]$, the first bracket { } on the left-hand side of (3.2.54) is negative on $[0,t^*]$ (note that we cannot have $F(t^*) = 0$ if $\dot{\overline{(F^{-\eta})}}(t^*) = 0$); then, by virtue of (3.2.54) the second bracket { } on the left-hand side of (3.2.54) is also negative on $[0,t^*]$, i.e.,

$$\dot{\overline{(F^{-\eta})}}(t^*) < -2\eta\nu\left[F^{-(2\gamma+1)}(t^*)\right]^{1/2} \leq 0 \tag{3.2.55}$$

so that (3.2.49) cannot hold and $\dot{\overline{F^{-\eta}}}(t) < 0$ for $0 \leq t \leq t_{\max}$, i.e., for $0 \leq t \leq t_{\max}$

$$\left[-\eta F^{-(\eta+1)}\dot{F}\right]^2 \geq \mathcal{C}_0 + 4\eta^2\nu^2 F^{-(2\eta+1)}(t) \tag{3.2.56}$$

As $-\eta F^{-(\eta+1)}\dot{F} < 0$, $0 \leq t \leq t_{\max}$, if we extract the square root on both sides of (3.2.56), we find that

$$\left|-\eta F^{-(\eta+1)}\dot{F}\right| \geq \left[\mathcal{C}_0 + 4\eta^2\nu^2 F^{-(2\eta+1)}\right]^{1/2} \tag{3.2.57}$$

which can be rewritten in the form

$$\dot{F}(t) \geq (4\nu^2 F(t) + \mathcal{C}_0\eta^{-2}F^{2(\eta+1)}(t))^{1/2} \tag{3.2.58}$$

for $0 \le t \le t_{\max}$. Integration of (3.2.58) now produces the result

$$\int_{F(0)}^{\infty} \frac{dy}{(4\nu^2 y + C_0\eta^{-2}y^{2(\eta+1)})^{1/2}} \ge t_{\max} \tag{3.2.59}$$

and the estimate (3.2.59) implies a finite time of existence for a classical solution of the initial-value problem, if $\delta < \infty$, because the integral on the left-hand side of (3.2.59) is convergent. The proof of Theorem 3.2 is now complete. □

Our last result in this section for the initial-value problem (3.2.5), (3.2.6), which makes use of the differential inequality type arguments introduced in § 3.1, is a growth estimate for classical solutions which is valid for $0 \le t < t_{\max}$; this result establishes that under the conditions on the initial data stated in Theorem 3.2, i.e., $\mathcal{E}(0) > 0$, $J(D_0, D_1) > 0$, and (3.2.46), $\|D(\cdot), t)\|_{L^2(-\infty,\delta)}$ must grow quadratically in time. Of course, given the hypotheses of Theorem 3.2, we have $t_{\max} < \infty$ if $\delta < \infty$, and, therefore, $\delta = \sup_{[0,t_{\max}]} \delta(t)$. Specifically we may state the following:

Theorem 3.3 ([24]): Under the hypotheses which prevail in Theorem 3.2, if $\delta < \infty$, then not only is $t_{\max} < \infty$, but for $t \in [0, t_{\max})$

$$\begin{aligned} 4\mu_0\delta^2 \|D(\cdot,t)\|^2_{L^2(-\infty,\delta)} &\ge I(D_0) \\ &+ 2^{3/2}\sqrt{\mathcal{E}(0)I(D_0)}t + 2\mathcal{E}(0)t^2 \end{aligned} \tag{3.2.60}$$

where $\delta = \sup_{[0,t_{\max})} \delta(t)$.

Proof: We begin, once again, with (3.2.47); our hypotheses imply that $\dot{F}(0) > 0$ and, thus, $\exists\, \theta > 0$ such that $\dot{F}(t) > 0$, $t \in [0, \theta)$. Multiplying (3.2.47) through by

$$-\eta\dot{\overline{(F^{-\eta})}}(t)\ddot{\overline{(F^{-(\eta+2)})}}(t),\ t \in [0,\theta)$$

and then integrating over $[0, t]$, $t < \theta$, we obtain

$$\begin{aligned} &\left[\dot{\overline{(F^{-\eta})}}(t)\right]^2 - 4\eta^2\nu^2 F^{-(2\eta+1)}(t) \\ &\ge \left[\dot{\overline{(F^{-\eta})}}(0)\right]^2 - 4\eta^2\nu^2 F^{-(2\eta+1)}(0) > 0 \end{aligned} \tag{3.2.61}$$

where the last inequality follows from the definition of $F(t)$ (we have set $\beta = 0$) and the hypotheses relative to the initial datum D_0 and D_1. We now factor both sides of the inequality (3.2.61), writing it in the form

$$\left[\overline{(F^{-\eta})}^{\cdot}(t) - 2\eta\nu F^{-(\eta+1/2)}(t)\right]\left[\overline{(F^{-\eta})}^{\cdot}(t) + 2\eta\nu F^{-(\eta+1/2)}(t)\right]$$
$$\geq \left[\overline{(F^{-\eta})}^{\cdot}(0) - 2\eta\nu F^{-(\eta+1/2)}(0)\right] \tag{3.2.62}$$
$$\times \left[\overline{(F^{-\eta})}^{\cdot}(0) + 2\eta\nu F^{-(\eta+1/2)}(0)\right]$$

Therefore, inasmuch as

$$\overline{(F^{-\eta})}^{\cdot}(t) = -\eta F^{-(\eta+1)}(t)\dot{F}(t) < 0, \tag{3.2.63}$$

for $0 \leq t < \theta$, it follows from (3.2.62) and our hypotheses relative to the initial data that, in fact,

$$\overline{(F^{-\eta})}^{\cdot}(t) < -2\eta\nu F^{-(\eta+1/2)}(t) \tag{3.2.64}$$

for $0 \leq t < \theta$. Combining (3.2.63), (3.2.64) we find that we cannot have $\dot{F}(\theta) = 0$ for any $\theta > 0$ once our hypotheses imply that $\dot{F}(0) > 0$. As $\dot{F}(t) > 0$, $0 \leq t < t_{\max}$, the differential inequality (3.2.64) also holds for $0 \leq t < t_{\max}$, and integration of (3.2.64) now yields the lower bound

$$F(t) \geq (\nu t + F^{1/2}(0))^2, \qquad 0 \leq t < t_{\max} \tag{3.2.65}$$

The stated quadratic growth estimate (3.2.60) is now a direct consequence of (3.2.65), the definition of the function $F(t)$, and the estimate (3.2.34) of Theorem 3.1. □

Before departing this section it may be worthwhile to note that the same techniques as those applied in this section to study the behavior of classical solutions to the initial-value problem (3.2.5), (3.2.6) on $[0, t_{\max})$ can (with appropriate hypotheses relative to the nonlinear damping term $\Sigma(D)_t$) be employed to study the behavior of initial-value problems for the nonlinear wave equation (2.6.21), although the author is not aware of any work along these lines.

3.3 Growth Estimates for Wave-Dielectric Interactions: Riemann Invariants

In our work in § 2.7 we considered the propagation of a plane wave pulse into a nonlinear dielectric under the assumption that the dielectric was not a perfect nonconductor and that the conduction vector $\boldsymbol{J}$ was given by a simple nonlinear Ohm's law; under these conditions it was demonstrated that singularities will form in the propagating electromagnetic wave, provided the initial gradients of the electromagnetic fields in the wave were positive and, pointwise, sufficiently large. The analysis of Chapter 2 pointed, albeit tentatively, to the fact that nonlinear conduction currents, in a nonlinear dielectric, may provide a natural dissipative mechanism in the plane wave-dielectric interaction problem; in this section we will take this line of thought one step further along the road to a global existence theorem (for smooth solutions to the initial-value problem associated with (2.6.21)) by deriving a lower bound for the life span of a classical solution in the case of periodic initial data. Our results in this section will be based on a Riemann invariants argument (akin to the analysis of § 2.7) as opposed to the differential inequality arguments used in the previous section for the undamped nonlinear problem.

In many ways, our work in this section is at least partially in the spirit of the recent body of work [82], [91], [84], [80], [83] on almost global existence for nonlinear wave equations where lower bounds are established for the life span of classical solutions to hyperbolic equations in three space dimensions. As noted in Bloom [20], some of the techniques employed in [82], [91], [84], [80], [83] may very well be applicable (or susceptible to extension) to classical solutions of initial-value problems for wave-dielectric interactions in various three dimensional situations, but this is another matter which remains to be addressed. The Riemann invariants argument which we present below appears to be self-limiting in the sense that it does not seem to be extendable in such a way so as to yield a global existence theorem for small data; while this may be formally accomplished, the conditions which would have to be imposed on the constitutive relation for $\boldsymbol{J}$ appear to be overly restrictive and, very likely, physically unrealizable. This barrier to the use of a Riemann invariants approach to obtain a global existence theorem for the wave-dielectric interaction problem, for the case of sufficiently small data, is not present in the corresponding problem for the nonlinear transmission line which will be studied in the next chapter; indeed, such arguments have been used successfully by both the author [19] and J. Wang [179]. The desired global existence theorem for the initial-value problem associated with the nonlinearly damped nonlinear wave equation (2.6.21) will follow in § 3.5 as a consequence of an energy argument.

The model to which we now turn is, in essence, the same one studied in § 2.7; the relevant quasilinear system for (D, B) is once again (2.6.16) and, if we again introduce the Riemann invariants defined by (2.7.3), this system assumes the form (2.7.48)-(2.7.50) where $\Phi(\kappa)$, $\kappa \in R^1$, is defined by (2.7.53), (2.7.54). We again define $\hat{\lambda}$ as in (2.7.59) and note that the characteristic curves in the x, t plane are defined by the solutions of the initial-value problems (2.7.60) so that along the characteristics our quasilinear system, in Riemann invariants form, reduces to the pair of nonlinear ordinary differential equations (2.7.61). We further assume, as in § 2.7, that associated with our system we have initial data of the form (2.7.62), in terms of the Riemann invariants w and z, with $w_0(\cdot)$ and $z_0(\cdot)$ of class C^1 on R^1 and periodic.

For the remainder of this section, the particular hypotheses which will apply to the constitutive functions $\mathsf{E}(\cdot)$ and $\tilde{\sigma}(\cdot)$ are as follows: we assume that $\mathsf{E}(\cdot) \in C^3(R^1)$ with $\mathsf{E}(0) = 0$ and $\mathsf{E}'(\zeta) \geq \epsilon_0$, $\forall\, \zeta \in R^1$, and some $\epsilon_0 > 0$, that $\tilde{\sigma}(\cdot) \in C^2(R^1)$, with $\tilde{\sigma} > 0$, and that

(i) For $j = 2, 3$, $\exists\, k_j > 0$ such that

$$\sup_{\zeta \in R^1} \left[|\mathsf{E}^{(j)}(\zeta)| \Big/ \mathsf{E}'(\zeta)^{1/2} \right] \leq k_j \tag{3.3.1}$$

(ii) For $j = 0, 1, 2$, $\exists\, K_j > 0$ such that

$$\sup_{\zeta \in R^1} \left| \frac{d^j}{d\zeta^j}(\zeta \tilde{\sigma}(\zeta)) \right| \leq K_j \tag{3.3.2}$$

(iii) For some $\sigma^* > 0$,

$$\sup_{\zeta \in R^1} \left| \mathsf{E}'^2(\zeta) \frac{d^2}{d\chi^2} \left[\chi \tilde{\sigma}(\chi) \right]_{\chi = E(\zeta)} \right| \leq \sigma^* \tag{3.3.3}$$

Given the (apparently) restrictive nature of hypotheses (i)-(iii) above, it is almost incumbent upon us to present a simple example of a physically reasonable set of constitutive hypotheses for $\mathsf{E}(\cdot)$ and $\tilde{\sigma}(\cdot)$ which are consistent with these restrictions.

Example: Consider the constitutive relations for $\mathsf{E}(\cdot)$ and $\tilde{\sigma}(\cdot)$ which are given, respectively, by

$$\mathsf{E}(\zeta) = (\epsilon_0 + \epsilon_2 \zeta^2)\zeta, \qquad \tilde{\sigma}(\zeta) = \sigma_0 e^{-\zeta^2}, \qquad \zeta \in R^1 \tag{3.3.4}$$

with ϵ_0, ϵ_2 and σ_0 all positive. Clearly $\mathsf{E}(\cdot) \in C^\infty(R^1)$, $\mathsf{E}(0) = 0$, and $\mathsf{E}'(\zeta) = \epsilon_0 + 3\epsilon_2\zeta^2 \geq \epsilon_0 > 0$, $\forall \zeta \in R^1$. Also, $\tilde{\sigma}(\cdot) \in C^\infty(R^1)$ with $\tilde{\sigma}(0) = \sigma_0 > 0$. Now,

$$\begin{aligned} |\mathsf{E}''(\zeta)| / \mathsf{E}'(\zeta)^{-1/2} &= 6\epsilon_2|\zeta| / \sqrt{\epsilon_0 + 3\epsilon_2\zeta^2} \\ &\equiv k_2(\zeta) \end{aligned} \tag{3.3.5a}$$

and

$$\begin{aligned} |\mathsf{E}'''(\zeta)| / \mathsf{E}'(\zeta)^{-1/2} &= 6\epsilon_2 / \sqrt{\epsilon_0 + 3\epsilon_2\zeta^2} \\ &\equiv k_3(\zeta) \end{aligned} \tag{3.3.5b}$$

Clearly, $k_2(\zeta) \to 0$, as $|\zeta| \to 0$, $k_2(\zeta) \to 2\sqrt{3\epsilon_2}$, as $|\zeta| \to \infty$, while $k_3(\zeta) \to 6\epsilon_2/\sqrt{\epsilon_0}$, as $|\zeta| \to 0$, and $k_3(\zeta) \to 0$, as $|\zeta| \to \infty$; the existence of the $k_j > 0$, $j = 2, 3$ in (3.3.1) is, thus, trivial to establish. From (3.3.4) we also have that

$$\zeta\tilde{\sigma}(\zeta) = \sigma_0\zeta e^{-\zeta^2} \equiv K_0(\zeta) \tag{3.3.6}$$

$$\frac{d}{d\zeta}[\zeta\tilde{\sigma}(\zeta)] = \sigma_0 e^{-\zeta^2}(1 - 2\zeta^2) \equiv K_1(\zeta) \tag{3.3.7}$$

and

$$\frac{d^2}{d\zeta^2}[\zeta\tilde{\sigma}(\zeta)] = -\sigma_0 e^{-\zeta^2}(6\zeta - 4\zeta^3) \equiv K_2(\zeta) \tag{3.3.8}$$

We therefore observe that $K_0(\zeta), K_1(\zeta)$, and $K_2(\zeta)$ are well-behaved as both $|\zeta| \to 0$, and $|\zeta| \to \infty$, nor can $K_0(\zeta), K_1(\zeta)$ or $K_2(\zeta)$ blow up for any finite value of ζ; thus, (3.3.2) is clearly satisfied for some positive constants $K_j, j = 0, 1, 2$ which we do not bother here to compute exactly. Finally, a straightforward calculation based on the pair of constitutive hypotheses (3.3.4) yields

$$\begin{aligned} &\mathsf{E}'^2(\zeta)\frac{d^2}{d\chi^2}[\chi\tilde{\sigma}(\chi)]\Big|_{\chi=\mathsf{E}(\zeta)} \\ &= -2\sigma_0 P(\zeta)\exp(-Q(\zeta)) \end{aligned} \tag{3.3.9}$$

where

$$\begin{aligned} P(\zeta) &\equiv \zeta(\epsilon_0 + \epsilon_2\zeta^2)(\epsilon_0 + 3\epsilon_2\zeta^2)^2 \\ &\qquad \times[3 - 2\zeta^2(\epsilon_0 + \epsilon_2\zeta^2)^2] \end{aligned} \tag{3.3.10a}$$

$$Q(\zeta) \equiv \zeta(\epsilon_0 + \epsilon_2\zeta^2)^2 \tag{3.3.10b}$$

A careful examination of (3.3.10a)-(3.3.10b) now shows that as $|\zeta| \to \infty$, $P(\zeta)\exp(-Q(\zeta)) \to 0$, while for $|\zeta| \to 0$, we also have $P(\zeta)\exp(-Q(\zeta)) \to 0$. As $P(\zeta)\exp(-Q(\zeta))$ is finite for each ζ, $0 < |\zeta| < \infty$, it follows that $\exists\, \sigma^* > 0$ such that (3.3.3) holds for $\mathsf{E}(\cdot)$, $\tilde{\sigma}(\cdot)$ given by (3.3.4). Therefore, the constitutive relations (3.3.4) are, indeed, consistent with the hypotheses (3.3.1)-(3.3.3).

The most important implication of the bounds expressed by (3.3.1)-(3.3.3) is now expressed by the following:

Lemma 3.3 With $\Phi(\cdot), \tilde{\lambda}(\cdot)$ as defined by (2.7.53), (2.7.54), and (2.7.59), and under the assumptions (3.3.1)-(3.3.3) we have

(i) $\Phi(0) = 0, \Phi'(0) < 0$, and $\sup_{\zeta \in R^1} |\Phi'(\zeta)| \le \bar{\Phi}$, for some $\bar{\Phi} > 0$

(ii) $\sup_{\zeta \in R^1} |\Phi''(\zeta)| \le \Phi^*$, for some $\Phi^* > 0$

(iii) $\exists$ constants $\lambda_* > 0, \lambda^* > 0$ such that $\forall \zeta \in R^1$:

$$|\hat{\lambda}(\zeta)| \ge \lambda_* \text{ and } \left|\frac{\hat{\lambda}'(\zeta)}{\hat{\lambda}(\zeta)}\right| \le \lambda^*$$

Proof: By virtue of the definition of $\Phi(\cdot)$ we have

$$\Phi(0) = -\frac{1}{\sqrt{\mu_0}} \cdot \sqrt{\mathsf{E}'(\hat{\eta}^{-1}(0))\Sigma(\hat{\eta}^{-1}(0))} \tag{3.3.11}$$

with $\hat{\eta}(\zeta) \equiv 2\int_0^\zeta \sqrt{\mathsf{E}'(\lambda)/\mu_0}\, d\lambda$. Thus $\hat{\eta}(\zeta) = 0 \Rightarrow \zeta = 0$ so that $\hat{\eta}^{-1}(0) = 0$. Therefore

$$\begin{aligned} \Phi(0) &= -\frac{1}{\sqrt{\mu_0}} \cdot \sqrt{\mathsf{E}'(0)\Sigma(0)} \\ &= -\frac{1}{\sqrt{\mu_0}} \cdot \sqrt{\mathsf{E}'(0)\tilde{\sigma}(\mathsf{E}(0))\mathsf{E}(0)} \\ &= 0 \end{aligned}$$

by virtue of the definition of $\Sigma(\cdot)$, i.e., (2.6.17) and our assumption that $\mathsf{E}(0) = 0$. Next, we note that we may, in view of our definitions of $\Phi(\cdot)$ and $\hat{\lambda}(\cdot)$, write $\Phi(\cdot)$ in the form

$$\Phi(\zeta) = \hat{\lambda}(\zeta)\Sigma(\hat{\eta}^{-1}(\zeta)), \qquad \zeta \in R^1 \tag{3.3.12}$$

and, therefore,

$$\begin{aligned}\Phi'(0) &= \hat{\lambda}'(0)\Sigma(\hat{\eta}^{-1}(0)) + \hat{\lambda}(0)\frac{d}{d\zeta}\Sigma(\hat{\eta}^{-1}(\zeta))\Big|_{\zeta=0}\\ &= \hat{\lambda}(0)\frac{d}{d\zeta}\Sigma(\hat{\eta}^{-1}(\zeta))\Big|_{\zeta=0}\end{aligned}$$

where we have again used the fact that $\Sigma(0) = 0$ when $\mathsf{E}(0) = 0$. Now $\hat{\lambda}(0) = -\sqrt{\mathsf{E}'(0)/\mu_0}$ while use of the easily verified relation

$$\frac{d}{d\zeta}\hat{\eta}^{-1}(\zeta) = \frac{1}{2}\sqrt{\frac{\mu_0}{\mathsf{E}'(\hat{\eta}^{-1}(\zeta))}} \tag{3.3.13}$$

produces

$$\frac{d}{d\zeta}\Sigma(\hat{\eta}^{-1}(\zeta)) = \frac{1}{2}\mu_0^{1/2}\Sigma'(\hat{\eta}^{-1}(\zeta))\Big/\sqrt{\mathsf{E}'(\hat{\eta}^{-1}(\zeta))} \tag{3.3.14}$$

Combining the results above we find that

$$\Phi'(0) = \frac{1}{2}\sqrt{\mu_0}\hat{\lambda}(0)\frac{\Sigma'(0)}{\sqrt{\mathsf{E}'(0)}} \equiv -\frac{1}{2}\Sigma'(0) \tag{3.3.15}$$

However, $\Sigma'(0) = \tilde{\sigma}(0)\mathsf{E}'(0)$ so, as $\tilde{\sigma}(0) > 0$, we have

$$\Phi'(0) = -\frac{1}{2}\tilde{\sigma}(0)\mathsf{E}'(0) < 0 \tag{3.3.16}$$

thus establishing the second statement in part (i) of the lemma. Now, directly from the definition of $\hat{\lambda}(\zeta)$, and the fact that $\mathsf{E}'(\zeta) \geq \epsilon_0$, $\forall\zeta \in R^1$, for some $\epsilon_0 > 0$, we find that

$$|\hat{\lambda}(\zeta)| = \sqrt{\frac{\mathsf{E}'(\hat{\eta}^{-1}(\zeta))}{\mu_0}} \geq \sqrt{\frac{\epsilon_0}{\mu_0}} \equiv \lambda_*, \quad \forall\zeta \in R^1 \tag{3.3.17a}$$

so the first statement in part (iii) of the lemma is trivially verified. Moreover, $\forall\,\zeta \in R^1$,

$$\begin{aligned}\hat{\lambda}'(\zeta) &= -\frac{1}{2}\mu_0^{-1/2}\left[(\mathsf{E}'(\kappa))^{-1/2}\mathsf{E}''(\kappa)\right]_{\kappa=\hat{\eta}^{-1}(\zeta)}\frac{d\hat{\eta}^{-1}(\zeta)}{d\zeta}\\ &= -\frac{1}{4}\mathsf{E}'(\hat{\eta}^{-1}(\zeta))^{-1}\mathsf{E}''(\hat{\eta}^{-1}(\zeta))\end{aligned} \tag{3.3.17b}$$

where we have used (3.3.13). Using the hypothesis (3.3.1), and the definition of λ_*, we now compute that $\forall\,\zeta \in R^1$,

$$\begin{aligned}\left|\frac{\hat{\lambda}'(\zeta)}{\hat{\lambda}(\zeta)}\right| &= \frac{1}{4}\mu_0^{1/2}\frac{|\mathsf{E}''(\hat{\eta}^{-1}(\zeta))|}{\mathsf{E}'(\hat{\eta}^{-1}(\zeta))^{3/2}}\\ &\leq \frac{1}{4}\sqrt{\frac{\mu_0}{\epsilon_0}}\frac{|\mathsf{E}''(\hat{\eta}^{-1}(\zeta))|}{\mathsf{E}'(\hat{\eta}^{-1}(\zeta))^{1/2}}\\ &\leq \frac{1}{4\lambda_*}\cdot k_2 \equiv \lambda^*\end{aligned} \tag{3.3.18}$$

In deriving the bound (3.3.18) we have, of course, again used the assumption that $\mathsf{E}'(\zeta) \geq \epsilon_0 > 0$, $\forall \zeta \in R^1$; the estimate (3.3.18) serves to establish the validity of the second statement in part (iii) of the lemma, so we now turn our attention to the content of part (ii) of the lemma and to the third statement in part (i) of the lemma.

To simplify matters somewhat, we begin by setting

$$\Phi^*(\kappa) = -\frac{1}{\sqrt{\mu_0}} \cdot \sqrt{\mathsf{E}'(\kappa)}\Sigma(\kappa), \qquad \kappa \in R^1 \tag{3.3.19}$$

so that $\Phi(\zeta) = \Phi^*(\hat{\eta}^{-1}(\zeta))$. Therefore,

$$\Phi'(\zeta) = \left.\frac{d\Phi^*}{d\kappa}\right|_{\kappa=\hat{\eta}^{-1}(\zeta)} \cdot \frac{d\hat{\eta}^{-1}(\zeta)}{d\zeta} \tag{3.3.20}$$

while

$$\begin{aligned} \Phi''(\zeta) &= \left.\frac{d^2\Phi^*}{d\kappa^2}\right|_{\kappa=\hat{\eta}^{-1}(\zeta)} \left(\frac{d\hat{\eta}^{-1}(\zeta)}{d\zeta}\right)^2 \\ &\quad + \left.\frac{d\Phi^*}{d\kappa}\right|_{\kappa=\hat{\eta}^{-1}(\zeta)} \frac{d^2\hat{\eta}^{-1}(\zeta)}{d\zeta^2} \end{aligned} \tag{3.3.21}$$

In view of (3.3.13), $\left(\frac{d\hat{\eta}^{-1}(\zeta)}{d\zeta}\right)^2 = \frac{1}{4}\mu_0 \mathsf{E}'(\hat{\eta}^{-1}(\zeta))^{-1}$, while

$$\begin{aligned} \frac{d^2\hat{\eta}^{-1}}{d\zeta^2} &= -\frac{1}{4}\mu_0^{1/2}\frac{\mathsf{E}''(\hat{\eta}^{-1}(\zeta))}{\mathsf{E}'(\hat{\eta}^{-1}(\zeta))^{3/2}} \cdot \frac{d\hat{\eta}^{-1}}{d\zeta} \\ &= -\frac{1}{8}\mu_0 \frac{\mathsf{E}''(\hat{\eta}^{-1}(\zeta))}{\mathsf{E}'^2(\hat{\eta}^{-1}(\zeta))} \end{aligned} \tag{3.3.22}$$

Combining the relations from (3.3.19) onward, and employing the hypotheses (3.3.1)-(3.3.3), we find that $\forall \zeta \in R^1$

$$\begin{aligned} \frac{8}{\sqrt{\mu_0}}|\Phi''(\zeta)| &\leq K_1\Bigg(\frac{1}{\mathsf{E}'(\hat{\eta}^{-1}(\zeta))^{3/2}}\left[\sup_{\zeta\in R^1}\frac{|\mathsf{E}''(\zeta)|}{\mathsf{E}'(\zeta)^{1/2}}\right]^2 \\ &\quad + \frac{1}{\mathsf{E}'(\hat{\eta}^{-1}(\zeta))}\left[\sup_{\zeta\in R^1}\frac{|\mathsf{E}''(\zeta)|}{\mathsf{E}'(\zeta)^{1/2}}\right]\Bigg) \\ &\quad + 3K_2 \sup_{\zeta\in R^1}\frac{|\mathsf{E}''(\zeta)|}{\mathsf{E}'(\zeta)^{1/2}} \\ &\quad + \frac{2}{\mathsf{E}'(\hat{\eta}^{-1}(\zeta))^{1/2}} \sup_{\zeta\in R^1}\left|\mathsf{E}'(\zeta)^2 \left.\frac{d^2}{d\chi^2}(\chi\tilde{\sigma}(\chi))\right|_{\chi=\mathsf{E}(\zeta)}\right| \\ &\leq K_1\left(\frac{k_2^2}{\epsilon_0^{3/2}} + \frac{k_3}{\epsilon_0}\right) + 3K_2k_2 + \frac{1}{\epsilon_0^{1/2}}\sigma^* \end{aligned}$$

or, that

$$\sup_{\zeta\in R^1} |\Phi''(\zeta)| \le \Phi^* \tag{3.3.23a}$$

with

$$\Phi^* \equiv \frac{\sqrt{\mu_0}}{8}\left[K_1\left(\frac{k_2^2}{\epsilon_0^{3/2}} + \frac{k_3}{\epsilon_0}\right) + 3K_2k_2 + \frac{2}{\epsilon_0^{1/2}}\sigma^*\right] \tag{3.3.23b}$$

and part (ii) of the lemma has been established. Also, by virtue of (3.3.19), (3.3.20), (3.3.13), and (2.6.17), we easily compute that

$$\begin{aligned} -\mu_0^{1/2}\Phi'(\zeta) &= \frac{d}{d\kappa}\left[\mathsf{E}'(\kappa)^{1/2}\Sigma(\kappa)\right]_{\kappa=\hat{\eta}^{-1}(\zeta)} \cdot \frac{d\hat{\eta}^{-1}(\zeta)}{d\zeta} \\ &= \frac{1}{4}\mu_0^{1/2}\mathsf{E}'(\hat{\eta}^{-1}(\zeta))^{-1}\tilde{\sigma}(\mathsf{E}(\hat{\eta}^{-1}(\zeta))\mathsf{E}(\hat{\eta}^{-1}(\zeta)) \\ &\quad + \frac{1}{2}\mu_0^{1/2}\frac{d}{d\kappa}\{\tilde{\sigma}(\mathsf{E}(\kappa))\mathsf{E}(\kappa)\}_{\kappa=\hat{\eta}^{-1}(\zeta)} \end{aligned}$$

and, therefore, $\forall\zeta \in R^1$

$$|\Phi'(\zeta)| \le \frac{1}{4\epsilon_0}K_0 + \frac{1}{2}K_1 \equiv \bar{\Phi} \tag{3.3.24}$$

thus establishing the third statement of part (i) of the lemma. □

We now want to rewrite the system (2.7.48) in (what will turn out to be) a more advantageous form which is suggested by Lemma 3.3. By virtue of the smoothness hypotheses relative to $\mathsf{E}(\cdot)$ and $\tilde{\sigma}(\cdot)$, the definition of $\Phi(\cdot)$, and part (i) of Lemma 3.3, we have, for any $\zeta \in R^1$, with $|\zeta|$ sufficiently small,

$$\Phi(\zeta) = \frac{1}{2}\alpha\zeta + \frac{1}{2}\Phi''(\theta(\zeta))\zeta^2 \tag{3.3.25}$$

where $\alpha = 2\Phi'(0) < 0$ while $0 < \theta(\zeta) < \zeta$, for $\zeta > 0$ and $\zeta < \theta(\zeta) < 0$, for $\zeta < 0$. We now set

$$\Psi(\zeta) = \frac{1}{2}\Phi''(\theta(\zeta))\zeta^2, \qquad \zeta \in R^1 \tag{3.3.26}$$

and rewrite our system (2.7.48) in Riemann invariant form as

$$\begin{cases} w' &= \frac{\alpha}{2}(w-z) + \Psi(w-z) \\ \grave{z} &= -\frac{\alpha}{2}(w-z) - \Psi(w-z) \end{cases} \tag{3.3.27}$$

with initial data of the form (2.7.62). We also make a useful observation at this juncture: If we set

$$\hat{\Phi}(\zeta) = \Psi'(\zeta), \qquad \zeta \in R^1 \tag{3.3.28}$$

then from (3.3.25), (3.3.26)

$$\hat{\Psi}(\zeta) = \Phi'(\zeta) - \Phi'(0) \equiv \Phi''(\hat{\theta}(\zeta))\zeta \tag{3.3.29}$$

with $0 < \hat{\theta}(\zeta) < \zeta$, for $\zeta > 0$, and $\zeta < \hat{\theta}(\zeta) < 0$, for $\zeta < 0$. Then, by part (ii) of Lemma 3.3

$$|\hat{\Psi}(\zeta)| \leq \Phi^{*}|\zeta|, \qquad \forall \zeta \in R^1 \tag{3.3.30}$$

With most of the required machinery in place, we begin our analysis of the system (3.3.27) by stating the following lemma, which is a direct consequence of extant results on local existence of solutions for initial-value problems associated with quasilinear symmetric hyperbolic systems, i.e., the results obtained, e.g., in [51], [88], [89] may be applied so as to conclude the following:

Lemma 3.4 Let $w_0(\cdot), z_0(\cdot)$ be periodic and of class C^1 with $\mathsf{E}(\cdot) \in C^3(R^1), \tilde{\sigma}(\cdot) \in C^2(R^1)$, and $\mathsf{E}'(\zeta) \geq \epsilon_0 > 0$, $\forall \zeta \in R^1$. Then $\exists t_\infty > 0$ such that the initial-value problem for (3.3.27) has a unique smooth solution $\{w, z\} \in C^1(R^1 \times [0, t_\infty))$. Furthermore, if $t_\infty < \infty$, then for some $x \in R^1$

$$\lim_{t \to t_\infty^-} [|w_x(x,t)| + |z_x(x,t)|] = +\infty \tag{3.3.31}$$

We now seek to derive a lower bound for the life span of a classical, i.e., C^1, solution to the initial-value problem for (3.3.27), with associated periodic initial data, and to relate that lower bound directly to the growth of the initial data in some appropriate fashion; the result that will be obtained (Theorem 3.4) is, in many ways, much more illuminating than the estimate which may be gleaned from (2.7.109), which was derived in the course of establishing singularity formation in the solution of the wave-dielectric interaction problem in § 2.7. We begin by noting that in view of Lemmas 2.2 and 3.4, and (3.3.24), we may state the following:

Lemma 3.5 Let $(w(x,t), z(x,t))$ be the unique local solution of the initial-value problem for (3.3.27), with associated periodic initial data, on $R^1 \times [0, t_\infty)$. Then, under hypotheses (3.3.1)-(3.3.3), relative to the constitutive functions $\mathsf{E}(\cdot)$ and $\tilde{\sigma}(\cdot)$, for each $x \in R^1$, and $t \in [0, t_\infty)$

$$|w(x,t)| + |z(x,t)| \leq (|w_0| + |z_0|)\, e^{\Gamma t_\infty} \tag{3.3.32}$$

where, as previously, $|w_0| = \sup_x |w_0(x)|$, $|z_0| = \sup_x |z_0(x)|$, and

$$\Gamma = \frac{K_0}{2\epsilon_0} + K_1 \tag{3.3.33}$$

Remarks: For the example defined by (3.3.4) it is easily checked that

$$\alpha = -\sigma_0\epsilon_0 \text{ and } \Gamma = \frac{e^{-1/2}}{2\sqrt{2}}\left(\frac{\sigma_0}{\epsilon_0}\right) + 2e^{-3/2}\sigma_0 \tag{3.3.34}$$

We are now in a position to state the main result of this section, namely,

Theorem 3.4 ([20]) Let $(w(x,t), z(x,t))$ be the unique local solution of the initial-value problem for (3.3.27), with associated periodic initial data, on $R^1 \times [0, t_\infty)$. Set

$$\begin{cases} W_0 &= \sup_x [|w_0(x)| + |z_0(x)|] \\ W_0^{\#} &= \sup_x [|w_0'(x)| + |z_0'(x)|] \end{cases} \tag{3.3.35}$$

and let $\alpha = 2\Phi'(0)$ satisfy

$$|\alpha| \geq \left(1 + \frac{2}{\lambda^*}\right)\Phi^* \tag{3.3.36}$$

with λ^*, Φ^* as in Lemma 3.3. Then either $t_\infty = \infty$ for $W_0 + W_0^{\#}$ sufficiently small or

$$t_\infty \geq \frac{1}{2\Gamma}\ln\left(\frac{1}{2\lambda^*}W_0^{-1}\right) \tag{3.3.37}$$

for all initial data with $W_0 + W_0^{\#}$ sufficiently small.

Remark: In the course of the proof we will use the assumption that if $t_\infty < \infty$, when (3.3.36) is satisfied, for all finite but arbitrarily small $W_0 + W_0^{\#}$, then

$$\lim_{W_0 + W_0^{\#} \to 0} W_0 e^{2\Gamma t_\infty} \neq (2\lambda^*)^{-1} \tag{3.3.38}$$

If (3.3.38) is (pathologically) not satisfied, then we may simply choose $\lambda^{**} > \lambda^*$ and replace (3.3.36) by

$$|\alpha| \geq \left(1 + \frac{2}{\lambda^{**}}\right)\Phi^*$$

in which case the growth estimate (3.3.7) for t_∞ holds with λ^{**} replacing λ^*. Without loss of generality, therefore, we will take (3.3.38) as being valid for the situation in which $t_\infty < \infty$ for arbitrarily small $W_0 + W_0^{\#}$.

Proof: (Theorem 3.4): We begin by differentiating the first relation in (3.3.27) with respect to x so as to obtain

$$\begin{aligned} w'_x + \hat{\lambda}_w w_x^2 + \hat{\lambda}_z w_x z_x \\ = \left(\frac{\alpha}{2} + \Psi'\right) w_x - \left(\frac{\alpha}{2} + \Psi'\right) z_x \end{aligned} \tag{3.3.39}$$

It is easily verified that both

$$w' + z' = 2\hat{\lambda} z_x \text{ and } \grave{z} - z' = -2\hat{\lambda} z_x$$

so that

$$\begin{aligned} w' - z' &= (w' - \grave{z}) + (\grave{z} - z') \\ &= \alpha(w - z) + 2\Psi - 2\hat{\lambda} z_x \end{aligned}$$

from which it follows that

$$z_x = \frac{\alpha}{2\hat{\lambda}}(w - z) + \frac{\Psi}{\hat{\lambda}} - \frac{1}{2\hat{\lambda}}(w' - z') \tag{3.3.40}$$

We now rewrite (3.3.39) in the form

$$\begin{aligned} w'_x + (\hat{\lambda}_w w_x - \frac{\alpha}{2} - \Psi) w_x \\ = (\hat{\lambda}_w w_x - \frac{\alpha}{2} - \hat{\Psi}) z_x \end{aligned} \tag{3.3.41}$$

using, in the process, the definition of $\hat{\Psi}$ and the fact that $\hat{\lambda}_z = -\hat{\lambda}_w$. Substituting for z_x in (3.3.41) from (3.3.40) we next obtain

$$\begin{aligned} & w'_x + (\hat{\lambda}_w w_x - \frac{\alpha}{2} - \hat{\Psi}) w_x \\ & = \left\{ \frac{\alpha \hat{\lambda}_w}{2\hat{\lambda}}(w - z) + \frac{\hat{\lambda}_w}{\hat{\lambda}} \Psi \right\} w_x - h' w_x \\ & - \left(\frac{\alpha}{2} + \Psi\right) \left\{ \frac{\alpha}{2\hat{\lambda}}(w - z) + \frac{\Psi}{\hat{\lambda}} - \frac{h'}{\hat{\lambda}_w} \right\} \end{aligned} \tag{3.3.42}$$

where

$$h(w - z) \equiv \int_0^{w-z} \frac{\hat{\lambda}'(\zeta)}{2\hat{\lambda}(\zeta)} \, d\zeta \tag{3.3.43}$$

so that $h' = \frac{\hat{\lambda}_w}{2\hat{\lambda}}(w' - z')$. We now make the additional definitions:

$$
\begin{aligned}
a(\alpha; w-z) &= \frac{\alpha}{2} + \hat{\Psi}(w-z) + \frac{\alpha\hat{\lambda}_w}{2\hat{\lambda}}(w-z) \\
&\quad + \frac{\hat{\lambda}_w}{\hat{\lambda}}\Psi(w-z)
\end{aligned}
\tag{3.3.44}
$$

$$
b(\alpha; w-z) = \frac{\alpha}{2} + \hat{\Psi}(w-z) \tag{3.3.45}
$$

$$
c(\alpha; w-z) = \frac{\alpha}{2\hat{\lambda}}(w-z) + \frac{\Psi(w-z)}{\hat{\lambda}} \tag{3.3.46}
$$

We note that $a = b + \hat{\lambda}_w c$ and that, in view of (3.3.44)-(3.3.46), we may rewrite (3.3.42) in the form

$$
\begin{aligned}
& w'_x + h'w_x + (\hat{\lambda}_w w_x - a)w_x \\
&= -b(c - \frac{1}{\hat{\lambda}_w}h')
\end{aligned}
$$

or, after multiplication by $\exp[h(w-z)]$, as

$$
\begin{aligned}
& (e^h w_x)' + (\hat{\lambda}_w w_x - a)e^h w_x \\
&= -b(c - \frac{1}{\hat{\lambda}_w}h')e^h
\end{aligned}
\tag{3.3.47}
$$

Setting

$$
d(\alpha; w-z) = b(\alpha; w-z)c(\alpha; w-z)\exp[h(w-z)] \tag{3.3.48}
$$

and

$$
g(\alpha; w-z) = \int_0^{w-z} \frac{b(\alpha;\zeta)f'(\zeta)}{\hat{\lambda}'(\zeta)}\,d\zeta; \quad f(\zeta) = e^{h(\zeta)}, \tag{3.3.49}
$$

we have

$$
g'(\alpha; w-z) = \frac{b(\alpha; w-z)}{\hat{\lambda}_w(w-z)}\exp[h(w-z)]h'(w-z) \tag{3.3.50}
$$

Moreover, equation (3.3.47) now becomes

$$
(e^h w_x)' + (\hat{\lambda}_w w_x - a)e^h w_x = g' - d \tag{3.3.51}
$$

It is our intention, at this point, to integrate (3.3.51) along the characteristic curve $x = x_1(t;\beta_1)$; to that end we now set

$$
\rho(t) = \left[\exp\{h(w(x,t) - z(x,t))\}\, w_x(x,t)\right]_{x = x_1(t,\beta_1)} \tag{3.3.52}
$$

$$\pi(t) = \Big[\hat{\lambda}_w(w(x,t)-z(x,t))w_x(x,t) \\ -a(\alpha; w(x,t)-z(x,t))\Big]\Big|_{x=x_1(t,\beta_1)} \tag{3.3.53}$$

$$\mu_1(t) = g(\alpha; w(x,t)-z(x,t))\Big|_{x=x_1(t,\beta_1)} \tag{3.3.54}$$

$$\mu_2(t) = -d(\alpha; w(x,t)-z(x,t))\Big|_{x=x_1(t,\beta_1)} \tag{3.3.55}$$

and

$$\mu(t) = \dot{\mu}_1(t) + \mu_2(t) \tag{3.3.56}$$

With the definitions (3.3.52)-(3.3.56), it is not difficult to see that along the characteristic $x = x_1(t,\beta_1)$ equation (3.3.51) becomes

$$\dot{\rho}(t) + \pi(t)\rho(t) = \mu(t), \qquad 0 \le t < t_\infty \tag{3.3.57}$$

If we integrate (3.3.57) along the characteristic $x = x_1(t,\beta_1)$, from $t = 0$ up to $t < t_\infty$, we then obtain

$$\rho(t) = \rho(0)\exp\left[-\int_0^t \pi(\tau)\,d\tau\right] \\ + \int_0^t \mu(s)\exp\left[-\int_0^t \pi(\tau)\,d\tau\right]\,ds \tag{3.3.58}$$

In order to be able to obtain an estimate from below for $\pi(t)$, valid for $0 \le t < t_\infty$, we first seek such an estimate (see (3.3.53)) for the quantity $-a(\alpha; w-z)$ along the characteristic $x = x_1(t,\beta_1)$. As $\alpha = 2\Phi'(0) < 0$,

$$-a(\alpha; w-z) = \frac{|\alpha|}{2}\left[1 + \frac{\hat{\lambda}_w}{\hat{\lambda}}(w-z)\right] \\ - \left(\frac{\hat{\lambda}_w}{\hat{\lambda}}\right)\Psi(w-z) - \hat{\Psi}(w-z) \tag{3.3.59}$$

However, for $0 \le t < t_\infty$,

$$\begin{aligned} \left|\left(\frac{\hat{\lambda}_w}{\hat{\lambda}}\right)\Psi(w-z)\right| &\le \frac{1}{2}\lambda^*\Phi^*(|w|+|z|)^2 \\ &\le \frac{1}{2}\lambda^*\Phi^* W_0^2\exp(2\Gamma t_\infty) \end{aligned} \tag{3.3.60}$$

and

$$|\hat{\Psi}(w-z)| \le \Phi^*(|w|+|z|) \le \Phi^* W_0 \exp(\Gamma t_\infty) \tag{3.3.61}$$

where we have used part (ii) of Lemma 3.3, (3.3.26), and (3.3.29). Assuming, without loss of generality, that $W_0 < 1$, we obtain from (3.3.60) and (3.3.61) the estimate

$$\begin{aligned} & -\left(\frac{\hat{\lambda}_w}{\hat{\lambda}}\right)\Psi(w-z) - \hat{\Psi}(w-z) \\ & \geq -\left(1+\frac{\lambda^*}{2}\right)\Phi^* W_0 \exp(2\Gamma t_\infty) \end{aligned} \tag{3.3.62}$$

Also,

$$\left|\left(\frac{\hat{\lambda}_w}{\hat{\lambda}}\right)(w-z)\right| \leq \lambda^* W_0 \exp(2\Gamma t_\infty) \tag{3.3.63}$$

so that, for $0 \leq t < t_\infty$,

$$\begin{aligned} -a(\alpha; w-z) \ \geq \ & \frac{|\alpha|}{2}\left[1 - \lambda^* W_0 \exp(2\Gamma t_\infty)\right] \\ & - \left(1+\frac{\lambda^*}{2}\right)\Phi^* W_0 \exp(2\Gamma t_\infty) \end{aligned} \tag{3.3.64}$$

In view of our hypothesis (3.3.36), however, $-\left(1+\frac{\lambda^*}{2}\right)\Phi^* \geq -|\alpha|/2\lambda^*$, so that (3.3.64) yields the lower bound

$$-a(\alpha; w-z) \geq \frac{|\alpha|}{2}\left[1 - 2\lambda^* W_0 \exp(2\Gamma t_\infty)\right] \tag{3.3.65}$$

for $0 \leq t < t_\infty$; as this last estimate clearly must hold along the characteristic $x = x_1(t;\beta_1)$, for $0 \leq t < t_\infty$, we may employ it in the definition (3.3.53) of $\pi(t)$ to obtain the lower bound

$$\begin{aligned} \pi(t) \geq \ & \hat{\lambda}_w(w(x,t) - z(x,t))w_x(x,t)\Big|_{x=x_1(t;\beta_1)} \\ & + \frac{1}{2}|\alpha|\left[1 - 2\lambda^* W_0 \exp(2\Gamma t_\infty)\right] \end{aligned} \tag{3.3.66}$$

Now, suppose that $t_\infty < \infty$, no matter how small $W_0 + W_0^\#$ is and that (3.3.37) is violated, i.e., that

$$t_\infty < \frac{1}{2\Gamma}\ln\left(\frac{1}{2\lambda^*}W_0^{-1}\right) \tag{3.3.67}$$

for arbitrarily small, but nonzero, $W_0 + W_0^\#$; we will establish a contradiction by proving that $\exists\, \epsilon > 0$ such that for $W_0 + W_0^\# < \epsilon$, the inequality (3.3.67) leads to the conclusion that $t_\infty = \infty$.

So, suppose that (3.3.67) holds for all initial data with $W_0 + W_0^\#$ sufficiently small; then for each such choice of the initial data, $\exists\, \alpha_0 > 0$ such that

$$1 - 2\lambda^* W_0 \exp(2\Gamma t_\infty) = 2\alpha_0/|\alpha|. \tag{3.3.68}$$

Using (3.3.17b), (3.3.1), and that fact that $\mathsf{E}'(\zeta) \geq \epsilon_0 > 0$, $\forall\, \zeta \in R^1$, we note that

$$\left|\hat{\lambda}_w(w - z)\right| \leq \frac{k_2}{4\sqrt{\epsilon_0}} \equiv \hat{\lambda}_\infty \tag{3.3.69}$$

Suppose now that on $R^1 \times [0, t_\infty]$

$$|w_x(x,t)| \leq \inf_{(w_0, z_0)} \left(\frac{\alpha_0}{2}\right) \cdot \frac{1}{\hat{\lambda}_\infty}, \tag{3.3.70}$$

a bound which will be verified below; by virtue of (3.3.68) and (3.3.38) we observe that $\inf\limits_{(w_0,z_0)} \alpha_0 > 0$. Therefore, by (3.3.68), (3.3.69), (3.3.70) and (3.3.66), we have $\pi(t) \geq \frac{1}{2}\alpha_0 \equiv \beta_0$, $0 \leq t < t_\infty$; employing this lower bound in (3.3.58) we then find that

$$\begin{aligned} \rho(t) \leq \rho(0)e^{-\beta_0 t} &+ \int_0^t \dot{\mu}_1(s) \exp[-\beta_0(t-s)]\, ds \\ &+ \int_0^t \mu_2(s) \exp[-\beta_0(t-s)]\, ds \end{aligned} \tag{3.3.71}$$

where we have also employed (3.3.56). Integrating the first integral in (3.3.71) by parts, we are led to the estimate

$$\begin{aligned} |\rho(t)| \leq \Big[|\rho(0)| + |\mu_1(0)|\Big] e^{-\beta_0 t} + |\mu_1(t)| \\ + \beta_0 e^{-\beta_0 t} \int_0^t |\mu_1(s)| e^{\beta_0 s}\, ds + e^{-\beta_0 t} \int_0^t |\mu_2(s)| e^{\beta_0 s}\, ds \end{aligned} \tag{3.3.72}$$

Now,

$$\rho(0) = \exp\left\{h[w_0(\beta_1) - z_0(\beta_1)] w_0'(\beta_1)\right\} \tag{3.3.73}$$

and by virtue of (3.3.43), and part (iii) of Lemma 3.3, we have

$$|h[w_0(\beta_1) - z_0(\beta_1)]| \leq \frac{1}{2}\lambda^* \left[|w_0(\beta_1)| + |z_0(\beta_1)|\right]$$

so that

$$|h[w_0(\beta_1) - z_0(\beta_1)]| \leq \frac{1}{2}\lambda^* W_0 \tag{3.3.74}$$

Therefore

$$\begin{aligned}|\rho(0)| &\leq \exp\left(\frac{\lambda^*}{2}W_0\right)\sup_{R^1}|w_0'(x)| \\ &\equiv C_1(\lambda^*;W_0)\sup_{R^1}|w_0'(x)|\end{aligned} \tag{3.3.75}$$

Also, by (3.3.49) and (3.3.54)

$$\mu_1(0) = \int_0^{w_0(\beta_1)-z_0(\beta_1)} b(\alpha;\zeta)\frac{f'(\zeta)}{\hat{\lambda}'(\zeta)}\,d\zeta$$

with

$$f'(\zeta) = \frac{d}{d\zeta}e^{h(\zeta)} \equiv e^{h(\zeta)}\frac{\hat{\lambda}'(\zeta)}{2\hat{\lambda}(\zeta)}$$

Thus

$$\begin{aligned}|\mu_1(0)| &\leq \int_0^{w_0(\beta_1)-z_0(\beta_1)}\frac{1}{2}\cdot\left|\frac{b(\alpha;\zeta)}{\hat{\lambda}(\zeta)}\right| e^{h(\zeta)}\,d\zeta \\ &\leq \frac{1}{2\lambda_*}\sup_{0\leq\zeta\leq W_0}\left\{|b(\alpha;\zeta)|e^{h(\zeta)}\right\}\cdot W_0 \\ &\leq C_2(|\alpha|,\Phi^*,\lambda_*,\lambda^*,W_0)\end{aligned} \tag{3.3.76}$$

where

$$C_2 = \frac{W_0}{2\lambda_*}\left(\frac{|\alpha|}{2}+\Phi^*W_0\right)\exp\left(\frac{\lambda^*}{2W_0}\right) \tag{3.3.77}$$

and we have employed (3.3.45) in conjunction with (3.3.30). Now, as

$$g(\alpha;w-z) = \int_0^{w-z}\frac{1}{2\hat{\lambda}(\zeta)}\left[\frac{\alpha}{2}+\hat{\Psi}(\zeta)\right]e^{h(\zeta)}\,d\zeta, \tag{3.3.78}$$

$$|g(\alpha;w-z)| \leq \frac{1}{2\lambda_*}\sup_{0\leq\zeta\leq(|w|+|z|)}\left\{\left[\frac{|\alpha|}{2}+|\hat{\Psi}(\zeta)|\right]e^{h(\zeta)}\right\}(|w|+|z|)$$

while

$$\begin{aligned}\sup_{0\leq\zeta\leq(|w|+|z|)}\left|\hat{\Psi}(\zeta)\right| &\leq \Phi^*(|w|+|z|) \\ &\leq \Phi^*W_0\exp(\Gamma t_\infty) \\ &\leq \frac{\Phi^*W_0}{\sqrt{2\lambda^*W_0}} = \frac{\Phi^*}{\sqrt{2\lambda^*}}\cdot\sqrt{W_0},\end{aligned} \tag{3.3.79}$$

and

$$|w|+|z| \leq W_0\exp(\Gamma t_\infty) < \sqrt{\frac{W_0}{2\lambda^*}} \tag{3.3.80}$$

where we have made use of (3.3.32) and (3.3.67). Furthermore,

$$\sup_{0\le\zeta\le(|w|+|z|)} \exp h(\zeta) \le \exp\left[\frac{1}{2}\lambda^*(|w|+|z|)\right] \le \exp\left(\sqrt{\frac{\lambda^* W_0}{8}}\right) \tag{3.3.81}$$

so if we now combine (3.3.78)-(3.3.81) we are led to the estimate

$$|g(\alpha; w-z)| < \frac{1}{2\lambda^*}\left(\frac{|\alpha|}{\sqrt{2\lambda^*}} + \frac{\Phi^*}{\lambda^*}\sqrt{W_0}\right)\sqrt{W_0}\exp\left(\sqrt{\frac{\lambda^* W_0}{8}}\right) \tag{3.3.82}$$

Making use of (3.3.82) in (3.3.54) we easily find that for $0 \le t < t_\infty$,

$$|\mu_1(t)| < C_3(|\alpha|, \Phi^*, \lambda_*, \lambda^*, W_0)\cdot\sqrt{W_0} \tag{3.3.83}$$

where C_3 has an obvious definition as implied by (3.3.82).

For our next set of estimates we appeal to (3.3.48); coupled with (3.3.81), (3.3.45), and (3.3.46), this relation and the estimates obtained to this point allow us to conclude that

$$|d(\alpha; w-z)| \le |c(\alpha; w-z)|\left(\frac{|\alpha|}{2} + \Phi^*\sqrt{\frac{W_0}{2\lambda^*}}\right)\exp\left(\sqrt{\frac{\lambda^* W_0}{8}}\right) \tag{3.3.84}$$

with

$$\begin{aligned} |c(\alpha; w-z)| &\le \frac{|\alpha|}{2\lambda^*}(|w|+|z|) + \frac{1}{\lambda^*}|\Psi(w-z)| \\ &\le \frac{|\alpha|}{2\lambda^*}(|w|+|z|) + \frac{\Phi^*}{2\lambda^*}(|w|+|z|)^2 \\ &\le \frac{|\alpha|}{2\lambda^*}W_0\exp(\Gamma t_\infty) + \frac{\Phi^*}{2\lambda^*}W_0^2\exp(2\Gamma t_\infty) \end{aligned}$$

or, by virtue of (3.3.67)

$$|c(\alpha; w-z)| < \frac{1}{2\lambda_*\sqrt{2\lambda_*}}\left(|\alpha| + \Phi^*\sqrt{\frac{W_0}{2\lambda^*}}\right)\sqrt{W_0} \tag{3.3.85}$$

By (3.3.55), (3.3.84), and (3.3.85) we now obtain as a bound for $\mu_2(t)$, $0 \le t < t_\infty$, an estimate of the form

$$|\mu_2(t)| < C_4(|\alpha|, \Phi^*, \lambda^*, W_0)\cdot\sqrt{W_0} \tag{3.3.86}$$

All the ingredients, i.e., (3.3.75), (3.3.76), (3.3.83), and (3.3.86), are now in place to enable us to deal effectively with (3.3.72) and, in fact, we easily determine that for $0 \le t < t_\infty$,

$$|\rho(t)| < C_1 \sup_{R^1} |W_0'(x)| + C_2 W_0 + \left(2C_3 + \frac{C_4}{\beta_0}\right) W_0^{1/2} \tag{3.3.87}$$

In order to obtain a bound on $|w_x(x,t)|$, we now return to the definition of $\rho(t)$, i.e., (3.3.52). From (3.3.43) we have, first of all, that

$$|h(w-z)| \le \frac{1}{2}\lambda^*(|w|+|z|) \le \frac{1}{2}\sqrt{\frac{\lambda^*}{2}} W_0^{1/2} \tag{3.3.88}$$

where we have again used part (iii) of Lemma 3.3 and the *a priori* estimate (3.3.32), coupled with (3.3.67). Thus

$$e^{-h(w-z)} \le \exp\left(\frac{1}{2}\sqrt{\frac{\lambda^*}{2}} W_0^{1/2}\right) \equiv C(\lambda^*, W_0) \tag{3.3.89}$$

Combining (3.3.87) and (3.3.89) with (3.3.52) we now obtain, for $0 \le t < t_\infty$, the estimate

$$\begin{aligned} |w_x(x,t)| &< C_1^* \sup_{R^1} |w_0'(x)| + C_2^* W_0 \\ &\le C^*(W_0 + W_0^\#) \end{aligned} \tag{3.3.90}$$

where $C_1^* = C_1 \cdot C$, $C_2^* = C(C_2 W_0^{1/2} + 2C_3 + \frac{C_4}{\beta_0})$, and $C^* = \max(C_1^*, C_2^*)$. In an analogous fashion, we may demonstrate the existence of $D^* = D^*(|\alpha|, \Phi^*, \lambda^*, W_0) > 0$ such that, for $0 \le t < t_\infty$,

$$|z_x(x,t)| \le D^*(W_0 + W_0^\#) \tag{3.3.91}$$

But, by virtue of (3.3.90) and (3.3.91), we readily conclude that

$$\lim_{t \to t_\infty^-} (|w_x(x,t)| + |z_x(x,t)|) < \infty \tag{3.3.92}$$

for $x \in R^1$, in which case $t_\infty = \infty$, by virtue of Lemma 3.4. Thus, if we can verify the assumption (3.3.70), then we will have the desired contradiction to the thesis that (3.3.67) is satisfied, for $W_0 + W_0^\#$ sufficiently small, if $t_\infty < \infty$; as $\beta_0 \equiv \frac{1}{2}\alpha_0$, (3.3.70) will be satisfied if we can show that for $x \in R^1$, and $W_0 + W_0^\#$ sufficiently small,

$$|w_x(x,t)| \le \hat{\beta}_0 / \hat{\lambda}_\infty, \qquad 0 \le t < t_\infty \tag{3.3.93}$$

where $\hat{\beta}_0 = \inf_{(w_0,z_0)} \beta_0 > 0$. Suppose, first of all, that $\sup_{R^1} |w_0'(x)| < \hat{\beta}_0/2\hat{\lambda}_\infty$; then $\exists \hat{t}$, $0 \le \hat{t} < t_\infty$, such that $|w_x(x,t)| < \hat{\beta}_0/\hat{\lambda}_\infty$ for $x \in R^1$ and $0 \le t \le \hat{t}$. From the estimate (3.3.90), we can, therefore, conclude that

$$|w_x(x,t)| < C_1^* \sup_{R^1} |w_0'(x)| + C_2^* W_0, \quad x \in R^1, \quad 0 \le t \le \hat{t} \tag{3.3.94}$$

By virtue of the definitions of C_1^*, C_2^*, we may now choose $\sup_{R^1} |w_0'(x)|$, and W_0, so small that

$$C_1^* \sup_{R^1} |w_0'(x)| + C_2^* W_0 \le \hat{\beta}_0/2\hat{\lambda}_\infty, \tag{3.3.95}$$

in which case, $\sup_{R^1} |w_x(x,\hat{t}| < \hat{\beta}_0/2\hat{\lambda}_\infty$. Taking $\hat{t}$ as the new initial time in the argument, and continuing in the same fashion, we are easily led to (3.3.93). In order to verify the validity of the estimate (3.3.91), we must, of course, begin with a bound, analogous to (3.3.70), for $|z_x(x,t)|$; this is easily established for $\sup_{R^1} |z_0'(x)|$ (and W_0) sufficiently small. Thus, both (3.3.70) and the similar bound for $|z_x(x,t)|$, on $R^1 \times [0, t_\infty)$, hold if $W_0 + W_0^\#$ is sufficiently small. The proof is now complete. □

Remarks: Theorem 3.4 does not allow us to distinguish between the situation in which the life span t_∞, of a C^1 solution to the initial-value problem for (3.3.27), is finite for arbitrarily small initial data (as measured by $W_0 + W_0^\#$) and then satisfies the lower bound given by (3.3.37), and the situation in which $t_\infty = \infty$ for $W_0 + W_0^\#$ chosen sufficiently small; it simply says that if we cannot achieve global (in time) existence of a C^1 solution by choosing the initial data sufficiently small then, at least, the life span of a C^1 solution must grow rapidly as the initial data shrinks. In fact, t_∞ must grow (at least) as fast as the estimate (3.3.37) dictates. In § 3.5 we will, in fact, show that for $W_0 + W_0^\#$ sufficiently small, i.e., for

$$\begin{cases} \tilde{W}_0 &= \sup_{x \in R^1} (|D_0(x)| + |B_0(x)|) \\ \tilde{W}_0^\# &= \sup_{x \in R^1} (|D_0'(x)| + |B_0'(x)|) \end{cases} \tag{3.3.96}$$

sufficiently small, the initial-value problem for the nonlinearly damped quasilinear system (2.6.16), governing the one-dimensional wave-dielectric interaction problem, possesses a C^1 solution (D,B) on $R^1 \times [0,\infty)$; this will be accomplished through the use of an energy argument, the basic nature of which is explained by means of a simple example in the next section.

3.4 Global Existence for Small Data: An Example

In § 2.2 we demonstrated that, in general, C^1 solutions of the Burger's equation (2.2.1) cannot exist for all $t > 0$; in fact, it follows from our work in that section that in many simple situations, e.g., $u_x(\bar{x}, 0) < 0$ at some $\bar{x} \in R^1$, the gradient $\nabla_{(x,t)}u = (u_x, u_t)$ must become unbounded in finite time t_∞ in the sense that at some $\bar{x} \in R^1$,

$$\lim_{t \to t_\infty^-} \left[u_x^2(\bar{x}, t)) + u_t^2(\bar{x}, t)\right] = \infty. \tag{3.4.1}$$

Suppose now that frictional damping is added to (2.2.1) so that the new initial-value problem is

$$\begin{cases} u_t + uu_x + \lambda u = 0, & -\infty < x < \infty, \quad t > 0 \\ u(x, 0) = u_0(x), & -\infty < x < \infty \end{cases} \tag{3.4.2}$$

with $\lambda > 0$ constant. The characteristics for (3.4.2) are determined, once again, by the solution curves of the nonlinear initial-value problem (2.2.2), but now, instead of (2.2.3), we find that along any characteristic $x = x(t)$,

$$\frac{du}{dt}(x(t), t) = -\lambda u(x(t), t) \tag{3.4.3}$$

Differentiating the equation in (3.4.2) through with respect to x, and then setting $w = u_x$, we observe that $w(x, t)$ satisfies

$$w_t + uw_x + w^2 + \lambda w = 0 \tag{3.4.4}$$

Therefore, along the characteristics determined by (2.2.2), we have for $\tilde{w}(t) = w(x(t), t)$,

$$\frac{dw}{dt} + w^2 + \lambda w = 0 \tag{3.4.5}$$

Of course, (3.4.5) is a Bernoulli equation and it is easily determined that $\tilde{w}(t) \equiv u_x(x(t), t)$ remains bounded for all $t > 0$ if $\tilde{w}(0) \geq -\lambda$, while for $\tilde{w}(0) < -\lambda$, $\tilde{w}(t)$ must blow up in finite time. Therefore, solutions of (3.4.2) which are of class C^1, for all $t > 0$, are obtained provided $\sup_{R^1} |u_x(x, 0)| \leq \lambda$.

In order to illustrate the energy argument that will be applied in § 3.5 to the initial-value problem associated with the quasilinear system (2.6.16), we now apply this argument to the simple problem represented by (3.4.2). Our task is, obviously, one of demonstrating that $\sup_{R^1} |u_x(x,t)|$ is bounded for all $t > 0$ and the method follows the analysis presented in [45]. We begin by recording the well-known fact that we may associate a local (in time) existence result with the initial-value problem (3.4.2) in the sense that for smooth initial data $u_0(\cdot)$, (3.4.2) has a unique solution $u \in C^1(R^1 \times [0, T_\infty))$, $T_\infty \leq +\infty$ such that for any $T \in [0, T_\infty)$:

$$u, u_x, u_{xx} \in L^\infty([0,T]; L^2(R^1)) \tag{3.4.6}$$

We begin by assuming that

$$\sup_{R^1} |u_x(x,0)| \leq \frac{2}{5}\lambda \tag{3.4.7}$$

Then $\exists\, s > 0$, $s < T$ such that

$$\sup_{R^1} |u_x(x,\tau)| \leq \frac{2}{5}\lambda, \qquad 0 \leq \tau < s \tag{3.4.8}$$

in which case for all $x \in R^1$,

$$u_x(x,\tau) > -\frac{2}{5}\lambda > -2\lambda, \qquad 0 \leq \tau < s \tag{3.4.9}$$

For $t \leq T$, we now differentiate (3.4.2) with respect to x, and then multiply the resulting equation through by $2u_x(x,t)$, so as to obtain

$$\frac{\partial}{\partial t}(u_x^2) + 2uu_xu_{xx} + 2u_x^3 + 2\lambda u_x^2 = 0 \tag{3.4.10}$$

or

$$\frac{\partial}{\partial t}(u_x^2) + \frac{\partial}{\partial x}(uu_x^2) + (2\lambda + u_x)u_x^2 = 0 \tag{3.4.11}$$

Next, we integrate (3.4.11) over $[0,s) \times R^1$ and then use the fact that (3.4.6) implies that $u(\pm\infty, t) = u_x(\pm\infty, t) = 0$, for $t \leq T$; we find that

$$\int_{-\infty}^{\infty} u_x^2(x,s)\,dx + \int_0^s \int_{-\infty}^{\infty} (2\lambda + u_x)u_x^2\,dx\,dt = \int_{-\infty}^{\infty} u_x^2(x,0)\,dx \tag{3.4.12}$$

Therefore, by virtue of (3.4.9), we obtain the bound

$$\|u_x(\cdot,s)\|_{L^2(R^1)} \leq \|u_x(\cdot,0)\|_{L^2(R^1)} \tag{3.4.13}$$

Our next step is to take the equation

$$u_{xt} + (uu_x)_x + \lambda u_x = 0 \tag{3.4.14}$$

obtained by differentiating (3.4.2) through with respect to x, differentiate again with respect to x, for $t \le T$, and then multiply the resulting equation by $2u_{xx}$ so as to obtain

$$\frac{\partial}{\partial t}(u_{xx}^2) + 2u_{xx}(uu_x)_{xx} + 2\lambda u_{xx}^2 = 0 \tag{3.4.15}$$

Inasmuch as

$$\begin{aligned} 2u_{xx}(uu_x)_{xx} &= 6u_x u_{xx}^2 + 2uu_x u_{xxx} \\ &= 5u_x u_{xx}^2 + \frac{\partial}{\partial x}(uu_{xx}^2) \end{aligned}$$

equation (3.4.15) can be rewritten in the form

$$\frac{\partial}{\partial t}(u_{xx}^2) + \frac{\partial}{\partial x}(uu_{xx}^2) + (2\lambda + 5u_x)u_{xx}^2 = 0 \tag{3.4.16}$$

We now integrate (3.4.16) over $[0, s) \times R^1$, use (3.4.9), and the fact that $u_{xx}^2(\pm\infty, t) = 0$, for $t \le T$, so as to arrive at the bound

$$\|u_{xx}(\cdot, s)\|_{L^2(R^1)} \le \|u_{xx}(\cdot, 0)\|_{L^2(R^1)} \tag{3.4.17}$$

However, for any $x \in R^1$,

$$\begin{aligned} u_x^2(x, s) &= \int_{-\infty}^{x} \frac{\partial}{\partial y}(u_x^2(y, s))\, dy \qquad (3.4.18) \\ &= 2\int_{-\infty}^{x} u_x(y, s)u_{xx}(y, s)\, dy \\ &\le 2\|u_x(\cdot, s)\|_{L^2(R^1)}\|u_{xx}(\cdot, s)\|_{L^2(R^1)} \end{aligned}$$

so combining the estimates (3.4.13), (3.4.17), and (3.4.18), we are led to the bound (for $x \in R^1$)

$$u_x^2(x, s) \le 2\|u_x(\cdot, 0)\|_{L^2(R^1)}\|u_{xx}(\cdot, 0)\|_{L^2(R^1)} \tag{3.4.19}$$

If, therefore, the initial data is chosen, consistent with (3.4.7), satisfying

$$\|u_x(\cdot, 0)\|_{L^2(R^1)}\|u_{xx}(\cdot, 0)\|_{L^2(R^1)} \le \frac{2}{25}\lambda^2, \tag{3.4.20}$$

then, by virtue of (3.4.19), we have

$$\sup_{R^1} |u_x(x, s)| \le \frac{2}{5}\lambda \tag{3.4.21}$$

We are now back to the position we were in at the starting point, i.e., with hypothesis (3.4.7). From (3.4.21) we infer the existence of an $s_1 > s$ such that

$$\sup_{R^1} |u_x(x,\tau)| \le \frac{2}{5}\lambda, \qquad s \le \tau < s_1 \tag{3.4.22}$$

Applying the local existence theorem cited above to the initial-value problem for (3.4.2), but with the initial time taken as $t = s$, and the "initial" data satisfying (3.4.20), we are led in successive steps from (3.4.22) to the conclusion that $\sup_{R^1} |u_x(x,t)| < \frac{2}{5}\lambda$ for $0 \le t < \infty$; this establishes the global existence, in time, of a C^1 solution for the initial-value problem (3.4.2) with associated small initial data in the sense of (3.4.19).

The simple energy argument employed above is, in essence, modeled on the one used by Matsumura [112] and Nishida [124], and may be traced back to earlier work of Courant, Friedrichs, and Lewy [36]; in various guises it has been used by Dafermos and Nohel [43], [42] and Hrusa and Nohel [71] to establish results on the global existence of solutions to problems arising in one-dimensional nonlinear viscoelasticity, by Hrusa [70] to prove similar results for materials with fading memory, by Slemrod [162] to establish the global existence, uniqueness, and asymptotic stability of classical smooth solutions for problems of one-dimensional, nonlinear thermoelasticity, and by Coleman, Hrusa, and Owen [35] to study the stability of equilibrium states for a nonlinear hyperbolic system which describes heat propagation in solids. The above list is not exhaustive and many other researchers have successfully applied the basic technique to other problems, in recent years, which have their origins in nonlinear continuum mechanics. In the next section we will describe how this basic type of energy argument can be applied to study the global existence and asymptotic stability of classical solutions to the initial-value problem associated with the wave-dielectric interaction system when the initial data is small in some appropriate sense.

3.5 Global Existence of Solutions in the Wave-Dielectric Interaction Problem

Our work in this section follows, closely, the recent paper of the author and H. Bellout [16], although we will often shortcut many of the calculations of a straightforward nature and will use a notation which conforms to the one already introduced in this book. As already noted in § 2.6, the relevant quasilinear system (2.6.16) assumes, whenever the indicated operations make sense, the form of the nonlinearly damped, nonlinear wave equation (2.6.21) which we here reproduce as

$$D_{tt} + \Sigma'(D)D_t = \mathsf{E}(D)_{xx} \tag{3.5.1}$$

where, without loss of generality, we have set $\mu_0 = 1$ and will use the subscript notation to indicate differentiation throughout the rest of this section. Equation (3.5.1) is to hold for $-\infty < x < \infty$ and $t > 0$ and we also have associated initial data of the form

$$D(x,0) = D_0(x), \qquad D_t(x,0) = D_1(x), \qquad -\infty < x < \infty \tag{3.5.2}$$

Note that, by virtue of the first relation in (2.6.16), with $\mu_0 = 1$,

$$D_1(x) = -B_0'(x) - \Sigma(D_0(x)), \qquad -\infty < x < \infty \tag{3.5.3}$$

The function $\Sigma(\cdot)$ is again given by (2.6.17), i.e.,

$$\Sigma(\zeta) = \tilde{\sigma}(\mathsf{E}(\zeta))\mathsf{E}(\zeta), \qquad \forall \zeta \in R^1 \tag{3.5.4}$$

where $\tilde{\sigma}(\cdot)$ is the constitutive function that determines the conduction current $\boldsymbol{J}$ for a given electric field $\boldsymbol{E}$.

For the balance of this section our basic hypotheses relative to the constitutive functions $\mathsf{E}(\cdot)$ and $\Sigma(\cdot)$ will be as follows:

$$\text{(i)} \quad \mathsf{E}(0) = 0,\ \Sigma(0) = 0, \tag{3.5.5a}$$

$$\text{(ii)} \quad \exists\, \epsilon_0 > 0,\ \Sigma_0 > 0 \text{ such that } \mathsf{E}'(0) \geq 2\epsilon_0,\ \Sigma'(0) \geq 2\Sigma_0, \tag{3.5.5b}$$

$$\text{(iii)} \quad \exists\, N > 0 \text{ such that } \mathsf{E} \in C^4[-N,N], \Sigma \in C^3[-N,N], \tag{3.5.5c}$$

and without loss of generality we will assume that $N \leq 1$. In view of (3.5.5b), (3.5.5c), it follows that $\exists\, \bar{\delta}$, $0 < \bar{\delta} \leq N$, such that

$$\mathsf{E}'(\zeta) \geq \epsilon_0, \qquad \Sigma'(\zeta) \geq \Sigma_0, \qquad |\zeta| \leq \bar{\delta} \tag{3.5.6}$$

We now define (for future use) the function

$$g(\zeta) = \begin{cases} \dfrac{2}{\zeta}\Sigma(\zeta) - \dfrac{2}{\zeta^2}\displaystyle\int_0^\zeta \Sigma(\lambda)\,d\lambda, \ \zeta \neq 0 \\ \Sigma'(0), \text{ at } \zeta = 0 \end{cases} \tag{3.5.7}$$

As a consequence of (3.5.5a)-(3.5.5c), $g(\zeta)$ is continuous on all of R^1 and $g(\zeta) \geq \Sigma_0$ for $|\zeta|$ chosen sufficiently small, i.e., $|\zeta| \leq \bar{\delta}$. Also

$$D^2(x,t)g(D(x,t))_t = \Sigma'(D(x,t))(D^2(x,t))_t \tag{3.5.8}$$

whenever both sides of (3.5.8) are well defined. In many of the calculations that follow, we will use the constant $\Lambda > 0$ which we define to be

$$\Lambda = \max_{\substack{1 \leq j \leq 4 \\ 1 \leq k \leq 3}} \left[\max_{[-N,N]} |\mathsf{E}^{(j)}(\cdot)|, \ \max_{[-N,N]} |\Sigma^{(k)}(\cdot)| \right] \tag{3.5.9}$$

and constants such as C, C', $\tilde{C}$, et al., or C_i, $i = 1, 2, \ldots$, when they appear, will be understood to depend only on Λ, ϵ_0, Σ_0, and, perhaps, constants λ_j, $j = 1, 2, \ldots$ which result from applying various elementary estimates such as the Cauchy-Schwarz inequality or the arithmetic-geometric mean inequality. Finally, we will assume that the initial data appearing in (3.5.2) satisfy (at least) the requirement that

$$D_0 \in H^3(R^1), \qquad D_1 \in H^2(R^1) \tag{3.5.10}$$

with $H^k(R^1)$, $k = 1, 2, 3, \ldots$ being the usual Sobolev spaces. For basic definitions and properties of these Sobolev spaces (which we shall use as required), the reader may consult a growing list of texts dealing with either functional analysis (e.g., Yoshida [187]) or the modern theory of partial differential equations (e.g., Friedman [58] or Nečas [123]).

Our first result is the analog of the local existence theorem expressed by Lemma 3.4 and is, in view of the hypotheses (3.5.5a)-(3.5.5c) and (3.5.10), a direct consequence of the fundamental results in, e.g., Kato [89]:

Lemma 3.6 The Cauchy problem (3.5.1), (3.5.2), subject to (3.5.5a)-(3.5.5c) and (3.5.10), has a unique solution $D \in C^2(R^1 \times [0, T_\infty)$ on a maximal interval $[0, T_\infty)$, $T_\infty \leq +\infty$, such that for any $t \in [0, T_\infty)$ the following holds:

$$D, D_t, D_x, D_{tt}, D_{xt}, D_{xx}, \ldots, D_{xxx} \in L^\infty([0,t]; L^2(R^1)) \tag{3.5.11}$$

and

$$\int_{-\infty}^{\infty} [D^2(x,t) + D_t^2(x,t) + D_x^2(x,t) + \cdots + D_{xxx}^2(x,t)]\, dx \to \infty, \tag{3.5.12}$$

as $t \to T_\infty^-$, if $T_\infty < \infty$.

Our goal in the remainder of this section will be to extend the local existence result, delineated in Lemma 3.5, to a global existence theorem (valid for $0 \le t < \infty$) and to study the behavior of the electric displacement field $D(x,t)$, $-\infty < x < \infty$, $t > 0$, and its derivatives, as $t \to \infty$. Following the general pattern of the example given in the previous section, our general strategy may be outlined as follows:

(i) We begin by defining a suitable energy functional $\mathcal{E}_0(t)$ on local solutions $D(x,t)$ of the Cauchy problem (3.5.1), (3.5.2); for $t < T_\infty$ such a functional has the form

$$\mathcal{E}_0(t) = \int_{-\infty}^{\infty} \left\{D^2 + D_t^2 + \cdots + D_{xxx}^2\right\}(x,t)\, dx \tag{3.5.13}$$

(ii) We also defined an enhanced energy functional $\mathcal{E}(t)$ as

$$\mathcal{E}(t) = \mathcal{E}_0(t) + \int_0^t \int_{-\infty}^{\infty} \left\{D_t^2 + D_x^2 + \cdots + D_{xxx}^2\right\}(x,\tau)\, dx\, d\tau \tag{3.5.14}$$

(iii) We attempt to show that if $\mathcal{E}_0(0)$ is sufficiently small, then for some nonnegative measure $\mathcal{I}_0$ (which involves only the $L^2(R^1)$ norms of the initial data, and certain derivatives of D at $t = 0$)

$$\mathcal{E}(t) \le C\mathcal{I}_0 \tag{3.5.15}$$

for some $C > 0$ and any $t < T_\infty$.

(iv) We show that the choice of $\mathcal{I}_0$ is such that for some $C' > 0$, if $\mathcal{E}_0(0)$ is sufficiently small, then

$$\mathcal{I}_0 \le C'\mathcal{E}_0(0) \tag{3.5.16}$$

Combining (3.5.13)-(3.5.16), we will then be led to the conclusion that for $\mathcal{E}_0(0)$ sufficiently small, $\exists\, \tilde{C} > 0$ such that

$$\mathcal{E}_0(t) \le \mathcal{E}(t) \le \tilde{C}\mathcal{E}_0(0), \qquad t < T_\infty \tag{3.5.17}$$

and the desired global existence theorem will follow as a direct consequence of the second part of Lemma 3.5, i.e., (3.5.12). From the same estimate (3.5.17) there will follow two basic results on the large time ($t \to \infty$) behavior of the classical solution $D(x,t)$ on $R^1 \times [0,\infty)$.

The various *a priori* estimates that will lead us to (3.5.17) will be stated and proved in a sequence of lemmas; whenever possible, we dispense with many of the straightforward intermediate calculations, as these may be found in the paper [16]. Before stating the first of our *a priori* estimates, we will define, below, a number of functionals, based on the initial data, that arise during the generation of the *a priori* estimates. Thus, for $k = 0, 1, \ldots, 7$ we set

$$J_0 = \int_{-\infty}^{\infty} g(D_0(x))D_0^2(x)\,dx + 2\left(\int_{-\infty}^{\infty} D_0^2(x)\,dx\right)^{1/2}\left(\int_{-\infty}^{\infty} D_1^2(x)\,dx\right)^{1/2} \tag{3.5.18a}$$

$$J_1 = \int_{-\infty}^{\infty} \left(D_1^2(x) + \mathsf{E}'(D_0(x))D_0'^2(x)\right)\,dx \tag{3.5.18b}$$

$$J_2 = \int_{-\infty}^{\infty} \left(D_1'^2(x) + \mathsf{E}'(D_0(x))D_0''^2(x)\right)\,dx \tag{3.5.18c}$$

$$J_3 = \int_{-\infty}^{\infty} \Sigma'(D_0(x))D_0'^2(x)\,dx \tag{3.5.18d}$$

$$J_4 = \int_{-\infty}^{\infty} \left(D_{tt}^2(x,0) + \mathsf{E}'(D_0(x))D_1'^2(x)\right)\,dx \tag{3.5.18e}$$

$$J_5 = \int_{-\infty}^{\infty} \left(D_1''^2(x) + \mathsf{E}'(D_0(x))D_0'''^2(x)\right)\,dx \tag{3.5.18f}$$

$$J_6 = \int_{-\infty}^{\infty} \left(D_{xtt}^2(x,0) + \mathsf{E}'(D_0(x))D_1''^2(x)\right)\,dx \tag{3.5.18g}$$

and

$$J_7 = \int_{-\infty}^{\infty} \left(D_0'^2(x) + D_0''^2(x) + D_0'''^2(x)\right)\,dx \tag{3.5.18h}$$

Calculations to be presented below will show that bounds for all derivatives of $D(x,t)$, at $t = 0$, which appear in the definitions (3.5.18a)-(3.5.18h), can be computed directly in terms of the initial data $D_0(\cdot)$ and $D_1(\cdot)$ if we make use of (3.5.1), e.g.,

$$\frac{1}{2}\int_{-\infty}^{\infty} D_{tt}^2(x,0)\,dx \leq \int_{-\infty}^{\infty} \Sigma'^2(D_0(x))D_1^2(x)\,dx + \int_{-\infty}^{\infty} \mathsf{E}'^2(D_0(x))D_0''^2(x)\,dx + \int_{-\infty}^{\infty} \mathsf{E}''^2(D_0(x))D_0'^4(x)\,dx \tag{3.5.19}$$

Suppose that the initial data $D_0(\cdot), D_1(\cdot)$ are sufficiently small in the sense that

$$\|D_0\|_{H^3}^2 + \|D_1\|_{H^2}^2 \leq \bar{\delta}^2$$

with $0 < \bar{\delta} \leq N$. Then it follows by the Sobolev embedding theorem, hypothesis (3.5.5c), and (3.5.19) that $\exists\, C > 0$ such that

$$\int_{-\infty}^{\infty} D_{tt}^2(x,0)\,dx \leq C\int_{-\infty}^{\infty}\left\{D_t^2 + D_x^2 + D_{xx}^2\right\}(x,0)\,dx \tag{3.5.20}$$

Various estimates which are entirely analogous to (3.5.20), for the integrals

$$\int_{-\infty}^{\infty} D_{xtt}^2(x,0)\,dx \quad \text{and} \quad \int_{-\infty}^{\infty} D_{ttt}^2(x,0)\,dx,$$

will be implicit in our subsequent calculations. As a consequence, it will be clear that for $\|D_0\|_{H^3}^2 + \|D_1\|_{H^2}^2$ chosen sufficiently small, we shall have

$$\mathcal{E}_0(0) \leq C^0\left(\|D_0\|_{H^3}^2 + \|D_1\|_{H^2}^2\right), \tag{3.5.21}$$

for some $C^0 > 0$, where the energy $\mathcal{E}_0$ is defined by (3.5.13).

If we begin by assuming that the initial data have been chosen so that for some $\delta,\ 0 < \delta \leq \bar{\delta}$, and some integer $k \geq 2$,

$$\mathcal{E}_0(0) \leq \delta^2/2^k \tag{3.5.22}$$

(i.e., by virtue of (3.5.21) we may choose $D_0(\cdot)$, $D_1(\cdot)$ so that $\|D_0\|_{H^3}^2 + \|D_1\|_{H^2}^2 \leq \delta^2/C^0 2^k$) then the definition of $\mathcal{E}_0(t)$, the Sobolev embedding theorem, and the fact that $k \geq 2$ imply that

$$\begin{aligned}\max\Big\{&|D(\cdot,0)|_\infty, |D_x(\cdot,0)|_\infty, |D_t(\cdot,0)|_\infty,\\ &|D_{tt}(\cdot,0)|_\infty, |D_{xt}(\cdot,0)|_\infty, |D_{xx}(\cdot,0)|_\infty\Big\}\\ &\leq \delta/2.\end{aligned} \tag{3.5.23}$$

where $|\cdot|_\infty$ denotes the $L^\infty(R^1)$ norm. Therefore, if $D(x,t)$ denotes the local solution of (3.5.1), (3.5.2) on $[0, T_\infty)$, $\exists\, s, 0 < s \leq T_\infty$, such that for any $t < s$,

$$\begin{aligned}\max\Big\{&|D(\cdot,t)|_\infty, |D_x(\cdot,t)|_\infty, |D_t(\cdot,t)|_\infty,\\ &|D_{tt}(\cdot,t)|_\infty, |D_{xt}(\cdot,t)|_\infty, |D_{xx}(\cdot,t)|_\infty\Big\} < \delta\end{aligned} \tag{3.5.24}$$

Our first estimate is expressed in the

Lemma 3.7 If $D(x,t)$ is the local solution of (3.5.1), (3.5.2) on $[0, T_\infty)$, hypotheses (3.5.5a)-(3.5.5c) are satisfied, and if $s \le T_\infty$ is such that (3.5.24) holds for any $t < s$, then $\exists\, C > 0$ such that

$$\int_{-\infty}^{\infty} (D^2(x,s) + D_t^2(x,s) + D_x^2(x,s))\,dx + \int_0^s \int_{-\infty}^{\infty} (D_t^2(x,t) + D_x^2(x,t))\,dx\,dt \le C(J_0 + J_1) \tag{3.5.25}$$

where J_0, J_1 are the initial datum functionals defined, respectively, by (3.5.18a), (3.5.18b).

Proof: We multiply (3.5.1) by $2D(x,t)$, $t < s$, integrate over $[0,s) \times (-\infty, \infty)$, and employ the identities

$$\begin{cases} 2DD_{tt} = (D^2)_{tt} - 2D_t^2 \\ (\mathsf{E}'(D)D_x)_x D = ((\mathsf{E}'(D)DD_x)_x - \mathsf{E}'(D)D_x^2 \end{cases}$$

so as to obtain

$$\begin{aligned} 2\int_0^s \int_{-\infty}^{\infty} &\mathsf{E}'(D(x,t))D_x^2(x,t)\,dx\,dt \\ &+ \int_0^s \int_{-\infty}^{\infty} \Sigma'(D(x,t))(D^2(x,t))_t\,dx\,dt \\ &= 2\int_0^s \int_{-\infty}^{\infty} D_t^2(x,t)\,dx\,dt - \int_0^s \int_{-\infty}^{\infty} (D^2(x,t))_{tt}\,dx\,dt \end{aligned} \tag{3.5.26}$$

where $\int_{-\infty}^{\infty} (\mathsf{E}'(D)DD_x)_x(x,t)\,dx = 0$ by virtue of the local existence theorem. Employing (3.5.24), (3.5.6), and (3.5.8) it now follows that

$$\begin{aligned} 2\epsilon_0 \int_0^s \int_{-\infty}^{\infty} &D_x^2(x,t)\,dx\,dt + \int_{-\infty}^{\infty} g(D(x,s))D^2(x,s)\,dx \\ &\le \int_{-\infty}^{\infty} g(D_0(x))D_0^2(x)\,dx + 2\left(\int_{-\infty}^{\infty} D_0^2(x)\,dx\right)^{1/2} \left(\int_{-\infty}^{\infty} D_1^2(x)\,dx\right)^{1/2} \\ &+2\int_0^s \int_{-\infty}^{\infty} D_t^2(x,t)\,dx\,dt + 2\int_{-\infty}^{\infty} |D(x,s)D_t(x,s)|\,dx \end{aligned} \tag{3.5.27}$$

However, $g(\zeta) \ge \Sigma_0$, for $|\zeta| \le \delta \le \bar{\delta}$, so if we once more make use of (3.5.24) and the inequality

$$|ab| \le \frac{1}{2\lambda_0}a^2 + \frac{\lambda_0}{2}b^2, \quad \text{with} \quad \lambda_0 > \frac{1}{\Sigma_0}$$

we obtain from (3.5.27) the estimate

$$\begin{aligned}\epsilon_0 \int_0^s \int_{-\infty}^{\infty} D_x^2(x,t)\,dx\,dt + \left(\Sigma_0 - \frac{1}{\lambda_0}\right) \int_{-\infty}^{\infty} D^2(x,s)\,dx \\ \le J_0 + \lambda_0 \int_{-\infty}^{\infty} D_t^2(x,s)\,dx + 2\int_0^s \int_{-\infty}^{\infty} D_t^2(x,t)\,dx\,dt\end{aligned} \tag{3.5.28}$$

Now, we return to (3.5.1), multiply through by $2D_t(x,t)$, integrate over $[0,s) \times (-\infty,\infty)$, and use the fact that $\Sigma'(\zeta) \ge \Sigma_0$ for $|\zeta| \le \delta \le \bar{\delta}$; in this manner we are led to the estimate

$$\begin{aligned}\int_{-\infty}^{\infty} D_t^2(x,s)\,dx + \Sigma_0 \int_0^s \int_{-\infty}^{\infty} D_t^2(x,t)\,dx\,dt \\ \le \int_{-\infty}^{\infty} D_1^2(x)\,dx + 2\int_0^s \int_{-\infty}^{\infty} D_t(x,t)\mathsf{E}(D(x,t))_{xx}\,dx\,dt\end{aligned} \tag{3.5.29}$$

Use of the identity

$$2D_t\mathsf{E}(D)_{xx} = 2(\mathsf{E}'(D)D_xD_t)_x - (\mathsf{E}'(D)D_x^2)_t + \mathsf{E}''(D)D_tD_x^2,$$

valid for $t < s \le T_\infty$, produces an estimate of the form

$$\begin{aligned}2\int_0^s \int_{-\infty}^{\infty} D_t\mathsf{E}(D)_{xx}\,dx\,dt \;\le\; & \int_{-\infty}^{\infty} \mathsf{E}'(D_0(x)){D_0'}^2(x)\,dx \\ & - \int_{-\infty}^{\infty} \mathsf{E}'(D(x,s))D_x^2(x,s)\,dx \\ & + \delta\Lambda \int_0^s \int_{-\infty}^{\infty} D_x^2(x,t)\,dx\,dt\end{aligned} \tag{3.5.30}$$

where Λ is given by (3.5.9). Thus, if we make use of (3.5.30) in (3.5.29), we obtain a bound of the form

$$\begin{aligned}\int_{-\infty}^{\infty} \left[D_t^2(x,s) + \epsilon_0 D_x^2(x,s)\right]\,dx \;+\; \Sigma_0 \int_0^s \int_{-\infty}^{\infty} D_t^2(x,t)\,dx\,dt \\ \le\; J_1 + \delta\Lambda \int_0^s \int_{-\infty}^{\infty} D_x^2(x,t)\,dx\,dt\end{aligned} \tag{3.5.31}$$

However, (3.5.28) implies that for some $C > 0$,

$$\begin{aligned}\delta\Lambda \int_0^s \int_{-\infty}^{\infty} D_x^2(x,t)\,dx\,dt \;\le\; & \delta C\Big(J_0 + \int_{-\infty}^{\infty} D_t^2(x,s)\,dx \\ & + \int_0^s \int_{-\infty}^{\infty} D_t^2(x,t)\,dx\,dt\Big)\end{aligned}$$

and, therefore, we may replace (3.5.31) with the estimate

$$\int_{-\infty}^{\infty}(D_t^2(x,s)+\epsilon_0 D_x^2(x,s))\,dx+\Sigma_0\int_0^s\int_{-\infty}^{\infty}D_t^2(x,t)\,dx\,dt \tag{3.5.32}$$
$$\leq (J_1+\delta C J_0)+\delta C\Big(\int_{-\infty}^{\infty}D_t^2(x,s)\,dx+\int_0^s\int_{-\infty}^{\infty}D_t^2(x,t)\,dx\,dt\Big)$$

We now choose $\delta \leq \bar{\delta}$ so small that both

$$1-\delta C \geq \frac{1}{2} \quad \text{and} \quad \Sigma_0-\delta C \geq \frac{1}{2}\Sigma_0$$

in which case, as $\delta \leq \bar{\delta} \leq N \leq 1$, we obtain, from (3.5.32), for some $C' > 0$, the bound

$$\int_{-\infty}^{\infty}(D_t^2(x,s)+\epsilon_0 D_x^2(x,s))\,dx \;+\; \Sigma_0\int_0^s\int_{-\infty}^{\infty}D_t^2(x,t)\,dx\,dt \tag{3.5.33}$$
$$\leq\; C'(J_0+J_1)$$

Next, we employ (3.5.33) in (3.5.28) and find that for some $\tilde{C} > 0$

$$\epsilon_0\int_0^s\int_{-\infty}^{\infty}D_x^2(x,t)\,dx\,dt \;+\; \Big(\Sigma_0-\frac{1}{\lambda_0}\Big)\int_{-\infty}^{\infty}D^2(x,s)\,dx \tag{3.5.34}$$
$$\leq\; \tilde{C}(J_0+J_1)$$

Finally, we combine our last two estimates (3.5.33) and (3.5.34) and we easily see that we are led to (3.5.25) for some $C > 0$. □

Our next set of estimates has as their goal the demonstration of the bound represented in

Lemma 3.8 Under the same conditions as those which prevail in Lemma 3.7,

$$\int_{-\infty}^{\infty}\left[D_{tt}^2(x,s)+D_{xt}^2(x,s)+D_{xx}^2(x,s)\right]\,dx \tag{3.5.35}$$
$$+\int_0^s\int_{-\infty}^{\infty}\left[D_{tt}^2+D_{xt}^2+D_{xx}^2\right](x,t)\,dx\,dt$$
$$\leq C\left(\sum_{k=0}^{4}J_k\right)$$

for some $C > 0$, where the $J_k, k=0,\ldots,4$ are given by (3.5.18a)-(3.5.18e).

Proof: For $t < s \le T_\infty$, we differentiate (3.5.1) through with respect to x, multiply the resulting equation by $2D_{xt}$, employ the identity

$$\begin{aligned} 2\Sigma(D)_{xt}D_{xt} &= 2\Sigma'(D)D_{xt}^2 + \left(\Sigma''(D)D_xD_t^2\right)_x \\ &- \left[\Sigma''(D)D_{xx} + \Sigma'''(D)D_x^2\right]D_t^2 \end{aligned}$$

and integrate over $[0, s) \times (-\infty, \infty)$; after using (3.5.24), and the fact that $\delta^2 \le \delta$ (as $\bar{\delta} \le 1$), we find that

$$\begin{aligned} &\int_{-\infty}^{\infty} D_{xt}^2(x,s)\,dx + \Sigma_0 \int_0^s \int_{-\infty}^{\infty} D_{xt}^2(x,t)\,dx\,dt \qquad (3.5.36) \\ &\le \int_{-\infty}^{\infty} D_{xt}^2(x,0)\,dx + 2\delta\Lambda \int_0^s \int_{-\infty}^{\infty} D_t^2(x,t)\,dx\,dt \\ &+2\int_0^s \int_{-\infty}^{\infty} \mathsf{E}(D(x,t))_{xxx}D_{xt}(x,t)\,dx\,dt \end{aligned}$$

However, for $t < s \le T_\infty$,

$$\begin{aligned} \mathsf{E}(D)_{xxx}D_{xt} &= \left[(\mathsf{E}'(D)D_{xx} + \mathsf{E}''(D)D_x^2)D_{xt}\right]_x \\ &- \frac{1}{2}(\mathsf{E}'(D)D_{xx}^2)_t + \frac{1}{2}\mathsf{E}''(D)D_tD_{xx}^2 \\ &- (\mathsf{E}''(D)D_x^2D_{xx})_t + \mathsf{E}'''(D)D_tD_{xx}D_x^2 \\ &+ 2\mathsf{E}''(D)D_xD_{xt}D_{xx} \end{aligned}$$

and it is a straightforward exercise to show that from this identity we may obtain an estimate of the form:

$$\begin{aligned} &\int_0^s \int_{-\infty}^{\infty} \mathsf{E}(D)_{xxx}D_{xt}\,dx\,dt \le \int_{-\infty}^{\infty} \mathsf{E}'(D_0(x))D_0'^2(x)\,dx \qquad (3.5.37) \\ &+2\delta\Lambda \int_{-\infty}^{\infty} D_0'^2(x)\,dx - \int_{-\infty}^{\infty} \mathsf{E}'(D(x,s))D_{xx}^2(x,s)\,ds \\ &+\delta C\left(\int_{-\infty}^{\infty} D_x^2(x,s)\,dx + \int_0^s \int_{-\infty}^{\infty} (D_x^2 + D_{xx}^2 + D_{xt}^2)\,dx\,dt\right) \end{aligned}$$

for some $C > 0$. By substituting (3.5.37) into (3.5.36) and, again, employing (3.5.6), we obtain

$$\begin{aligned} &\int_{-\infty}^{\infty} D_{xt}^2(x,s) + \epsilon_0 D_{xx}^2(x,s))\,ds + \Sigma_0 \int_0^s \int_{-\infty}^{\infty} D_{xt}^2(x,t)\,dx\,dt \qquad (3.5.38) \\ &\le J_2 + 2\delta\Lambda \int_{-\infty}^{\infty} D_0'^2(x)\,dx + \delta\bar{C}\Big(\int_{-\infty}^{\infty} D_x^2(x,s)\,ds \\ &+\int_0^s \int_{-\infty}^{\infty} (D_x^2 + D_t^2 + D_{xx}^2 + D_{xt}^2)(x,t)\,dx\,dt\Big) \end{aligned}$$

for some $\bar{C} > 0$, with J_2 as given by (3.5.18c). Choosing δ so small that

$$\Sigma_0 - \delta\bar{C} \geq \frac{1}{2}\Sigma_0$$

and making use of our first fundamental estimate, i.e., (3.5.25), we now see that for some $C' > 0$,

$$\begin{aligned}\int_{-\infty}^{\infty} D_{xt}^2(x,s) + \epsilon_0 D_{xx}^2(x,s))\,dx \;&+\; \frac{1}{2}\Sigma_0 \int_0^s \int_{-\infty}^{\infty} D_{xt}^2(x,t)\,dx\,dt \\ \leq J_2 + 2\delta\Lambda \int_{-\infty}^{\infty} D_0'^2(x)\,dx \;&+\; \delta C'(J_0 + J_1) \\ &+\; \delta\bar{C} \int_0^s \int_{-\infty}^{\infty} D_{xx}^2(x,t)\,dx\,dt\end{aligned} \tag{3.5.39}$$

In view of (3.5.39), it should be clear that an estimate for $\int_0^s \int_{-\infty}^{\infty} D_{xx}^2(x,t)\,dx\,dt$ is now in order; such an estimate may be generated by rewriting (3.5.1) (for $t < s \leq T_\infty$) in the form

$$\mathsf{E}'(D)D_{xx} + \mathsf{E}''(D)D_x^2 = D_{tt} + \Sigma'(D)D_t,$$

multiplying through by D_{xx}, integrating over $[0,s)\times(-\infty,\infty)$, and employing (3.5.24) so as to obtain the bound

$$\begin{aligned}\epsilon_0 \int_0^s \int_{-\infty}^{\infty} D_{xx}^2\,dx\,dt &\leq \delta\Lambda \int_0^s \int_{-\infty}^{\infty} D_x^2\,dx\,dt \\ &+ \int_0^s \int_{-\infty}^{\infty} \Sigma'(D)D_t D_{xx}\,dx\,dt + \frac{1}{2\lambda_1}\int_0^s \int_{-\infty}^{\infty} D_{xx}^2\,dx\,dt \\ &+ \frac{\lambda_1}{2}\int_0^s \int_{-\infty}^{\infty} D_{tt}^2\,dx\,dt\end{aligned} \tag{3.5.40}$$

which is valid for any $\lambda_1 > 0$. Choosing λ_1 so large that $\epsilon_0 - \dfrac{1}{2\lambda_1} > 0$, and employing the identity

$$\begin{aligned}\Sigma'(D)D_t D_{xx} \;&=\; [\Sigma'(D)D_t D_x]_x - \frac{1}{2}\Sigma''(D)D_t D_x^2 \\ &\quad - \frac{1}{2}(\Sigma'(D)D_x^2)_t,\end{aligned}$$

we are easily led from (3.5.40) to the estimate

$$\begin{aligned}\frac{1}{2}\Sigma_0 \int_{-\infty}^{\infty} D_x^2(x,s)\,dx + \left(\epsilon_0 - \frac{1}{2\lambda_1}\right)\int_0^s \int_{-\infty}^{\infty} D_{xx}^2\,dx\,dt \\ \leq \frac{1}{2}\int_{-\infty}^{\infty} \Sigma'(D_0(x))D_0'^2(x)\,dx + \frac{\lambda_1}{2}\int_0^s \int_{-\infty}^{\infty} D_{tt}^2\,dx\,dt \\ + \frac{3}{2}\delta\Lambda \int_0^s \int_{-\infty}^{\infty} D_x^2\,dx\,dt\end{aligned} \tag{3.5.41}$$

Now, we employ (3.5.25) again so as to obtain from (3.5.41) the final estimate in this set, namely, for some $\hat{C}, \tilde{C} > 0$,

$$\int_0^s \int_{-\infty}^{\infty} D_{xx}^2 \, dx \, dt \le \hat{C}(J_0 + J_1 + J_3) + \tilde{C} \int_0^s \int_{-\infty}^{\infty} D_{tt}^2 \, dx \, dt \tag{3.5.42}$$

with J_3 as given by (3.5.18d). A truly useful estimate for $\int_0^s \int_{-\infty}^{\infty} D_{xx}^2 \, dx \, dt$ will result, however, only if an appropriate bound is found for $\int_0^s \int_{-\infty}^{\infty} D_{tt}^2 \, dx \, dt$, which we now set about establishing. To accomplish this, we first differentiate (3.5.1) through with respect to t, for $t < s$, multiply the resulting equation by $2u_{tt}$, and then integrate over $[0, s) \times (-\infty, \infty)$, thus achieving an estimate of the form

$$\begin{aligned} &\int_{-\infty}^{\infty} D_{tt}^2(x, s) \, dx + \Sigma_0 \int_0^s \int_{-\infty}^{\infty} D_{tt}^2 \, dx \, dt \\ &\quad \le \int_{-\infty}^{\infty} D_{tt}^2(x, 0) \, dx + 2 \int_0^s \int_{-\infty}^{\infty} \mathsf{E}'(D) D_{xxt} D_{tt} \, dx \, dt \\ &\quad +\delta C \int_0^s \int_{-\infty}^{\infty} (D_t^2 + D_x^2 + D_{tt}^2 + D_{xt}^2 + D_{xx}^2)(x, t) \, dx \, dt \end{aligned} \tag{3.5.43}$$

for some $C > 0$. By using the identity

$$\begin{aligned} \mathsf{E}'(D) D_{xxt} D_{xt} &= (\mathsf{E}'(D) D_{xt} D_{tt})_x - \mathsf{E}''(D) D_x D_{tt} D_{xt} \\ &\quad - \frac{1}{2} (\mathsf{E}'(D) D_{xt}^2)_t + \frac{1}{2} \mathsf{E}''(D) D_t D_{xt}^2, \end{aligned}$$

and choosing δ sufficiently small, we are now able to achieve an estimate of the form

$$\begin{aligned} &\int_{-\infty}^{\infty} (D_{tt}^2(x, s) + \epsilon_0 D_{xt}^2(x, s)) \, dx + \frac{\Sigma_0}{4} \int_0^s \int_{-\infty}^{\infty} D_{tt}^2 \, dx \, dt \\ &\quad \le J_4 + \delta C_0 \int_0^s \int_{-\infty}^{\infty} (D_t^2 + D_x^2 + D_{xt}^2 + D_{xx}^2) \, dx \, dt \end{aligned} \tag{3.5.44}$$

for some $C_0 > 0$, where J_4 is given by (3.5.18e). If we now combine (3.5.42) with (3.5.39) we find that for some $C_1, C_2, C_3 > 0$,

$$\begin{aligned} &\int_{-\infty}^{\infty} (D_{xt}^2(x, s) + \epsilon_0 D_{xx}^2(x, s)) \, dx + \frac{1}{2} \Sigma_0 \int_0^s \int_{-\infty}^{\infty} D_{xt}^2 \, dx \, dt \\ &\quad \le J_2 + \delta C_1 \int_{-\infty}^{\infty} D_0'^2(x) \, dx + \delta C_2 (J_0 + J_1 + J_3) \\ &\quad +\delta C_3 \int_0^s \int_{-\infty}^{\infty} D_{tt}^2 \, dx \, dt \end{aligned} \tag{3.5.45}$$

while combining (3.5.44) with (3.5.42) yields an estimate of the form

$$\int_{-\infty}^{\infty}(D_{tt}^2(x,s)+\epsilon_0 D_{xt}^2(x,s))\,dx+\frac{1}{8}\Sigma_0\int_0^s\int_{-\infty}^{\infty}D_{tt}^2\,dx\,dt \tag{3.5.46}$$
$$\leq J_4+\delta C_4(J_0+J_1+J_3)+\delta C_0\int_0^s\int_{-\infty}^{\infty}(D_t^2+D_x^2+D_{xt}^2)\,dx\,dt$$

for some $C_4>0$ and δ chosen sufficiently small. Adding the estimates (3.5.45) and (3.5.46), choosing δ so small that both

$$\frac{1}{2}\Sigma_0-\delta C_0\geq\frac{1}{12}\Sigma_0 \quad\text{and}\quad \frac{1}{8}\Sigma_0-\delta C_3\geq\frac{1}{12}\Sigma_0,$$

and making use, once more, of (3.5.25), as well as the fact that

$$\int_{-\infty}^{\infty}D_0'^2(x)\,dx\leq C'J_1,\quad\text{for some } C'>0,$$

we now find that for some $C^*>0$,

$$\int_{-\infty}^{\infty}(D_{tt}^2(x,s)+D_{xt}^2(x,s)+D_{xx}^2(x,s))\,dx \tag{3.5.47}$$
$$+\int_0^s\int_{-\infty}^{\infty}(D_{xt}^2+D_{tt}^2)\,dx\,dt\leq C^*\left(\sum_{k=0}^{4}J_k\right)$$

However, by virtue of (3.5.42) and (3.5.47), we also have

$$\int_0^s\int_{-\infty}^{\infty}D_{xx}^2\,dx\,dt\leq C_*\left(\sum_{k=0}^{4}J_k\right) \tag{3.5.48}$$

for some $C_*>0$, and the required result, i.e., (3.5.35), now follows directly by summing the estimates (3.5.47) and (3.5.48). □

In order to proceed further down the path which we have been following, we now need to obtain an estimate for the third order derivatives of $D(x,t)$ which is similar in nature to that represented by (3.5.35); however, the local solution D on $(0,T_\infty)$ is simply not smooth enough to justify our formally continuing the procedure which has been used to arrive at the estimates in Lemmas 3.7 and 3.8. In order to continue with the same kind of computations as those which produced the two previous lemmas, we would need the (local) existence of various fourth order derivates of $D(x,t)$ and such derivatives are not guaranteed by the local existence theorem to exist as *bona fide* functions. Fortunately, it turns out that we may proceed in precisely such a formal manner as that indicated above, because the estimates obtained from such a formal procedure can, in fact, be derived in a rigorous fashion by using a standard approximation technique, i.e., forward differencing operators. This observation appears in an analogous context in the recent work of Hrusa and Nohel [71] where, indeed, only the formal derivations of the relevant energy estimates are presented. Hrusa and Nohel [71] (and we, as well) refer the interested reader to the paper of Dafermos and Nohel [42] for details on how to apply forward differencing operators in a situation entirely analogous to the one we are confronted with now at this junction in our analysis. For other similar applications of this approximation technique, the reader may also consult the papers of Hrusa [70] and Coleman, Hrusa, and Owen [35]. With the aforementioned considerations in mind, we now are in a position to state and prove the following:

Lemma 3.9 Under the same conditions as those which prevail in Lemmas 3.7 and 3.8,

$$\begin{aligned}\int_{-\infty}^{\infty}(D_{ttt}^2 + D_{xxt}^2 + D_{xtt}^2 + D_{xxx}^2)(x,s)\,dx \\ + \int_0^s\int_{-\infty}^{\infty}(D_{ttt}^2 + D_{xxt}^2 + D_{xtt}^2 + D_{xxx}^2)\,dx\,dt \\ \leq C\left(\sum_{k=0}^{7} J_k\right)\end{aligned} \tag{3.5.49}$$

for some $C > 0$ where the $J_k, k = 0, \ldots, 7$ are defined by (3.5.18a)-(3.5.18f).

Proof: Proceeding formally, for $t < s$ we differentiate (3.5.1) through with respect to x and then use (3.5.6) and (3.5.24) to obtain estimates of the form

$$\begin{aligned}\int_{-\infty}^{\infty} D_{xxx}^2(x,s)\,dx \leq \int_{-\infty}^{\infty} D_{xtt}^2(x,s)\,dx \\ +C\int_{-\infty}^{\infty}(D_{xt}^2(x,s) + D_x^2(x,s))\,dx\end{aligned} \tag{3.5.50a}$$

and

$$\int_0^s\int_{-\infty}^{\infty} D_{xxx}^2\,dx\,dt \leq \int_0^s\int_{-\infty}^{\infty} D_{xtt}^2\,dx\,dt + C\int_0^s\int_{-\infty}^{\infty}(D_{xt}^2 + D_x^2)\,dx\,dt \tag{3.5.50b}$$

for some $C > 0$. In a similar manner, differentiation of (3.5.1) with respect to t, for $t < s$, produces the two estimates

$$\int_{-\infty}^{\infty} D_{ttt}^2\,dx \leq \bar{C}\int_{-\infty}^{\infty}(D_t^2 + D_{xt}^2 + D_{tt}^2 + D_{xxt}^2)(x,s)\,dx \tag{3.5.51a}$$

and

$$\int_0^s \int_{-\infty}^{\infty} D_{ttt}^2\,dx\,dt \le \bar{C}\int_0^s \int_{-\infty}^{\infty} (D_t^2 + D_{xt}^2 + D_{tt}^2 + D_{xxt}^2)\,dx \tag{3.5.51b}$$

for some $\bar{C} > 0$. In order to obtain our next estimate, we differentiate (3.5.1) twice in succession with respect to x, for $t < s$, multiply the resulting equation by $2D_{xxt}$, and then integrate over $[0,s) \times (-\infty,\infty)$ so as to obtain, with the aid of (3.5.6) again, the estimate

$$\begin{aligned}
&\int_{-\infty}^{\infty} D_{xxt}^2(x,s)\,dx + \Sigma_0 \int_0^s \int_{-\infty}^{\infty} D_{xxt}^2\,dx\,dt \qquad (3.5.52)\\
&\le \int_{-\infty}^{\infty} D_{xxt}^2(x,0)\,dx - \int_0^s \int_{-\infty}^{\infty} \Sigma''(D)(1+D_x)(D_{xt}^2)_x\,dx\,dt\\
&-2\int_0^s \int_{-\infty}^{\infty} \Sigma'''(D)D_t D_x^2 D_{xxt}\,dx\,dt\\
&-2\int_0^s \int_{-\infty}^{\infty} \Sigma'''(D)D_t D_{xx} D_{xxt}\,dx\,dt\\
&+2\int_0^s \int_{-\infty}^{\infty} \mathsf{E}(D)_{xxxx} D_{xxt}\,dx\,dt
\end{aligned}$$

Elementary calculations employing (3.5.5a)-(3.5.5c), and (3.5.24), then yield the existence of constants $C_i > 0$, $i = 1,2,3$, such that

$$-\int_0^s \int_{-\infty}^{\infty} \Sigma''(D)(1+D_x)(D_{xt}^2)_x\,dx\,dt \le C_1 \int_0^s \int_{-\infty}^{\infty} D_{xt}^2\,dx\,dt \tag{3.5.53a}$$

$$-2\int_0^s \int_{-\infty}^{\infty} \Sigma'''(D)D_t D_x^2 D_{xxt}\,dx\,dt \le \delta C_2 \int_0^s \int_{-\infty}^{\infty} (D_t^2 + D_{xxt}^2)\,dx\,dt \tag{3.5.53b}$$

and

$$-2\int_0^s \int_{-\infty}^{\infty} \Sigma''(D)D_t D_{xx} D_{xxt}\,dx\,dt \le \delta C_3 \int_0^s \int_{-\infty}^{\infty} (D_t^2 + D_{xxt}^2)\,dx\,dt \tag{3.5.53c}$$

Employing (3.5.53a)-(3.5.53c) in (3.5.52), and choosing (in a now familiar fashion) δ sufficiently small, we are led to an estimate of the form

$$\begin{aligned}
&\int_{-\infty}^{\infty} D_{xxt}^2(x,s)\,dx + \frac{1}{2}\Sigma_0 \int_0^s \int_{-\infty}^{\infty} D_{xxt}^2\,dx\,dt \qquad (3.5.54)\\
&\le \int_{-\infty}^{\infty} D_{xxt}^2(x,0)\,dx + C\int_0^s \int_{-\infty}^{\infty} (D_t^2 + D_{xt}^2)\,dx\,dt\\
&+2\int_0^s \int_{-\infty}^{\infty} \mathsf{E}(D)_{xxxx} D_{xxt}\,dx\,dt
\end{aligned}$$

for some $C > 0$. Next, we integrate, by parts, the last integral in (3.5.54) over $[0, s) \times (-\infty, \infty)$ first in x, and then in t, and substitute the result back into (3.5.54), thus generating the estimate

$$\begin{aligned}
&\int_{-\infty}^{\infty} (D_{xxt}^2(x,s) + \epsilon_0 D_{xxx}^2(x,s))\, dx + \frac{1}{2}\Sigma_0 \int_0^s \int_{-\infty}^{\infty} D_{xxt}^2\, dx\, dt \qquad (3.5.55)\\
&\quad \le J_5 + C \int_0^s \int_{-\infty}^{\infty} (D_t^2 + D_{xt}^2)\, dx\, dt + \delta C' \int_0^s \int_{-\infty}^{\infty} D_{xxx}^2\, dx\, dt\\
&\qquad - 6 \int_0^s \int_{-\infty}^{\infty} \mathsf{E}''(D) D_x D_{xx} D_{xxxt}\, dx\, dt\\
&\qquad - 2 \int_0^s \int_{-\infty}^{\infty} \mathsf{E}'''(D) D_x^3 D_{xxxt}\, dx\, dt
\end{aligned}$$

for some $C' > 0$, where $C > 0$ is as in (3.5.54) and J_5 is given by (3.5.18f). Our task now is to estimate, from above, the last two integrals in (3.5.55); computing both of these integrals, by integrating twice by parts, first in t, and then in x, we obtain for some constants $C_i > 0$, $i = 1, \ldots, 6$, $\tilde{C} > 0$, and $\hat{C} > 0$, bounds of the form

$$\begin{aligned}
&- 6 \int_0^s \int_{-\infty}^{\infty} \mathsf{E}''(D) D_x D_{xx} D_{xxxt}\, dx\, dt \qquad (3.5.56a)\\
&\quad \le C_1 \int_{-\infty}^{\infty} (D_x^2 + D_{xx}^2 + D_{xxx}^2)(x,0)\, dx\\
&\qquad + C_2 \int_{-\infty}^{\infty} D_{xx}^2(x,s)\, dx + C_3 \int_0^s \int_{-\infty}^{\infty} D_{xx}^2\, dx\, dt\\
&\qquad + \delta \tilde{C} \int_0^s \int_{-\infty}^{\infty} (D_{xxt}^2 + D_{xxx}^2)\, dx\, dt
\end{aligned}$$

and

$$\begin{aligned}
&- 2 \int_0^s \int_{-\infty}^{\infty} \mathsf{E}'''(D) D_x^3 D_{xxxt}\, dx\, dt \qquad (3.5.56b)\\
&\quad \le C_4 \int_{-\infty}^{\infty} (D_x^2(x,0) + D_{xxx}^2(x,0))\, dx\\
&\qquad + C_5 \int_{-\infty}^{\infty} (D_x^2 + D_{xx}^2)(x,s)\, dx + C_6 \int_0^s \int_{-\infty}^{\infty} (D_t^2 + D_x^2)\, dx\, dt\\
&\qquad + \delta \hat{C} \int_0^s \int_{-\infty}^{\infty} D_{xxx}^2\, dx\, dt
\end{aligned}$$

Combining (3.5.56a), (3.5.56b) with (3.5.55), taking δ so small that $\Sigma_0 - 2\delta\tilde{C} \ge \frac{1}{2}\Sigma_0$, then employing both the estimates (3.5.25) and (3.5.35), and relabeling the generic constants, we obtain the estimate

$$\int_{-\infty}^{\infty}(D_{xxt}^2(x,s)+\epsilon_0 D_{xxx}^2(x,s))\,dx+\frac{1}{4}\Sigma_0\int_0^s\int_{-\infty}^{\infty}D_{xxt}^2\,dx\,dt \tag{3.5.57}$$

$$\leq C_0\left(\sum_{k=0}^{5}J_k\right)+C\int_{-\infty}^{\infty}(D_x^2+D_{xx}^2+D_{xxx}^2)(x,0)\,dx$$

$$+\delta\hat{C}\int_0^s\int_{-\infty}^{\infty}D_{xxx}^2\,dx\,dt$$

for some positive $C_0, C, \hat{C}$ with $J_k, k=0,\ldots,5$ as given by (3.5.18a)-(3.5.18f). At this point it is clear that we need an estimate for $\int_0^s\int_{-\infty}^{\infty}D_{xxx}^2\,dx\,dt$; however, some reflection on the structure of the estimates generated up to now indicates that in order to avail ourselves of such an estimate, we must first produce a bound for $\int_0^s\int_{-\infty}^{\infty}D_{xtt}^2\,dx\,dt$, which we now proceed to do.

To begin with, we differentiate (3.5.1), first with respect to x, and then with respect to t; we then multiply the resulting equation by $2D_{xtt}$ (all of this, of course, for $t<s$) and integrate over $[0,s)\times(-\infty,\infty)$, so as to produce

$$\int_{-\infty}^{\infty}D_{xtt}^2(x,s)\,dx+2\int_0^s\int_{-\infty}^{\infty}\Sigma(D)_{xtt}D_{xtt}\,dx\,dt \tag{3.5.58}$$

$$=\int_{-\infty}^{\infty}D_{xtt}^2(x,0)\,dx+2\int_0^s\int_{-\infty}^{\infty}\mathsf{E}(D)_{xxxt}D_{xtt}\,dx\,dt$$

Expanding the second integral on the left-hand side of (3.5.58), and integrating by parts, we obtain from (3.5.58), with the help of (3.5.5c), an estimate of the form

$$\int_{-\infty}^{\infty}D_{xtt}^2(x,s)\,dx+\Sigma_0\int_0^s\int_{-\infty}^{\infty}D_{xtt}^2\,dx\,dt \tag{3.5.59}$$

$$\leq\int_{-\infty}^{\infty}D_{xtt}^2(x,0)\,dx+2\int_0^s\int_{-\infty}^{\infty}\mathsf{E}(D)_{xxxt}D_{xtt}\,dx\,dt$$

$$+C\int_0^s\int_{-\infty}^{\infty}(D_t^2+D_x^2)\,dx\,dt+\delta C'\int_0^s\int_{-\infty}^{\infty}D_{xtt}^2\,dx\,dt$$

for some $C, C'>0$. For δ so small that $\Sigma_0-\delta C'\geq\frac{1}{2}\Sigma_0$ we have, therefore, achieved an estimate of the form

$$\int_{-\infty}^{\infty}D_{xtt}^2(x,s)\,dx+\frac{1}{2}\Sigma_0\int_0^s\int_{-\infty}^{\infty}D_{xtt}^2\,dx\,dt \tag{3.5.60}$$

$$\leq\int_{-\infty}^{\infty}D_{xtt}^2(x,0)\,dx+\int_0^s\int_{-\infty}^{\infty}(D_t^2+D_x^2)\,dx\,dt$$

$$+2\int_0^s\int_{-\infty}^{\infty}\mathsf{E}(D)_{xxxt}D_{xtt}\,dx\,dt$$

Judiciously estimating the last integral on the right-hand side of (3.5.60), and choosing δ to be sufficiently small, now produces the bound

$$\int_{-\infty}^{\infty} (D_{xtt}^2(x,s)\,dx + \epsilon_0 D_{xxt}^2(x,s))\,dx + \frac{1}{8}\Sigma_0 \int_0^s \int_{-\infty}^{\infty} D_{xtt}^2\,dx\,dt \tag{3.5.61}$$
$$\leq J_6 + C_1 \int_0^s \int_{-\infty}^{\infty} (D_t^2 + D_x^2 + D_{xt}^2)\,dx\,dt + \delta C_2 \int_0^s \int_{-\infty}^{\infty} (D_{xxx}^2 + D_{xxt}^2)\,dx\,dt$$

for some $C_1, C_2 > 0$, with J_6 as defined by (3.5.18g). Using the bounds (3.5.25), (3.5.35), and the estimate (3.5.50b), in conjunction with (3.5.61), we easily obtain from this latter estimate the result that

$$\int_{-\infty}^{\infty} (D_{xtt}^2(x,s)\,dx + \epsilon_0 D_{xxt}^2(x,s))\,dx + \frac{1}{8}\Sigma_0 \int_0^s \int_{-\infty}^{\infty} D_{xtt}^2\,dx\,dt \tag{3.5.62}$$
$$\leq C_* \left(\sum_{k=0}^{6} J_k\right) + \delta\hat{C} \int_0^s \int_{-\infty}^{\infty} D_{xtt}^2\,dx\,dt + \delta C^* \int_0^s \int_{-\infty}^{\infty} D_{xxt}^2\,dx\,dt$$

which is valid for some $C_*, C^*, \hat{C} > 0$. Thus, by choosing δ so small that $\Sigma_0 - 2\delta\hat{C} \geq \frac{1}{2}\Sigma_0$, we have, effectively, produced the estimate

$$\int_{-\infty}^{\infty} (D_{xtt}^2(x,s)\,dx + \epsilon_0 D_{xxt}^2(x,s))\,dx + \frac{1}{16}\Sigma_0 \int_0^s \int_{-\infty}^{\infty} D_{xtt}^2\,dx\,dt \tag{3.5.63}$$
$$\leq C_* \left(\sum_{k=0}^{6} J_k\right) + \delta C^* \int_0^s \int_{-\infty}^{\infty} D_{xxt}^2\,dx\,dt$$

On the other hand, by virtue of (3.5.57), (3.5.25), (3.5.35), and (3.5.50b), we clearly have the estimate

$$\int_{-\infty}^{\infty} (D_{xxt}^2(x,s)\,dx + \epsilon_0 D_{xxx}^2(x,s))\,dx + \frac{1}{4}\Sigma_0 \int_0^s \int_{-\infty}^{\infty} D_{xxt}^2\,dx\,dt \tag{3.5.64}$$
$$\leq C_0 \left(\sum_{k=0}^{5} J_k\right) + C \int_{-\infty}^{\infty} (D_x^2 + D_{xx}^2 + D_{xxx}^2)(x,0)\,dx + \delta C' \int_0^s \int_{-\infty}^{\infty} D_{xtt}^2\,dx\,dt$$

for some $C_0, C, C' > 0$. By adding the inequalities (3.5.63) and (3.5.64) there now results an estimate of the form

$$\epsilon_0^* \int_{-\infty}^{\infty} (D_{xtt}^2 + D_{xxt}^2 + D_{xxx}^2)(x,s)\,dx \tag{3.5.65}$$
$$+\frac{1}{16}\Sigma_0 \int_0^s \int_{-\infty}^{\infty} D_{xtt}^2\,dx\,dt + \frac{1}{4}\Sigma_0 \int_0^s \int_{-\infty}^{\infty} D_{xxt}^2\,dx\,dt$$
$$\leq \tilde{C} \left(\sum_{k=0}^{7} J_k\right) + \delta C^* \int_0^s \int_{-\infty}^{\infty} D_{xxt}^2\,dx\,dt$$
$$+\delta C' \int_0^s \int_{-\infty}^{\infty} D_{xtt}^2\,dx\,dt$$

for some $\tilde{C} > 0$, where $\epsilon_0^* = \min(1, \epsilon_0)$ and J_7 is given by (3.5.18h). Choosing δ so small that both

$$\Sigma_0 - 16\delta C' \geq \frac{1}{2}\Sigma_0 \quad \text{and} \quad \Sigma_0 - 4\delta C^* \geq \frac{1}{8}\Sigma_0$$

we now obtain from (3.5.65) the bound

$$\int_{-\infty}^{\infty} (D_{xtt}^2 + D_{xxt}^2 + D_{xxx}^2)(x,s)\,dx \qquad (3.5.66)$$

$$+ \int_0^s \int_{-\infty}^{\infty} (D_{xtt}^2 + D_{xxt}^2)\,dx\,dt \leq C\left(\sum_{k=0}^{7} J_k\right)$$

for some $C > 0$. However, from (3.5.50b) and (3.5.66), it follows that $\exists\, C_1 > 0$ such that

$$\int_0^s \int_{-\infty}^{\infty} D_{xxx}^2\,dx\,dt \leq C_1\left(\sum_{k=0}^{7} J_k\right) \qquad (3.5.67)$$

while by (3.5.51a), (3.5.51b), (3.5.25), (3.5.35), and (3.5.66) it follows that $\exists\, C_2 > 0$ such that

$$\int_{-\infty}^{\infty} D_{ttt}^2(x,s)\,dx + \int_0^s \int_{-\infty}^{\infty} D_{ttt}^2\,dx\,dt \leq C_2\left(\sum_{k=0}^{7} J_k\right) \qquad (3.5.68)$$

The required estimate, i.e., (3.5.49), which is valid for some $C > 0$, provided δ is chosen sufficiently small, is now obtained by combining (3.5.66), (3.5.67), and (3.5.68). The proof of the lemma is now complete. □

Using the estimates established in Lemmas 3.7, 3.8, and 3.10, we are now in a position to synthesize the desired global existence theorem. We begin by adding the estimates (3.5.25), (3.5.35), and (3.5.49); by making note of the definitions (3.5.13) and (3.5.14) of $\mathcal{E}_0(t)$ and $\mathcal{E}(t)$, respectively, we deduce that for some $C > 0$,

$$\mathcal{E}(s) \leq C\left(\sum_{k=0}^{7} J_k\right) \equiv C\mathcal{I}_0 \qquad (3.5.69)$$

provided that $\mathcal{E}_0(0) \leq \frac{\delta^2}{2^k}$ with $k \geq 2$ and δ is chosen sufficiently small, $\delta \leq \bar{\delta} \leq N < 1$. However, for $\mathcal{E}_0(0) \leq \frac{\delta^2}{2^k}$, with $k \geq 2$, and δ sufficiently small, $\delta \leq \bar{\delta} \leq N$, we have $\mathcal{I}_0 \leq C'\mathcal{E}_0(0)$, for some $C' > 0$ and, therefore, for some $C > 0$

$$\mathcal{E}(s) \leq C\mathcal{E}_0(0) \leq \frac{C}{2^k}\delta^2 \qquad (3.5.70)$$

Choosing $k \geq 2$ so large that $C \leq 2^k/4$, it then follows from (3.5.70) that

$$\mathcal{E}(s) \leq \delta^2/4 \tag{3.5.71}$$

provided $\mathcal{E}_0(0) \leq \frac{\delta^2}{2^k}$, with $k \geq 2$ sufficiently large and $\delta > 0$ sufficiently small. However, by virtue of the bound (3.5.21), which is valid for some $C^0 > 0$, whenever $\|D_0\|^2_{H^3} + \|D_1\|^2_{H^2}$ is sufficiently small, it follows that we will have $\mathcal{E}_0(0) \leq \frac{\delta^2}{2^k}$, provided $(\|D_0\|^2_{H^3} + \|D_1\|^2_{H^2}) \leq \delta^2/2^k C^0$ and, in this situation, with δ sufficiently small and $k \geq 2$ sufficiently large, (3.5.71) holds. From (3.5.71), the definition of $\mathcal{E}(t)$, and the Sobolev embedding theorem, we infer that

$$\begin{aligned} &\max\{\|D(\cdot,s)\|_\infty, \|D_x(\cdot,s)\|_\infty, \|D_t(\cdot,s)\|_\infty, \\ &\qquad \|D_{tt}(\cdot,s)\|_\infty, \|D_{xt}(\cdot,s)\|_\infty, \|D_{xx}(\cdot,s)\|_\infty\} \leq \delta/2 \end{aligned} \tag{3.5.72}$$

in which case it follows that $\exists\, s_1,\ s_1 > s$, such that (3.5.24) holds for $s \leq t < s_1$. Continuing this procedure, we find that for $\|D_0\|^2_{H^3} + \|D_1\|^2_{H^2}$ chosen sufficiently small, $\exists\, C_\infty, C'_\infty$ such that

$$\mathcal{E}(t) \leq C_\infty \mathcal{I}_0 \leq C'_\infty \left(\|D_0\|^2_{H^3} + \|D_1\|^2_{H^2}\right) \tag{3.5.73}$$

for any $t < T_\infty$, with $\mathcal{I}_0 \equiv \sum_{k=0}^{7} J_k$, and the J_k, $k = 0, \ldots, 7$ defined by (3.5.18a)-(3.5.18h). However, (3.5.73) implies that $\mathcal{E}(t)$ must remain finite as $t \to T_\infty^-$ and, therefore, by the second part of the local existence theorem for the Cauchy problem (3.5.1), (3.5.2), i.e., by Lemma 3.6, it follows that $T_\infty = +\infty$. We have now established the following result on global existence of solutions to (3.5.1), (3.5.2) with small initial data:

Theorem 3.5 Under hypotheses (3.5.5a)-(3.5.5c) and (3.5.10), for $\|D_0\|^2_{H^3} + \|D_1\|^2_{H^2}$ sufficiently small there exists a unique solution $D \in C^2(R^1 \times [0,\infty))$ of the Cauchy problem (3.5.1), (3.5.2) which, for any $t < \infty$, satisfies the estimates (3.5.73) for some $C_\infty, C'_\infty > 0$. Furthermore, from (3.5.73) and the definition of $\mathcal{E}(t)$ it follows that

$$\begin{aligned} &D, D_t, D_x, D_{tt}, D_{xt}, D_{xx}, D_{ttt}, D_{xxt}, \\ &\qquad D_{xtt}, \text{ and } D_{xxx} \in L^\infty([0,\infty); L^2(R^1)) \end{aligned} \tag{3.5.74}$$

and

$$D_t, D_x, D_{tt}, D_{xt}, D_{xx}, D_{ttt}, D_{xxt}, \tag{3.5.75}$$

$$D_{xtt}, \text{ and } D_{xxx} \in L^2([0,\infty); L^2(R^1))$$

As a direct consequence of (3.5.74), (3.5.75) we also have the following asymptotic (large time) result for the unique solution of the Cauchy problem (3.5.1), (3.5.2) governing the one-dimensional electromagnetic wave-dielectric interaction:

Corollary 3.1 Under the conditions which prevail in Theorem 3.5, we have

$$D, D_t, D_x, D_{tt}, D_{xt}, \text{ and } D_{xx} \to 0, \tag{3.5.76}$$

$$\text{as } t \to \infty, \text{ uniformly on } R^1$$

and

$$D(\cdot,t), D_t(\cdot,t), D_x(\cdot,t), D_{tt}(\cdot,t), \tag{3.5.77}$$

$$D_{xt}(\cdot,t), \text{ and } D_{xx}(\cdot,t) \to 0,$$

$$\text{as } t \to \infty, \text{ in } L^2(R^1).$$

Moreover, as $t \to \infty$,

$$\begin{cases} D(\cdot,t) \rightharpoonup 0, & \text{in } H^3(R^1) \\ D_t(\cdot,t) \rightharpoonup 0, & \text{in } H^2(R^1) \\ D_{tt}(\cdot,t) \rightharpoonup 0, & \text{in } H^2(R^1) \end{cases} \tag{3.5.78}$$

where the $\rightharpoonup$ denotes, as is customary, weak convergence.

The conclusion that $D \in L^\infty([0,\infty); L^2(R^1))$ (in Theorem 3.5) as well as the asymptotic results $D(x,t) \to 0$, uniformly on R^1, as $t \to \infty$, and $D(\cdot,t) \to 0$, in $L^2(R^1)$, as $t \to \infty$, are a distinct consequence of the structure of the right-hand side of (3.5.1), i.e., the fact that $\mathsf{E}(D)_{xx}$ appears here in lieu of $\mathsf{E}(D_x)_x$; in particular, we may note that the conclusions just cited above, for the electromagnetic displacement field $D(\cdot,t)$, $t > 0$, do not necessarily hold even for solutions of the linearly damped nonlinear wave equation

$$u_{tt} + \alpha u_t = \mathsf{E}(u_x)_x, \qquad \alpha > 0$$

which was treated by Nishida [124].

To this point we have displayed, for the Cauchy problem associated with the quasilinear system (2.6.16), governing plane electromagnetic wave-nonlinear dielectric interactions (or, for the equivalent initial-value problem associated with the nonlinearly damped, nonlinear wave equation (2.6.21)), the following results: breakdown of classical solutions for sufficiently *large* data, long time growth behavior of classical solutions, and existence of classical smooth solutions for sufficiently *small* data; the meaning of the terms *large* and *small* in this context is made precise in the statements of Theorems 2.2, 3.4, and 3.5. Using the methods of compensated compactness and weak convergence, it is also possible to prove a theorem concerning the existence of weak, i.e., $L^\infty(R^1)$, solutions for the initial value problem associated with the quasilinear system (2.6.16), for the case of arbitrarily large initial data; however, such a result for the system (2.6.16) will also follow by analogy from the similar result that will be proven in the next chapter for the somewhat more general quasilinear hyperbolic system which governs the evolution of the current and voltage in a distributed parameter nonlinear transmission line. In the next section we turn to a study of the behavior of solutions to the initial-value problem associated with (2.6.21) when $\mu \neq \mu_0$, indeed to a study of the behavior of solutions as $\mu \to 0^+$.

3.6 A Model of Superconductivity in a Nonlinear Dielectric

In the past few sections of this book, we have been concerned with the propagation of a linearly polarized plane electromagnetic wave of the form

$$\begin{cases} \boldsymbol{E} = (0, E(x,t), 0), & \boldsymbol{D} = (0, D(x,t), 0) \\ \boldsymbol{B} = (0, 0, B(x,t)), & \boldsymbol{H} = (0, 0, H(x,t)) \end{cases}$$

through an unbounded (cylindrical) domain $\Omega \subseteq R^3$ that is filled with a nonlinear dielectric substance which conforms to the constitutive laws

$$E = \mathsf{E}(D), \quad B = \mu H, \text{ and } J = \sigma(E)E$$

where $\mathsf{E}'(D) > 0$, $\mu > 0$, and $\sigma(E) > 0$. We were led, as a consequence, to deal with the nonlinearly damped, nonlinear wave equation (3.5.1), i.e., for $\mu \equiv 1$, with

$$D_{tt}(x,t) + \Sigma(D(x,t))_t = \mathsf{E}(D(x,t))_{xx}$$

in which $\Sigma(\zeta) = \sigma(\mathsf{E}(\zeta))\mathsf{E}(\zeta)$, $\forall \zeta \in R^1$.

In any dielectric, the magnetic permeability μ is given by the relation

$$\mu = \mu_0(1 + \chi_m) > 0 \tag{3.6.1}$$

with μ_0, of course, the permeability of free space and χ_m the magnetic susceptibility. The quantity χ_m may be a constant, but it may also be spatially inhomogeneous, field dependent, temperature dependent, or a combination of some (or all) of these. In a superconductor, it is observed that as the temperature is increased, $\chi_m \to -1$, in which case the magnetic permeability $\mu \to 0^+$; at the same time, in a superconducting material it is also observed that the conductivity $\sigma \to +\infty$. We may idealize the above considerations by assuming that Ω is filled with a nonlinear dielectric substance in which

$$E = \mathsf{E}(D), \quad B = \mu H, \text{ and } J = \sigma_\mu E \tag{3.6.2}$$

with $\sigma_\mu \to +\infty$, as $\mu \to 0^+$. While we could consider here the more general situation in which $J = \sigma_\mu(E)E$ with $\sigma_\mu(\zeta) \to \infty$, as $\mu \to 0^+$, for each fixed $\zeta \in R^1$, we will limit ourselves, in this section, to the assumption that μ is spatially homogeneous, with

$$\sigma_\mu = \sigma/\mu, \quad \sigma > 0 \text{ (const.)}, \tag{3.6.3}$$

and that $\mathsf{E}(\cdot)$ is given by the constitutive hypothesis

$$\mathsf{E}(\zeta) = \lambda_0\zeta + \lambda_2\zeta^3, \ \forall \zeta \in R^1 \tag{3.6.4}$$

for some $\lambda_0, \lambda_2 > 0$. With any loss of generality for the analysis which follows below, we may set $\sigma \equiv 1$ in (3.6.3). If the domain Ω is, for example, maintained at a uniform termperature θ, which we can envision as being steadily increased in a quasistatic fashion, and the magnetic susceptibility $\chi_m = \chi_m(\theta)$, then the problem to be studied here can be thought of as conforming to the situation depicted below:

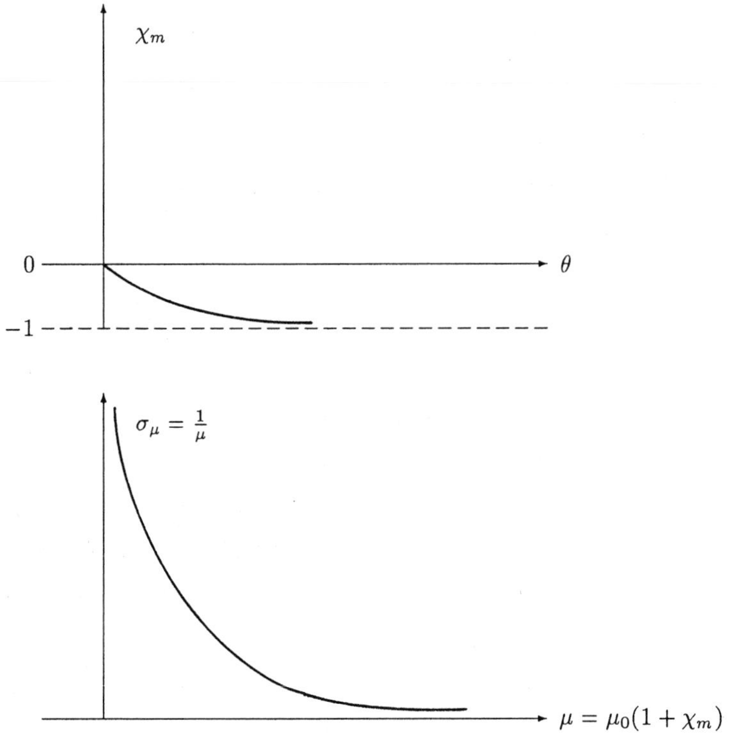

Figure 3.1

If we combine our constitutive relations (3.6.2) with the assumption of linearly polarized plane wave propagation, and Maxwell's equations, we are easily led (for $\sigma = 1$) to the nonlinear wave equation (for $-\infty < x < \infty$, $t > 0$)

$$\mu D_{tt}(x,t) + \mathsf{E}(D(x,t))_t = \mathsf{E}(D(x,t))_{xx} \tag{3.6.5}$$

Even though (3.6.5), for fixed $\mu > 0$, is but a special case of (3.5.1), the methods delineated in the past few sections (most notably § 3.5) and the estimates obtained therein, do not lend themselves to a study of the initial-value problem for (3.6.5) as $\mu \to 0^+$; in fact, we do not wish here to limit the initial data to be, *a priori*, small in any particular norm.

As $\mathsf{E}'(\zeta) > 0$, $\forall \zeta \in R^1$, the constitutive relation $E = \mathsf{E}(D)$ is globally invertible and we may write that

$$D = \mathsf{E}^{-1}(E) \equiv \mathcal{D}(E) \tag{3.6.6}$$

In so doing, we may avail ourselves of (3.6.6) to rewrite (3.6.5) in the equivalent form

$$\mu\mathcal{D}(E)_{tt} + E_t = E_{xx},\ -\infty < x < \infty,\ t > 0 \tag{3.6.7}$$

The nonlinear wave equation (3.6.7) will turn out to be a far more advantageous form to work with, in the limit as $\mu \to 0^+$, than the equivalent equation (3.6.5) and we will, from now on, denote solutions of (3.6.7) as $E^\mu(x,t)$. Of course, associated with (3.6.7) we will have initial data of the form

$$E^\mu(x,0) = E_0(x),\ E_t^\mu(x,0) = E_1(x),\ -\infty < x < \infty \tag{3.6.8}$$

for all $\mu > 0$.

Our basic goal here is to show that a wave which propagates with a finite speed of propagation, for fixed $\mu > 0$, according to (3.6.7), (3.6.8), approaches, in the limit as $\mu \to 0^+$, one which propagates with an infinite speed of propagation; such a result tends to lend a measure of credibility to our intention of characterizing the above (albeit somewhat naïve) model as being one appropriate to a superconducting nonlinear dielectric. To be more precise, on this account, we want to prove that, for each fixed $\mu > 0$, there exists a solution $E^\mu \in L^\infty([0,T];L^2(R^1))$ of (3.6.7), (3.6.8), for any $T > 0$, which has the property that as $\mu \to 0^+$, $E^\mu(x,t) \to E(x,t)$ in $L^2(Q_T)$, $Q_T = R^1 \times [0,T]$, where $E(x,t)$ is the unique classical solution of (3.6.7), for $\mu = 0$, which satisfies the initial condition $E(x,0) = E_0(x)$, $-\infty < x < \infty$; for such a result to hold, we will need to impose a suitable compatability assumption on the initial data functions $E_0(\cdot)$ and $E_1(\cdot)$ in (3.6.8). The specific assumption that we will make is that $(E_0(\cdot), E_1(\cdot))$ is to be restricted to that subset of the space $H^2(R^1)\times(H^1(R^1)\cap L^4(R^1))$ in which

$$E_1(x) = E_0''(x),\ x \in R^1 \tag{3.6.9}$$

is satisfied. All indicated derivatives here are to be understood, of course, as weak derivatives; by virtue of the common device of approximating $L^2(R^1)$ functions by elements of $C_0^\infty(R^1)$ we may operate, in the remainder of this section, as though indicated differentiations are applied to sufficiently smooth functions, i.e., we always have in mind a standard density argument. The rationale behind the compatability condition (3.6.9) will become clear during the course of the derivation of the various *a priori* estimates we will need in order to prove Theorem 3.6 below.

From the constitutive hypothesis (3.6.4), and the definition of $\mathcal{D}(\cdot)$, it is immediate that $\exists\, M > 0$ such that

$$\max\{\|\mathcal{D}'(\cdot)\|_\infty, \|\mathcal{D}''(\cdot)\|_\infty, \|\mathcal{D}'''(\cdot)\|_\infty\} \le M \tag{3.6.10}$$

where $\|\cdot\|_\infty = \|\cdot\|_{L^\infty(R^1)}$. Appealing yet once more to the results of Kato [89] for the relevant local existence theorem, we may state the following:

Lemma 3.10 With $E_0(\cdot) \in H^2(R^1)$ and $E_1(\cdot) \in H^1(R^1)$, the Cauchy problem (3.6.7), (3.6.8) has, for each $\mu > 0$, a unique solution $E^\mu(x,t)$, defined on a maximal time interval $[0, T_\infty)$, $T_\infty \le +\infty$, such that for any $T \in [0, T_\infty)$

$$\begin{gathered} E^\mu, E^\mu_t, E^\mu_x, E^\mu_{tt}, E^\mu_{xt}, \text{ and} \\ E^\mu_{xx} \in L^\infty([0,T]; L^2(R^1)) \end{gathered} \tag{3.6.11}$$

Furthermore, if $T_\infty < \infty$, then as $t \to T_\infty^-$,

$$\begin{aligned} \int_{-\infty}^{\infty} \Big[&(E^\mu(x,t))^2 + (E^\mu_t(x,t))^2 + (E^\mu_x(x,t))^2 \\ &+ (E^\mu_{tt}(x,t))^2 + (E^\mu_{xt}(x,t))^2 + (E^\mu_{xx}(x,t))^2 \Big]\, dx \to \infty \end{aligned} \tag{3.6.12}$$

Now, as a consequence of (3.6.11), it follows that for any T, $0 \le T < T_\infty$,

$$\begin{aligned} J_\mu(T) \equiv \int_0^T \int_{-\infty}^{\infty} \Big[&(E^\mu(x,t))^2 + (E^\mu_t(x,t))^2 + (E^\mu_x(x,t))^2 \\ &+ (E^\mu_{tt}(x,t))^2 + (E^\mu_{xt}(x,t))^2 + (E^\mu_{xx}(x,t))^2 \Big]\, dx\, dt \end{aligned} \tag{3.6.13}$$

is bounded. Our goal will be to obtain estimates for $J_\mu(T)$, for $T < T_\infty$, which are independent of μ and which will enable us to conclude that for each $\mu > 0$

$$\lim_{T \to T_\infty^-} J_\mu(T) < \infty \tag{3.6.14}$$

Employing (3.6.14), we will be able to extend the local weak solution of (3.6.7), (3.6.8), for any $\mu > 0$, to a global weak solution on $Q_T = R^1 \times [0,T]$, for any $T < \infty$; then the estimates which we will have for $J_\mu(T)$, $T < \infty$, which are μ-independent, will enable us to extract the weak limit (in $L^2(Q_T)$), of the globally defined weak solution E^μ, as $\mu \to 0^+$.

The first estimate, which is relevant to the process described above, is now delineated in the following:

Lemma 3.11 Let $\mathcal{I}_1(\mu; E_0, E_1)$ be defined by

$$\mathcal{I}_1 = \frac{1}{2}\mu \int_{-\infty}^{\infty} \mathcal{D}'(E_0(x))E_1^2(x)\,dx + \frac{1}{2}\int_{-\infty}^{\infty} (E_0'(x))^2\,dx \tag{3.6.15}$$

Then for any $t < T_\infty$, the local solution $E^\mu(x,t)$ of (3.6.7), (3.6.8) satisfies

$$\frac{1}{2}\mu \int_{-\infty}^{\infty} \mathcal{D}'(E^\mu(x,t))(E_t^\mu(x,t))^2\,dx + \int_0^t \int_{-\infty}^{\infty} (E_t^\mu(x,\tau))^2\,dx\,d\tau + \frac{1}{2}\int_{-\infty}^{\infty} (E_t^\mu(x,t))^2\,dx \le \mathcal{I}_1 + \frac{1}{2}\mu \int_0^t \int_{-\infty}^{\infty} |\mathcal{D}''(E^\mu)|\,|E_t^\mu|^3\,dx\,d\tau \tag{3.6.16}$$

Proof: Choosing $t < T_\infty$, we multiply (3.6.7) by E_t^μ and integrate over $R^1 \times [0,t]$ so as to obtain

$$\mu \int_0^t \int_{-\infty}^{\infty} \mathcal{D}(E^\mu)_{tt} E_t^\mu\,dx\,d\tau + \int_0^t \int_{-\infty}^{\infty} (E_t^\mu)^2\,dx\,d\tau - \int_0^t \int_{-\infty}^{\infty} E_{xx}^\mu E_t^\mu\,dx\,d\tau = 0 \tag{3.6.17}$$

Expanding the first integral on the left-hand side of (3.6.17) and, then, integrating by parts, we have

$$\int_0^t \int_{-\infty}^{\infty} \mathcal{D}(E^\mu)_{tt} E_t^\mu\,dx\,d\tau = \frac{1}{2}\int_0^t \int_{-\infty}^{\infty} \mathcal{D}''(E^\mu)(E_t^\mu)^3\,dx\,d\tau + \frac{1}{2}\int_{-\infty}^{\infty} \mathcal{D}'(E^\mu(x,t))(E_t^\mu(x,t))^2\,dx - \frac{1}{2}\int_{-\infty}^{\infty} \mathcal{D}'(E_0(x))E_1^2(x)\,dx \tag{3.6.18}$$

while

$$\int_0^t \int_{-\infty}^{\infty} E_{xx}^\mu E_t^\mu\,dx\,d\tau = -\frac{1}{2}\int_{-\infty}^{\infty} (E_x^\mu(x,t))^2\,dx + \frac{1}{2}\int_{-\infty}^{\infty} (E_0'(x))^2\,dx \tag{3.6.19}$$

Combining (3.6.17)-(3.6.19), we easily find that

$$\frac{1}{2}\mu \int_0^t \int_{-\infty}^{\infty} \mathcal{D}''(E^\mu)(E_t^\mu)^3\,dx\,d\tau + \frac{1}{2}\mu \int_{-\infty}^{\infty} \mathcal{D}'(E^\mu(x,t))(E_t^\mu(x,t))^2\,dx + \int_0^t \int_{-\infty}^{\infty} (E_t^\mu)^2\,dx\,d\tau + \frac{1}{2}\int_{-\infty}^{\infty} (E_x^\mu(x,t))^2\,dx = \mathcal{I}_1 \tag{3.6.20}$$

where $\mathcal{I}_1 = \mathcal{I}_1(\mu; E_0, E_1)$ is given by (3.6.15); the required estimate, i.e., (3.6.16), is now a direct consequence of (3.6.20). □

Lemma 3.12 Let $\mathcal{I}_2(\mu; E_0, E_1)$ be defined by

$$\begin{aligned}\mathcal{I}_2 &= \frac{1}{2}\mu \int_{-\infty}^{\infty} \mathcal{D}'(E_0(x))(E_{tt}^{\mu}(x,0))^2\,dx \\ &\quad + \frac{1}{2}\int_{-\infty}^{\infty} (E_1'(x))^2\,dx\end{aligned} \tag{3.6.21}$$

Then for any $t < T_\infty$, the local solution $E^\mu(x,t)$ of (3.6.7), (3.6.8) satisfies

$$\begin{aligned}&\frac{1}{2}\int_{-\infty}^{\infty} \mathcal{D}'(E^\mu(x,t))(E_{tt}^{\mu}(x,t))^2\,dx + \int_0^t\int_{-\infty}^{\infty} (E_{tt}^{\mu})^2\,dx\,d\tau \\ &\quad + \frac{1}{2}\int_{-\infty}^{\infty} (E_{xt}^{\mu}(x,t))^2\,dx \le \mathcal{I}_2 + \mu\int_0^t\int_{-\infty}^{\infty} |\mathcal{D}'''(E^\mu)|\,|E_t^\mu|^3\,|E_{tt}^\mu|\,dx\,d\tau \\ &\quad + \frac{5}{2}\int_0^t\int_{-\infty}^{\infty} |\mathcal{D}''(E_{tt}^\mu)|\,|E_{tt}^\mu|^2\,|E_t^\mu|\,dx\,d\tau\end{aligned} \tag{3.6.22}$$

Proof: To generate the estimate (3.6.22), we begin by differentiating (3.6.7) through with respect to t, for $t < T_\infty$; then we multiply the resulting equation by E_{tt}^μ and subsequently integrate over $R^1 \times [0,t]$ so as to obtain

$$\begin{aligned}&\mu\int_0^t\int_{-\infty}^{\infty} \mathcal{D}(E^\mu)_{ttt}E_{tt}^\mu\,dx\,d\tau + \int_0^t\int_{-\infty}^{\infty} (E_{tt}^\mu)^2\,dx\,d\tau \\ &\quad - \int_0^t\int_{-\infty}^{\infty} E_{xxt}^\mu E_{tt}^\mu\,dx\,d\tau = 0\end{aligned} \tag{3.6.23}$$

However,

$$\begin{aligned}&\int_0^t\int_{-\infty}^{\infty} \mathcal{D}(E^\mu)_{ttt}E_{tt}^\mu\,dx\,d\tau \\ &\quad = \int_0^t\int_{-\infty}^{\infty} \mathcal{D}'''(E^\mu)(E_t^\mu)^3E_{tt}^\mu\,dx\,d\tau \\ &\quad + \frac{5}{2}\int_0^t\int_{-\infty}^{\infty} \mathcal{D}''(E^\mu)(E_{tt}^\mu)^2E_t^\mu\,dx\,d\tau \\ &\quad + \frac{1}{2}\int_{-\infty}^{\infty} \mathcal{D}'(E^\mu(x,t))(E_{tt}^\mu(x,t))^2\,dx - \frac{1}{2}\int_{-\infty}^{\infty} \mathcal{D}'(E_0(x))(E_{tt}^\mu(x,0))^2\,dx\end{aligned} \tag{3.6.24}$$

and

$$\begin{aligned}\int_0^t\int_{-\infty}^{\infty} E_{xxt}^\mu E_{tt}^\mu\,dx\,d\tau &= -\frac{1}{2}\int_{-\infty}^{\infty} (E_{xt}^\mu(x,t))^2\,dx \\ &\quad + \frac{1}{2}\int_{-\infty}^{\infty} (E_1'(x))^2\,dx\end{aligned} \tag{3.6.25}$$

Combining, now, (3.6.23)-(3.6.25), we readily find that

$$\begin{aligned}
&\mu \int_0^t \int_{-\infty}^{\infty} \mathcal{D}'''(E^\mu)(E_t^\mu)^3 E_{tt}^\mu \, dx \, d\tau \\
&+ \frac{5}{2}\mu \int_0^t \int_{-\infty}^{\infty} \mathcal{D}''(E^\mu)(E_{tt}^\mu)^2 E_t^\mu \, dx \, d\tau \\
&+ \frac{1}{2}\mu \int_{-\infty}^{\infty} \mathcal{D}'(E^\mu(x,t))(E_{tt}^\mu(x,t))^2 \, dx \\
&+ \int_0^t \int_{-\infty}^{\infty} (E_{tt}^\mu)^2 \, dx \, d\tau + \frac{1}{2} \int_{-\infty}^{\infty} (E_{xt}^\mu(x,t))^2 \, dx = \mathcal{I}_2
\end{aligned} \tag{3.6.26}$$

where $\mathcal{I}_2 = \mathcal{I}_2(\mu; E_0, E_1)$ is given by (3.6.21); the stated estimate, i.e., (3.6.22), is now a straightforward consequence of (3.6.26). □

Remarks: If we set $t = 0$ in (3.6.7), then we find that for any $x \in R^1$

$$\begin{aligned}
E_{tt}(x,0) \;\; = \;\; & \frac{1}{\mu \mathcal{D}'(E_0(x))} \{E_0''(x) - E_1(x)\} \\
& - \left\{ \frac{\mathcal{D}''(E_0(x))}{\mathcal{D}'(E_0(x))} \right\} E_1^2(x).
\end{aligned} \tag{3.6.27}$$

Therefore, if $E_0''(x) \neq E_1(x)$, a.e. on R^1, we would deduce that $E_{tt}^2(x,0) \sim \mu^{-2}$ as $\mu \to 0^+$; this would, in view of the form (3.6.21) of $\mathcal{I}_2(\mu; E_0, E_1)$ in the estimate of Lemma 3.12, present us with an entirely untenable situation. The relation (3.6.27), and the form of $\mathcal{I}_2(\mu; E_0, E_1)$ are, thus, the motivating factors behind the compatability hypothesis (3.6.9). Therefore, when (3.6.9) is satisfied, we will have

$$E_{tt}(x,0) = -\left\{ \frac{\mathcal{D}''(E_0(x))}{\mathcal{D}'(E_0(x))} \right\} E_1^2(x), \qquad x \in R^1 \tag{3.6.28}$$

We also make the following observations:

(i) For $\mu \le 1$ we have

$$\begin{aligned}
\mathcal{I}_1 + \mathcal{I}_2 \le \mathcal{I}_0 \equiv \frac{1}{2} \bigg\{ & \int_{-\infty}^{\infty} \mathcal{D}'(E_0(x)) E_1^2(x) \, dx \\
& + \int_{-\infty}^{\infty} (E_0'(x))^2 \, dx + \int_{-\infty}^{\infty} (E_1'(x))^2 \, dx \\
& + \int_{-\infty}^{\infty} \left\{ \frac{\mathcal{D}''^2(E_0(x))}{\mathcal{D}'(E_0(x))} \right\} E_1^4(x) \, dx \bigg\}
\end{aligned} \tag{3.6.29}$$

(ii) Also,

$$\frac{\mathcal{D}''^2(\zeta)}{\mathcal{D}'(\zeta)} = \mathsf{E}'^3(\mathcal{D}(\zeta))\mathsf{E}''^2(\mathcal{D}(\zeta)), \qquad \zeta \in R^1 \tag{3.6.30}$$

So, by virtue of (3.6.4), and the definition of $\mathcal{D}(\cdot)$, we have $\mathcal{D}''^2(E_0(x))/\mathcal{D}'(E_0(x))$ bounded and strictly positive $\forall\, x \in R^1$. Our previous estimates and the observations (i), (ii) cited above can now be combined so as to yield

Lemma 3.13 Let $E^\mu(x,t)$ be the local solution of (3.6.7), (3.6.8) on $R^1 \times [0,T]$, for any $T < T_\infty$. Then for all μ sufficiently small

$$E_t^\mu, E_{tt}^\mu, E_x^\mu, \text{ and } E_{xt}^\mu \text{ are bounded} \tag{3.6.31a}$$
$$\text{in } L^2(Q_T), \text{ independent of } \mu$$

and

$$\mu\mathcal{D}(E^\mu)_{tt}, E_{xx}^\mu, \text{ and } E^\mu \text{ are bounded} \tag{3.6.31b}$$
$$\text{independent of } \mu$$

Proof: We begin by defining the quantities

$$A_1(\mu,T) = \left(\int_0^T \int_{-\infty}^{\infty} (E_t^\mu)^2\, dx\, d\tau\right)^{1/2} \tag{3.6.32a}$$

$$A_2(\mu,T) = \left(\int_0^T \int_{-\infty}^{\infty} (E_{tt}^\mu)^2\, dx\, d\tau\right)^{1/2} \tag{3.6.32b}$$

$$A_3(\mu,T) = \sup_{[0,T]} \left(\int_{-\infty}^{\infty} (E_{xt}^\mu(x,t))^2\, dx\right)^{1/2} \tag{3.6.32c}$$

$$A_4(\mu,T) = \sup_{[0,T]} \left(\int_{-\infty}^{\infty} (E_x^\mu(x,t))^2\, dx\right)^{1/2} \tag{3.6.32d}$$

Next, we add the estimates (3.6.16) and (3.6.22) and then employ (3.6.29) so as to obtain, for any $t \in [0, T_\infty)$, and any $\mu \le 1$:

$$\frac{1}{2}\mu \int_{-\infty}^{\infty} \mathcal{D}'(E^\mu(x,t))(E_t^\mu(x,t))^2\, dx + \int_0^T \int_{-\infty}^{\infty} (E_t^\mu)^2\, dx\, d\tau \tag{3.6.33}$$
$$+\frac{1}{2}\mu \int_{-\infty}^{\infty} \mathcal{D}'(E^\mu(x,t))(E_{tt}^\mu(x,t))^2\, dx + \int_0^T \int_{-\infty}^{\infty} (E_{tt}^\mu)^2\, dx\, d\tau$$

$$+\frac{1}{2}\int_{-\infty}^{\infty}(E_{xt}^{\mu}(x,t))^2\,dx+\frac{1}{2}\int_{-\infty}^{\infty}(E_{x}^{\mu}(x,t))^2\,dx$$
$$\leq \mathcal{I}_0+\frac{1}{2}\mu\int_0^t\int_{-\infty}^{\infty}|\mathcal{D}''(E^{\mu})|\,|E_t^{\mu}|^3\,dx\,d\tau$$
$$+\mu\int_0^t\int_{-\infty}^{\infty}|\mathcal{D}''(E^{\mu})|\,|E_t^{\mu}|^3\,|E_{tt}^{\mu}|\,dx\,d\tau$$
$$+\frac{5\mu}{2}\int_0^t\int_{-\infty}^{\infty}|\mathcal{D}''(E^{\mu})|\,|E_{tt}^{\mu}|^2\,|E_t^{\mu}|\,dx\,d\tau$$

Using (3.6.10), it is a straightforward matter to show that the estimate (3.6.33) implies that

$$\frac{1}{2}\mu\int_{-\infty}^{\infty}\mathcal{D}'(E^{\mu}(x,t))(E_t^{\mu}(x,t))^2\,dx+\int_0^T\int_{-\infty}^{\infty}(E_t^{\mu})^2\,dx\,d\tau \tag{3.6.34}$$
$$+\frac{1}{2}\mu\int_{-\infty}^{\infty}\mathcal{D}'(E^{\mu}(x,t))(E_{tt}^{\mu}(x,t))^2\,dx+\int_0^T\int_{-\infty}^{\infty}(E_{tt}^{\mu})^2\,dx\,d\tau$$
$$+\frac{1}{2}\int_{-\infty}^{\infty}(E_{xt}^{\mu}(x,t))^2\,dx+\frac{1}{2}\int_{-\infty}^{\infty}(E_{x}^{\mu}(x,t))^2\,dx$$
$$\leq \mathcal{I}_0+\frac{\mu M}{2}\left(5\|E_t^{\mu}\|_{L^\infty(Q_t)}+\|E_t^{\mu}\|^2_{L^\infty(Q_T)}\right)$$
$$\times\left(\int_0^T\int_{-\infty}^{\infty}(E_t^{\mu})^2\,dx\,d\tau+\int_0^T\int_{-\infty}^{\infty}(E_{tt}^{\mu})^2\,dx\,d\tau\right)$$

However, standard estimates (which are direct consequences of the usual Sobolev embedding results) provide the bounds

$$\|E_t^{\mu}\|_{L^\infty(R^1)} \leq C_0\left[\left(\int_{-\infty}^{\infty}(E_t^{\mu}(x,t))^2\,dx\right)^{1/2}+\left(\int_{-\infty}^{\infty}(E_{xt}^{\mu}(x,t))^2\,dx\right)^{1/2}\right] \tag{3.6.35}$$

and

$$\left(\int_{-\infty}^{\infty}(E_t^{\mu}(x,t))^2\,dx\right)^{1/2} \leq C_1\left(\int_0^T\int_{-\infty}^{\infty}(E_t^{\mu})^2\,dx\,d\tau\right)^{1/2}+C_2\left(\int_0^T\int_{-\infty}^{\infty}(E_{tt}^{\mu})^2\,dx\,d\tau\right)^{1/2} \tag{3.6.36}$$

which are valid for some positive constants C_0, C_1, and C_2, all of which are independent of t. Combining (3.6.35) and (3.6.36), we see that for some $k>0$, which is independent of t,

$$\|E_t^\mu\|_{L^\infty(R^1)} \le k\left[\left(\int_0^t\int_{-\infty}^{\infty}(E_{tt}^\mu)^2\,dx\,d\tau\right)^{1/2}\right. \tag{3.6.37}$$

$$\left. + \left(\int_0^t\int_{-\infty}^{\infty}(E_t^\mu)^2\,dx\,d\tau\right)^{1/2} + \left(\int_{-\infty}^{\infty}(E_{xt}^\mu(x,t))^2\,dx\right)^{1/2}\right]$$

Returning to (3.6.34), we observe that we may rewrite this estimate in the form

$$\frac{1}{2}\mu\int_{-\infty}^{\infty}\mathcal{D}'(E^\mu(x,t))(E_t^\mu(x,t))^2\,dx \tag{3.6.38}$$

$$+\frac{1}{2}\mu\int_{-\infty}^{\infty}\mathcal{D}'(E^\mu(x,t))(E_{tt}^\mu(x,t))^2\,dx$$

$$+\frac{1}{2}\int_{-\infty}^{\infty}(E_{xt}^\mu(x,t))^2\,dx + \frac{1}{2}\int_{-\infty}^{\infty}(E_x^\mu(x,t))^2\,dx$$

$$+\gamma(\mu,M;t)\int_0^t\int_{-\infty}^{\infty}\left[(E_t^\mu)^2 + (E_{tt}^\mu)^2\right]\,dx\,d\tau \le \mathcal{I}_0$$

with

$$\gamma(\mu,M;t) = 1 - \frac{1}{2}\mu M\left[5\|E_t^\mu\|_{L^\infty(Q_t)} + \|E_t^\mu\|^2_{L^\infty(Q_t)}\right] \tag{3.6.39}$$

Defining the $A_i(\mu;T)$, $i = 1,\ldots,4$ as in (3.6.32a)-(3.6.32d), we now obtain directly from (3.6.37) the fact that

$$\|E_t^\mu\|_{L^\infty(Q_t)} \le k\sum_{j=1}^{4}A_j(\mu;T) \equiv kB(\mu;T) \tag{3.6.40}$$

If we set

$$k^* = \frac{1}{2}M\cdot\max\left\{5k, k^2\right\} \tag{3.6.41}$$

then

$$\frac{1}{2}\mu M\left[5\|E_t^\mu\|_{L^\infty(Q_T)} + \|E_t^\mu\|^2_{L^\infty(Q_T)}\right] \le \mu k^*(B(\mu;T) + B^2(\mu;T))$$

in which case

$$\gamma(\mu,M;T) \ge 1 - \mu k^*\left[B(\mu;T) + B^2(\mu;T)\right] \tag{3.6.42}$$

We now return to the estimate (3.6.38), which is valid for $t \le T < \infty$, and $\mu \le 1$; we delete the first two terms on the left-hand side of this inequality, employ the lower bound (3.6.42), and the definitions (3.6.32a)-(3.6.32d), so as to obtain the estimate

$$\begin{aligned}\frac{1}{2}(A_3^2(\mu,T) + A_4^2(\mu,T))& \\ +(1-\mu k^*[B(\mu;T)+B^2(\mu;T)])& \\ \times(A_1^2(\mu;T)+A_2^2(\mu;T)) \le \mathcal{I}_0&\end{aligned} \tag{3.6.43}$$

or

$$\left(1-\mu k^*[B(\mu;T)+B^2(\mu;T)]\right)\sum_{j=1}^{4} A_j^2(\mu;T) \le 2\mathcal{I}_0 \tag{3.6.44}$$

for $\mu \le 1$ and $T < T_\infty$. However

$$B^2(\mu;T) \le 4\sum_{j=1}^{4} A_j^2(\mu;T)$$

so that (3.6.44) implies that

$$\left(1-\mu k^*[B(\mu;T)+B^2(\mu;T)]\right) B^2(\mu;T) \le 8\mathcal{I}_0 \tag{3.6.45}$$

At this point, we fix $\mu \le 1$ and $T < T_\infty$ and examine the function

$$f(B) = B^2 - \mu k^*(B^3 + B^4) \tag{3.6.46}$$

whose graph is sketched in Figure 3.2 in the domain $B \ge 0$; upon this graph we have also superimposed a sketch of the graph of $g(B) = \frac{1}{2}B^2$. A straightforward calculation now establishes the following

(i) $f'(B) > 0$ for $B < B_\mu = -\dfrac{3}{8} + \dfrac{1}{2}\sqrt{\dfrac{9}{16} + \dfrac{2}{\mu k^*}}$

(ii) $g(B)$ intersects $f(B)$ at $B^\mu = -\dfrac{1}{2} + \dfrac{1}{2}\sqrt{1 + \dfrac{2}{\mu k^*}}$
We also note that

(iii) B_μ and $B^\mu \to +\infty$, as $\mu \to 0^+$ and

(iv) $B \le B^\mu \Rightarrow \dfrac{1}{2}B^2 \le f(B)$

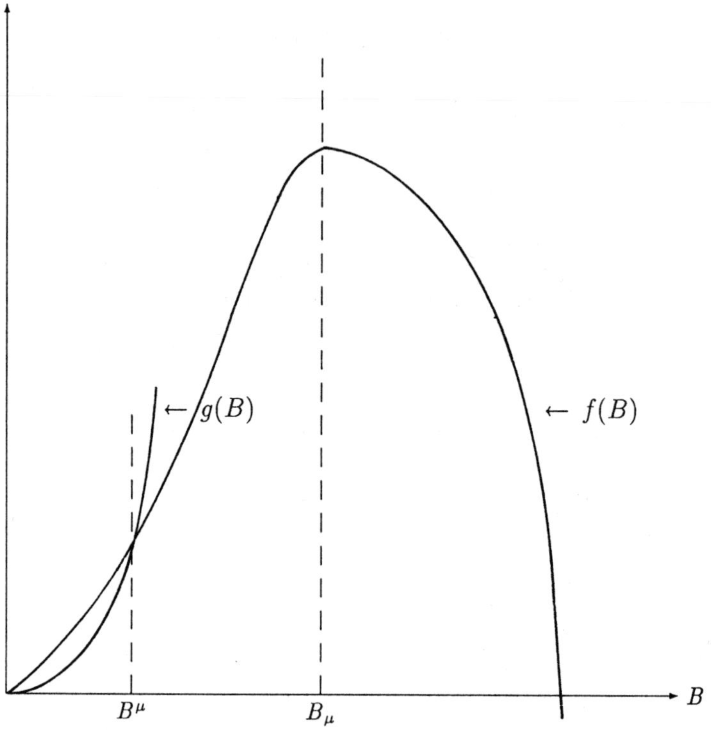

Figure 3.2

Therefore, by virtue of (3.6.45), for any $t < T_\infty$

$$B(\mu; t) \leq B^{\mu} \Rightarrow B(\mu; t) \leq 4\sqrt{\mathcal{I}_0} \tag{3.6.47}$$

In view of the definitions (3.6.32a)-(3.6.32d) of the $A_j(\mu; T)$, $j = 1, \ldots, 4$, the definition of $B(\mu; T)$, and (3.6.29), we have

$$B^2(\mu; 0) \leq 2\int_{-\infty}^{\infty} \left\{E_0'^2(x) + E_1'^2(x)\right\}\, dx \;\leq 4\sqrt{\mathcal{I}_0} \tag{3.6.48}$$

and, therefore, for μ sufficiently small, say, $\mu \leq \mu_0$,

$$B(\mu; 0) \leq 2\sqrt{\mathcal{I}_0} \leq \frac{1}{4}B^{\mu} \tag{3.6.49}$$

so that by continuity $\exists\, t_1 > 0$ such that

$$B(\mu; t_1) \leq B^{\mu} \text{ (for } \mu \leq \mu_0) \tag{3.6.50}$$

Combining (3.6.47) and (3.6.49) we now find that

$$B(\mu;t_1) \leq 4\sqrt{\mathcal{I}_0} \leq \frac{1}{2}B^\mu \text{ (for } \mu \leq \mu_0) \tag{3.6.51}$$

in which case $\exists\, t_2 > t_1$ such that

$$B(\mu;t_2) \leq B^\mu \text{ (for } \mu \leq \mu_0) \tag{3.6.52}$$

Continuing in this manner, we obtain an increasing sequence $\{t_n\}$, without an accumulation point, in the interval $(0, T_\infty)$, such that for each n we have $B(\mu;t_n) \leq 4\sqrt{\mathcal{I}_0}$ provided that $\mu \leq \mu_0$. Thus, as a direct consequence of (3.6.45), we have established that

$$B(\mu;T) \leq 4\sqrt{\mathcal{I}_0}; \qquad \mu \leq \mu_0,\ T < T_\infty \tag{3.6.53}$$

Referring once more to the definition of $B(\mu;T)$, and employing (3.6.53) for $\mu \leq \mu_0$ and $T < T_\infty$, we now observe the estimate

$$\sum_{j=1}^{4} A_j^2(\mu;T) \leq 4B^2(\mu;T) \leq 64\mathcal{I}_0$$

or

$$\int_0^T \int_{-\infty}^{\infty} \left[(E_t^\mu)^2 + (E_{tt}^\mu)^2\right] dx\, dt$$
$$+ \sup_{[0,T]} \int_{-\infty}^{\infty} \left[(E_x^\mu(x,t))^2 + (E_{xt}^\mu(x,t))^2\right] dx \leq 64\mathcal{I}_0 \tag{3.6.54}$$

From (3.6.54), we may now infer directly the validity of the first part of the lemma, i.e., of (3.6.31a), for $\mu \leq \mu_0$. We remark that bounds for the $L^2(Q_T)$ norms of both E_x^μ and E_{xt}^μ grow, at worst, linearly in T; these norms are, therefore, finite as $T \to T_\infty < \infty$.

For the proof of the second part of the lemma, we observe that straightforward calculations yield the series of estimates

$$\int_0^T \int_{-\infty}^{\infty} (\mu\mathcal{D}(E^\mu)_{tt})^2\, dx\, dt$$
$$\leq 2\mu^2 \int_0^T \int_{-\infty}^{\infty} \left[\mathcal{D}'^2(E^\mu)(E_{tt}^\mu)^2 + \mathcal{D}''^2(E^\mu)(E_t^\mu)^4\right] dx\, dt$$
$$\leq 2M^2\mu^2 \int_0^T \int_{-\infty}^{\infty} (E_{tt}^\mu)^2\, dx\, dt$$

$$+2M^2\mu^2\|E_t^\mu\|_{L^\infty(Q_T)}\int_0^T\int_{-\infty}^{\infty}(E_t^\mu)^2\,dx\,dt$$

$$\leq 2M^2\mu^2\int_0^T\int_{-\infty}^{\infty}(E_{tt}^\mu)^2\,dx\,dt$$

$$+2k^2M^2\mu^2B^2(\mu;T)\int_0^T\int_{-\infty}^{\infty}(E_t^\mu)^2\,dx\,dt$$

where, in the last estimate, we have used (3.6.40). Thus, for $\mu \leq \mu_0$, we have, as a consequence of (3.6.53), the estimate

$$\int_0^T\int_{-\infty}^{\infty}(\mu\mathcal{D}(E^\mu)_{tt})^2\,dx\,dt \tag{3.6.55}$$

$$\leq 2M^2\mu_0^2\left[\int_0^T\int_{-\infty}^{\infty}(E_{tt}^\mu)^2\,dx\,dt\right.$$

$$\left.+16k^2\mathcal{I}_0\int_0^T\int_{-\infty}^{\infty}(E_t^\mu)^2\,dx\,dt\right]$$

Combining (3.6.55) with (3.6.54) we see that for $\mu \leq \mu_0$,

$$\mu\mathcal{D}(E^\mu)_{tt} \text{ is bounded in } L^2(Q_T) \text{ independently of } \mu \tag{3.6.56}$$

However, by virtue of (3.6.31a) and (3.6.56), it follows directly from the evolution equation (3.6.7) that for all $\mu \leq \mu_0$,

$$E_{xx}^\mu \text{ is bounded in } L^2(Q_T) \text{ independently of } \mu \tag{3.6.57}$$

It then follows directly from the elementary estimate

$$\int_0^T\int_{-\infty}^{\infty}(E^\mu)^2\,dx\,dt \leq k^*T^2\int_0^T\int_{-\infty}^{\infty}(E_t^\mu)^2\,dx\,dt + 2T\int_{-\infty}^{\infty}E_0^2(x)\,dx \tag{3.6.58}$$

that, for $\mu \leq \mu_0$,

$$E^\mu \text{ is bounded in } L^2(Q_T) \text{ independently of } \mu \tag{3.6.59}$$

and all elements of the second part of the lemma, i.e., (3.6.31b), have now been established. □

As a consequence of Lemmas 3.10 and 3.13 we may now state the following result relative to the global existence of solutions for the Cauchy problem (3.6.7), (3.6.8) when $\mu > 0$ is sufficiently small:

Theorem 3.6 Suppose that $E_0(\cdot) \in H^2(R^1)$, that $E_1(\cdot) \in H^1(R^1) \cap L^4(R^1)$, and that the compatability condition (3.6.9) is satisfied. Then $\exists\, \mu_0 > 0$ such that, for each $\mu \leq \mu_0$, the Cauchy problem (3.6.7), (3.6.8) has a unique solution $E^\mu \in L^2(Q_T)$, for each $T < \infty$; for each such solution, E^μ_x, E^μ_t, E^μ_{xx}, E^μ_{xt}, and E^μ_{tt} also belong to $L^2(Q_T)$, for each $T < \infty$.

Remarks: As (3.6.55) holds for all $\mu \leq \mu_0$, we see that the global solution of (3.6.7), (3.6.8) satisfies

$$\begin{aligned}
&\int_0^T \int_{-\infty}^{\infty} (\mathcal{D}(E^\mu)_{tt})^2\,dx\,dt \\
&\qquad \leq 2M^2 \left[\int_0^T \int_{-\infty}^{\infty} (E^\mu_{tt})^2\,dx\,dt \right.\\
&\qquad\qquad \left. + 16k^2\mathcal{I}_0 \int_0^T \int_{-\infty}^{\infty} (E^\mu_t)^2\,dx\,dt\right]
\end{aligned}$$

for any $T < \infty$ provided $\mu \leq \mu_0$. Therefore, not only does (3.6.56) hold but, also,

$$\begin{gathered}\mathcal{D}(E^\mu)_{tt} \text{ is bounded in } L^2(Q_T) \\ \text{independently of } \mu\end{gathered} \tag{3.6.60}$$

so that, in fact,

$$\mu\mathcal{D}(E^\mu)_{tt} \to 0, \quad \text{in } L^2(Q_T) \tag{3.6.61}$$

as $\mu \to 0^+$. We are now in a position to state and prove the main result of this section, namely,

Theorem 3.7 Under the same conditions as those which hold in Theorem 3.6, the weak solution $E^\mu \in L^2(Q_T)$ of (3.6.7), (3.6.8), for $\mu \leq \mu_0$, converges weakly in $L^2(Q_T)$, as $\mu \to 0^+$, to the unique classical solution of the initial-value problem:

$$\left\{\begin{aligned} E_t(x,t) &= E_{xx}(x,t), \quad \text{on } R^1 \times [0,T] \\ E(x,0) &= E_0(x), \quad x \in R^1 \end{aligned}\right. \tag{3.6.62}$$

Proof: By virtue of the boundedness of E^μ in $L^2(Q_T)$, for any $T < \infty$, with $\mu \le \mu_0$, it follows that there exists a subsequence $E^{\mu_n(T)}$, for each fixed $T < \infty$, such that

$$E^{\mu_n(T)} \rightharpoonup E, \text{ in } L^2(Q_T) \tag{3.6.63}$$

as $\mu \to 0^+$. As we assume that $T < \infty$ is fixed, we will now suppress displaying the explicit dependence of sequences such as $\{\mu_n(T)\}$ on T. Now, inasmuch as $E_t^{\mu_n(T)}$ is bounded in $L^2(Q_T)$, for $\mu_n \le \mu_0$, we may infer the existence of $E^* \in L^2(Q_T)$ such that

$$E_t^{\mu_{n_k}} \rightharpoonup E^*, \text{ in } L^2(Q_T) \tag{3.6.64}$$

as $\mu_{n_k} \to 0^+$, for some subsequence $\{\mu_{n_k}\}$. Now, let $\mathcal{H} = C_0^\infty(Q_T)$ denote, in the usual way, the space of infinitely differentiable functions with compact support in Q_T; we also denote the dual space to $C_0^\infty(Q_T)$, i.e., the space of distributions, by $\mathcal{H}'$. Then, because weak convergence in L^2 implies convergence in the sense of distributions, i.e., in $\mathcal{H}'$, we also have

$$E_t^{\mu_{n_k}} \longrightarrow E^*, \text{ in } \mathcal{H}'(Q_T) \tag{3.6.65}$$

as $\mu_{n_k} \to 0^+$. However, in view of (3.6.63),

$$E^{\mu_{n_k}} \rightharpoonup E, \text{ in } L^2(Q_T) \tag{3.6.66}$$

as $\mu_{n_k} \to 0^+$, so

$$E^{\mu_{n_k}} \longrightarrow E, \text{ in } \mathcal{H}'(Q_T) \tag{3.6.67}$$

From (3.6.67), it follows that

$$E_t^{\mu_{n_k}} \longrightarrow E_t, \text{ in } \mathcal{H}'(Q_T) \tag{3.6.68}$$

By comparing (3.6.65) with (3.6.68), we see that $E^* = E_t$ so that, in fact, as $\mu_{n_k} \to 0^+$,

$$E_t^{\mu_{n_k}} \rightharpoonup E_t, \text{ in } L^2(Q_T) \tag{3.6.69}$$

Now, $E_{xx}^{\mu_{n_k}}$ is also bounded in $L^2(Q_T)$, for $\mu_{n_k} \le \mu_0$, so there exists a subsequence $\{\mu_{\tilde{n}_k}\}$ of $\{\mu_{n_k}\}$ such that for some $E^{**} \in L^2(Q_T)$,

$$E_{xx}^{\mu_{\tilde{n}_k}} \rightharpoonup E^{**}, \text{ in } L^2(Q_T) \tag{3.6.70}$$

as $\mu_{\bar{n}_k} \to 0^+$. For this subsequence $\{\mu_{\bar{n}_k}\}$ we must still have, of course, that

$$\begin{cases} E^{\mu_{\bar{n}_k}} \rightharpoonup E, & \text{in } L^2(Q_T) \\ E_t^{\mu_{\bar{n}_k}} \rightharpoonup E_t, & \text{in } L^2(Q_T) \end{cases} \tag{3.6.71}$$

as $\mu_{\bar{n}_k} \to 0^+$. By virtue of (3.6.70) we have

$$E_{xx}^{\mu_{\bar{n}_k}} \longrightarrow E^{**}, \text{ in } \mathcal{H}'(Q_T) \tag{3.6.72}$$

as $\mu_{\bar{n}_k} \to 0^+$. However, from (3.6.71),

$$E^{\mu_{\bar{n}_k}} \longrightarrow E, \text{ in } \mathcal{H}'(Q_T) \tag{3.6.73}$$

as $\mu_{\bar{n}_k} \to 0^+$, so that

$$E_{xx}^{\mu_{\bar{n}_k}} \longrightarrow E_{xx}, \text{ in } \mathcal{H}'(Q_T) \tag{3.6.74}$$

and, therefore, $E^{**} = E_{xx}$. Thus,

$$E_{xx}^{\mu_{\bar{n}_k}} \rightharpoonup E_{xx}, \text{ in } L^2(Q_T) \tag{3.6.75}$$

as $\mu_{\bar{n}_k} \to 0^+$. From the evolution equation (3.6.7), with $\mu = \mu_{\bar{n}_k}$, and $E = E^{\mu_{\bar{n}_k}}$, i.e.,

$$\mu_{\bar{n}_k} \mathcal{D}(E^{\mu_{\bar{n}_k}})_{tt} + E_t^{\mu_{\bar{n}_k}} = E_{xx}^{\mu_{\bar{n}_k}} \tag{3.6.76}$$

we now find, as a consequence of (3.6.71), (3.6.75), and (3.6.61) that, for $T < \infty$, $E^{\mu_{\bar{n}_k}} \longrightarrow E$ in $L^2(Q_T)$, as $\mu_{\bar{n}_k} \to 0^+$, with $E(x,t)$ a weak (and, therefore, a classical) solution of the Cauchy problem (3.6.62).

We know that if $\{\mu_m\}$, $\mu_m \to 0^+$, is any other sequence with the property that for some subsequence $\{\mu_{m_k}\}$:

$$\begin{cases} E^{\mu_{m_k}} \rightharpoonup \tilde{E}, & \text{in } L^2(Q_T) \\ E_t^{\mu_{m_k}} \rightharpoonup \tilde{E}_t, & \text{in } L^2(Q_T) \\ E_{xx}^{\mu_{m_k}} \rightharpoonup \tilde{E}_{xx}, & \text{in } L^2(Q_T) \end{cases} \tag{3.6.77}$$

as $\mu_{m_k} \to 0^+$, then we must have that $\tilde{E}(x,t)$ is also a classical solution of the Cauchy problem (3.6.62). By the uniqueness of solutions to the Cauchy problem for the heat equation, we then have that $E(x,t) \equiv \tilde{E}(x,t)$, on $R^1 \times [0,\infty)$; in other words, all subsequences $\{\mu_n\}$ yield sequences $\{E^{\mu_n}\}$ such that, as $\mu_n \to 0^+$, $E^{\mu_n} \to E$, in $L^2(Q_T)$, with $E(x,t)$ being the unique classical solution of the Cauchy problem (3.6.62). □

Remarks: The convergence Theorem 3.7 established above shows that a wave which travels with finite speed of propagation according to (3.6.7), for fixed $\mu > 0$, tends in the limit as $\mu \to 0^+$ to one which propagates with infinite speed of propagation. The very simplistic nature of the model notwithstanding, this conclusion lends substantial credence to our earlier characterization of the model as one representing a superconducting nonlinear dielectric. The essential content of what we have presented here may be found in the paper [11] by the author and his colleagues H. Bellout and J. Nečas; an improved model might, for example, allow for an explicit dependence of μ on the temperature and would then require our coupling a heat equation to (3.6.7) and employing an associated Fourier heat conduction law for the heat flux which could involve the electric field component E as well as the temperature gradient.

3.7 Multidimensional and Nonstrictly Hyperbolic Problems Governing Wave-Dielectric Interactions

Equation (2.8.2) is the two-dimensional analog of the nonlinearly damped, nonlinear wave equation (3.5.1) which governs the evolution of the electric displacement field in the one space dimension, wave-dielectric interaction problem; for this latter equation, we have established the existence of a global (in time) smooth solution of the associated Cauchy problem, with sufficiently small initial data, by using, in § 3.5, an energy argument which is due, in essence, to Matsumura [112] and Nishida [124]. The arguments in [112], [124], however, are not restricted to problems posed in one space dimension. Central, however, to the analysis presented in § 3.5 was the fact that we were able to establish, for the locally defined solutions of (3.5.1), an estimate of the form

$$\|D\|_{H^3} \leq C_0 + \delta \|\frac{\partial^2 D}{\partial x^2}\|_{L^\infty} \tag{3.7.1}$$

for δ sufficiently small, with C_0 depending only on the initial data, and that in one space dimension we have, for some $C_1 > 0$

$$\|\frac{\partial^2 D}{\partial x^2}\|_{L^\infty} \leq C_1 \|D\|_{H^3} \tag{3.7.2}$$

Then (3.7.1), (3.7.2) combine to yield

$$\|D\|_{H^3} \leq C_0 + \delta C_1 \|D\|_{H^3}, \tag{3.7.3}$$

which, for δ sufficiently small, yields a bound for $\|D\|_{H^3}$ on $[0, T_\infty)$, in terms of the initial data. However, the analog of (3.7.2) is no longer true in two space dimensions; what does hold in two space dimensions is the analog of the estimate

$$\|\frac{\partial^2 D}{\partial x^2}\|_{L^\infty} \leq C_2 \|D\|_{H^4} \tag{3.7.4}$$

for some $C_2 > 0$. If the analysis of § 3.5, for the one-dimensional situation, can be carried further, so as to yield an estimate of the form

$$\|D\|_{H^4} \leq C_0 + \delta \|\frac{\partial^2 D}{\partial x^2}\|_{L^\infty} \tag{3.7.5}$$

then it is likely that the analog of (3.7.5), for solutions of the Cauchy problem associated with (2.8.2), can be established as well. Such a result, in conjunction with the two-dimensional analog of the embedding displayed in (3.7.4), would then yield the desired global existence theorem for (2.8.2) with (sufficiently small) associated Cauchy data.

We will close out this chapter by offering some remarks concerning nonstrictly hyperbolic problems and phase transitions. In all of our work on wave propagation in nonlinear dielectrics, in Chapters 2 and 3, we have assumed that $\mathsf{E}'(D) > 0$. However, monotonicity of $\mathsf{E}(\cdot)$ is not necessarily applicable in all problems involving nonlinear dielectrics and, in many situations, phase transitions may take place. For example, in [63] Grindlay considers equations of state of the form

$$E = \beta D + \xi D^3 + \zeta D^5 \tag{3.7.6}$$

with $\beta = A(\theta - \theta_0)$; here θ is the temperature of the dielectric and A, θ_0, ζ are positive constants, whereas ξ may be a positive or negative constant. Equation (3.7.6) provides a good representation of the dielectric behavior of barium titanate ($\xi < 0$) in the vicinity of the Curie-Weiss temperature θ_0. Introducing dimensionless variables t, e, d, with $t = 4A\zeta(\theta - \theta_0)/\xi^2$, (3.7.6) may be put in the form

$$e = td + 2d^3 + d^5 \tag{3.7.7}$$

As $e(d,t)$ is an odd function of d, for $t < 0$, we say that the dielectric possesses a reversible spontaneous dipole moment because, in this case, the non-zero roots of $e = 0$ occur in pairs of opposite sign and equal magnitudes. Because there is no spontaneous dipole moment for $t \geq 0$, the system is said to undergo a *phase transition* at $t = 0$, i.e., at the Curie-Weiss temperature θ_0. The problem, in this case, is to study the behavior of solutions to the Cauchy problem for the quasilinear system

$$D_t + \frac{1}{\mu}B_x = \Sigma(D), \qquad B_t + E_x = 0 \tag{3.7.8}$$

with nonmonotone $E = \mathsf{E}(D)$ of the form (3.7.6).

With $\Sigma = 0$, there is an analog of the nonhyperbolic system (3.7.8) in the theory of compressible fluid flow, i.e., the equations

$$u_t = -p(w)_x, \quad w_t = u_x \tag{3.7.9}$$

$$p(w) = \frac{R\theta_0}{w-b} - \frac{a}{w^2}; \; 0 < b < w < \infty; \; R, a > 0 \tag{3.7.10}$$

governing the motion of a Van der Waals fluid, which has been considered by many authors ([159], [158], [160], [161], [65], [142], [155], [156], [64]) recently as a model for phase transitions in fluids. In (3.7.9), u is the fluid velocity, $w = \rho^{-1}$ is the specific volume (ρ is the density), and p is the pressure. For (3.7.9), $\exists\, w_\alpha, w_\beta > 0$ such that (i) $p'(w) < 0$, $0 < b < w < w_\alpha$, $w_\beta < w$, (ii) $p'(w_\alpha) = p'(w_\beta) = 0$, (iii) $p'(w) > 0$ if $w_\alpha < w < w_\beta$. The domains (b, w_α) and (w_β, ∞) are called the *α-phase* and *β-phase*, respectively, the α-phase corresponding to the fluid being liquid, while the β-phase corresponds to the fluid being vapor. With the above constitutive hypothesis, the equations (3.7.9) become a mixed hyperbolic-elliptic system and solutions of associated initial-value problems which lie in the elliptic regime exhibit a classical Hadamard instability. Weak solutions of the Cauchy problem for (3.7.9) may possess propagating singular surfaces, termed *phase boundaries*, which separate states in one hyperbolic domain from states in another hyperbolic domain. More precisely, if $\Gamma : x = \gamma(t)$ denotes a singular surface, i.e., a C^1 curve across which u, w experience jump discontinuities and u_+, w_+, u_-, w_- are the limits from right and left, for u, w, as $(x,t) \to (\gamma(\bar{t}), \bar{t})$ on Γ, then Γ is called a *phase boundary* (shock) if for every point on Γ, w_+, w_- lie in a different (the same) phase. As with shocks, it is necessary to discuss the formulation of admissibility criteria for systems which admit propagating phase boundaries. For the system (3.7.9), Serrin [155] revived the Korteweg theory of interfacial capillarity [94] according to which the stress τ has the form

$$\tau = -p(w) + D(w)w_x^2 - C(w)w_{xx} + \mu(w)w_x \tag{3.7.11}$$

with $p(w)$ given by (3.7.10); the problem now is to determine whether solutions of (3.7.9) with $u_t = \tau_x$ and τ given by (3.7.11), which tend to a limit bounded, a.e., as $\mu(w) \to 0$, $D(w) \to 0$, $C(w) \to 0$, converge to solutions of the (same) initial-value problem for the Van der Waals fluid. Slemrod [158] sets $C(w) = \mu_0^2 A$, $D(w) = 0$, $\mu(w) = \mu_0$ and looks for travelling wave solutions of the resulting system, i.e.,

$$\begin{cases} u_t = (-p(w) - \mu_0^2 A w_{xx} + \mu_0 u_x)_x \\ w_t = u_x \end{cases} \tag{3.7.12}$$

of the form $u(x,t) = \hat{u}(\xi)$, $w(x,t) = \hat{w}(\xi)$, $\xi = \dfrac{x - Ut}{\mu_0}$ $\left(U = \left(\dfrac{[\tau]}{[w]}\right)^{1/2}\right.$ being the velocity of a specific propagating phase boundary Γ). It follows that $\hat{u}, \hat{w}$ must satisfy

$$\begin{cases} -U\hat{u}' &= (-p(\hat{w}) - A\hat{w}'' + \hat{u}')' \\ -U\hat{w}' &= \hat{u}' \end{cases} \tag{3.7.13}$$

$$(\hat{u}(-\infty), \hat{w}(-\infty), \hat{u}(+\infty), \hat{w}(+\infty)) = (u_-, w_-, u_+, w_+) \tag{3.7.14}$$

The singular surface Γ is then said to satisfy the *viscosity-capillarity admissibility condition* if there exists $\hat{u}, \hat{w}$ such that (3.7.13), (3.7.14) are satisfied at all points $(\gamma(t), t) \in \Gamma$. For $A > 0$, Slemrod [158] establishes the equivalence of the above admissibility criterion and the standard Lax shock condition for the case of a strictly hyperbolic, genuinely nonlinear system; he also establishes conditions under which, for a propagating phase boundary Γ, a state "on the left" in the liquid phase determines a state "on the right" so as to make Γ (and the weak solution) admissible according to the viscosity-capillarity criteria. This latter result differs from the standard shock wave theorem, for materials which do not undergo phase transitions, where one state can be connected by a shock to any other state, as long as the Rankine-Hugoniot conditions are satisfied and the shock is compressive.

Our system (3.7.8), with nonmonotone $\mathbb{E}(\cdot)$ given by (3.7.6), is not a conservation law if $\Sigma \neq 0$. Also, there are no analogies to the Korteweg constitutive theory

(3.7.11) which are available in the electromagnetic theory literature and which permit one to discuss either stabilization of solutions for (3.7.8), (3.7.6) or admissibility of propagating shocks and phase boundaries, even in the case $\Sigma = 0$. The goals in studying (3.7.6), (3.7.8) should, therefore, be the following:

(i) to formulate a physically relevant and justifiable modification of the constitutive relation (3.7.6) which will enable one to discuss the problem of global existence of solutions for the Cauchy problem associated with (3.7.8) and

(ii) to formulate (and justify) a modification of (3.7.6) which will enable one to prove, along the lines of the analysis in [158], the admissibility of propagating shocks and phase boundaries.

We conjecture that a modification of the constitutive relation between E and D to allow for the influence of (electric) displacement gradients may provide the stabilization sought in (i); a related analysis of this type has been effected by Pego [132], who considers a viscosity-type perturbation of the Van der Waals constitutive function and derives both global existence results for classical solutions, as well as admissibility criteria for weak solutions, with appropriate assumptions on the initial data.

Chapter 4

DISTRIBUTED PARAMETER NONLINEAR TRANSMISSION LINES I: Shock Formation and Existence of Smooth Solutions

4.0 Introduction

In this chapter we begin our study of the initial-value problem for a distributed parameter nonlinear transmission line with periodic initial values for the current and charge distribution in the line; we allow for the possibility of a voltage-dependent capacitance, which tends to destabilize signals propagating in the line, as well as a leakage conductance, per unit length of the line, which, under appropriate circumstances, can be shown to provide a dissipative mechanism in the line. In § 4.1 we begin by deriving the quasilinear hyperbolic system which governs the evolution of the current and charge distribution in the nonlinear transmission line and some analogies are drawn with certain aspects of the equations that govern the propagation, in one dimension, of plane waves in the wave-dielectric interaction problem studied in Chapters 2 and 3. The Riemann invariants germane to the discussion of the nonlinear transmission line equations are then introduced here and the equations are rewritten as a set of nonlinear ordinary differential equations for these Riemann invariants along the appropriate set of characteristic curves. *A priori* estimates are derived for the Riemann invariants, assuming the initial data to be periodic, and these estimates are then employed to prove that smooth (C^1) solutions of the equations must break down in finite-time; these results on finite-time breakdown of classical solutions are then compared with those of a similar nature which have been derived by other authors. In § 4.2 we take advantage of that term in the nonlinear transmission line equations which arises as a consequence of the presence of the resistance element in the line, in order to prove, using the Riemann invariants formulation of the problem, that a global C^1 solution will exist if the leakage conductance in the line is sufficiently dissipative in a sense which is made precise in the statement of hypothesis $(\tilde{G})$ of this section. Some drawbacks with respect to the structure of hypothesis $(\tilde{G})$ are indicated and, immediately following the proof of global (in time) existence based on $(\tilde{G})$, we offer the reader the essence of an alternative proof, still grounded in the Riemann invariants formulation, but based on a far less restrictive condition than $(\tilde{G})$, which we call (G^*), that was proposed to the author by J. Wang. All of the work in § 4.2 is, of course, based on the assumption that the initial data (with respect to an appropriate measure) is sufficiently small. Continuing in § 4.3 with the assumption of sufficiently small data, we show that the energy argument which was employed in § 3.5 to establish the global existence, in time, of smooth solutions to the equations governing the one-space dimension wave-dielectric interaction problem, can also be adapted, with an appropriate set of hypotheses, to prove the global existence of C^1 solutions to the nonlinear transmission line equations; an example is provided of a set of constitutive relations for the voltage and leakage conductance which conform to the hypotheses used in this section. Aside from the global existence theorem obtained in this section, we also obtain what appears to be the strongest results to date on the asymptotic (or large time) behavior of the charge distribution in the nonlinear transmission line as $t \to \infty$. A word is in order about what is not covered in this chapter, i.e., we do not attempt here a discussion of the behavior of nonlinear transmission lines which are dispersive in nature. While most of the results presented in this chapter for the nonlinear, nondispersive line are quite recent, and have not appeared before in book form, the book by A. C. Scott [151] contains an excellent discussion of dispersive effects in nonlinear transmission lines and it would appear to be the case that little (if anything) new on this important problem has appeared in the mathematical literature since the publication of [151]; there are exceptions, of course, the most notable of which might be the excellent papers of Peng and Landauer [98] and Landauer [96]. There are also, of course, some striking similarities between the equations governing the nonlinear transmission line problem and those which govern the wave-dielectric interaction problem in one-space dimension; the effects of dispersion on the propagation of electromagnetic waves in nonlinear dielectrics are covered in most texts on nonlinear optics, the best of which, from a mathematical standpoint, would seem to be the recent monograph by Moloney and Newell [114]. Other classic texts which cover dispersive effects in the realm of nonlinear wave propagation in electromagnetic theory would include the volumes by Yariv [186] and Bloembergen [18].

4.1 Shock Formation in Nonlinear Transmission Lines

In § 1.4 we surveyed, for the reader, some of the basic concepts connected with the topic of transmission lines. In particular, we discussed the ideas of series and parallel connection in a line and several of the constituent components of a circuit were examined in detail, e.g., capacitors, inductors, and resistors. A detailed discussion of the basic Kirchhoff's laws was also provided in § 1.4 and both the transient and steady-state behavior of a simple R-L-C circuit were examined, the operative linear differential equation in this simple situation being (1.4.11). This latter equation reflects the absence of a leakage conductance component in the transmission line, as well as the fact that R, L, and C have all been assumed constant. However, in many practical applications of transmission line theory, the line does contain a leakage conductance and, also, $Q = Q(v)$, where $v = v(x,t)$ is the voltage at a point x units distant from an origin chosen in the line and $Q(v)$ is the voltage-dependent charge per unit length of the line. In such a situation, we must have a capacitance $C = C(v) = Q'(v)$ which is both voltage-dependent and nonlinear (in general) in v.

For a distributed parameter nonlinear transmission line with $C = C(v)$, but R, L, and G all positive constants, $\frac{1}{G}$ being the leakage conductance in the line, an application of both Ohm's and Kirchhoff's laws to an element of the transmission line (see, e.g., Kataev [50]) yields the system of equations

$$\begin{cases} L\dfrac{\partial i}{\partial t} + \dfrac{\partial v}{\partial x} + Ri = 0 \\ C(v)\dfrac{\partial v}{\partial t} + \dfrac{\partial i}{\partial x} + Gv = 0 \end{cases} \tag{4.1.1}$$

with $i = i(x,t)$ being the current in the line at a point x units distant from the chosen origin in the line; the situation is depicted in Figure 4.1.

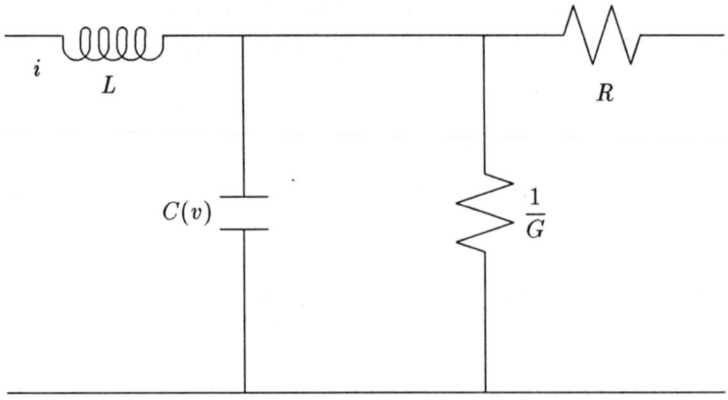

Figure 4.1

More interesting (and complicated) models of distributed parameter nonlinear transmission lines than that represented by the equations (4.1.1) are also of interest in applications, although we shall have little to say about such models in this book. For example, we could very well consider the model introduced in Kataev [50], for which the relevant transmission line equations are given as

$$\left\{ \begin{array}{l} \dfrac{\partial \Phi}{\partial t} + \dfrac{\partial v}{\partial x} = 0 \\ \dfrac{\partial Q}{\partial t} + \dfrac{\partial i}{\partial x} = -j \end{array} \right. \tag{4.1.2}$$

in which Φ, the magnetic flux per unit length between the conductors, is given as $\Phi = \Phi(i)$, so that $L = L(i) = \partial\Phi/\partial i$, while $Q = Q(v)$ and j, the leakage current, is given by a relation of the form $j = j(i, v) = G(i, v) \cdot v$. A direct comparison of (4.1.2), with the associated constitutive assumptions as indicated above, and the system (4.1.1), shows that (4.1.1) may be obtained from the more general model by taking $L(i) = L =$ const. and $G(i, v) = G =$ const.

In this section we will first present the work of the author [22] on shock formation in solutions of initial-value problems associated with the quasilinear system (4.1.1); then, we will compare and contrast these results, on electromagnetic shock wave formation in nonlinear distributed parameter transmission lines, with earlier work of Kataev [50], A. Jeffrey [74], Cumberbatch [38], Landauer [97],[96], and Peng and Landauer [98]; as noted in this last reference, the fact that one of the primary concepts of nonlinear wave propagation, i.e., that various parts of a wave travel with different velocities and that wave fronts (or tails) can sharpen into shock waves, carries over into the analysis of electromagnetic wave propagation in transmission lines, seems to have been noted for the first time by Salinger [146]. Other basic papers in this area include the works of Riley [135], Mullick [118], Ostrovskii [128], and A. C. Scott [150], while the volume by Scott [151] appears to be the last book in this area which presents a serious treatment of shock formation in nonlinear distributed parameter transmission lines. Most of the work which we delineate in this section appeared, of course, some considerable time after the publication of the monograph [151].

We begin our analysis by noting that (4.1.1) has the form of an inhomogeneous quasilinear system, i.e.,

$$\begin{pmatrix} i \\ v \end{pmatrix}_{,t} + \begin{bmatrix} 0 & \frac{1}{L} \\ \frac{1}{C(v)} & 0 \end{bmatrix} \begin{pmatrix} i \\ v \end{pmatrix}_{,x} = \begin{bmatrix} -\frac{R}{L}i \\ -\frac{Gv}{C(v)} \end{bmatrix} \tag{4.1.3}$$

and if $C(v) = \dfrac{dQ}{dv} > 0$, then (4.1.3) is clearly hyperbolic with the associated characteristics defined by the nonlinear ordinary differential equations

$$\frac{dx}{dt} = \pm\frac{1}{\sqrt{LC(v)}} \tag{4.1.4}$$

As a special case, consider the situation in which C is independent of v, i.e., $C = C_0 = \text{const.}$; then it is easily shown that (Jeffrey [74])

$$\frac{d}{dt}\left[i\left(\frac{L}{C_0}\right)^{1/2} + v\right] = -\left(\frac{R}{\sqrt{LC_0}}i + \frac{G}{C_0}v\right) \tag{4.1.5}$$

along a solution of $\dfrac{dx}{dt} = -\dfrac{1}{\sqrt{LC_0}}$ while

$$\frac{d}{dt}\left[i\left(\frac{L}{C_0}\right)^{1/2} - v\right] = -\left(\frac{R}{\sqrt{LC_0}}i - \frac{G}{C_0}v\right) \tag{4.1.6}$$

along a solution of $\dfrac{dx}{dt} = +\dfrac{1}{\sqrt{LC_0}}$.

From (4.1.5), (4.1.6) it follows, for example, that both the current i and the voltage v propagate along this linear transmission line with an attenuation factor of $-G/C_0$, provided we have $R/L = G/C_0$.

In order to rewrite the system (4.1.3) as an inhomogeneous quasilinear system in "conservation form", we note that inasmuch as Q is monotone increasing in v, $\exists\mathcal{V}$

such that $v = \mathcal{V}(Q)$ with $\mathcal{V} \in C^1$ when $Q \in C^1$. Our basic system (4.1.3) now assumes the form

$$\begin{cases} \dfrac{\partial i}{\partial t} + \dfrac{1}{L}\mathcal{V}'(Q)\dfrac{\partial Q}{\partial x} = -\dfrac{R}{L}i \\ \dfrac{\partial Q}{\partial t} + \dfrac{\partial i}{\partial x} = -G\mathcal{V}(Q) \end{cases} \tag{4.1.7}$$

or

$$\begin{pmatrix} i \\ Q \end{pmatrix}_{,t} + \begin{pmatrix} 0 & \frac{1}{L}\mathcal{V}(Q) \\ 1 & 0 \end{pmatrix} \begin{pmatrix} i \\ Q \end{pmatrix}_{,x} = -\begin{pmatrix} \frac{R}{L}i \\ G\mathcal{V}(Q) \end{pmatrix} \tag{4.1.8}$$

Because $\dfrac{d\mathcal{V}}{dQ} = 1/Q'(v) > 0$, the system (4.1.7) is hyperbolic with associated characteristics defined by $\dfrac{dx}{dt} = \pm\dfrac{1}{\sqrt{L}} \cdot \sqrt{\mathcal{V}'(Q)}$. If $R \equiv 0$, then we may draw a direct analogy between the system (4.1.7) and the system (2.6.16) governing the evolution of the fields (D, B) in a linearly polarized plane wave propagating in a nonlinear dielectric material; this analogy, for the case $R \equiv 0$, is made by replacing $Q \to D$, $i \to \dfrac{1}{\mu_0}B$, $G\mathcal{V}(Q) \to \Sigma(D)$, and $\dfrac{1}{L}\mathcal{V}(Q) \to \dfrac{1}{\mu_0}\mathsf{E}(D)$. Our interest in this chapter, therefore, will be in the situation where $R \neq 0$. We also note that if $G \equiv 0$ (so that there does not exist leakage current between the conductors), then the system (4.1.7) reduces to

$$\begin{cases} \dfrac{\partial Q}{\partial t} + \dfrac{\partial i}{\partial x} = 0 \\ \dfrac{\partial i}{\partial t} + \dfrac{1}{L}(\mathcal{V}(Q))_{,x} = -\dfrac{R}{L}i \end{cases}$$

which is directly comparable to the systems treated by Slemrod [163] and Nishida [125]. Therefore, we will also assume that $G \neq 0$ in all that follows. Our fundamental system, as pertains to the problem of electromagnetic wave propagation in a distributed parameter nonlinear transmission is, thus, (4.1.7) subject to $R^2 + G^2 \neq 0$.

As with the analysis of the system (2.6.16), governing the wave-dielectric interaction problem in one-space dimension in Chapter 2, we begin by writing the system (4.1.7) in Riemann invariants form; the appropriate Riemann invariants for the problem at hand are easily seen to be given by

$$r(x,t) = -i(x,t) + \int_0^{Q(x,t)} \sqrt{\frac{\mathcal{V}'(\zeta)}{L}}\,d\zeta \tag{4.1.9a}$$

and

$$s(x,t) = -i(x,t) - \int_0^{Q(x,t)} \sqrt{\frac{\mathcal{V}'(\zeta)}{L}}\,d\zeta \tag{4.1.9b}$$

We now set

$$\lambda = -\sqrt{\frac{\mathcal{V}'(Q)}{L}}, \qquad \nu = \sqrt{\frac{\mathcal{V}'(Q)}{L}} \tag{4.1.10}$$

and define the directional derivatives in the characteristic directions in the usual way, i.e.,

$$' = \frac{\partial}{\partial t} + \lambda(x,t)\frac{\partial}{\partial x}, \qquad \grave{} = \frac{\partial}{\partial t} + \nu(x,t)\frac{\partial}{\partial x} \tag{4.1.11}$$

As many of our basic calculations are similar to those involved in the proof of Theorem 2.2, we will proceed at a fairly rapid pace. We begin by noting that

$$r(x,t) - s(x,t) = 2\int_0^{Q(x,t)} \sqrt{\frac{\mathcal{V}'(\zeta)}{L}}\,d\zeta \equiv \eta(x,t) \tag{4.1.12}$$

so that $\eta = \hat{\eta}(Q(x,t))$ satisfies

$$\frac{d\hat{\eta}}{dQ} = 2\sqrt{\frac{\mathcal{V}'(Q)}{L}} > 0 \tag{4.1.13}$$

in which case $\exists\, \hat{\eta}^{-1}$ such that

$$Q(x,t) = \hat{\eta}^{-1}[r(x,t) - s(x,t)] \tag{4.1.14}$$

Combining (4.1.10) with (4.1.4) we see that we may write

$$\begin{aligned} \lambda(x,t) &= \hat{\lambda}(r(x,t) - s(x,t)) \\ \nu(x,t) &= \hat{\nu}(r(x,t) - s(x,t)) \end{aligned} \tag{4.1.15}$$

with the obvious definitions for $\hat{\lambda}$ and $\hat{\nu}$. The system (4.1.7) can now be written in terms of the Riemann invariants r, s as

$$\begin{cases} r' &= -\frac{\alpha}{2}(r+s) + \Phi(r-s) \\ \grave{s} &= -\frac{\alpha}{2}(r+s) - \Phi(r-s) \end{cases} \tag{4.1.16}$$

where $\alpha = R/L > 0$ and

$$\Phi(r-s) = -G\mathcal{V}[\hat{\eta}^{-1}(r-s)]\sqrt{\frac{\mathcal{V}'(\hat{\eta}^{-1}(r-s))}{L}} \tag{4.1.17}$$

The system (4.1.16) reduces, of course, to a pair of nonlinear ordinary differential equations along the characteristics $x_1(t,\beta_1)$, $x_2(t,\beta_2)$ defined by the solutions of the initial-value problems

$$\frac{dx_1}{dt} = \hat{\lambda}[r(x_1,t) - s(x_1,t)]; \quad x_1(0,\beta_1) = \beta_1 \tag{4.1.18a}$$

$$\frac{dx_2}{dt} = \hat{\nu}[r(x_2,t) - s(x_2,t)]; \quad x_2(0,\beta_2) = \beta_2 \tag{4.1.18b}$$

namely,

$$\frac{d}{dt}r(x_1(t,\beta_1),t) = -\frac{\alpha}{2}\{r(x_1(t,\beta_1),t) + s(x_1(t,\beta_1),t)\} + \Phi(r(x_1(t,\beta_1),t) - s(x_1(t,\beta_1),t)) \tag{4.1.19a}$$

and

$$\frac{d}{dt}s(x_2(t,\beta_2),t) = -\frac{\alpha}{2}\{r(x_2(t,\beta_2),t) + s(x_2(t,\beta_2),t)\} - \Phi(r(x_2(t,\beta_2),t) - s(x_2(t,\beta_2),t)) \tag{4.1.19b}$$

By virtue of (4.1.12) and (4.1.17)

$$\hat{\eta}(0) = 0 \quad \text{and} \quad \Phi(0) = -G\mathcal{V}(0)\sqrt{\frac{\mathcal{V}'(0)}{L}} = 0 \tag{4.1.20}$$

so that

$$\Phi(r-s) = \int_0^{r-s} \Phi'(\eta)\,d\eta$$

in which case

$$|\Phi(r-s)| \le \sup_\eta |\Phi'(\eta)|\,(|r| + |s|) \tag{4.1.21}$$

We now make the following basic constitutive assumption relative to the voltage $\mathcal{V}(\cdot)$:

The constitutive function $\mathcal{V}(\cdot)$ satisfies

$$\sup_\eta \left|\frac{d\Phi(\eta)}{d\eta}\right| = M < \infty \tag{4.1.22}$$

where for $\eta \in R^1$

$$\Phi(\eta) = -G\mathcal{V}(\hat{\eta}^{-1}(\eta))\sqrt{\frac{\mathcal{V}'(\hat{\eta}^{-1}(\eta))}{L}} \tag{4.1.23}$$

Remarks: The constitutive hypothesis (4.1.22) is, of course, directly analogous to the assumption (2.7.63) made in the analysis of shock formation in the wave-dielectric interaction problem in one-space dimension and serves the same role, in what follows, as (2.7.63) did in Chapter 2. A concrete example of a (reasonable) voltage function $\mathcal{V}(\cdot)$, which satisfies (4.1.22), is presented in § 4.2. As a consequence of the hypothesis (4.1.22), and (4.1.21), we have, of course,

$$\Phi(r-s) \leq M(|r|+|s|) \tag{4.1.24}$$

The following lemma is now the direct analog, for the problem at hand, of Lemma 2.2:

Lemma 4.1 Let (r,s) be a C^1 solution, for $0 \leq t \leq T$, $T < \infty$, of the initial-value problem for (4.1.16) with associated initial data

$$r(x,0) = r_0(x), \qquad s(x,0) = s_0(x) \tag{4.1.25}$$

where $r_0(\cdot)$ and $s_0(\cdot)$ are assumed to be periodic on R^1 and of class C^1. Then if the voltage $\mathcal{V}(\cdot)$ satisfies (4.1.22) for some $M > 0$,

$$|r(x,t)| + |s(x,t)| \leq (|r_0| + |s_0|)e^{2MT} \tag{4.1.26}$$

for all $t < T$, where $|r_0| = \sup_x |r_0(x)|$ and $|s_0| = \sup_x |s_0(x)|$.

Proof: In view of the close similarity of the proof to that of Lemma 2.2, we will proceed in a concise fashion: We begin by integrating (4.1.19a), (4.1.19b) along the respective characteristic curves and applying (4.1.24) so as to obtain the pair of estimates for $0 \leq \tau \leq t \leq T$:

$$\begin{aligned}\exp\left(\frac{\alpha}{2}t\right)|r(x_1,t)| \leq{}& |r_0(\beta_1)| \\ &+ \frac{(\alpha+2M)}{2}\int_0^t \exp\left(\frac{\alpha}{2}\tau\right)|s(x_1,\tau)|\,d\tau \\ &+ M\int_0^t \exp\left(\frac{\alpha}{2}\tau\right)|r(x_1,\tau)|\,d\tau\end{aligned} \tag{4.1.26a}$$

$$\begin{aligned}\exp\left(\frac{\alpha}{2}t\right)|s(x_2,t)| \leq{}& |s_0(\beta_1)| \\ &+ \frac{(\alpha+2M)}{2}\int_0^t \exp\left(\frac{\alpha}{2}\tau\right)|r(x_2,\tau)|\,d\tau \\ &+ M\int_0^t \exp\left(\frac{\alpha}{2}\tau\right)|r(x_2,\tau)|\,d\tau\end{aligned} \tag{4.1.26b}$$

Following the pattern established in the proof of Lemma 2.2, we now set

$$\begin{cases} R(t) = \sup_x \exp\left(\frac{\alpha}{2}t\right)|r(x,t)| \\ S(t) = \sup_x \exp\left(\frac{\alpha}{2}t\right)|s(x,t)| \end{cases} \tag{4.1.27}$$

Using the definitions (4.1.27) we obtain from (4.1.26a), (4.1.26b):

$$\exp\left(\frac{\alpha}{2}t\right)|r(x_1,t)| \leq |r_0| + \left(\frac{\alpha+2M}{2}\right)\int_0^t S(\tau)\,d\tau + M\int_0^t R(\tau)\,d\tau \tag{4.1.28a}$$

$$\exp\left(\frac{\alpha}{2}t\right)|s(x_2,t)| \leq |r_0| + \left(\frac{\alpha+2M}{2}\right)\int_0^t R(\tau)\,d\tau + M\int_0^t S(\tau)\,d\tau \tag{4.1.28b}$$

As $r(x,t)$ and $s(x,t)$ are periodic in x, for each $t \leq T$, for $t = \hat{t} \leq T$, $\exists\, \hat{x}_1, \hat{x}_2$ such that

$$R(\hat{t}) = \exp\left(\frac{\alpha}{2}\hat{t}\right)|r(\hat{x}_1,\hat{t})|, \qquad S(\hat{t}) = \exp\left(\frac{\alpha}{2}\hat{t}\right)|s(\hat{x}_2,\hat{t})|$$

Choosing $\hat{\beta}_1 = \beta_1(\hat{t})$, $\hat{\beta}_2 = \beta_2(\hat{t})$ such that $\hat{x}_1 = x_1(\beta_1(\hat{t}),\hat{t})$, $\hat{x}_2 = x_2(\beta_2(\hat{t}),\hat{t})$ we obtain

$$R(\hat{t}) = \exp\left(\frac{\alpha}{2}\hat{t}\right)|r(x_1(\beta_1(\hat{t}),\hat{t}),\hat{t})| \tag{4.1.29a}$$

$$S(\hat{t}) = \exp\left(\frac{\alpha}{2}\hat{t}\right)|s(x_2(\beta_2(\hat{t}),\hat{t}),\hat{t})| \tag{4.1.29b}$$

We now choose $x_1 = x_1(\beta_1(t),t)$ and $x_2 = x_2(\beta_2(t),t)$, respectively, on the left-hand sides of the estimates (4.1.28a), (4.1.28b) so as to obtain for each $t \leq T$

$$R(t) \leq |r_0| + \left(\frac{\alpha+2M}{2}\right)\int_0^t S(\tau)\,d\tau + M\int_0^t R(\tau)\,d\tau \tag{4.1.30a}$$

$$S(t) \leq |s_0| + \left(\frac{\alpha+2M}{2}\right)\int_0^t R(\tau)\,d\tau + M\int_0^t S(\tau)\,d\tau \tag{4.1.30b}$$

Setting $W(t) = R(t) + S(t)$ and adding the estimates (4.1.30a), (4.1.30b), we now obtain, for $t \leq T$,

$$W(t) \leq |r_0| + |s_0| + \left(\frac{\alpha+4M}{2}\right)\int_0^t W(\tau)\,d\tau \tag{4.1.31}$$

so that by virtue of Gronwall's inequality,

$$W(t) \leq (|r_0| + |s_0|)\exp\left[\frac{\alpha+4M}{2}t\right], \qquad t \leq T \tag{4.1.32}$$

However, in view of the definitions of $R(t)$, $S(t)$, and $W(t)$,

$$W(t) \geq \exp\left(\frac{\alpha}{2}t\right)\left[|r(x,t)| + |s(x,t)|\right], \qquad t \leq T \tag{4.1.33}$$

and the requisite result, i.e., (4.1.26) now follows directly from (4.1.32) and (4.1.33). □

With the aid of Lemma 4.1 we are now in a position to prove that C^1 solutions $(r(x,t), s(x,t))$ of the initial-value problem (4.1.16), (4.1.25) cannot exist for all $t > 0$ when $\mathcal{V}(\cdot)$ satisfies the constitutive hypothesis (4.1.22), as well as the additional hypothesis of genuine nonlinearity that we delineate below; the analysis is, of course, similar to that of the proof of Theorem 2.2, so once again we proceed at a fairly rapid pace. We will assume that there exists a C^1 solution $(r(x,t), s(x,t))$ of (4.1.16), (4.1.25) for $0 \leq t \leq T$ with $T > 0$ arbitrary and will then derive a contradiction to this assumption by showing that for $|r_0|, |s_0|$ sufficiently small, and $r_x(x,0)$ sufficiently large, for some $x \in R^1$, $|r_x(x,t)| \to \infty$ as $t \to t_\infty < T$. Therefore, by differentiating the first of the pair of equations (4.1.16) through with respect to x, using the definition of the directional derivative` in (4.1.11), and the fact that

$$\frac{\partial \hat{\lambda}}{\partial x} = \frac{\partial \hat{\lambda}}{\partial r}\left(\frac{\partial r}{\partial x} - \frac{\partial s}{\partial x}\right)$$

we obtain

$$\left(\frac{\partial r}{\partial x}\right)' + \frac{\partial \hat{\lambda}}{\partial r}\frac{\partial s}{\partial x}\frac{\partial r}{\partial x} = \frac{\partial \hat{\lambda}}{\partial r}\left(\frac{\partial r}{\partial x}\right)^2 + \left(\Psi - \frac{\alpha}{2}\right)\frac{\partial r}{\partial x} - \left(\Psi + \frac{\alpha}{2}\right)\frac{\partial s}{\partial x} \tag{4.1.34}$$

with $\Psi(\zeta) = \Phi'(\zeta)$, $\zeta \in R^1$. However,

$$\frac{\partial s}{\partial x} = -\frac{\Phi}{\hat{\lambda}} + \frac{(r' - s')}{2\hat{\lambda}}$$

so that

$$\begin{aligned}\left(\frac{\partial r}{\partial x}\right)' + \frac{\partial \hat{\lambda}}{\partial r}\cdot\frac{(r' - s')}{2\hat{\lambda}}\frac{\partial r}{\partial x} - \frac{\Phi}{\hat{\lambda}}\frac{\partial \hat{\lambda}}{\partial r}\frac{\partial r}{\partial x} \\ = \frac{\partial \hat{\lambda}}{\partial r}\left(\frac{\partial r}{\partial x}\right)^2 + \left(\Psi - \frac{\alpha}{2}\right)\frac{\partial r}{\partial x} \\ - \left(\Psi + \frac{\alpha}{2}\right)\left(\frac{(r' - s')}{2\hat{\lambda}}\right) + \frac{\Phi}{\hat{\lambda}}\left(\Psi + \frac{\alpha}{2}\right)\end{aligned} \tag{4.1.35}$$

Using the fact that

$$(\ln \hat{\lambda})' = \frac{1}{\hat{\lambda}}\frac{\partial \hat{\lambda}}{\partial r}(r' - s'),$$

and multiplying (4.1.35) through by $\hat{\lambda}^{1/2}$ we now obtain

$$\chi' = \hat{\lambda}^{-1/2}\frac{\partial \hat{\lambda}}{\partial r}\chi^2 + \left(\Psi + \frac{\Phi}{\hat{\lambda}}\frac{\partial \hat{\lambda}}{\partial r} - \frac{\alpha}{2}\right)\chi \tag{4.1.36}$$

$$- \left(\frac{\Psi}{2} + \frac{\alpha}{4}\right)\hat{\lambda}^{-1/2}(r' - s') + \hat{\lambda}^{-1/2}\Phi\left(\Psi + \frac{\alpha}{2}\right)$$

where we have set $\chi = \hat{\lambda}^{1/2}\left(\frac{\partial r}{\partial x}\right)$. By making the further definitions

$$F_\alpha(r - s) = -\frac{1}{\alpha}\int_0^{r-s}\left\{\frac{\Psi(\zeta)}{2} + \frac{\alpha}{4}\right\}\hat{\lambda}^{-1/2}(\zeta)\,d\zeta \tag{4.1.37}$$

$$\phi_\alpha = \Psi + \frac{\Phi}{\hat{\lambda}}\frac{\partial \hat{\lambda}}{\partial r} - \frac{\alpha}{2} \tag{4.1.38}$$

and

$$\pi_\alpha = \hat{\lambda}^{-1/2}\Phi\left(\Psi + \frac{\alpha}{2}\right) \tag{4.1.39}$$

it is an easy exercise to show that (4.1.36) assumes the form

$$\chi' = \hat{\lambda}^{-1/2}\frac{\partial \hat{\lambda}}{\partial r}\chi^2 + \phi_\alpha\chi + \pi_\alpha + \alpha F_\alpha' \tag{4.1.40}$$

Finally, we set $\theta = \chi - \alpha F_\alpha$ and we have

$$\theta' = \hat{\lambda}^{-1/2}\frac{\partial \hat{\lambda}}{\partial r}(\theta - \alpha \cdot F_\alpha)^2 + \phi_\alpha(\theta + \alpha \cdot F_\alpha) + \pi_\alpha \tag{4.1.41}$$

$$= \hat{\lambda}^{-1/2}\frac{\partial \hat{\lambda}}{\partial r}\theta^2 + 2\alpha\hat{\lambda}^{-1/2}\frac{\partial \hat{\lambda}}{\partial r}\theta F_\alpha$$

$$+ \hat{\lambda}^{-1/2}\frac{\partial \hat{\lambda}}{\partial r}\alpha^2 F_\alpha^2 + \phi_\alpha\theta + \alpha\phi_\alpha F_\alpha + \pi_\alpha$$

$$\geq \hat{\lambda}^{-1/2}\frac{\partial \hat{\lambda}}{\partial r}\theta^2 + (\phi_\alpha + 2\alpha\hat{\lambda}^{-1/2}\frac{\partial \hat{\lambda}}{\partial r}F_\alpha)\theta + \pi_\alpha$$

whenever $\hat{\lambda}^{-1/2}\frac{\partial \hat{\lambda}}{\partial r} > 0$; we now examine the meaning of this last condition.

Suppose that, in addition to our previous hypotheses, we require that the voltage $\mathcal{V}(\cdot)$ satisfy $\mathcal{V}(\cdot) \in C^2(R^1)$ with $\mathcal{V}''(0) > 0$; then for $|\zeta|$ sufficiently small, $\mathcal{V}''(\zeta) > 0$. As a consequence of the fact that

$$Q(x,t) = \hat{\eta}^{-1}(r(x,t) - s(x,t)),$$

and Lemma 4.1, we see that for $|r_0| = \sup_x |r(x,0)|$ and $|s_0| = \sup_x |s(x,0)|$ sufficiently small,

$$\mathcal{V}''(Q(x,t)) > 0, \qquad 0 \le t \le T \tag{4.1.42}$$

However, a direct computation produces

$$\hat{\lambda}^{-1/2}\frac{\partial\hat{\lambda}}{\partial r} = \frac{1}{4}L^{1/4}\left\{\mathcal{V}''(Q)\big/[\mathcal{V}'(Q)]^{5/4}\right\} \tag{4.1.43}$$

so that for $|r_0|$, $|s_0|$ sufficiently small, the condition $\hat{\lambda}^{-1/2}\frac{\partial\hat{\lambda}}{\partial r} > 0$, for $0 \le t \le T$, is a direct consequence of the genuine nonlinearity assumption $\mathcal{V}''(0) > 0$.

With the definition

$$\gamma_\alpha = \phi_\alpha + 2\alpha\hat{\lambda}^{-1/2}\left(\frac{\partial\hat{\lambda}}{\partial r}\right)F_\alpha$$

the differential inequality (4.1.41) assumes the form

$$\theta' \ge \hat{\lambda}^{-1/2}\frac{\partial\hat{\lambda}}{\partial r}\theta^2 + \gamma_\alpha\theta + \pi_\alpha \tag{4.1.44}$$

Now each of the coefficients on the right-hand side of (4.1.44), i.e., $\hat{\lambda}^{-1/2}\frac{\partial\hat{\lambda}}{\partial r}$, γ_α, π_α is a function of $r - s$; we are assuming that (r,s) is a C^1 solution of the initial-value problem for $t \in [0,T]$, $T > 0$ and that the estimate (4.1.26) applies on $[0,T]$. As $\mathcal{V}(\cdot) \in C^2(R^1)$, $\exists\, \rho > 0$ such that for $|r-s| < \rho$ we have $\mathcal{V}''(\hat{\eta}^{-1}(r-s)) > 0$ and the condition $|r-s| < \rho$ on $[0,T]$ follows by taking $|r_0|$, $|s_0|$ to be sufficiently small. We now define the set

$$S_\rho = \{\eta \equiv r - s|\, |\eta| < \rho,\, 0 \le t \le T\}.$$

Clearly, for $|r_0|$, $|s_0|$ sufficiently small, we have that $\eta \equiv (r-s) \in S_\rho$ and with the definitions

$$\Lambda^* = \inf_{S_\rho} \hat{\lambda}^{-1/2}\frac{\partial\hat{\lambda}}{\partial r} > 0 \tag{4.1.45a}$$

$$\Gamma_\alpha = \sup_{S_\rho} |\gamma_\alpha| \tag{4.1.45b}$$

$$\Pi_\alpha = \inf_{S_\rho} \pi_\alpha \tag{4.1.45c}$$

we obtain from (4.1.44) the estimate

$$\theta' \ge \Lambda^*\theta^2 + \Gamma_\alpha\theta + \Pi_\alpha, \qquad 0 \le t \le T \tag{4.1.46}$$

which holds on $[0,T]$ provided the initial data are chosen so as to satisfy

$$|r_0| + |s_0| < \rho e^{-2MT}$$

Without loss of generality we may assume that $\Pi_\alpha < 0$ (otherwise, simply strengthen (4.1.46) by dropping Π_α on the right-hand side of (4.1.46)). Setting

$$\mathcal{I}_\alpha = \left[\left(\frac{\Gamma_\alpha}{\Lambda^*}\right)^2 - \frac{2}{\Lambda^*}\Pi_\alpha\right]^{1/2} \tag{4.1.47}$$

we easily find that (4.1.46) implies that

$$\theta' \geq \frac{1}{2}\Lambda^*(\theta^2 - \mathcal{I}_\alpha)^2, \qquad 0 \leq t \leq T \tag{4.1.48}$$

We also note that with the differential inequality (4.1.48) for $\theta(x,t)$, $x \in R^1$, $0 \leq t \leq T$, we have, associated, the initial condition on R^1:

$$\begin{aligned} \theta(x,0) &= \hat{\lambda}^{-1/2}(r_0(x) - s_0(x))\frac{\partial r}{\partial x}(x,0) \\ &+ \frac{1}{2}\int_0^{r_0(x)-s_0(x)} \{\Psi(\zeta) + \alpha/2\}\, \hat{\lambda}^{-1/2}(\zeta)\, d\zeta \\ &\equiv \bar{\theta}(x) \end{aligned} \tag{4.1.49}$$

We now want to compare the solution of the initial-value problem (4.1.48), (4.1.49) with that for the problem

$$\tilde{\theta}' = \frac{1}{2}\Lambda^*(\tilde{\theta}^2 - \mathcal{I}_\alpha)^2, \quad 0 \leq t \leq T \tag{4.1.50}$$

$$\tilde{\theta}(x,0) = \bar{\theta}(x). \tag{4.1.51}$$

Employing, as in Chapter 2, a standard comparison theorem [95] we have that

$$\theta(x,t) \geq \tilde{\theta}(x,t), \qquad x \in R^1, 0 \leq t \leq T \tag{4.1.52}$$

However, it is easily seen that the solution of the initial-value problem (4.1.50), (4.1.51) satisfies

$$\begin{aligned} \frac{1}{\tilde{\theta}(x,t) + \mathcal{I}_\alpha} &= \frac{\exp(\mathcal{I}_\alpha \Lambda^* t)}{\bar{\theta}(x) + \mathcal{I}_\alpha} \\ &+ \frac{1}{2\mathcal{I}_\alpha}[1 - \exp(\mathcal{I}_\alpha \Lambda^* t)] \end{aligned} \tag{4.1.53}$$

for $0 \le t \le T$. As a consequence of (4.1.53) it is clear that $\exists t_\infty < \infty$ such that $|\tilde{\theta}(x,t)| \to \infty$ as $t \to t_\infty$, provided for such t_∞

$$\lim_{t \to t_\infty} \left[\frac{\exp(\mathcal{I}_\alpha \Lambda^* t)}{\bar{\theta}(x) + \mathcal{I}_\alpha} + \frac{1}{2\mathcal{I}_\alpha}(1 - \exp(\mathcal{I}_\alpha \Lambda^* t)) \right] = 0$$

and this latter condition is satisfied, for some $t_\infty < \infty$, if and only if

$$\frac{2\mathcal{I}_\alpha}{\bar{\theta}(x) + \mathcal{I}_\alpha} = 1 - \exp(-\mathcal{I}_\alpha \Lambda^* t_\infty) \tag{4.1.54}$$

Because $0 < 1 - \exp(-\mathcal{I}_\alpha \Lambda^* t) < 1$, $\forall t > 0$, (4.1.54) is satisfied for some $t_\infty < \infty$ if and only if $\bar{\theta}(x)$ is chosen so as to satisfy

$$0 < \frac{2\mathcal{I}_\alpha}{\bar{\theta}(x) + \mathcal{I}_\alpha} < 1 \tag{4.1.55}$$

An examination of the structure of the initial function $\bar{\theta}(\cdot)$, i.e., of (4.1.49), shows that (4.1.55) is going to be satisfied if at some $x \in R^1$, $\frac{\partial r}{\partial x}(x,0)$ is positive and sufficiently large, in which case the time $t_\infty < \infty$ such that $\lim_{t \to t_\infty} |\tilde{\theta}(x,t)| = +\infty$ is then given by

$$t_\infty = -\frac{1}{\mathcal{I}_\alpha \Lambda^*} \ln \left[1 - \frac{2\mathcal{I}_\alpha}{\bar{\theta}(x) + \mathcal{I}_\alpha} \right] \tag{4.1.56}$$

Of course, t_∞ as given by (4.1.56) is positive as a consequence of (4.1.55); what is crucial at this point is the fact that the blow-up time t_∞, which is given by (4.1.56), will also satisfy $t_\infty < T$ provided that $\frac{\partial r}{\partial x}(x,0) > 0$ is chosen so large that at $x \in R^1$

$$\bar{\theta}(x) > \frac{2\mathcal{I}_\alpha}{1 - \exp(-\mathcal{I}_\alpha \Lambda^* T)} - \mathcal{I}_\alpha \tag{4.1.57}$$

Combining all of our results, we have now established the existence of a time $t_\infty < T$ such that $\lim_{t \to t_\infty} |\theta(x,t)| = +\infty$ whenever the initial data is chosen so that $\frac{\partial r}{\partial x}(x,0) > 0$, at some $x \in R^1$, with $\frac{\partial r}{\partial x}(x,0)$ so large that (4.1.57) holds. Tracing our way back through the various definitions introduced above, we find that $\left|\frac{\partial r}{\partial x}(x,t)\right| \to +\infty$, at some $x \in R^1$, if $\frac{\partial r}{\partial x}$ is both positive and sufficiently large at $(x,0)$; an analogous result holds, of course, for $\frac{\partial s}{\partial x}$, thus contradicting the assumed existence of a C^1 solution $(r(x,t), s(x,t))$ to the initial-value problem (4.1.16), (4.1.25) on $[0,T]$. In terms of the finite-time breakdown of class C^1 solutions for the original quasilinear hyperbolic system (4.1.7), which governs the evolution of charge and current in our nonlinear, distributed parameter, transmission line, we may state the following result:

Theorem 4.1 [22] Consider the inhomogeneous quasilinear system (4.1.7) with associated periodic initial-data $i(x,0) = i_0(x)$, $Q(x,0) = Q_0(x)$, $x \in R^1$. Assume that the voltage function $\mathcal{V}(\cdot)$ satisfies $\mathcal{V}'(\zeta) > 0$, $\forall \zeta \in R^1$, $\mathcal{V}''(0) > 0$, and (4.1.22), where Φ is defined by (4.1.17). Then a C^1 solution $(i(x,t), Q(x,t))$ cannot exist for all $t > 0$ if $\sup_x |i_0(x)|$, $\sup_x |Q_0(x)|$ are chosen sufficiently small while, at some $x \in R^1$,

$$\mu(x) \equiv \sqrt{\frac{\mathcal{V}'(Q_0(x))}{L}} Q_0'(x) - i_0'(x) \tag{4.1.58}$$

is chosen so as to be positive and sufficiently large.

Remarks: It should be clear that for $\sup_x |i_0(x)|$, $\sup_x |Q_0(x)|$ chosen sufficiently small, we may replace the hyperbolicity requirement $\mathcal{V}'(\zeta) > 0$, $\forall \zeta \in R^1$, in Theorem 4.1, by the milder condition of local hyperbolicity, i.e., $\mathcal{V}'(0) > 0$.

We now want to compare the result stated above, in Theorem 4.1, with earlier results on the finite-time breakdown of smooth solutions to initial-value problems associated with the quasilinear system (4.1.1) governing the evolution of the current and voltage in a distributed parameter nonlinear transmission line. For Q monotone increasing in the voltage v, this system is, of course, equivalent to (4.1.7) and we maintain our previous assumptions that the resistance R, inductance L, and leakage conductance $1/G$ are all constant. Jeffrey [74] has considered the system (4.1.1); the primary focus in this work was aimed at making use of the author's previous work [76] (see also Jeffrey and Taniuti [78]) on the development of jump discontinuities in solutions to certain hyperbolic quasilinear systems with Lipschitz continuous initial data. The results of [76] are used, specifically, in [74] in order to determine the time t_c and the distance x_c at which a continuous wave propagating down a distributed parameter nonlinear transmission line, having a voltage-dependent capacitance, first forms an electromagnetic shock wave. In examining electromagnetic shock wave formation in the transmission line, the author [74] assumes that the wave is propagating into the state in which $i = 0$ and $v = 0$; the critical time t_c at which the electromagnetic shock wave first forms corresponds to the first time at which the Jacobian of the so-called (ϕ, t') transformation vanishes on the C_0^+ (forward) characteristic bounding the constant state $i = 0$, $v = 0$, where $\phi(x,t) = 0$ corresponds to the wavefront trace and $t' = t'(t)$ is a single-valued function of t. If we denote by a zero subscript the value of the indicated expression, in the constant state immediately ahead of the wavefront trace, and by $\tilde{i}_x$ the limiting value of $\dfrac{\partial i}{\partial x}$ taken along the wavefront trace C_0^+ as $t' \to 0$, then the essential content of the main result in [74] is the following: If either

$$\text{(i)} \quad \tilde{i}_x \left(\frac{\partial C}{\partial v} \right)_0 < 0 \tag{4.1.59a}$$

or

$$\text{(ii)} \quad \tilde{i}_x \left(\frac{\partial C}{\partial v} \right)_0 > 0 \text{ and}$$

$$\frac{C_0^2}{\tilde{i}_x \left(\frac{\partial C}{\partial v} \right)_0} \left(\frac{R}{L} + \frac{G}{C_0} \right) < 1 \tag{4.1.59b}$$

then there exists a (positive) critical time t_c for the formation of an electromagnetic shock wave in the transmission line which is given by

$$t_c = \frac{2}{\left(\frac{R}{L} + \frac{G}{C_0} \right)} \ln \left\{ 1 - \frac{C_0^2}{\tilde{i}_x \left(\frac{\partial C}{\partial v} \right)_0} \left(\frac{R}{L} + \frac{G}{C_0} \right) \right\}^{-1} \tag{4.1.60}$$

As the wave is assumed in [74] to be propagating into a constant state with speed $(LC_0)^{-1/2}$, the distance travelled by the current and voltage wavefronts from the time $t = 0$ to the time $t = t_c$ is

$$x_c = t_c \big/ (LC_0)^{1/2} \tag{4.1.61}$$

For a non-dissipative line in which $R = G = 0$ we may take the limit in (4.1.60) as the quantity

$$\lambda = \left(\frac{R}{L} + \frac{G}{C_0} \right) \longrightarrow 0$$

in which case we obtain for the critical time until the formation of an electromagnetic shock in the line

$$t_c = 2C_0^2 / \tilde{i}_x \left(\frac{\partial C}{\partial v} \right)_0 \tag{4.1.62}$$

The condition (4.1.59b) ties in nicely with the conclusion of Theorem 4.1, in the sense that for $\tilde{i}_x \left(\frac{\partial C}{\partial v} \right)_0 > 0$, this criterion requires that $\tilde{i}_x \left(\frac{\partial C}{\partial v} \right)_0$ be sufficiently large; moreover, when (4.1.59b) is satisfied, the expression (4.1.60) clearly indicates that the critical time t_c to the development of an electromagnetic shock is shortened by increasing the value of the product $\tilde{i}_x \left(\frac{\partial C}{\partial v} \right)_0$.

Nonlinear effects in transmission lines, including shock formation, were also studied in a fundamental paper by Cumberbatch [38]. In [38] the author considers the system (4.1.1), but allows for the more general scenario in which we can have $L = L(i)$ and $G = G(v)$ in addition to $C = C(v)$; for such a situation, but with $R = G = 0$, he first exhibits the existence of simple wave solutions of the transmission line equations which assume, in this special case, the form

$$\frac{d}{dt}\left[\int \sqrt{L(i)}\,di \pm \int \sqrt{C(v)}\,dv\right] = 0 \tag{4.1.63}$$

on the characteristic curves $C^{\pm}$ given by

$$\frac{dx}{dt} = \pm 1 \Big/ \sqrt{C(v)L(i)} \tag{4.1.64}$$

Labelling the C^{+} characteristics by the time $t = \phi$ at which they emanated from the point $x = 0$ in the line, these simple wave solutions for the case $C = C(v)$, $L = L(i)$, $R = G = 0$ have the form

$$i = i(\phi), \qquad v = v(\phi), \qquad t = \phi + \sqrt{C(v)L(i)}x \tag{4.1.65}$$

With the assumption of a boundary condition of the form

$$v(0,t) = \bar{v}\sin nt \tag{4.1.66}$$

the simple wave solutions obtained in [38] look like

$$v(x,t) = \bar{v}\sin n(t - \sqrt{C(v)L(i)}x) \tag{4.1.67}$$

and the position at which a shock wave forms is then given by the intersection of neighboring characteristics, i.e., where

$$\frac{\partial t}{\partial \phi} = 1 + \frac{x}{2\sqrt{CL}}(LC'(v)v'(\phi) + CL'(i)i'(\phi)) = 0 \tag{4.1.68}$$

For $L = L_0 =$ const., and for a wave propagating into the constant state $i = v = 0$, it is also demonstrated in [38] that when it is assumed that the capacitance $C(v)$ has, near $v = 0$, an expansion of the form

$$C(v) = C_0 + C_1 v + C_2 v^2 + \ldots, \tag{4.1.69}$$

the position at which a shock is first formed at the wave front $\phi = 0$ occurs when $C'(v)|_{v=0} \cdot v'(\phi)|_{\phi=0} \neq 0$ and, for $C_1 \neq 0$, and the simple wave solution (4.1.67), this results in

$$x_c = -2\sqrt{\frac{C_0}{L_0}} \Big/ C_1 \bar{v} n \tag{4.1.70}$$

The expression (4.1.70) requires that $C_1\bar{v} < 0$ in order to obtain a positive shock distance. The remainder of the analysis in [38] revolves around the concept of a relatively undistorted wave (e.g., Varley and Cumberbatch [174]); the author uses this concept in order to obtain an asymptotic solution to the transmission line equations, for small dissipation parameters R and G, which describes waves that propagate in a slowly-varying manner. For these slowly-varying solutions of the distributed parameter nonlinear transmission line equations, the analysis in [38] provides expressions for the location of the points x_c, in the line, at which shocks first form, which are based on linearized approximations of the solutions. If we again denote by a zero subscript the values of the indicated expression in the constant state ($i = 0, v = 0$), immediately ahead of the wavefront trace, then the requirement in [38] that the analysis yield a positive time t_c, for the (first) formation of an electromagnetic shock, assumes the form

$$\text{(i)} \qquad \omega = n\left(\frac{G_0}{C_0} + \frac{R_0}{L_0}\right)^{-1} > -C_0/\bar{v}C_1 \tag{4.1.71a}$$

and

$$\text{(ii)} \qquad \omega > -2C_0/\bar{v}^2 C_2 \tag{4.1.71b}$$

As the author points out [38], the conditions (4.1.71a), (4.1.71b) indicate that shock waves will form on the transmission line provided the product of amplitude ($\bar{v}$) and frequence ($\bar{n}$) of the input voltage is sufficiently large.

Prior to the work which appeared in the mid 1960's (as exemplified by [74] and [38]), there appeared a series of papers by R. Landauer of IBM and his colleagues which considered the development and propagation of shock waves in nonlinear transmission lines and their effect on parametric amplification; a prime example of this work is the paper [97] in which it is shown that, in general, a shock wave will form on the line, but in a distance that is too short for the purpose of parametric amplification. It should be pointed out that much of the analysis in these earlier works, including [97], dealt with the nonlinear transmission line equations (4.1.1) but with $R = G = 0$, and that attention was focused primarily on simple wave solutions of the form $v = v(x - \frac{1}{\sqrt{LC(v)}}t)$; however, the author [97] does pursue an analysis of the propagation of the shock subsequent to its formation on the line showing that asymptotically, as $t \to \infty$, the shock amplitude decreases inversely as the distance along the line. Although the analysis in [97] leads to the conclusion that parametric amplification of a signal cannot be achieved on transmission lines which are relatively dispersionless, for the purposes of harmonic generation, wave shaping, intermittent amplification accompanied by signal compression in time, or for the control of transit time through transmission systems, lines governed by (4.1.1), with $R = G = 0$, do appear to have potential. For subsequent work by Landauer and his colleagues (some of which we discuss in § 6, below) on the nonlinear transmission line problem, both with and without dispersion, as well as for further references to some of the earlier work in this area, the reader may consult the references [96] and [98]. Other aspects of the problem of shock formation in distributed parameter nonlinear transmission lines (some of which will also be covered in § 6, below) are treated in the monographs by Kataev [50] and A. C. Scott [151], as well as in the book by Khokhlov [147].

4.2 Global Existence with Small Data: Riemann Invariants Arguments

In this section we begin our study of the existence problem for the system (4.1.7) with associated initial data

$$i(x,0) = i_0(x), \qquad Q(x,0) = Q_0(x), \qquad x \in R^1 \tag{4.2.1}$$

We begin with a Riemann invariants argument based on the system (4.1.16), where the Riemann invariants r, s are given by (4.1.9a), (4.1.9b); our results in this case will be valid under an appropriate set of "smallness" assumptions relative to the associated initial data (4.1.25). For the wave-dielectric interaction problem, in one-space dimension, we indicated, in Chapter 3, that such a Riemann invariants argument was not feasible with regard to the matter of proving existence of global smooth solutions, even for sufficiently small data. In our efforts here, however, our work is aided by the presence of the additional strong dissipative term which is generated by the presence of the resistance element in the transmission line, a term which has no direct counterpart in the quasilinear system (2.6.16) which governs the evolution of the fields D and B in the one-dimensional wave-dielectric interaction problem. Even with the presence of the rather strong damping induced by the resistance element in the nonlinear transmission line, the initial effort in this area by the author, i.e., [19], made use of a rather strong condition on the (inverse) of the leakage conductance G which we now assume to be a nonnegative function of the voltage v; this condition (which we label below as $(\tilde{G})$) is easily seen to be superfluous with respect to the analysis presented if G is a constant, as was the case with the shock formation analysis delineated in the previous section. Moreover, the global condition $(\tilde{G})$ has subsequently been weakened, in recent work of J. Wang [179], [177], to the condition which we denote later on in our discussion as (G^*); this is a requirement which is much easier to verify than the original condition $(\tilde{G})$ employed by this author in [19]. In order to convey the full flavor of the development in this area, we have chosen to first present the essence of the existence argument, for the system (4.1.7), with $G = G(v)$, as originally given in [19], and then to discuss, in some detail, the substantial improvement which is embodied in the more recent work [179], [177] of Wang. Our work on the existence problem for the system (4.1.7), governing the distributed parameter nonlinear transmission problem, then continues in § 4.2, where we present an analysis (again for sufficiently small data) which is grounded in the same basic type of energy argument that was employed in § 3.5 for the wave-dielectric interaction problem. Following this, we offer, in § 4.3, a brief survey of those ideas and concepts which are central to the notions of compensated compactness and the Young measure. An argument based on compensated compactness and weak convergence is then employed in § 4.4 in order to prove the existence of weak solutions for the system (4.1.7) with arbitrarily large initial data.

For the situation in which we allow for a voltage dependence on the part of the leakage conductance, the system (4.1.7) assumes the form

$$\begin{pmatrix} i \\ Q \end{pmatrix}_{,t} + \begin{pmatrix} 0 & \frac{1}{L}\mathcal{V}'(Q) \\ 1 & 0 \end{pmatrix} \begin{pmatrix} i \\ Q \end{pmatrix}_{,x} = \begin{pmatrix} -\frac{R}{L}i \\ -G(\mathcal{V}(Q))\mathcal{V}(Q) \end{pmatrix} \tag{4.2.2}$$

and we take the initial data in (4.2.1) to be both periodic and of class C^1. As indicated above, the appropriate Riemann invariants are once again given by (4.1.9a), (4.1.9b) and the definitions and relations embodied in (4.1.10)-(4.1.15) will apply in this section as well—in particular, the charge distribution $Q(x,t)$ is given in terms of $r(x,t)$, $s(x,t)$ by (4.1.14). If we set

$$\tilde{\Phi}(Q) = -G(\mathcal{V}(Q))\mathcal{V}(Q)\sqrt{\frac{\mathcal{V}'(Q)}{L}} \tag{4.2.3}$$

then in place of (4.1.16) we have

$$\begin{cases} r' = -\dfrac{\alpha}{2}(r+s) + \tilde{\Phi}(r-s) \\ \grave{s} = -\dfrac{\alpha}{2}(r+s) - \tilde{\Phi}(r-s) \end{cases} \tag{4.2.4}$$

with $\tilde{\Phi}(r-s) = \Phi(\hat{\eta}^{-1}(r-s))$ and $\alpha = \dfrac{R}{L} > 0$. Simply for the purpose of having a more convenient notation with which to work below, we will often write $Q = \tilde{Q}(\eta)$, where $\eta(x,t)$ is defined by (4.1.12), instead of the equivalent relation (4.1.14). We also set $\tilde{\Psi}(\eta) = \tilde{\Phi}'(\eta)$, i.e.,

$$\tilde{\Psi}(\eta) = \left.\frac{d\tilde{\Phi}}{dQ}\right|_{Q=\tilde{Q}(\eta)} \cdot \frac{d\tilde{Q}}{d\eta} \tag{4.2.5}$$

and with λ, ν as given by (4.1.10) we will find it useful to use, in place of the notation in (4.1.15), the notation

$$\tilde{\lambda}(\eta) = \lambda(\tilde{Q}(\eta)) \text{ and } \tilde{\nu}(\eta) = \nu(\tilde{Q}(\eta))$$

If we set

$$\Phi_*(Q) = \mathcal{V}(Q)\lambda(Q) \tag{4.2.6}$$

then

$$\tilde{\Phi}(Q) = G(\mathcal{V}(Q))\Phi_*(Q) \tag{4.2.7}$$

From (4.1.12), with $Q = \tilde{Q}(\eta)$, we have

$$\frac{d\tilde{Q}}{d\eta} = \frac{\sqrt{L}}{2} \cdot \frac{1}{\sqrt{\mathcal{V}'(\tilde{Q}(\eta))}} \tag{4.2.8}$$

and as

$$\lambda'(Q) = -\frac{1}{2\sqrt{L}}\frac{\mathcal{V}''(Q)}{\sqrt{\mathcal{V}'(Q)}}$$

$$\tilde{\lambda}'(\eta) \equiv \left.\frac{d\lambda}{dQ}\right|_{Q=\tilde{Q}(\eta)} \frac{d\tilde{Q}}{d\eta} = -\frac{\mathcal{V}''(\tilde{Q}(\eta))}{4\mathcal{V}'(\tilde{Q}(\eta))} \tag{4.2.9a}$$

$$\frac{\tilde{\lambda}'(\eta)}{\tilde{\lambda}(\eta)} = \frac{\sqrt{L}}{4} \frac{\mathcal{V}''(\tilde{Q}(\eta))}{\mathcal{V}'(\tilde{Q}(\eta))\sqrt{\mathcal{V}'(\tilde{Q}(\eta))}} \tag{4.2.9b}$$

provided $\mathcal{V}''$ is well-defined, and

$$\frac{\tilde{\lambda}'(\eta)}{\tilde{\lambda}(\eta)} \cdot \tilde{\Phi}(\eta) = -\frac{G\mathcal{V}(\tilde{Q}(\eta))}{4} \frac{\mathcal{V}(\tilde{Q}(\eta))\mathcal{V}''(\tilde{Q}(\eta))}{\mathcal{V}'(\tilde{Q}(\eta))} \tag{4.2.9c}$$

A direct calculation yields

$$\frac{d\Phi_*}{dQ} = -\frac{\mathcal{V}(Q)\mathcal{V}''(Q) - 2\mathcal{V}'^2(Q)}{2\sqrt{L}\sqrt{\mathcal{V}'(Q)}} \tag{4.2.10}$$

and

$$\tilde{\Psi}(\eta) = \left.\left(G(\mathcal{V}(Q))\frac{d\Phi_*}{dQ} + G'(\mathcal{V}(Q))\mathcal{V}'(Q)\Phi_*(Q)\right)\right|_{Q=\tilde{Q}(\eta)} \times \frac{d\tilde{Q}}{d\eta}$$

so that

$$\begin{aligned}\tilde{\Psi}(\eta) &= -\frac{G(\mathcal{V}(\tilde{Q}(\eta))}{4} \cdot \frac{\mathcal{V}(\tilde{Q}(\eta))\mathcal{V}''(\tilde{Q}(\eta)) - 2\mathcal{V}'^2(\tilde{Q}(\eta))}{\mathcal{V}'(\tilde{Q}(\eta))} \\ &\quad - \frac{1}{2}G'(\mathcal{V}(\tilde{Q}(\eta))\mathcal{V}'(\tilde{Q}(\eta))\mathcal{V}(\tilde{Q}(\eta))\end{aligned} \tag{4.2.11}$$

Finally, for future reference, we record the fact that

$$\tilde{\Psi}(\eta) - \frac{\tilde{\lambda}'(\eta)}{\tilde{\lambda}(\eta)} \cdot \tilde{\Phi}(\eta) = -\frac{\mathcal{V}'(\tilde{Q}(\eta))}{2} \left.\frac{d}{d\mathcal{V}}(\mathcal{V}G(\mathcal{V}))\right|_{\mathcal{V}=\mathcal{V}(\tilde{Q}(\eta))} \tag{4.2.12}$$

We now want to delineate the basic constitutive hypotheses that will be employed throughout the remainder of this section; they are as follows:

(i) For all $\zeta \in R^1$, $\frac{d}{d\zeta}(\zeta G(\zeta))$ exists and is bounded, i.e., $\exists\, G_1 > 0$ such that

$$|(\zeta G(\zeta))'| \le G_1, \ \forall \zeta \in R^1 \tag{4.2.13}$$

A simple example of a leakage conductance which conforms to the requirement expressed by (4.2.13) might be $G(\zeta) = \exp(-|\zeta|)$, in which case

$$H(\zeta) = \frac{d}{d\zeta}(\zeta G(\zeta)) = \begin{cases} e^{-\zeta}(1-\zeta), \zeta \ge 0 \\ e^{\zeta}(1+\zeta), \zeta < 0 \end{cases} \tag{4.2.14}$$

the graph of which is depicted below:

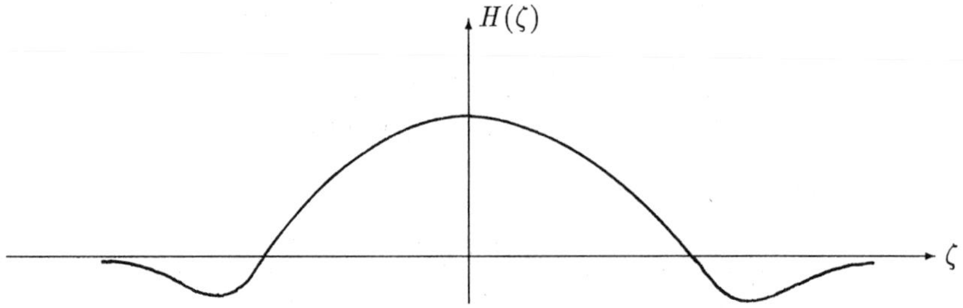

Figure 4.2

If we define, in the obvious manner, $\tilde{\Phi}_*(\eta) = \Phi_*(\tilde{Q}(\eta))$ so that $\tilde{\Phi}(\eta) = \tilde{G}(\eta)\tilde{\Phi}_*(\eta)$ with $\tilde{G}(\eta) = G(\mathcal{V}(\tilde{Q}(\eta))$, then we could write

$$\tilde{\Phi}(\eta) = \tilde{G}(\eta) \int_0^\eta \frac{d\tilde{\Phi}_*}{d\zeta}(\zeta)\, d\zeta \tag{4.2.15}$$

if $\tilde{\Phi}_*(0) = \Phi_*(\tilde{Q}(0)) = 0$; as a consequence of the fact that

$$\eta = 2 \int_0^{\tilde{Q}(\eta)} \sqrt{\frac{\mathcal{V}'(\zeta)}{L}}\, d\zeta$$

however,

$$\int_0^{\tilde{Q}(0)} \sqrt{\frac{\mathcal{V}'(\zeta)}{L}}\, d\zeta = 0 \tag{4.2.16}$$

from which it follows that $\tilde{Q}(0) = 0$. Also, by virtue of (4.2.6)

$$\Phi_*(0) \equiv \mathcal{V}(0)\lambda(0) = -\mathcal{V}(0)\sqrt{\frac{\mathcal{V}'(0)}{L}} = 0 \tag{4.2.17}$$

provided $\mathcal{V}(0) = 0$; thus, (4.2.15) applies if $\mathcal{V}(0) = 0$. Our first hypothesis relative to the voltage function $\mathcal{V}(\cdot)$ is, therefore, as follows:

(ii) $\mathcal{V} = \mathcal{V}(\zeta)$ satisfies $\mathcal{V}(\cdot) \in C^1(R^1)$ with

$$\mathcal{V}(0) = 0 \text{ and } \mathcal{V}_0 \geq \mathcal{V}'(\zeta) \geq \epsilon > 0,\ \forall\, \zeta \in R^1 \tag{4.2.18}$$

for some $\mathcal{V}_0, \epsilon > 0$. In view of (4.2.15),

$$|\tilde{\Phi}(\eta)| \leq \tilde{G}(\eta)|\eta| \sup_\eta |\tilde{\Phi}'_*(\eta)|,\ \forall\, \eta \in R^1 \tag{4.2.19}$$

Also, by (4.2.8) and (4.2.10)

$$\tilde{\Phi}'_*(\eta) \equiv \left.\frac{d\Phi_*}{dQ}\right|_{Q=\tilde{Q}(\eta)} \cdot \frac{d\tilde{Q}}{d\eta} = -\frac{1}{4} \cdot \frac{\mathcal{W}(\tilde{Q}(\eta))}{\mathcal{V}'(\tilde{Q}(\eta))} \tag{4.2.20}$$

where we have set

$$\mathcal{W}(\zeta) = \mathcal{V}(\zeta)\mathcal{V}''(\zeta) + 2\mathcal{V}'(\zeta)^2, \; \forall \zeta \in R^1 \tag{4.2.21}$$

In order that $\tilde{\Phi}'_*(\eta)$, $\eta \in R^1$, be well-defined, we will need the first part of hypothesis (iii) below; the second part of the hypothesis will be used in an integral fashion in the proof of the global existence theorem. It is worth noting that neither part of hypothesis (iii) requires that $\mathcal{V}''(\zeta)$ exist $\forall \zeta \in R^1$.

(iii)

(a) For all $\zeta \in R^1$ the product $\mathcal{V}(\zeta)\mathcal{V}''(\zeta)$ exists and is finite; furthermore,

$$\sup_{\zeta} \left| \frac{\mathcal{W}(\zeta)}{\mathcal{V}'(\zeta)} \right| < \infty \tag{4.2.22}$$

i.e., $\exists B_1 > 0$ such that $|\mathcal{W}(\zeta)| \le B_1 |\mathcal{V}'(\zeta)|$, $\forall \zeta \in R^1$.

(b) $\mathcal{V}''(\zeta)$ exists for all ζ, with $|\zeta|$ sufficiently small, and for all such ζ, $\exists B_2 > 0$ such that $|\mathcal{V}''(\zeta)| \le B_2 \mathcal{V}'(\zeta)$.

The complete set of constitutive hypothesis consists, therefore, of hypothesis (i) relative to the leakage conductance $1/G$ and hypotheses (ii) and (iii) concerning the nature of the dependence of the voltage $\mathcal{V}$ on the charge Q. It will be necessary, below, to make one additional hypothesis $(\tilde{G})$ concerning the behavior of the leakage conductance and this hypothesis is the one which will subsequently be weakened to the condition (G^*) as introduced in the work of Wang [179], [177]. Before proceeding, however, we want to offer an example of a voltage function $\mathcal{V}(\cdot)$ whose dependence on Q is such as to satisfy all the requirements of hypotheses (ii) and (iii) above:

Example. For arbitrary $\epsilon > 0$, $\mathcal{V}_1 > 0$, define, for $Q \in R^1$

$$\mathcal{V}(Q) = \begin{cases} \epsilon Q + \mathcal{V}_1(1 - e^{-Q}), & Q \ge 0 \\ \epsilon Q + \mathcal{V}_1(e^{Q} - 1), & Q < 0 \end{cases} \tag{4.2.23}$$

Clearly, $\mathcal{V}(0) = 0$, $\mathcal{V}(Q) > 0$ for $Q > 0$, and $\mathcal{V}(Q) < 0$ for $Q < 0$. Also, $\mathcal{V}(Q) = -\mathcal{V}(-Q)$ and $\mathcal{V} \to \pm\infty$ as $Q \to \pm\infty$. Next, $\mathcal{V}'(Q) = \epsilon + \mathcal{V}_1 e^{-|Q|}$ so that $\forall Q \in R^1$,

$\epsilon < \mathcal{V}'(Q) \leq \epsilon + \mathcal{V}_1$ and $\mathcal{V}'(Q) \to \epsilon$ as $|Q| \to \infty$. A direct computation produces

$$\mathcal{V}''(Q) = \begin{cases} -\mathcal{V}_1 e^{-Q}, & Q > 0 \\ \mathcal{V}_1 e^{Q}, & Q < 0 \end{cases} \tag{4.2.24}$$

so that $\mathcal{V}''(Q) > 0$ for $Q < 0$, $\mathcal{V}''(Q) < 0$ for $Q > 0$, and

$$\lim_{Q \to 0+} \mathcal{V}''(Q) = -\mathcal{V}_1 \neq \lim_{Q \to 0-} \mathcal{V}''(Q) = \mathcal{V}_1 \tag{4.2.25}$$

Thus $\mathcal{V}''(0)$ does not exist if $\mathcal{V}_1 \neq 0$; however,

$$\mathcal{V}(Q)\mathcal{V}''(Q) = \mathcal{V}_1^2(e^{-Q} - 1)e^{-Q} - \epsilon \mathcal{V}_1 Q e^{-Q}, \quad \text{for } Q \geq 0 \tag{4.2.25a}$$

while

$$\mathcal{V}(Q)\mathcal{V}''(Q) = \mathcal{V}_1^2(e^{Q} - 1)e^{Q} + \epsilon \mathcal{V}_1 Q e^{Q}, \quad \text{for } Q < 0 \tag{4.2.25b}$$

so that $\mathcal{V}(Q)\mathcal{V}''(Q)$ does exist and is continuous $\forall\, Q \in R^1$ with $\mathcal{V}\mathcal{V}''\Big|_{Q=0} = 0$. Also, for $Q \neq 0$ we clearly have $\mathcal{V}(Q)\mathcal{V}''(Q) < 0$, as a direct consequence of (4.2.25a), (4.2.25b). Moreover,

$$\mathcal{V}'^2(Q) = \begin{cases} \mathcal{V}_1^2 e^{-2Q} + 2\epsilon \mathcal{V}_1 e^{-Q} + \epsilon^2, & Q \geq 0 \\ \mathcal{V}_1^2 e^{2Q} + 2\epsilon \mathcal{V}_1 e^{Q} + \epsilon^2, & Q < 0 \end{cases} \tag{4.2.26}$$

Therefore, employing the definition (4.2.22) of $\mathcal{W}(\cdot)$, we compute that for $Q \geq 0$,

$$\begin{aligned} \mathcal{W}(Q) &= 3\mathcal{V}_1^2 e^{-2Q} + (4\epsilon \mathcal{V}_1 - \mathcal{V}_1^2)e^{-Q} \\ &\quad - \epsilon \mathcal{V}_1 Q e^{-Q} + 2\epsilon^2 \end{aligned} \tag{4.2.27a}$$

while for $Q < 0$,

$$\begin{aligned} \mathcal{W}(Q) &= 3\mathcal{V}_1^2 e^{2Q} + (4\epsilon \mathcal{V}_1 - \mathcal{V}_1^2)e^{Q} \\ &\quad + \epsilon \mathcal{V}_1 Q e^{Q} + 2\epsilon^2 \end{aligned} \tag{4.2.27b}$$

The results (4.2.27a), (4.2.27b) may be rewritten in the more compact form as

$$\mathcal{W}(Q) = \mathcal{A} e^{-2|Q|} + \mathcal{B}(Q) e^{-|Q|} + \mathcal{C} \tag{4.2.28}$$

with

$$\mathcal{A} = 3\mathcal{V}_1^2, \quad \mathcal{B} = 4\epsilon\mathcal{V}_1 - \mathcal{V}_1^2 - \epsilon\mathcal{V}_1|Q|, \quad \mathcal{C} = 2\epsilon^2 \tag{4.2.29}$$

An elementary calculation now produces the estimates

$$|\mathcal{W}(Q)| = 6\mathcal{V}_1^2 + 4\epsilon^2 + \frac{1}{e}(\epsilon\mathcal{V}_1) \leq a\mathcal{V}_1^2 + b\epsilon^2 \tag{4.2.30}$$

where $a = 6 + (2e)^{-1}$ and $b = 4 + (2e)^{-1}$. Then, as $\mathcal{V}'(Q) > \epsilon$, $\forall\, Q \in R^1$, we have

$$\sup_{Q \in R^1} |\mathcal{W}(Q)\,/\mathcal{V}'(Q)| \leq \frac{1}{\epsilon}(a\mathcal{V}_1^2 + b\epsilon^2) \equiv B_1 \tag{4.2.31}$$

with B_1 now the constant whose existence insures the validity of part (a) of hypothesis (iii), i.e., of (4.2.22). A sketch of the graph of $\mathcal{V}(Q)$ appears below:

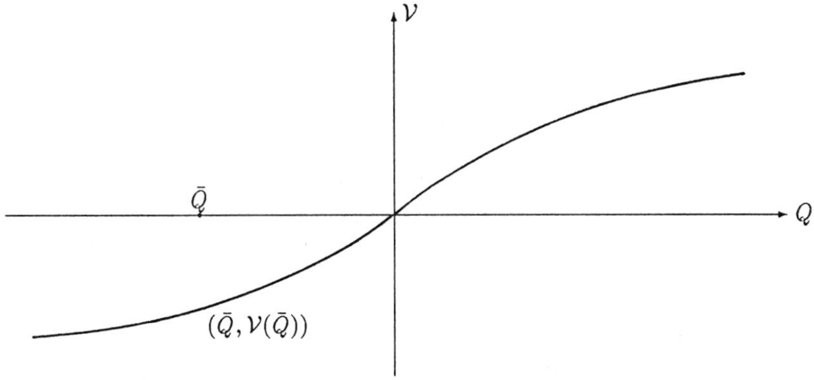

Figure 4.3

Now, as indicated above in Figure 4.3, let $\bar{Q} < 0$ so that $(\bar{Q}, \mathcal{V}(\bar{Q})$ lies on the graph of $\mathcal{V}(\cdot)$ in the third quadrant of the $(Q, \mathcal{V})$ plane. Define

$$\bar{\mathcal{V}}(Q) = \mathcal{V}(Q + \bar{Q}) - \mathcal{V}(\bar{Q}), \quad Q \in R^1 \tag{4.2.32}$$

Then we clearly have $\bar{\mathcal{V}}(0) = 0$, $\epsilon < \bar{\mathcal{V}}'(Q) \leq \epsilon + \mathcal{V}_1$, $\forall\, Q \in R^1$, while $\bar{\mathcal{V}}''(Q)$ exists and is continuous $\forall\, Q$ such that $Q < |\bar{Q}|$ with $\bar{\mathcal{V}}''(Q) > 0$ for $|Q|$ sufficiently small. It is easy to see that we again obtain a bound of the form

$$\sup_{Q} \left|\bar{\mathcal{W}}(Q)\,/\bar{\mathcal{V}}'(Q)\right| \leq \bar{B}_1 \tag{4.2.33}$$

for some $\bar{B}_1 > 0$ where, obviously, $\bar{\mathcal{W}}(\cdot)$ is given by (4.2.21) with $\bar{\mathcal{V}}(\cdot)$ replacing $\mathcal{V}(\cdot)$. As concerns the second part of hypothesis (iii), we note that for $Q \neq 0$,

$$\frac{\mathcal{V}''(Q)}{\mathcal{V}'(Q)} = -\mathrm{sgn}Q \cdot \frac{\mathcal{V}_1 e^{-|Q|}}{\epsilon + \mathcal{V}_1 e^{-|Q|}}$$

so that for $Q < |\bar{Q}|$

$$\frac{\bar{\mathcal{V}}''(Q)}{\bar{\mathcal{V}}'(Q)} = \frac{\mathcal{V}_1 e^{Q-|\bar{Q}|}}{\epsilon + \mathcal{V}_1 e^{Q-|\bar{Q}|}} \tag{4.2.34}$$

in which case it follows that for $|Q|$ sufficiently small, $\bar{\mathcal{V}}''(Q)/\bar{\mathcal{V}}'(Q)$ is, indeed, well-defined with $|\bar{\mathcal{V}}''(Q)/\bar{\mathcal{V}}'(Q)| < 1$.

We now introduce our last hypothesis, the hypothesis which we have referred to, above, as $(\tilde{G})$; the introduction of this hypothesis will have the effect of allowing us to strengthen the statement of Lemma 4.1, relative to the initial-value problem (4.1.16), (4.1.25) in Riemann invariants form, to a degree which will permit locally defined smooth solutions to persist for all $t > 0$. The existence of a local C^1 (in (x,t)) solution for the quasilinear hyperbolic system of interest here, i.e., for (4.1.7), with associated C^1 periodic initial data (4.2.1) follows in the usual way from the work, e.g., of Kato [89] and implies the corresponding local existence result for the initial-value problem (4.1.16), (4.1.25). More precisely, $\exists\, t_m > 0$ such that a C^1 solution $(r(x,t), s(x,t))$ of (4.1.16), (4.1.25) exists for $0 \le t < t_m \le \infty$ and, if $t_m < \infty$, then

$$\lim_{t\to t_m^-} \sqrt{r_x^2(x,t) + s_x^2(x,t)} = \infty \tag{4.2.35}$$

at some $x \in R^1$ so that a shock forms at (x, t_m) when the C^1 solution is not globally defined. Our additional hypothesis will be seen to be equivalent to the requirement that, for the local C^1 solution, the leakage conductance $1/G$, per unit length of the line, grows sufficiently fast on $[0, t_m)$.

From (4.2.19) and (4.2.20), coupled with hypothesis (iii), above, we have, $\forall\, \eta \in R^1$

$$\begin{aligned} |\tilde{\Phi}(\eta)| &\le \frac{1}{4}\tilde{G}(\eta)|\eta| \sup_{\eta} \left| \frac{\mathcal{W}(\tilde{Q}(\eta))}{\mathcal{V}'(\tilde{Q}(\eta))} \right| \\ &\le \frac{B_1}{4}\tilde{G}(\eta)|\eta| \sup_{\eta} \left| \frac{\mathcal{V}''(\tilde{Q}(\eta))}{\mathcal{V}'(\tilde{Q}(\eta))} \right| \\ &\le \frac{\tilde{B}_1}{4}(\tilde{G}(\eta)|\eta|) \end{aligned} \tag{4.2.36}$$

with $\tilde{G}(\eta) = G(\mathcal{V}(\tilde{Q}(\eta)))$ and $\tilde{B}_1 = B_1 B_2$. If we now denote by $(r(x,t), s(x,t))$ the unique, locally defined solution of (4.1.16), (4.1.25) on $[0, t_m)$ and recall that $\eta(x,t) = r(x,t) - s(x,t)$, then our last hypothesis is the following:

$(\tilde{G})$: For some $G_0 > 0$ and some $\delta > \alpha = R/L$,

$$\sup_{x \in R^1} \tilde{G}(\eta(x,t)) \leq G_0 e^{-\delta t}, \qquad 0 \leq t < t_m \tag{4.2.37}$$

As we have already indicated at several points in the text, we will, subsequently, weaken the condition $(\tilde{G})$ to a condition (G^*) which will be a condition relative to the leakage conductance that is independent of the locally defined solution (r,s) on $[0,t_m)$.

With hypotheses (i)-(iii), and $(\tilde{G})$, in hand, we are now in a position to state and prove the strengthened version of Lemma 4.1 which will be needed in the global existence theorem that follows:

Lemma 4.2 ([19]): Suppose that $\mathcal{V}(\cdot), G(\cdot)$ satisfy hypotheses (i)-(iii) and that, for the uniquely defined local C^1 solution of (4.2.4), (4.1.25) on $[0,t_m)$, (4.2.37) is satisfied for some $G_0 > 0$ and some $\delta > \alpha$; then $\exists\, C_1 = C_1(\alpha)$ such that

$$|r(x,t)| + |s(x,t)| \leq C_1(|r_0| + |s_0|) \tag{4.2.38}$$

for all $x \in R^1$ and $0 \leq t \leq t_m$.

Proof: The basic scheme for the proof follows, of course, the same pattern as in the proofs of Lemmas 4.1 and (2.2) with a few important differences. First of all, as a direct consequence of (4.2.36) and (4.2.37), we have, for $0 \leq t < t_m$

$$|\tilde{\Phi}(r(x,t) - s(x,t))| \leq M_0 e^{-\delta t}(|r(x,t)| + |s(x,t)|) \tag{4.2.39}$$

with $M_0 = 1/4\tilde{B}_1 G_0$. For β_1, $\beta_2 \in R^1$ we again denote by $x_1(t,\beta_1)$ and $x_2(t,\beta_2)$, respectively, the characteristics which pass through the points $(\beta_1, 0)$ and $(\beta_2, 0)$, i.e., $x_1(t,\beta_1)$ and $x_2(t,\beta_2)$ satisfy (4.1.18a), (4.1.18b) with $\hat{\lambda}, \hat{\nu}$ replaced by $\tilde{\lambda}, \tilde{\nu}$. Of course, $r(x_1(t,\beta_1),t)$ and $s(x_2(t,\beta_2),t)$ then satisfy, along their respective characteristics, the nonlinear ordinary differential equations (4.1.19a), (4.1.19b) with Φ replaced by $\tilde{\Phi}$; integrating these equations along the characteristics $x_1(t,\beta_1)$ and $x_2(t,\beta_2)$, respectively, and taking absolute values on both sides of each equation, we find that for $t \in [0,t_m)$,

$$\exp\left(\frac{\alpha}{2}t\right)|r(x_1,t)| \leq |r_0(\beta_1)|$$

$$+\frac{\alpha}{2}\int_0^t \exp\left(\frac{\alpha}{2}\tau\right)|s(x_1,\tau)|\,d\tau$$
$$+\int_0^t \exp\left(\frac{\alpha}{2}\tau\right)|\tilde{\Phi}(r(x_1,\tau)-s(x_1,\tau))|\,d\tau$$

and

$$\exp\left(\frac{\alpha}{2}t\right)|s(x_2,t)| \le |s_0(\beta_2)|$$
$$+\frac{\alpha}{2}\int_0^t \exp\left(\frac{\alpha}{2}\tau\right)|r(x_2,\tau)|\,d\tau$$
$$+\int_0^t \exp\left(\frac{\alpha}{2}\tau\right)|\tilde{\Phi}(r(x_2,\tau)-s(x_2,\tau))|\,d\tau$$

By virtue of the hypothesis $(\tilde{G})$, for $0 \le t < t_m$, $\tilde{\Phi}$ satisfies the estimate (4.2.39) for any $x \in R^1$ and, in particular, $\tilde{\Phi}$ satisfies (4.2.39) along the characteristics $x_1(t,\beta_1)$ and $x_2(t,\beta_2)$, for $0 \le t < t_m$. Therefore, as $\delta > \alpha$, if we employ (4.2.39) in the last two estimates, we easily find that

$$\exp\left(\frac{\alpha}{2}t\right)|r(x_1,t)| \le |r_0(\beta_1)| + \frac{\alpha}{2}\int_0^t \exp\left(\frac{\alpha}{2}\tau\right)|s(x_1,\tau)|\,d\tau \tag{4.2.40a}$$
$$+M_0\int_0^t \exp\left(-\frac{\alpha}{2}\tau\right)(|r(x_1,\tau)|+|s(x_1,\tau)|)\,d\tau$$

and

$$\exp\left(\frac{\alpha}{2}t\right)|s(x_2,t)| \le |s_0(\beta_2)| + \frac{\alpha}{2}\int_0^t \exp\left(\frac{\alpha}{2}\tau\right)|r(x_2,\tau)|\,d\tau \tag{4.2.40b}$$
$$+M_0\int_0^t \exp\left(-\frac{\alpha}{2}\tau\right)(|r(x_2,\tau)|+|s(x_2,\tau)|)\,d\tau$$

Setting

$$R(t) = \sup_{R^1}\exp\left(\frac{\alpha}{2}t\right)|r(x,t)| \tag{4.2.41a}$$

and

$$S(t) = \sup_{R^1}\exp\left(\frac{\alpha}{2}t\right)|s(x,t)| \tag{4.2.41b}$$

and using the definitions of $|r_0|, |s_0|$, we have

$$\exp\left(\frac{\alpha}{2}t\right)|r(x_1,t)| \le |r_0| + \frac{\alpha}{2}\int_0^t S(\tau)\,d\tau \tag{4.2.42a}$$
$$+M_0\int_0^t \exp(-\alpha\tau)[R(\tau)+S(\tau)]\,d\tau$$

and

$$\exp\left(\frac{\alpha}{2}t\right)|s(x_2,t)| \le |s_0| + \frac{\alpha}{2}\int_0^t R(\tau)\,d\tau \qquad (4.2.42b)$$
$$+M_0\int_0^t \exp(-\alpha\tau)[R(\tau)+S(\tau)]\,d\tau$$

Using the periodicity of the initial data, and thus of the solution $(r(x,t), s(x,t))$ an argument entirely analogous to the one which led us from (4.1.28a), (4.1.28b) to (4.1.30a), (4.1.30b) now allows us to obtain from (4.2.42a), (4.2.42b) the estimates

$$R(t) \le |r_0| + \frac{\alpha}{2}\int_0^t S(\tau)\,d\tau \qquad (4.2.43a)$$
$$+M_0\int_0^t \exp(-\alpha\tau)[R(\tau)+S(\tau)]\,d\tau$$
$$S(t) \le |s_0| + \frac{\alpha}{2}\int_0^t R(\tau)\,d\tau \qquad (4.2.43b)$$
$$+M_0\int_0^t \exp(-\alpha\tau)[R(\tau)+S(\tau)]\,d\tau$$

which are valid for $0 \le t < t_m$. Setting

$$\begin{cases} W(t) = R(t) + S(t) \\ W_0 = |r_0| + |s_0| \end{cases} \qquad (4.2.44)$$

and

$$m(t) = \frac{1}{2}\alpha + 2M_0\exp(-\alpha\tau) \qquad (4.2.45)$$

and adding the estimates (4.2.43a), (4.2.43b), we now obtain

$$W(t) \le W_0 + \int_0^t m(\tau)W(\tau)\,d\tau, \qquad 0 \le t < t_m \qquad (4.2.46)$$

As $m(t) > 0$, the standard Gronwall inequality (Hale [66], Corollary 6.6) may be applied to the estimate (4.2.46) so as to yield

$$W(t) \le W_0 \exp\left(\frac{\alpha}{2}t\right)\left[2M_0\int_0^t \exp(-\alpha\tau)\,d\tau\right] \qquad (4.2.47)$$

for $0 \le t < t_m$. However,

$$W(t) \ge \exp\left(\frac{\alpha}{2}t\right)(|r(x,t)| + |s(x,t)|)$$

and, therefore, for $0 \le t < t_m$,

$$|r(x,t)| + |s(x,t)| \le W_0\left\{\frac{2M_0}{\alpha}[1-\exp(-\alpha t)]\right\} \qquad (4.2.48)$$

so that the lemma follows with $C_1 = \exp(2M_0/\alpha)$. □

With Lemma 4.2 in hand, we are now in a position to state and prove the main result of this section, namely, that any C^1 local solution of the initial-value problem associated with the system (4.2.4), which satisfies the hypothesis $(\tilde{G})$, may be extended to a global C^1 solution on $[0, \infty)$. Our approach will follow the scheme in Nishida [124], i.e., we will derive an appropriate set of *a priori* bounds on the gradients $|r_x(x,t)|, |s_x(x,t)|$ of the Riemann continuation argument; it is sufficient, of course, to derive the required *a priori* estimate for $|r_x(x,t)|$ as an entirely analogous argument will then yield the similar estimate for $|s_x(x,t)|$.

Theorem 4.2 [19] Let (r, s) be the local C^1 solution of the initial-value problem (4.2.4), (4.1.25) on $[0, t_m)$ and assume that hypotheses (i)-(iii) are satisfied. There exist positive constants $\alpha_0, k_1, k_2(\alpha)$ such that if $\alpha \geq \alpha_0$ and condition $(\tilde{G})$ is satisfied, for some $G_0 > 0$, and $\delta > \alpha$, then for $|r_0|, |s_0|$, and $\sup_{R^1} |r_0'(x)|$ chosen sufficiently small,

$$|r_x(x,t)| \leq k_1 \sup_{R^1} |r_0'(x)| + k_2(\alpha)(|r_0| + |s_0|) \tag{4.2.49}$$

for $x \in R^1$ and $0 \leq t \leq t_m$.

Proof: We begin by differentiating the first equation in (4.2.4) through with respect to x so as to obtain

$$r_x' + \tilde{\lambda}_r r_x^2 + \tilde{\lambda}_s r_x s_x = \left(-\frac{\alpha}{2} + \tilde{\Phi}'\right) r_x - \left(\frac{\alpha}{2} + \tilde{\Phi}'\right) s_x \tag{4.2.50}$$

An elementary calculation shows that

$$r' - s' = 2\tilde{\Phi} - 2\tilde{\lambda} s_x \tag{4.2.51}$$

and if we now substitute for r' in (4.2.51), from (4.2.4), we have that

$$s_x = \frac{s'}{2\tilde{\lambda}} + \frac{\alpha}{4\tilde{\lambda}}(r+s) + \frac{\tilde{\Phi}}{2\tilde{\lambda}} \tag{4.2.52}$$

We now define

$$\tilde{h}(r-s) = \frac{1}{2}\ln[-\tilde{\lambda}(r-s)] \tag{4.2.53}$$

so that $\tilde{h}' = \tilde{h}_r(r' - s')$. However, $\tilde{h}_r = \frac{1}{2}\tilde{\lambda}_r/\tilde{\lambda}$ so if we take note of (4.2.51) we find that

$$\tilde{h}' = \left(\frac{\tilde{\lambda}_r}{\tilde{\lambda}}\right)\tilde{\Phi} - \tilde{\lambda}_r s_x \tag{4.2.54}$$

Substituting in (4.2.54) for s_x, as given by (4.2.52), we obtain

$$\tilde{h}' = -\frac{\alpha\tilde{\lambda}_r}{4\tilde{\lambda}}(r+s) - \frac{\tilde{\lambda}_r}{2\tilde{\lambda}}s' + \frac{\tilde{\lambda}_r}{2\tilde{\lambda}}\tilde{\Phi} \tag{4.2.55}$$

We now return to (4.2.50) which we write in the form

$$r'_x + \left(\tilde{\lambda}_r r_x + \frac{\alpha}{2} + \tilde{\Psi}\right) r_x = -\left(\tilde{\lambda}_s r_x + \frac{\alpha}{2} + \tilde{\Psi}\right) s_x \tag{4.2.56}$$

where $\tilde{\Psi}(\eta) = \tilde{\Psi}'(\eta)$. Substituting for s_x in (4.2.56) from (4.2.52), employing (4.2.55), and using the fact that $\tilde{\lambda}_r = -\tilde{\lambda}_s$, we obtain the equation

$$\begin{aligned} r'_x + \tilde{h}' r_x + \left(\tilde{\lambda}_r r_x + \frac{\alpha}{2} + \tilde{\Psi} - \frac{\tilde{\lambda}_r}{\tilde{\lambda}}\tilde{\Phi}\right) r_x & \\ = -\frac{\alpha}{2}\left[\frac{s'}{2\tilde{\lambda}} + \frac{\alpha}{4\tilde{\lambda}}(r+s)\right] - \frac{\alpha}{4\tilde{\lambda}}\tilde{\Phi} & \\ -\tilde{\Psi}\left(\frac{s'}{2\tilde{\lambda}} + \frac{\alpha}{4\tilde{\lambda}}(r+s) + \frac{\tilde{\Phi}}{2\tilde{\lambda}}\right) & \end{aligned} \tag{4.2.57}$$

We now define

$$\tilde{z}(r-s) = \int_0^{r-s} \frac{\alpha e^{\tilde{h}(\zeta)}}{4\tilde{\lambda}(\zeta)}\, d\zeta \tag{4.2.58}$$

so that $\tilde{z}' = \frac{\alpha}{4\tilde{\lambda}} e^{\tilde{h}}(r' - s')$ or,

$$\frac{2}{\alpha}\tilde{z}' - \frac{e^{\tilde{h}}}{2\tilde{\lambda}}\tilde{\Phi} = -e^{\tilde{h}}\left[\frac{\alpha}{4\tilde{\lambda}}(r+s) + \frac{s'}{2\tilde{\lambda}}\right] \tag{4.2.59}$$

where we have used (4.2.4) once again to substitute for r'. Now we multiply (4.2.57) through by $\exp(\tilde{h})$, employ (4.2.59), and simplify so as to obtain

$$\begin{aligned} (e^{\tilde{h}} r_x)' + \left(\tilde{\lambda}_r r_x + \frac{\alpha}{2} + \tilde{\Psi} - \frac{\tilde{\lambda}_r}{\tilde{\lambda}}\tilde{\Phi}\right) e^{\tilde{h}} r_x & \\ = \left(\frac{\alpha}{2} + \tilde{\Psi}\right)\left(\frac{2}{\alpha}\tilde{z}' - \frac{e^{\tilde{h}}}{\tilde{\lambda}}\tilde{\Phi}\right) & \end{aligned} \tag{4.2.60}$$

At this point we want to integrate (4.2.60) along the characteristic curve $x = x_1(t, \beta_1)$; to expedite matters, we will first define the following quantities:

$$\rho(t) = \exp[\tilde{h}(r-s)] r_x(x_1(t,\beta_1), t) \tag{4.2.61}$$

$$\mu(t) = \left[\frac{\alpha}{2} + \tilde{\Psi}(r-s)\right]\left[\frac{2}{\alpha}\tilde{z}'(r-s) - \frac{e^{\tilde{h}}(r-s)}{\tilde{\lambda}(r-s)}\tilde{\Phi}(r-s)\right] \tag{4.2.62}$$

and

$$\begin{aligned}\pi(t) &= \tilde{\lambda}_r(r-s)\cdot r_x(x_1(t,\beta_1),t)+\frac{\alpha}{2}+\tilde{\Psi}(r-s) \\ &- \frac{\tilde{\lambda}_r(r-s)}{\tilde{\lambda}(r-s)}\tilde{\Phi}(r-s)\end{aligned} \tag{4.2.63}$$

where, in (4.2.61)-(4.2.63), $r-s=r(x_1(t,\beta_1),t)-s(x_1(t,\beta_1),t)$, and $0\le t<t_m$. Employing the definitions (4.2.61)-(4.2.63), in order to rewrite (4.2.60), and then integrating along the characteristic $x=x_1(t,\beta_1)$, we easily obtain

$$\rho(t)=\rho(0)\exp\left[-\int_0^t \pi(\tau)\,d\tau\right]+\int_0^t \mu(s)\exp\left[-\int_0^t \pi(\tau)\,d\tau\right]\,ds \tag{4.2.64}$$

for $0\le t<t_m$. For the sake of convenience, we now set

$$\tilde{\Sigma}(\eta)=\tilde{\Psi}(\eta)-\frac{\tilde{\lambda}'(\eta)}{\tilde{\lambda}(\eta)}\tilde{\Phi}(\eta), \qquad \eta\in R^1$$

and recall (4.2.12), so that for $\eta\in R^1$,

$$\tilde{\Sigma}(\eta)=-\frac{1}{2}\mathcal{V}'(\tilde{Q}(\eta))\frac{d}{d\mathcal{V}}(\mathcal{V}G(\mathcal{V}))\Big|_{\mathcal{V}=\mathcal{V}(\tilde{Q}(\eta))} \tag{4.2.65}$$

In lieu of (4.2.63) we then have

$$\pi(t)=\tilde{\lambda}_r(r-s)r_x(x_1(t,\beta_1),t)+\frac{\alpha}{2}+\tilde{\Sigma}(r-s) \tag{4.2.66}$$

with $\eta\equiv r-s$ evaluated along $x=x_1(t,\beta_1)$, $0\le t<t_m$. But, in view of hypothesis (i), i.e., (4.2.13)

$$\left|\frac{d}{d\mathcal{V}}(\mathcal{V}G(\mathcal{V}))\Big|_{\mathcal{V}=\mathcal{V}(\tilde{Q}(\eta))}\right|\le G_1,\ \forall\eta\in R^1 \tag{4.2.67}$$

while, by hypothesis (ii), we have $\mathcal{V}'(\tilde{Q}(\eta))\le\mathcal{V}_0$, $\forall\eta\in R^1$. Therefore, along the characteristic curve $x=x_1(t,\beta_1)$. for $0\le t<t_m$

$$|\tilde{\Sigma}(r-s)|\le\frac{1}{2}\mathcal{V}_0G_1\equiv\Sigma_1 \tag{4.2.68}$$

from which it follows that for $0\le t<t_m$,

$$\tilde{\Sigma}(r(x_1(t,\beta_1),t)-s(x_1(t,\beta_1),t))\ge-\Sigma_1 \tag{4.2.69}$$

From (4.2.66) we now obtain the lower bound

$$\pi(t)\ge\tilde{\lambda}_r(r-s)r_x(x_1(t,\beta_1),t)+\frac{\alpha}{2}-\Sigma_1,\ 0\le t<t_m \tag{4.2.70}$$

along $x = x_1(t, \beta_1)$. We now set $\alpha_0 = 4\Sigma_1$; then, for $\alpha \geq \alpha_0$

$$\pi(t) \geq \tilde{\lambda}_r(r-s) r_x(x_1(t,\beta_1),t) + \frac{\alpha}{4} \equiv \pi_1(t) \tag{4.2.71}$$

along $x = x_1(t, \beta_1)$, $0 \leq t < t_m$. Thus, for $0 \leq s \leq t < t_m$,

$$\exp\left[-\int_s^t \pi(\tau)\,d\tau\right] \leq \exp\left[-\int_s^t \pi_1(\tau)\,d\tau\right] \tag{4.2.72}$$

We now note that as a consequence of (4.2.9a)-(4.2.9c)

$$|\tilde{\lambda}_r(r-s) r_x| = |\tilde{\lambda}'(\eta)|\,|r_x| = \frac{1}{4}\left|\frac{\mathcal{V}''(\tilde{Q}(\eta))}{\mathcal{V}'(\tilde{Q}(\eta))}\right| |r_x| \tag{4.2.73}$$

However, as a consequence of Lemma 4.2, it follows that $|\tilde{Q}(\eta)|$, $\eta \equiv r - s$, may be made arbitrarily small if $|r_0|$ and $|s_0|$ are both chosen sufficiently small, in which case by part (b) of hypothesis (iii), $\exists\, B_2 > 0$

$$\left|\frac{\mathcal{V}''(\tilde{Q}(\eta))}{\mathcal{V}'(\tilde{Q}(\eta))}\right| \leq B_2 \tag{4.2.74}$$

for $|r_0|, |s_0|$ chosen sufficiently small. Combining (4.2.73) and (4.2.74), we deduce the existence of a constant $B_3 > 0$ such that

$$|\tilde{\lambda}_r(r-s) r_x| \leq B_3 |r_x|, \ 0 \leq t < t_m \tag{4.2.75}$$

holds whenever $|r_0|$ and $|s_0|$ are chosen sufficiently small. Let us now assume that for $0 \leq t < t_m$ we have, in fact,

$$|r_x| \leq \frac{\alpha - 4\epsilon_1}{4B_3} \tag{4.2.76}$$

for a fixed but arbitrary $\epsilon_1 < \frac{\alpha}{4}$. A standard argument, which we present towards the end of the proof, will justify the assumption (4.2.76). By virtue of the assumed bound (4.2.76), for $0 \leq t < t_m$, (4.2.71), and (4.2.75), we clearly have

$$\pi_1(t) = \tilde{\lambda}_r(r-s) r_x(x_1(t,\beta_1),t) + \frac{\alpha}{4} \geq \epsilon_1 \tag{4.2.77}$$

for $0 \leq t < t_m$. We now decompose $\mu(t)$, as defined by (4.2.62), into $\mu(t) = \mu_1(t) + \mu_2(t)$ where

$$\mu_1(t) = \left[1 + \frac{2}{\alpha}\tilde{\Psi}(r-s)\right] \tilde{z}'(r-s) \tag{4.2.78a}$$

and

$$\mu_2(t) = -\frac{e^{\tilde{h}}(r-s)}{\tilde{\lambda}(r-s)}\tilde{\Phi}(r-s)\left[\frac{\alpha}{2} + \tilde{\Psi}(r-s)\right] \tag{4.2.78b}$$

In (4.2.78a), (4.2.78b), of course, $\eta = r - s$ is evaluated along the characteristic $x = x_1(t, \beta_1)$. Setting

$$\tilde{\Lambda}(\eta) = \int_0^\eta \left[1 + \frac{2}{\alpha}\tilde{\Psi}(s)\right] \frac{d\tilde{z}}{ds}\, ds \tag{4.2.79}$$

we have

$$\tilde{\Lambda}' = \frac{d\tilde{\Lambda}}{d\eta}\eta' = \left[1 + \frac{2}{\alpha}\tilde{\Psi}(\eta)\right] \tilde{z}'(\eta)$$

or

$$\tilde{\Lambda}'(r - s)\Big|_{x=x_1(t,\beta_1)} = \mu_1(t) \tag{4.2.80}$$

Therefore,

$$\mu(t) = \tilde{\Lambda}'(r - s)\Big|_{x=x_1(t,\beta_1)} + \mu_2(t) \tag{4.2.81}$$

Also,

$$\begin{aligned}
&\int_0^t \mu_1(s) \exp\left[-\int_s^t \pi(\tau)\, d\tau\right] ds \\
&= \int_0^t \frac{d\tilde{\Lambda}}{ds}(\eta(x_1(s, \beta_1), s)) \exp\left[-\int_s^t \pi(\tau)\, d\tau\right] ds \\
&= \tilde{\Lambda}(\eta(x_1(t, \beta_1), t) - \tilde{\Lambda}(\eta(x_1(0, \beta_1), 0) \exp\left[-\int_0^t \pi(\tau)\, d\tau\right] \\
&\quad - \int_0^t \tilde{\Lambda}(\eta(x_1(s, \beta_1), s)\pi(s) \exp\left[-\int_s^t \pi(\tau)\, d\tau\right]
\end{aligned}$$

Employing this last result in (4.2.64) we have, for $0 \leq t < t_m$,

$$\begin{aligned}
\rho(t) = \rho(0) \exp\left[-\int_0^t \pi(\tau)\, d\tau\right] + \int_0^t \mu_2(s) \exp\left[-\int_s^t \pi(\tau)\, d\tau\right] \\
+ \tilde{\Lambda}_{\beta_1}(t) - \tilde{\Lambda}_{\beta_1}(0) \exp\left[-\int_0^t \pi(\tau)\, d\tau\right] \\
- \int_0^t \tilde{\Lambda}_{\beta_1}(s)\pi(s) \exp\left[-\int_s^t \pi(\tau)\, d\tau\right]
\end{aligned} \tag{4.2.82}$$

where we have set

$$\tilde{\Lambda}_{\beta_1}(t) = \tilde{\Lambda}(\eta(x_1(t, \beta_1), t) \tag{4.2.83}$$

so that

$$\tilde{\Lambda}_{\beta_1}(0) = \tilde{\Lambda}(\eta(x_1(0, \beta_1), 0) \equiv \tilde{\Lambda}(r_0(\beta_1) - s_0(\beta_1)) \tag{4.2.84}$$

We now obtain from (4.2.72) and (4.2.82) the estimate

$$\begin{aligned}|\rho(t)| &\leq |\rho(0)| \exp\left[-\int_0^t \pi_1(\tau)\,d\tau\right] \\ &+\int_0^t |\mu_2(s)| \exp\left[-\int_s^t \pi_1(\tau)\,d\tau\right] \\ &+|\tilde{\Lambda}_{\beta_1}(t)| + |\tilde{\Lambda}_{\beta_1}(0)| \exp\left[-\int_0^t \pi_1(\tau)\,d\tau\right] \\ &+\int_0^t |\tilde{\Lambda}_{\beta_1}(s)|\,|\pi(s)| \exp\left[-\int_0^t \pi_1(\tau)\,d\tau\right] ds\end{aligned} \tag{4.2.85}$$

However, by virtue of the lower bound expressed by (4.2.77), we have $-\pi_1(t) \leq -\epsilon_1$, for $0 \leq t < t_m$, and, therefore,

$$\begin{aligned}|\rho(t)| &\leq |\rho(0)| \exp(-\epsilon_1 t) \\ &+\int_0^t |\mu_2(s)| \exp[-\epsilon_1(t-s)]\,ds \\ &+|\tilde{\Lambda}_{\beta_1}(t)| + |\tilde{\Lambda}_{\beta_1}(0)| \exp(-\epsilon_1 t) \\ &+\int_0^t |\tilde{\Lambda}_{\beta_1}(s)|\,|\pi(s)| \exp[-\epsilon_1(t-s)]\,ds\end{aligned}$$

or

$$\begin{aligned}|\rho(t)| &\leq |\tilde{\Lambda}_{\beta_1}(t)| + \left\{|\rho(0)| + |\tilde{\Lambda}_{\beta_1}(0)|\right\} \exp(-\epsilon_1 t) \\ &+\int_0^t \left(|\mu_2(s)| + |\tilde{\Lambda}_{\beta_1}(s)|\,|\pi(s)|\right) \exp[-\epsilon_1(t-s)]\,ds\end{aligned} \tag{4.2.86}$$

for $0 \leq t < t_m$.

Now, in view of the relations (4.2.66), (4.2.68), and (4.2.75), we have, for $0 \leq t < t_m$

$$|\pi(t)| \leq B_3 |r_x(x,(t,\beta_1),t)| + \frac{\alpha}{2} + \Sigma_1$$

However, by virtue of the (tentative) assumption (4.2.76), for $0 \leq t < t_m$, we obtain from our last estimate, above, the upper bound on $[0, t_m)$:

$$|\pi(t)| \leq \left(\frac{\alpha - 4\epsilon_1}{4}\right) + \frac{\alpha}{2} + \Sigma_1 = \frac{3\alpha}{4} + (\Sigma_1 - \epsilon_1) \equiv \pi_0 \tag{4.2.87}$$

Setting $\tilde{\Gamma}(t) = |\mu_2(t)| + \pi_0 |\tilde{\Lambda}_{\beta_1}(t)|$ we now find, as a consequence of (4.2.86), the estimate

$$\begin{aligned}|\rho(t)| &\leq |\rho(0)| + 2 \sup_{0\leq t<t_m} |\tilde{\Lambda}_{\beta_1}(t)| + \frac{1}{\epsilon_1} \sup_{0\leq t<t_m} \tilde{\Gamma}(t) \\ &\leq |\rho(0)| + \frac{1}{\epsilon_1} \sup_{0\leq t<t_m} |\mu_2(t)| + \left(2 + \frac{\pi_0}{\epsilon_1}\right) \sup_{0\leq t<t_m} |\tilde{\Lambda}_{\beta_1}(t)|\end{aligned} \tag{4.2.88}$$

for $0 \le t < t_m$. We are now faced with the task of estimating $|\mu_2(t)|$ and $|\tilde{\Lambda}_{\beta_1}(t)|$ on $[0, t_m)$. We begin by noting that (see (4.2.78b))

$$|\mu_2(t)| \le \frac{\exp[\tilde{h}(r-s)]|\tilde{\Phi}(r-s)|}{|\tilde{\lambda}(r-s)|}\left[\frac{\alpha}{2} + |\tilde{\Psi}(r-s)|\right] \tag{4.2.89}$$

where, as previously, $\eta = r - s$ is evaluated along $x = x_1(t, \beta_1)$ for $0 \le t < t_m$. In view of the definition of $\tilde{h}(r-s)$, i.e., (4.2.53)

$$\exp[\tilde{h}(r-s)] = [-\tilde{\lambda}(r-s)]^{1/2} \equiv \left[\frac{\mathcal{V}'(\tilde{Q}(r-s))}{L}\right]^{1/4} \tag{4.2.90}$$

so that by hypothesis (ii)

$$\exp[\tilde{h}(r-s)] \le (\mathcal{V}_0/L)^{1/4} \tag{4.2.91}$$

Also, by virtue of hypothesis (ii)

$$\frac{\exp[\tilde{h}(r-s)]}{|\tilde{\lambda}(r-s)|} \le \left(\frac{L}{\epsilon}\right)^{1/2}\left(\frac{\mathcal{V}_0}{L}\right)^{1/4} \equiv \frac{1}{\sqrt{\epsilon}}(L\mathcal{V}_0)^{1/4} \tag{4.2.92}$$

Employing (4.2.39), and using Lemma 4.2, we find that

$$|\tilde{\Phi}(r-s)| \le M_0 C_1(\alpha)(|r_0| + |s_0|) \tag{4.2.93}$$

with this last estimate holding, in particular, along the characteristic $x = x_1(t, \beta_1)$ for $0 \le t < t_m$. By (4.2.12) and (4.2.9c) we have

$$\begin{aligned} \tilde{\Psi}(r-s) &= -\frac{G(\mathcal{V}(\tilde{Q}(r-s))}{4} \cdot \frac{\mathcal{V}(\tilde{Q}(r-s))\mathcal{V}''(\tilde{Q}(r-s))}{\mathcal{V}'(\tilde{Q}(r-s))} \\ &\quad - \frac{\mathcal{V}'(\tilde{Q}(r-s))}{2}\frac{d}{d\mathcal{V}}(\mathcal{V}G(\mathcal{V}))\Big|_{\mathcal{V}=\mathcal{V}(\tilde{Q}(r-s))} \end{aligned} \tag{4.2.94}$$

Therefore, if we employ the condition $(\tilde{G})$, hypotheses (i) and (ii), and the elementary fact that $\forall \tau \in R^1$

$$\begin{aligned} \frac{\mathcal{V}(\tau)\mathcal{V}''(\tau)}{\mathcal{V}'(\tau)} &< \frac{\mathcal{V}(\tau)\mathcal{V}''(\tau)}{\mathcal{V}'(\tau)} + 2\mathcal{V}'(\tau) \\ &= \frac{\mathcal{W}(\tau)}{\mathcal{V}'(\tau)} \le \left|\frac{\mathcal{W}(\tau)}{\mathcal{V}'(\tau)}\right| \le B_1, \end{aligned}$$

with $B_1 > 0$ the constant whose existence is implied by (4.2.22), we obtain from (4.2.94) the estimate

$$|\tilde{\Psi}(r-s)| \le \frac{B_1 G_0}{4} + \frac{\mathcal{V}_0 G_1}{2} \equiv \Psi_0 \tag{4.2.95}$$

which is, of course, valid along the characteristic $x = x_1(t, \beta_1)$ for $0 \le t < t_m$. Combining the relations (4.2.89), (4.2.92), (4.2.93), and (4.2.95), we find for $|\mu_2(t)|$ the estimate

$$|\mu_2(t)| \le \frac{1}{\sqrt{\epsilon}}\left(\frac{\alpha}{2} + \Psi_0\right) \cdot (L\mathcal{V}_0)^{1/4} M_0 C_1(\alpha)[\,|r_0| + |s_0|\,] \tag{4.2.96}$$

for $0 \le t < t_m$. Our last set of estimates concern $\tilde{\Lambda}_{\beta_1}(t)$; clearly, by virtue of (4.2.83), (4.2.79), and (4.2.58) we have, as a first estimate, that

$$|\tilde{\Lambda}_{\beta_1}(t)| \le \frac{\alpha}{4}\int_0^{|\eta|}\left[1 + \frac{2}{\alpha}|\tilde{\Psi}(s)|\right]\frac{\exp[\tilde{h}(s)]}{|\tilde{\lambda}(s)|}\,ds \tag{4.2.97}$$

for $0 \le t < t_m$, where $|\eta| = |r(x_1(t, \beta_1), t) - s(x_1(t, \beta_1), t)|$. Applying the estimates (4.2.92) and (4.2.95), it now follows that for $0 \le t < t_m$

$$|\tilde{\Lambda}_{\beta_1}(t)| \le \frac{\alpha}{4\sqrt{\epsilon}}\left(1 + \frac{2}{\alpha}\Psi_0\right)(L\mathcal{V}_0)^{1/4} C_1(\alpha)[\,|r_0| + |s_0|\,] \tag{4.2.98}$$

By combining (4.2.88), (4.2.96), and (4.2.98), we now find that, for $0 \le t < t_m$,

$$|\rho(t)| \le |\rho(0)| + k(\alpha) C_1(\alpha)[\,|r_0| + |s_0|\,] \tag{4.2.99}$$

where

$$k(\alpha) = \left(\frac{\alpha}{2} + \Psi_0\right)\frac{(L\mathcal{V}_0)^{1/4}}{\sqrt{\epsilon}}\left(1 + \frac{\pi_0}{2\epsilon_1} + \frac{M_0}{\epsilon_1}\right) \tag{4.2.100}$$

Of course, in (4.2.100), the constants ϵ_1 and π_0 also depend on α. By virtue of the definition of $\rho(t)$, i.e., (4.2.61) and (4.2.91), it is clear that

$$|\rho(0)| \le \left(\frac{\mathcal{V}_0}{L}\right)^{1/4}|r_x(\beta, 0)| \le \left(\frac{\mathcal{V}_0}{L}\right)^{1/4}\sup_x |r_0'(x)| \tag{4.2.101}$$

Also, along $x = x_1(t, \beta_1)$, for $0 \le t < t_m$

$$\exp[-\tilde{h}(r - s)] = \left[\frac{L}{\mathcal{V}'(\tilde{Q}(r - s))}\right]^{1/4} \le \left(\frac{L}{\epsilon}\right)^{1/4} \tag{4.2.102}$$

Employing both (4.2.101) and (4.2.102) in the estimate (4.2.100), and, once again, using the definition of $\rho(t)$, we now find that for $0 \le t < t_m$,

$$\begin{aligned} |r_x(x_1(t, \beta_1), t)| \;\le\; & \left(\frac{L}{\epsilon}\right)^{1/4}\left\{\left(\frac{\mathcal{V}_0}{L}\right)^{1/4}\sup_x |r_0'(x)| \right. \\ & \left. + k(\alpha) C_1(\alpha)[\,|r_0| + |s_0|\,]\right\} \end{aligned} \tag{4.2.103}$$

or

$$|r_x(x_1(t,\beta_1),t)| \leq k_1 \sup_x |r_0'(x)| + k_2(\alpha)(|r_0| + |s_0|) \tag{4.2.104}$$

with

$$\begin{cases} k_1 = \left(\dfrac{\mathcal{V}_0}{\epsilon}\right)^{1/4} \\ k_2(\alpha) = \left(\dfrac{L}{\epsilon}\right)^{1/4} k(\alpha)C_1(\alpha) \end{cases} \tag{4.2.105}$$

The required estimate, i.e., (4.2.49), now follows, with k_1, $k_2(\alpha)$ as in (4.2.105), from the periodicity of the initial data (through the use of an argument analogous to the one employed in the proof of Lemma 4.2) if we can verify the condition (4.2.76), for $0 \leq t < t_m$, and $\epsilon_1 < \alpha/4$; in order to see that (4.2.76) does indeed hold, we begin by choosing $\epsilon_1 < \alpha/4$, where $\alpha \geq \alpha_0 = 4\Sigma_1$, and set

$$R_1 = \frac{\alpha - 4\epsilon_1}{4B_1} \tag{4.2.106}$$

We now choose $r_0(x)$ such that

$$\sup_x |r_0'(x)| < R_1 \tag{4.2.107}$$

Then, by the continuity of $r_x(x,t)$ on $[0,t_m)$, it follows that $\exists t_1 \leq t_m$ such that $|r_x(x,t)| \leq R_1$ for $t \in [0,t_1)$; the estimate of the theorem, i.e., (4.2.49) will now apply on $[0,t_1]$. Choosing $\sup_x |r_0'(x)|$ even smaller, if necessary, and $|r_0| + |s_0|$ sufficiently small, we obtain from (4.2.49) the conclusion that $|r_x(x,t)| < R_1$ on $[0,t_1]$, in which case there must exist a t_2, $t_1 < t_2 \leq t_m$, such that $|r_x(x,t)| \leq R_1$ for $t \in [0,t_2)$; repetition of this (standard) argument establishes the fact that (4.2.49) holds for all $x \in R^1$, $0 \leq t < t_m$, provided that $\sup_x |r_0'(x)|$, $|r_0|$, and $|s_0|$ are all chosen sufficiently small. □

As a direct consequence of Theorem 4.2, we can now state the following corollary, which guarantees the existence of a global smooth solution for the initial-value problem associated with the system (4.2.4) and, therefore, also implies the existence of a global smooth solution for the initial-value problem associated with the quasilinear system (4.1.7) governing the evolution of the charge and current distributions in a distributed parameter nonlinear transmission line:

Corollary 4.1 [19] Given hypotheses (i)-(iii), above, condition $(\tilde{G})$ implies that there exists a global C^1 (in (x,t)) solution for the system (4.2.4), with associated periodic initial data $r_0(\cdot)$, $s_0(\cdot)$, provided α is chosen sufficiently large and $|r_0|$, $|s_0|$, $\sup_x |r_0'(x)|$, and $\sup_x |s_0'(x)|$ are all chosen sufficiently small.

We now want to describe, in some detail, the important improvement on the global existence result cited above which is embodied in the work of J. Wang [179], [177]; as we indicated previously, the basic contribution here involves a substantial weakening of the restrictive condition $(\tilde{G})$ to the condition which we denote by (G^*) below. In Wang's work, the condition that $|r_0|$ and $|s_0|$ be sufficiently small is also replaced by a condition which, in essence, requires that the oscillation in the initial data (in a sense which we make precise below) be sufficiently small; the basic goal of the arguments in [179] and [177] is to reproduce the essential content of Lemma 4.2 under the auspices of a condition on the leakage conductance in the transmission line which is somewhat more realistic and easier to verify. Thus, consider again the transmission line problem in Riemann invariants form, i.e., the system (4.2.4) with associated periodic initial data $r_0(\cdot)$, $s_0(\cdot)$, of class C^1 on R^1; as was the case leading up to Lemma 4.2, we suppose that $\mathcal{V}(\cdot)$ and $G(\cdot)$ satisfy hypotheses (i)-(iii) and we work with the uniquely defined local C^1 solution on $[0, t_m)$. In lieu of the hypothesis $(\tilde{G})$, we now assume that $\forall\, Q \in R^1$, $G(Q) \geq 0$ (i.e., the leakage conductance is nonnegative) and

(G^*): For some $G_* > 0$, sufficiently small,

$$\sup_{Q \in R^1} G(Q) \leq G_* \tag{4.2.108}$$

We also assume that the following small oscillation condition holds, namely, for $\beta_1, \beta_2 \in R^1$,

$$\sup_{\beta_2 \leq \beta_1} |r_0(\beta_2) - s_0(\beta_1)| \leq \mu_0 \tag{4.2.109}$$

with $\mu_0 > 0$ sufficiently small. Given (4.2.108), (4.2.109), we will show that the same basic estimate which prevailed in Lemma 4.2, i.e., (4.2.38), is still valid for the local C^1 solution on $[0, t_m)$. Throughout the remainder of this section, the curves $x = x_1(t, \beta_1)$, $x = x_2(t, \beta_2)$ will again denote, respectively, the characteristic curves defined by (4.1.18a), (4.1.18b)—with $\hat{\lambda}, \hat{\nu}$ replaced by $\tilde{\lambda}, \tilde{\nu}$—which emanate from the points $(\beta_1, 0)$, $(\beta_2, 0)$ on the x-axis.

We begin by referring to the situation depicted in Figure 4.4 below.

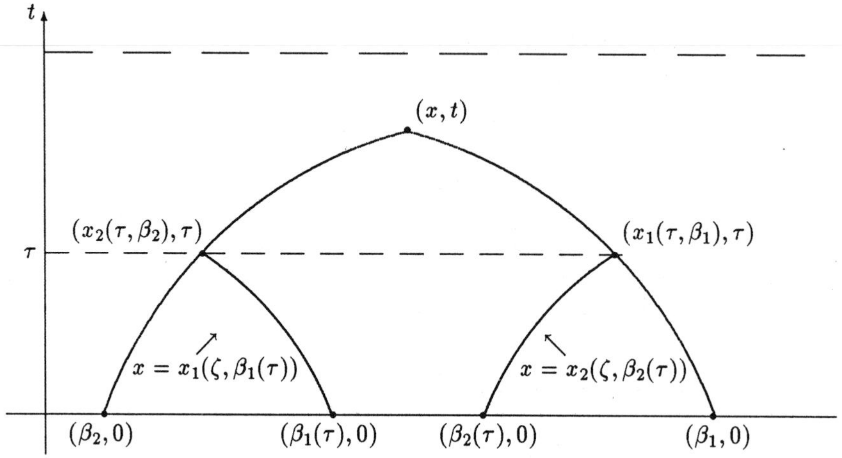

Figure 4.4

As indicated above, we first choose a point (x,t), $x \in R^1$, $t < t_m$ in the upper half plane; from this point we draw the x_1 and x_2 characteristics, i.e., $x = x_1(\tau,\beta_1)$ and $x = x_2(\tau,\beta_2)$ backwards until they intersect, respectively, the x axis at $(\beta_1,0)$ and $(\beta_2,0)$. Next we fix $\tau > 0$, such that $\tau \leq t$, and locate the points $(x_1(\tau,\beta_1),\tau)$ and $(x_2(\tau,\beta_2),\tau)$. From the point $(x_1(\tau,\beta_1),\tau)$, we draw the x_2 characteristic $x = x_2(\zeta,\beta_2(\tau))$, which intersects the x-axis at the point $(\beta_2(\tau),0)$, i.e., $x_2(0,\beta_2(\tau)) = \beta_2(\tau)$ and from the point $(x_2(\tau,\beta_2),\tau)$ we draw the x_1 characteristic $x = x_1(\zeta,\beta_1(\tau))$, which intersects the x-axis at the point $(\beta_1(\tau),0)$, i.e., $x_1(0,\beta_1(\tau)) = \beta_1(\tau)$. Clearly, $\beta_1(t) = \beta_1$, $\beta_1(0) = \beta_2$, $\beta_2(t) = \beta_2$, and $\beta_2(0) = \beta_1$. As the locally defined solution on $[0,t_m)$ is of class C^1, all the functions indicated in Figure 4.4 are continuously differentiable and, therefore, both $\beta_1(\tau)$ and $\beta_2(\tau)$ are differentiable.

Next, we integrate the equations in the system (4.2.4), respectively, along the characteristic curves $x = x_2(\zeta,\beta_2(\tau))$, from $(\beta_2(\tau),0)$ to $(x_1(\tau,\beta_1),\tau)$, and $x = x_1(\zeta,\beta_1(\tau))$, from $(\beta_1(\tau),0)$ to $(x_2(\tau,\beta_2),\tau)$; we find that

$$\begin{aligned} r(x_1(\tau,\beta_1),\tau) &= r_0(\beta_2(\tau)) \\ &\quad - \frac{\alpha}{2}\int_0^\tau (r+s)(x_2(\zeta,\beta_2(\tau)),\zeta)\,d\zeta \end{aligned} \tag{4.2.110}$$

$$+ \int_0^\tau \tilde{\Phi}(r-s)(x_2(\zeta,\beta_2(\tau)),\zeta)\,d\zeta$$

and

$$\begin{aligned} s(x_2(\tau,\beta_2),\tau) &= s_0(\beta_1(\tau)) \\ &- \frac{\alpha}{2}\int_0^\tau (r+s)(x_1(\zeta,\beta_1(\tau)),\zeta)\,d\zeta \\ &+ \int_0^\tau \tilde{\Phi}(r-s)(x_1(\zeta,\beta_1(\tau)),\zeta)\,d\zeta \end{aligned} \tag{4.2.111}$$

Now, we again integrate the equations in the system (4.2.4), but this time along $x = x_2(\zeta,\beta_2)$ from $(\beta_2,0)$ to $(x_2(\tau,\beta_2),\tau)$ and along $x = x_1(\zeta,\beta_1)$ from $(\beta_1,0)$ to $(x_1(\tau,\beta_1),\tau)$; we obtain

$$\begin{aligned} r(x_2(\tau,\beta_2),\tau) = r_0(\beta_2) - \frac{\alpha}{2}\int_0^\tau (r+s)(x_2(\zeta,\beta_2),\zeta)\,d\zeta \\ + \int_0^\tau \tilde{\Phi}(r-s)(x_2(\zeta,\beta_2),\zeta)\,d\zeta \end{aligned} \tag{4.2.112}$$

and

$$\begin{aligned} s(x_1(\tau,\beta_1),\tau) = s_0(\beta_1) - \frac{\alpha}{2}\int_0^\tau (r+s)(x_1(\zeta,\beta_1),\zeta)\,d\zeta \\ + \int_0^\tau \tilde{\Phi}(r-s)(x_1(\zeta,\beta_1),\zeta)\,d\zeta \end{aligned} \tag{4.2.113}$$

From (4.2.110) and (4.2.113) we obtain

$$\begin{aligned} \frac{d}{d\tau}(r-s)(x_1(\tau,\beta_1),\tau) = r_0'(\beta_2(\tau))\beta_2'(\tau) \\ + \frac{\alpha}{2}(r+s)(x_1(\tau,\beta_1),\tau) + \tilde{\Phi}(r-s)(x_1(\tau,\beta_1),\tau) \\ - \frac{\alpha}{2}\frac{d}{d\tau}\int_0^\tau (r+s)(x_2(\zeta,\beta_2(\tau)),\zeta)\,d\zeta \\ + \frac{d}{d\tau}\int_0^\tau \tilde{\Phi}(r-s)(x_2(\zeta,\beta_2(\tau)),\zeta)\,d\zeta \end{aligned} \tag{4.2.114}$$

while (4.2.111) and (4.2.112) yield

$$\begin{aligned} \frac{d}{d\tau}(r-s)(x_2(\tau,\beta_2),\tau) = -s_0'(\beta_1(\tau))\beta_1'(\tau) \\ - \frac{\alpha}{2}(r+s)(x_2(\tau,\beta_2),\tau) + \tilde{\Phi}(r-s)(x_2(\tau,\beta_2),\tau) \\ + \frac{\alpha}{2}\frac{d}{d\tau}\int_0^\tau (r+s)(x_1(\zeta,\beta_1(\tau)),\zeta)\,d\zeta \\ + \frac{d}{d\tau}\int_0^\tau \tilde{\Phi}(r-s)(x_1(\zeta,\beta_1(\tau)),\zeta)\,d\zeta \end{aligned} \tag{4.2.115}$$

In an entirely analogous fashion, we also find that

$$\frac{d}{d\tau}(r+s)(x_1(\tau,\beta_1),\tau) = r_0'(\beta_2(\tau))\beta_2'(\tau)$$
$$-\frac{\alpha}{2}(r+s)(x_1(\tau,\beta_1),\tau) - \tilde{\Phi}(r-s)(x_1(\tau,\beta_1),\tau)$$
$$-\frac{\alpha}{2}\frac{d}{d\tau}\int_0^\tau (r+s)(x_2(\zeta,\beta_2(\tau)),\zeta)\,d\zeta$$
$$+\frac{d}{d\tau}\int_0^\tau \tilde{\Phi}(r-s)(x_2(\zeta,\beta_2(\tau)),\zeta)\,d\zeta$$

or, in view of (4.2.114), that

$$\begin{aligned}\frac{d}{d\tau}(r+s)(x_1(\tau,\beta_1),\tau) &= \frac{d}{d\tau}(r-s)(x_1(\tau,\beta_1),\tau) \\ &- \alpha(r+s)(x_1(\tau,\beta_1),\tau) \\ &- 2\tilde{\Phi}(r-s)(x_1(\tau,\beta_1),\tau)\end{aligned} \tag{4.2.116}$$

Also,

$$\begin{aligned}\frac{d}{d\tau}(r+s)(x_2(\tau,\beta_2),\tau) &= \frac{d}{d\tau}(r-s)(x_2(\tau,\beta_2),\tau) \\ &- \alpha(r+s)(x_2(\tau,\beta_2),\tau) \\ &+ 2\tilde{\Phi}(r-s)(x_2(\tau,\beta_2),\tau)\end{aligned} \tag{4.2.117}$$

Making note of the fact that

$$(r+s)(x_1(t,\beta_1),t) = (r+s)(x_2(t,\beta_2),t) = (r+s)(x,t)$$

we now integrate (4.2.116) along the characteristic $x = x_1(\tau,\beta_1)$ from $\tau = 0$ to $\tau = t$ so as to obtain

$$\begin{aligned}(r+s)(x,t) &= e^{-\alpha t}\Big[r_0(\beta_1) + s_0(\beta_1) \\ &+ \int_0^t \frac{d}{d\tau}[(r-s)(x_1(\tau,\beta_1),\tau)]e^{\alpha\tau}\,d\tau \\ &-2\int_0^t \tilde{\Phi}(r-s)(x_1(\tau,\beta_1),\tau)e^{\alpha\tau}\,d\tau\Big]\end{aligned} \tag{4.2.118}$$

In a similar manner, we integrate (4.2.117) along the characteristic $x = x_2(\tau,\beta_2)$ from $\tau = 0$ to $\tau = t$ and find that

$$(r+s)(x,t) = e^{-\alpha t}\Big[r_0(\beta_2) + s_0(\beta_2) \tag{4.2.119}$$

$$-\int_0^t \frac{d}{d\tau}[(r-s)(x_2(\tau,\beta_2),\tau)]e^{\alpha\tau}\,d\tau$$

$$+2\int_0^t \tilde{\Phi}(r-s)(x_2(\tau,\beta_2),\tau)e^{\alpha\tau}\,d\tau\Big]$$

Subtracting (4.2.119) from (4.2.118), we now produce the relation

$$\begin{aligned}
&e^{-\alpha t}[r_0(\beta_1)+s_0(\beta_1)-r_0(\beta_2)-s_0(\beta_2)]\\
&+e^{-\alpha t}\Big[\int_0^t\{\frac{d}{d\tau}(r-s)(x_1(\tau,\beta_1),\tau)\\
&\qquad+\frac{d}{d\tau}(r-s)(x_2(\tau,\beta_2),\tau)\}e^{\alpha\tau}\,d\tau\\
&-2\int_0^t\{\tilde{\Phi}(r-s)(x_1(\tau,\beta_1),\tau)+\tilde{\Phi}(r-s)(x_2(\tau,\beta_2),\tau)\}e^{\alpha\tau}\,d\tau\Big]\\
&=0
\end{aligned}$$

from which, by an obvious integration by parts, we deduce the equation

$$\begin{aligned}
(r-s)(x,t) &= e^{-\alpha t}\Big[r_0(\beta_2)-s_0(\beta_1)\\
&+\frac{\alpha}{2}\int_0^t\{(r-s)(x_1(\tau,\beta_1),\tau)+(r-s)(x_2(\tau,\beta_2),\tau)\}e^{\alpha\tau}\,d\tau\\
&+2\int_0^t\{\tilde{\Phi}(r-s)(x_1(\tau,\beta_1),\tau)+\tilde{\Phi}(r-s)(x_2(\tau,\beta_2),\tau)\}e^{\alpha\tau}\,d\tau\Big]
\end{aligned}\tag{4.2.120}$$

Now, by virtue of (4.2.15), (4.2.20), (4.2.22), and hypotheses (ii) and (iii), it follows that for $|r-s|$ sufficiently small, the sign of $\tilde{\Phi}(\eta)$ is determined solely by the sign of $\eta = r-s$ with sgn $\tilde{\Phi}(\eta) = -$sgn η (essentially because $\tilde{\Phi}'_*(\eta) < 0$ by hypotheses (ii) and (iii) if $|\eta|$ is sufficiently small). Returning to (4.2.120) we now rewrite it in the form

$$\begin{aligned}
[(r-s)(x,t)]e^{\alpha t} &= \Big[r_0(\beta_2)-s_0(\beta_1)\\
&+\int_0^t\{\frac{\alpha}{2}(r-s)(x_1(\tau,\beta_1),\tau)+2\tilde{\Phi}(r-s)(x_1(\tau,\beta_1),\tau)\}e^{\alpha\tau}\,d\tau\\
&+\int_0^t\{\frac{\alpha}{2}(r-s)(x_2(\tau,\beta_2),\tau)+2\tilde{\Phi}(r-s)(x_2(\tau,\beta_2),\tau)\}e^{\alpha\tau}\,d\tau\Big]
\end{aligned}\tag{4.2.121}$$

By virtue of (4.2.108), i.e., condition (G^*) with $G_* > 0$ sufficiently small, (4.2.15), and the fact that sgn $\tilde{\Phi}(\eta) = -$sgn η,

$$\left|\int_0^t\left\{\left[\frac{\alpha}{2}(r-s)+2\tilde{\Phi}(r-s)\right](x_1(\tau,\beta_1),\tau)\right\}e^{\alpha\tau}\,d\tau\right| \tag{4.2.122}$$

$$\leq \frac{k_1}{2} \int_0^t |(r-s)(x_1(\tau, \beta_1), \tau)| e^{\alpha\tau} \, d\tau$$

for some k_1, $0 \leq k_1 \leq \alpha$. In an analogous fashion we have, for some k_2, $0 \leq k_2 \leq \alpha$,

$$\left| \int_0^t \left\{ \left[\frac{\alpha}{2}(r-s) + 2\tilde{\Phi}(r-s) \right] (x_2(\tau, \beta_2), \tau) \right\} e^{\alpha\tau} \, d\tau \right| \tag{4.2.123}$$

$$\leq \frac{k_2}{2} \int_0^t |(r-s)(x_2(\tau, \beta_2), \tau)| e^{\alpha\tau} \, d\tau$$

We now set, for $\tau > 0$

$$Z(\tau) = \sup_{x \in R^1} |(r-s)(x, \tau)| e^{\alpha\tau} \tag{4.2.124}$$

Then, taking absolute values on both sides of (4.2.121), and using the estimates (4.2.122) and (4.2.123), we obtain

$$Z(t) \leq M + k \int_0^t Z(\tau) \, d\tau \tag{4.2.125}$$

where

$$M = \sup_{\beta_1 \leq \beta_2} |r_0(\beta_2) - s_0(\beta_1)| \tag{4.2.126a}$$

$$k = \frac{1}{2}(k_1 + k_2) \leq \alpha \tag{4.2.126b}$$

From the standard Gronwall inequality and (4.2.124), it then follows that for all $x \in R^1$,

$$|(r-s)(x,t)| \leq M e^{(k-\alpha)t} \tag{4.2.127}$$

whenever G_* and $|\eta|$ are sufficiently small. Of course, in a manner analogous to the argument employed towards the conclusion of the proof of Theorem 4.2, the hypothesis that $|\eta| = |r - s|$ be sufficiently small may be guaranteed by choosing M sufficiently small, i.e., $M \leq \mu_0$ as in (4.2.109), and, also, choosing $|r_0|$ and $|s_0|$ sufficiently small. Having established (4.2.127), we now go back and add (4.2.118) and (4.2.119) and then integrate by parts so as to obtain

$$\begin{aligned}(r+s)(x,t) = e^{-\alpha t} \Big[& r_0(\beta_2) + s_0(\beta_1) \\ & - \frac{\alpha}{2} \int_0^t \{(r-s)(x_1(\tau, \beta_1), \tau) - (r-s)(x_2(\tau, \beta_2), \tau)\} e^{\alpha\tau} \, d\tau \\ & + 2 \int_0^t \{\tilde{\Phi}(r-s)(x_2(\tau, \beta_2), \tau) - \tilde{\Phi}(r-s)(x_1(\tau, \beta_1), \tau)\} e^{\alpha\tau} \, d\tau \Big]\end{aligned} \tag{4.2.128}$$

Employing (4.2.127) in (4.2.128) and using condition (G^*), hypothesis (iii), and the estimate (4.2.19), we obtain

$$\begin{aligned} |(r+s)(x,t)| \;\le\; & e^{-\alpha t}\left|\left[M^* - \frac{M}{k}(\alpha + B_1 G_*)\right]\right| \\ & + \frac{M}{k}(\alpha + B_1 G_*)e^{(k-\alpha)t} \end{aligned} \tag{4.2.129}$$

where

$$M^* = \sup_{\beta_2 \le \beta_1} |r_0(\beta_2) + s_0(\beta_1)| \tag{4.2.130}$$

Finally, combining (4.2.129) with (4.2.127), we are led to the following upper and lower bounds for the Riemann invariants r and s at the point (x,t):

$$\begin{aligned} & -e^{-\alpha t}\left|M^* - \frac{M}{k}(\alpha + B_1 G_*)\right| - Me^{(k-\alpha)t}\left(\frac{\alpha}{k} + \frac{B_1 G_*}{k} + 1\right) \\ & \le r \le e^{-\alpha t}\left|M^* - \frac{M}{k}(\alpha + B_1 G_*)\right| + Me^{(k-\alpha)t}\left(\frac{\alpha}{k} + \frac{B_1 G_*}{k} + 1\right) \end{aligned} \tag{4.2.131}$$

and

$$\begin{aligned} & -e^{-\alpha t}\left|M^* - \frac{M}{k}(\alpha + B_1 G_*)\right| - Me^{(k-\alpha)t}\left(\frac{\alpha}{k} + \frac{B_1 G_*}{k} + 2\right) \\ & \le s \le e^{-\alpha t}\left|M^* - \frac{M}{k}(\alpha + B_1 G_*)\right| + Me^{(k-\alpha)t}\left(\frac{\alpha}{k} + \frac{B_1 G_*}{k} + 2\right) \end{aligned} \tag{4.2.132}$$

From (4.2.131) and (4.2.132) we now easily extract the following alternative to Lemma 4.2, namely,

Lemma 4.3 ([179],[177]): Suppose that $\mathcal{V}(\cdot)$, $G(\cdot)$ satisfy hypotheses (i)-(iii) and that $(r(x,t), s(x,t))$ is the locally defined C^1 solution of the initial-value problem (4.2.4), (4.1.25) on $[0, t_m)$. If condition (G^*) holds, with $G_* > 0$ sufficiently small, and the initial data satisfies the small oscillation condition (4.2.109), with $\mu_0 > 0$ sufficiently small, then $\exists\, C^* > 0$ independent of (x,t), $C^* = C^*(|r_0|, |s_0|, \alpha)$, with $C^* \to 0$ as $(|r_0| + |s_0|) \to 0$, such that

$$|r(x,t)| + |s(x,t)| \le C^*, \qquad x \in R^1,\ 0 \le t < t_m \tag{4.2.133}$$

Remarks: If both $k_1, k_2 < \alpha$, so that $k < \alpha$, once Lemma 4.3 has been used (again) to establish the global existence Theorem 4.2, it follows from (4.2.131) and (4.2.132) that both $|r|, |s| \to 0$ as $t \to +\infty$. Stronger asymptotic decay results for the unique global smooth solution of the system (4.1.7), governing the evolution of charge and current in a distributed parameter nonlinear transmission line, under the assumption of small initial data, will follow from the energy type estimates of the next section, to which we now turn our attention.

4.3 Global Existence with Small Data: Energy Estimates

In the last section we studied the problem of global existence of smooth solutions for the initial-value problem associated with the quasilinear system (4.1.7) which governs the evolution of the current and charge distribution in a nonlinear transmission line; in studying (4.1.7), we assumed that G was voltage dependent and nonnegative, i.e., $G = G(\mathcal{V}(Q)) \geq 0$, $\forall\, Q \in R^1$. The existence results generated in § 4.2 apply to the case of sufficiently "small" initial data, as made precise, e.g., in Theorem 4.2, and are dependent on a Riemann invariants argument which is inherently tied to a one-space dimension situation. In this section we will again establish the global (in time) existence of smooth solutions to the initial-value problem associated with (4.1.7), for the case of sufficiently small initial data, but will employ an energy argument which is analogous to the one used in § 3.5 to prove the existence of global smooth solutions to the one-space dimension wave-dielectric interaction problem; our analysis will employ a somewhat weaker set of conditions on the relevant constitutive functions than those employed in the previous section and will, at the same time, yield the strongest results known, to date, on the large time behavior of the charge distribution in a nonlinear distributed parameter transmission line. The analysis described below is based on the joint study [15] by H. Bellout and the author.

We begin by eliminating among the first order equations in (4.1.7) so as to obtain the following wave equation for the evolution of the charge distribution in the line:

$$Q_{tt} + \alpha Q_t + \psi(Q)_t + \alpha\psi(Q) = \mathcal{F}(Q)_{xx} \tag{4.3.1}$$

In (4.3.1), $\alpha = \dfrac{R}{L} > 0$, as was the case in § 4.2, while

$$\psi(Q) = G(\mathcal{V}(Q))\mathcal{V}(Q); \qquad \mathcal{F}(Q) = \frac{1}{L}\mathcal{V}(Q) \tag{4.3.2}$$

Clearly, (4.3.1) is a second-order damped nonlinear wave equation which possesses both strong and weakly nonlinear damping terms, as well as both semilinear and quasilinear terms which tend to destabilize smooth solutions; also (4.3.1) represents a somewhat nontrivial generalization of the nonlinearly damped, nonlinear wave equation (3.5.1), governing the evolution of the electric induction field D in the wave-dielectric interaction problem, if $\alpha \neq 0$, due to the presence of new stabilizing (i.e., αQ_t) and destabilizing (i.e., $\alpha\psi(Q)$) terms. In spite of these complications, for $\alpha > 0$ and G voltage dependent, the initial-value problem associated with (4.3.1) is still amenable to an energy argument of the same basic type as that which was employed in § 3.5; in fact, as will be seen below, the resulting global existence proof can be established without invoking the strong restrictions on the leakage conductance $1/G$ which were employed in § 4.2.

Throughout the remainder of this section, the following essential hypotheses will apply relative to the functions ψ and $\mathcal{F}$ as defined in (4.3.2):

(i) $\psi(0) = 0,\ \mathcal{F}(0) = 0$ (4.3.3a)

(ii) $\exists\, \psi_0,\ f_0 > 0$ such that $\psi'(0) = 2\psi_0,\ \mathcal{F}'(0) = 2f_0$ (4.3.3b)

(iii) $\exists\, N > 0$ such that $\psi \in C^3[-N, N],\ \mathcal{F} \in C^4[-N, N]$ (4.3.3c)

Without loss of generality we assume that $N \leq 1$. As a consequence of our second hypothesis, above, it follows that $\exists\, \eta,\ 0 < \eta \leq N$, such that

$$\psi'(\zeta) \geq \psi_0 > 0, \quad \mathcal{F}'(\zeta) \geq f_0 > 0, \quad \text{for } |\zeta| \leq \eta \tag{4.3.4}$$

We also note the obvious fact that $\mathcal{F}(0) = 0 \Leftrightarrow \mathcal{V}(0) = 0$ so that $\psi(0) = 0$ provided $G(0)$ is finite. We now define, in analogy with (3.5.7), for $\zeta \in R^1$,

$$\Psi(\zeta) = \begin{cases} \dfrac{2}{\zeta}\psi(\zeta) - \dfrac{2}{\zeta^2}\displaystyle\int_0^\zeta \psi(\lambda)\, d\lambda, & \zeta \neq 0 \\ \psi'(0), \quad \zeta = 0 \end{cases} \tag{4.3.5}$$

As a consequence of hypotheses (i)-(iii), it follows that $\Psi(\zeta)$ is continuous at $\zeta = 0$ and that for $|\zeta| \leq \eta$ we have $\Psi(\zeta) \geq \psi_0$. Also, by virtue of the definition (4.3.5), it is easily verified that for any sufficiently differentiable function w :

$$[w^2(x,t)\Psi(w(x,t))]_t = \psi'(w(x,t))(w^2(x,t))_t \tag{4.3.6}$$

As an example of constitutive functions for the voltage and leakage conductance which conform to the hypotheses (i)-(iii), consider $\mathcal{V}(\cdot)$ and $G(\cdot)$ as defined by

$$\mathcal{V}(\zeta) = \zeta(v_0 + v_1\zeta^2), \quad \zeta \in R^1 \tag{4.3.7a}$$

and

$$G(\zeta) = g_0 + g_1\zeta^2, \quad \zeta \in R^1 \tag{4.3.7b}$$

where $v_0, v_1, g_0, g_1 > 0$. For $\zeta \in R^1$, we set

$$\hat{G}(\zeta) = G(\mathcal{V}(\zeta)) \equiv g_0 + g_1\zeta^2(v_0 + v_1\zeta^2)^2 \tag{4.3.8}$$

so that $\hat{G}'(0) = 0$. It then follows that

$$\begin{cases} \psi(0) = \hat{G}(0)\mathcal{V}(0) = 0 \\ \psi'(0) = \hat{G}'(0)\mathcal{V}(0) + \hat{G}(0)\mathcal{V}'(0) = g_0 v_0 \\ \mathcal{F}(0) = \dfrac{1}{L}\mathcal{V}(0) = 0 \\ \mathcal{F}'(\zeta) = \dfrac{1}{L}(v_0 + 3v_1\zeta^2) \geq \dfrac{v_0}{L}, \quad \forall \zeta \in R^1 \end{cases} \tag{4.3.9}$$

so that hypotheses (i)-(iii) are clearly satisfied with $\psi_0 = \frac{1}{2}g_0 v_0 > 0$, $f_0 = \frac{v_0}{2L} > 0$, and any (arbitrary) $N > 0$.

In a manner similar to the definition (3.5.9), we set

$$\Lambda = \max_{\substack{1 \lesssim j \lesssim 4 \\ 1 \lesssim k \lesssim 3}} \left(\max_{[-N,N]} |\mathcal{F}^{(j)}(\cdot)|, \max_{[-N,N]} |\psi^{(k)}(\cdot)| \right) \tag{4.3.10}$$

and will denote by the symbols $C, C', \tilde{C}$, etc., or $C_i, i = 1, 2, \ldots$ generic constants which depend, at most, on $\Lambda, f_0, \psi_0, \alpha$ and various constants that arise from employing, in the estimates that follow, the Cauchy-Schwarz or arithmetic-geometric mean inequalities.

We now associate, with the nonlinear wave equation (4.3.1) for $Q(x,t)$, periodic initial data

$$Q(x,0) = Q_0(x), \quad Q_t(x,0) = Q_1(x), \quad x \in R^1 \tag{4.3.11}$$

which is assumed to satisfy

$$Q_0 \in H^3(R^1), \qquad Q_1 \in H^2(R^1) \tag{4.3.12}$$

Under the set of hypotheses (i)-(iii), and (4.3.12), we may apply the local existence results of Kato [89] so as to be able to conclude that for the initial-value problem (4.3.1), (4.3.12) the results of Lemma 3.5 apply, without change, provided we exchange, of course, $Q(x,t)$ for $D(x,t)$. As in § 3.5, therefore, our goal will be to extend the local existence result to a global existence theorem valid for $0 \le t < \infty$ and to study the behavior of the charge distribution $Q(x,t)$, $-\infty < x < \infty$, $t > 0$, and its derivatives, as $t \to \infty$. As in § 3.5, we again define functions $\mathcal{E}_0(t)$ and $\mathcal{E}(t)$, given by (3.5.13) and (3.5.14), respectively, but with Q in lieu of D, and attempt to follow the strategy highlighted by the series of estimates (3.5.15)-(3.5.17). Many of the pertinent estimates, of course, will have to be somewhat different now from those employed in § 3.5, because of the presence in (4.3.1) of the strong damping term αQ_t and the semilinear term $\alpha\psi(Q)$, e.g., (3.5.20) with Q replacing D no longer applies but, rather, as a consequence of (4.3.1) it is easily seen that

$$\begin{aligned}\frac{1}{4}\int_{-\infty}^{\infty} Q_{tt}^2(x,0)\,dx &\le \alpha^2\int_{-\infty}^{\infty} Q_t^2(x,0)\,dx \\ &+\int_{-\infty}^{\infty}\psi'^2(Q_0(x))Q_t^2(x,0)\,dx+\alpha^2\int_{-\infty}^{\infty}\psi^2(Q_0(x))\,dx \\ &+2\int_{-\infty}^{\infty}\mathcal{F}''^2(Q_0(x))Q_x^4(x,0)\,dx+2\int_{-\infty}^{\infty}\mathcal{F}'^2(Q_0(x))Q_{xx}^2(x,0)\,dx\end{aligned} \tag{4.3.13}$$

so that as a consequence of hypotheses (i) and (iii), above, and the Sobolev theorem, it follows that for

$$\|Q_0\|_{H^3}^2+\|Q_1\|_{H^2}^2 \le \delta^2, \quad 0<\delta\le\eta\le N \tag{4.3.14}$$

$\exists C > 0$ such that (in lieu of (3.5.20) with Q in place of D) we have the estimate

$$\int_{-\infty}^{\infty} Q_{tt}^2(x,0)\,dx \le C\int_{-\infty}^{\infty}\left\{Q^2+Q_t^2+Q_x^2+Q_{xx}^2\right\}(x,0)\,dx \tag{4.3.15}$$

Also, as was the case in § 3.5, various functionals based on the initial data will arise in the estimates to follow; these are formally equivalent to the functionals delineated in (3.5.18a)-(3.5.18h), except for one which arises as a consequence of the presence of the semilinear expression $\psi(Q)$ in (4.3.1), so for the sake of ease of exposition we now list these as follows:

$$K_0 = 2\int_{-\infty}^{\infty}\int_0^{Q_0(x)}\psi(\lambda)\,d\lambda\,dx \tag{4.3.16a}$$

$$K_1 = \int_{-\infty}^{\infty} \Psi(Q_0(x))Q_0^2(x)\,dx + 2\left(\int_{-\infty}^{\infty} Q_0^2(x)\,dx\right)^{1/2}\left(\int_{-\infty}^{\infty} Q_1^2(x)\,dx\right)^{1/2} \tag{4.3.16b}$$

$$K_2 = \int_{-\infty}^{\infty} \left(Q_1^2(x) + \mathcal{F}'(Q_0(x))Q_0'^2(x)\right)\,dx \tag{4.3.16c}$$

$$K_3 = \int_{-\infty}^{\infty} \left(Q_1'^2(x) + \mathcal{F}'(Q_0(x))Q_0''^2(x)\right)\,dx \tag{4.3.16d}$$

$$K_4 = \int_{-\infty}^{\infty} \psi'(Q_0(x))\left\{Q_0'^2(x) + Q_1^2(x)\right\}\,dx \tag{4.3.16e}$$

$$K_5 = \int_{-\infty}^{\infty} \left(Q_{tt}^2(x,0) + \mathcal{F}'(Q_0(x))Q_1'^2(x)\right)\,dx \tag{4.3.16f}$$

$$K_6 = \int_{-\infty}^{\infty} \left(Q_1''^2(x) + \mathcal{F}'(Q_0(x))Q_0'''^2(x)\right)\,dx \tag{4.3.16g}$$

$$K_7 = \int_{-\infty}^{\infty} \left(Q_{xtt}^2(x,0) + \mathcal{F}'(Q_0(x))Q_1''^2(x)\right)\,dx \tag{4.3.16h}$$

and

$$K_8 = \int_{-\infty}^{\infty} \left(Q_0'^2(x) + Q_0''^2(x) + Q_0'''^2(x)\right)\,dx \tag{4.3.16i}$$

We now assume that the initial data Q_0, Q_1 are chosen to be small in the sense that (4.3.14) holds for some δ, $0 < \delta \le \eta \le N$ (we may, in fact, again take $\delta = \eta/2^{k/2}$, with k sufficiently large, as in § 3.5); then, by virtue of the definition of $\mathcal{E}_0(t)$, i.e., (3.5.13) with Q replacing D, we have, first of all, the existence of $\mathcal{C}_0 > 0$ such that

$$\mathcal{E}_0(0) \le \left(\|Q_0\|_{H^3}^2 + \|Q_1\|_{H^2}^2\right) \tag{4.3.17}$$

and, therefore,

$$\mathcal{E}_0(0) \le \frac{\mathcal{C}_0\eta^2}{2^k} \le \frac{\eta^2}{4} \tag{4.3.18}$$

for k sufficiently large. Using the definition of $\mathcal{E}_0(t)$ again, and the Sobolev embedding theorem, it now follows that

$$\max\Big\{\|Q_0(\cdot)\|_{L^\infty}, \|Q_0'(\cdot)\|_{L^\infty}, \|Q_1(\cdot)\|_{L^\infty}, \|Q_{tt}(\cdot,0)\|_{L^\infty}, \|Q_1'(\cdot)\|_{L^\infty}, \|Q_0''(\cdot)\|_{L^\infty}\Big\} \le \eta/2. \tag{4.3.19}$$

so for the local solution $Q(x,t)$ of (4.3.1), (4.3.12) on $[0, T_\infty)$, $\exists\, s$, $0 < s \le T_\infty$, such that for any $t < s$,

$$\max\Big\{\|Q(\cdot,t)\|_{L^\infty}, \|Q_x(\cdot,t)\|_{L^\infty}, \|Q_t(\cdot,t)\|_{L^\infty}, \tag{4.3.20}$$

$$\|Q_{tt}(\cdot,t)\|_{L^\infty}, \|Q_{xt}(\cdot,t)\|_{L^\infty}, \|Q_{xx}(\cdot,t)\|_{L^\infty}\Big\} \le \eta.$$

Our first set of estimates will serve to establish the following result, which is the natural counterpart to Lemma 3.6 of § 3.5, with the important exception that a new term, i.e.,

$$\int_0^s \int_{-\infty}^{\infty} Q^2(x,\tau)\,dx\,d\tau$$

appears on the left-hand side of the estimate as a consequence of the presence of the semilinear term $\alpha\psi(Q)$ in (4.3.1).

Lemma 4.4 Let $Q(x,t)$ be the local solution of (4.3.1), (4.3.12) on $[0,T_\infty)$ and suppose that hypotheses (i)-(iii) are satisfied (i.e., (4.3.3a)-(4.3.3c)). If $s \le T_\infty$ is such that (4.3.20) holds, for any $t < s$, and η is sufficiently small, then $\exists\, \mathcal{C}^\# > 0$ such that

$$\int_{-\infty}^{\infty} (Q^2 + Q_x^2 + Q_t^2)(x,s)\,dx \tag{4.3.21}$$

$$+\int_0^s \int_{-\infty}^{\infty} (Q^2 + Q_x^2 + Q_t^2)(x,\tau)\,dx\,d\tau \le \mathcal{C}^\# \sum_{j=0}^{2} K_j$$

with K_0, K_1, and K_2 the functionals defined by (4.3.16a)-(4.3.16c).

Proof: We begin by multiplying (4.3.1) through by $2Q(x,t)$, $t < s$, assuming, of course, that the initial data satisfies (4.3.14) with δ of the form $\delta = \eta/2^{k/2}$ and k chosen so large that (4.3.18) holds; thus (4.3.19) applies, as does (4.3.20), for $t < s \le T_\infty$. The result of the operation cited above is the relation

$$(Q^2)_{tt} - 2Q_t^2 + \alpha(Q^2)_t + \psi'(Q)(Q^2)_t \tag{4.3.22}$$

$$+ 2\alpha Q\psi(Q) = 2(\mathcal{F}'(Q)QQ_x)_x - 2\mathcal{F}'(Q)Q_x^2$$

for $t < s$. By virtue of hypotheses (i) and (ii), and (4.3.20), for η sufficiently small we have

$$\begin{aligned} Q\psi(Q) &= \psi'(0)Q^2 + \mathcal{O}(Q^3) \\ &= 2\psi_0 Q^2 + \mathcal{O}(Q^3) \\ &\ge \psi_0 Q^2 \end{aligned} \tag{4.3.23}$$

Also, in view of (4.3.6)

$$\psi'(Q)(Q^2)_t = (Q^2\Psi(Q))_t \tag{4.3.24}$$

with $\Psi(Q) \geq \psi_0 > 0$ for $|Q|$ sufficiently small. Thus, for sufficiently small η we obtain from (4.3.22) the estimate

$$\begin{aligned}(Q^2)_{tt} - 2Q_t^2 + \alpha(Q^2)_t + (\psi'(Q)Q^2)_t \\ + 2\alpha\psi_0 Q^2 \leq 2(\mathcal{F}'(Q)QQ_x)_x - 2\mathcal{F}'(Q)Q_x^2\end{aligned} \tag{4.3.25}$$

valid for $t < s$. Integrating (4.3.25) over $[0, s) \times (-\infty, \infty)$, and using the fact that $\mathcal{F}'(Q) \geq f_0 > 0$, for $|Q|$ sufficiently small, we obtain, for small η, the estimate

$$\begin{aligned}&2f_0 \int_0^s \int_{-\infty}^{\infty} Q_x^2(x,\tau)\,dx\,d\tau + 2\alpha\psi_0 \int_0^s \int_{-\infty}^{\infty} Q^2(x,\tau)\,dx\,d\tau \\ &\quad +(\alpha + \psi_0) \int_{-\infty}^{\infty} Q^2(x,s)\,dx \leq 2\int_0^s \int_{-\infty}^{\infty} Q_t^2(x,\tau)\,dx\,d\tau \\ &\quad - \int_{-\infty}^{\infty} \int_0^s (Q^2)_{tt}(x,\tau)\,d\tau\,dx + \int_{-\infty}^{\infty} (Q_0^2(x) + \Psi(Q_0(x))Q_0^2(x)\,dx\end{aligned} \tag{4.3.26}$$

By applying the arithmetic-geometric mean inequality to the second integral on the right-hand side of (4.3.26), we find that for any $\lambda_0 > 0$

$$\begin{aligned}&2f_0 \int_0^s \int_{-\infty}^{\infty} Q_x^2(x,\tau)\,dx\,d\tau + 2\alpha\psi_0 \int_0^s \int_{-\infty}^{\infty} Q^2(x,\tau)\,dx\,d\tau \\ &\quad +(\alpha + \psi_0) \int_{-\infty}^{\infty} Q^2(x,s)\,dx \\ &\quad \leq \left\{2\int_{-\infty}^{\infty} Q_0(x)Q_1(x)\,dx + \int_{-\infty}^{\infty} (Q_0^2(x) + \Psi(Q_0(x))Q_0^2(x)\,dx\right\} \\ &\quad +2\int_0^s \int_{-\infty}^{\infty} Q_t^2(x,\tau)\,dx\,d\tau + \lambda_0 \int_{-\infty}^{\infty} Q^2(x,s)\,dx + \frac{1}{\lambda_0}\int_{-\infty}^{\infty} Q_t^2(x,s)\,dx\end{aligned} \tag{4.3.27}$$

Therefore, with $\lambda_0 < \alpha + \psi_0$ it follows from the definition of the initial data functional K_1 that

$$\begin{aligned}&2f_0 \int_0^s \int_{-\infty}^{\infty} Q_x^2(x,\tau)\,dx\,d\tau + 2\alpha\psi_0 \int_0^s \int_{-\infty}^{\infty} Q^2(x,\tau)\,dx\,d\tau \\ &\quad +(\alpha + \psi_0 - \lambda_0) \int_{-\infty}^{\infty} Q^2(x,s)\,dx \leq K_1 \\ &\quad +2\int_0^s \int_{-\infty}^{\infty} Q_t^2(x,\tau)\,dx\,d\tau + \frac{1}{\lambda_0}\int_{-\infty}^{\infty} Q_t^2(x,s)\,dx\end{aligned} \tag{4.3.28}$$

For the next set of estimates we multiply (4.3.1) by $2Q_t(x,t)$, $t < s$, integrate over $[0, s) \times (-\infty, \infty)$, integrate by parts, and apply the definition of Λ; as $|Q| \leq \eta \leq N$

we obtain

$$\int_{-\infty}^{\infty} Q_t^2(x,s)\,dx + 2(\alpha+\psi_0)\int_0^s\int_{-\infty}^{\infty} Q_t^2(x,\tau)\,dx\,d\tau \tag{4.3.29}$$
$$+2\alpha\int_0^s\int_{-\infty}^{\infty} Q_t(x,\tau)\psi(Q(x,\tau))\,dx\,d\tau$$
$$\leq \int_{-\infty}^{\infty}(Q_1^2(x)+\mathcal{F}'(Q_0(x))Q_0'^2(x))\,dx$$
$$-\int_{-\infty}^{\infty}\mathcal{F}'(Q(x,s))Q_x^2(x,s)\,dx + \eta\Lambda\int_0^s\int_{-\infty}^{\infty} Q_x^2(x,\tau)\,dx\,d\tau$$

In view of the fact that $\mathcal{F}'(Q(x,s)) \geq f_0$, if we apply the estimate (4.3.28) to (4.3.29) we obtain

$$\int_{-\infty}^{\infty}(f_0Q_x^2(x,s)+Q_t^2(x,s))\,dx + 2(\alpha+\psi_0)\int_0^s\int_{-\infty}^{\infty} Q_t^2(x,\tau)\,dx\,d\tau \tag{4.3.30}$$
$$\leq K_2 + \eta\mathcal{C}\left(K_1 + 2\int_0^s\int_{-\infty}^{\infty} Q_t^2(x,\tau)\,dx\,d\tau + \frac{1}{\lambda_0}\int_{-\infty}^{\infty} Q_t^2(x,s)\,dx\right)$$

for some $\mathcal{C} > 0$, where K_2 is defined by (4.3.16c). For η sufficiently small we find, therefore, that

$$\int_{-\infty}^{\infty}(f_0Q_x^2(x,s)+\frac{1}{2}Q_t^2(x,s))\,dx + (\alpha+\psi_0)\int_0^s\int_{-\infty}^{\infty} Q_t^2(x,\tau)\,dx\,d\tau \tag{4.3.31}$$
$$+2\alpha\int_0^s\int_{-\infty}^{\infty} Q_t\psi(Q)\,dx\,d\tau \leq K_2 + \eta\mathcal{C}K_1$$

As $Q\psi(Q) = \dfrac{\partial}{\partial t}\displaystyle\int_0^{Q(x,t)}\psi(\zeta)\,d\zeta$ we have

$$\int_0^s Q_t(x,\tau)\psi(Q(x,\tau))\,d\tau = \int_0^{Q(x,s)}\psi(\zeta)\,d\zeta - \int_0^{Q_0(x)}\psi(\zeta)\,d\zeta \tag{4.3.32a}$$

$$\int_0^s\int_{-\infty}^{\infty} Q_t\psi(Q)\,dx\,d\tau = \int_{-\infty}^{\infty}\int_0^{Q(x,s)}\psi(\zeta)\,d\zeta\,dx - \int_{-\infty}^{\infty}\int_0^{Q_0(x)}\psi(\zeta)\,d\zeta\,dx \tag{4.3.32b}$$

However, for small ζ, $\psi(\zeta) = \psi'(0)\zeta + \mathcal{O}(\zeta^2)$ so

$$\int_0^Q \psi(\zeta)\,d\zeta = \psi_0 Q^2 + \mathcal{O}(Q^3) > 0 \tag{4.3.33}$$

for $|Q(x,s)|$ sufficiently small. Employing the identities (4.3.32a), (4.3.32b) in the estimate (4.3.31), and using (4.3.33), we now obtain, for η sufficiently small

$$\int_{-\infty}^{\infty}(f_0Q_x^2(x,s)+\frac{1}{2}Q_t^2(x,s))\,dx + (\alpha+\psi_0)\int_0^s\int_{-\infty}^{\infty} Q_t^2(x,\tau)\,dx\,d\tau \tag{4.3.34}$$
$$\leq K_2 + \eta\mathcal{C}K_1 + 2\alpha\int_{-\infty}^{\infty}\int_0^{Q_0(x)}\psi(\zeta)\,d\zeta\,dx$$

so that with η chosen so small that $\eta\mathcal{C} < 1$ we have

$$\int_{-\infty}^{\infty}(f_0 Q_x^2(x,s) + \frac{1}{2}Q_t^2(x,s))\,dx + (\alpha + \psi_0)\int_0^s \int_{-\infty}^{\infty} Q_t^2(x,\tau)\,dx\,d\tau \tag{4.3.35}$$
$$\leq \alpha K_0 + K_1 + K_2$$

with K_0 as defined by (4.3.16a). From (4.3.35) it is immediate that for some $\mathcal{C}_1 > 0$

$$\int_{-\infty}^{\infty} Q_x^2\,dx + \int_{-\infty}^{\infty} Q_t^2\,dx + \int_0^s \int_{-\infty}^{\infty} Q_t^2\,dx\,d\tau \leq \mathcal{C}_1 \sum_{j=0}^{2} K_j \tag{4.3.36}$$

Finally, employing (4.3.36) in the estimate (4.3.28) we infer the existence of $\mathcal{C}_2 > 0$ such that

$$\int_{-\infty}^{\infty} Q^2\,dx + \int_0^s \int_{-\infty}^{\infty} Q^2\,dx\,d\tau + \int_0^s \int_{-\infty}^{\infty} Q_x^2\,dx\,d\tau \leq \mathcal{C}_2 \sum_{j=0}^{2} K_j \tag{4.3.37}$$

The required result, i.e., (4.3.21) now follows upon combining the estimates (4.3.36) and (4.3.37). □

Remarks: As noted just prior to the statement of the lemma, the bound on $\int_0^s \int_{-\infty}^{\infty} Q^2(x,\tau)\,dx\,d\tau$ which is implicit in (4.3.37) has no counterpart in the estimates derived in § 3.5 for the solution of the one-space dimension wave-dielectric interaction problem; this is, of course, a direct consequence of the presence of the semilinear term $\psi(Q)$ in (4.3.1), as is easily seen by tracing through the estimates which led to (4.3.37).

Our second set of estimates are now geared towards establishing the following

Lemma 4.5 Under the same conditions as apply in Lemma 4.4, for any $t < s$, and η sufficiently small, $\exists\,\mathcal{C}^* > 0$ such that

$$\int_{-\infty}^{\infty}(Q_{xx}^2 + Q_{xt}^2 + Q_{tt}^2)(x,s)\,dx \tag{4.3.38}$$
$$+\int_0^s \int_{-\infty}^{\infty}(Q_{xx}^2 + Q_{xt}^2 + Q_{tt}^2)(x,\tau)\,dx\,d\tau \leq \mathcal{C}^* \sum_{j=0}^{5} K_j$$

Proof: We initiate this series of estimates by differentiating (4.3.1) through with respect to x and multiplying the resulting equation by $2Q_{xt}$; for $0 \leq t < T_\infty$, $-\infty < x < \infty$, we obtain

$$(Q_{xt}^2)_t + 2\alpha Q_{xt}^2 + 2\psi(Q)_{xt} Q_{xt} \tag{4.3.39}$$
$$+2\alpha\psi(Q)_x Q_{xt} = 2\mathcal{F}(Q)_{xxx} Q_{xt}$$

Using the identity

$$\begin{aligned} 2\psi(Q)_{xt}Q_{xt} &= 2\psi'(Q)Q_{xt}^2 + (\psi''(Q)Q_xQ_t^2)_x \\ &- \left\{\psi''(Q)Q_{xx} + \psi'''(Q)Q_x^2\right\} Q_t^2 \end{aligned}$$

in (4.3.39), integrating over $[0,s) \times (-\infty,\infty)$, and applying the definition of Λ and (4.3.20), we obtain the estimate

$$\begin{aligned} &\int_{-\infty}^{\infty} Q_{xt}^2(x,s)\,dx + 2(\alpha+\psi_0)\int_0^s\int_{-\infty}^{\infty} Q_{xt}^2(x,\tau)\,dx\,d\tau \\ &+2\alpha\int_0^s\int_{-\infty}^{\infty}\psi(Q(x,\tau))Q_{xt}(x,\tau)\,dx\,d\tau \\ &\le \int_{-\infty}^{\infty} Q_{xt}^2(x,0)\,dx + 2\int_0^s\int_{-\infty}^{\infty}\mathcal{F}(Q(x,\tau))_{xxx}Q_{xt}(x,\tau)\,dx\,d\tau \\ &+2\eta\Lambda\int_0^s\int_{-\infty}^{\infty} Q_t^2(x,\tau)\,dx\,d\tau \end{aligned} \tag{4.3.40}$$

As a consequence of the identity

$$\begin{aligned} \mathcal{F}(Q)_{xxx}Q_{xt} &= \left(\left[\mathcal{F}'(Q)Q_{xx} + \mathcal{F}''(Q)Q_x^2\right]Q_{xt}\right)_x \\ &-\frac{1}{2}(\mathcal{F}'(Q)Q_{xx}^2)_t + \frac{1}{2}\mathcal{F}''(Q)Q_tQ_{xx}^2 \\ &-(\mathcal{F}''(Q)Q_x^2Q_{xx})_t + \mathcal{F}'''(Q)Q_tQ_{xx}Q_x^2 \\ &+2\mathcal{F}''(Q)Q_xQ_{xt}Q_{xx} \end{aligned}$$

it is easy to show that $\exists \mathcal{C} > 0$ such that

$$\begin{aligned} &2\int_0^s\int_{-\infty}^{\infty}\mathcal{F}(Q)_{xxx}Q_{xt}\,dx\,d\tau \le \int_{-\infty}^{\infty}\mathcal{F}'(Q_0(x)){Q_0'}^2(x)\,dx \\ &+2\eta\Lambda\int_{-\infty}^{\infty}{Q_0'}^2(x)\,dx - \int_{-\infty}^{\infty}\mathcal{F}'(Q(x,s))Q_{xx}^2(x,s)\,dx \\ &+\eta\mathcal{C}\left(\int_{-\infty}^{\infty}Q_x^2(x,s)\,dx + \int_0^s\int_{-\infty}^{\infty}(Q_x^2+Q_{xx}^2+Q_{xt}^2)(x,\tau)\,dx\,d\tau\right) \end{aligned} \tag{4.3.41}$$

Using, once again, the lower bound $\mathcal{F}'(Q(x,s)) \ge f_0$ for $-\infty < x < \infty$, and employing (4.3.41) in (4.3.40), we are led to an estimate of the form

$$\begin{aligned} &\int_{-\infty}^{\infty}(Q_{xt}^2(x,s) + f_0Q_{xx}^2(x,s))\,dx + 2(\alpha+\psi_0)\int_0^s\int_{-\infty}^{\infty}Q_{xt}^2\,dx\,d\tau \\ &+2\alpha\int_0^s\int_{-\infty}^{\infty}\psi(Q)_xQ_{xt}\,dx\,d\tau \le K_3 + 2\eta\Lambda\int_{-\infty}^{\infty}{Q_0'}^2(x)\,dx \\ &+\eta\bar{\mathcal{C}}\left(\int_{-\infty}^{\infty}Q_x^2(x,s)\,dx + \int_0^s\int_{-\infty}^{\infty}(Q_x^2+Q_t^2+Q_{xx}^2+Q_{xt}^2)(x,\tau)\,dx\,d\tau\right) \end{aligned} \tag{4.3.42}$$

for some $\bar{\mathcal{C}} > 0$ where K_3 is defined by (4.3.16d). However, we may now estimate the integrals $\int_{-\infty}^{\infty} Q_x^2\,dx$ and $\int_0^s \int_{-\infty}^{\infty} (Q_x^2 + Q_t^2)\,dx\,d\tau$ on the right-hand side of (4.3.42) by using the result of Lemma 4.4, i.e. (4.3.21), in which case we obtain, for η sufficiently small, and some $\hat{\mathcal{C}} > 0$

$$\begin{aligned} &\int_{-\infty}^{\infty} (Q_{xt}^2(x,s) + f_0 Q_{xx}^2(x,s))\,dx \\ &\quad +2(\alpha+\psi_0)\int_0^s \int_{-\infty}^{\infty} Q_{xt}^2\,dx\,d\tau \\ &\quad +2\alpha \int_0^s \int_{-\infty}^{\infty} \psi(Q)_x Q_{xt}\,dx\,d\tau \le K_3 + \eta\hat{\mathcal{C}} \sum_{j=0}^{2} K_j \\ &\quad +2\eta\Lambda \int_{-\infty}^{\infty} Q_0'^2(x)\,dx + \eta\bar{\mathcal{C}} \int_0^s \int_{-\infty}^{\infty} (Q_{xt}^2 + Q_{xx}^2)\,dx\,d\tau \end{aligned} \tag{4.3.43}$$

Choosing η even smaller, if necessary, we now absorb the last integral on the right-hand side of (4.3.43), using the like expressions on the left-hand side of this estimate, and we are led to the inequality

$$\begin{aligned} &\int_{-\infty}^{\infty} (Q_{xt}^2(x,s) + f_0 Q_{xx}^2(x,s))\,dx + (\alpha+\psi_0)\int_0^s \int_{-\infty}^{\infty} Q_{xt}^2\,dx\,d\tau \\ &\quad +2\alpha \int_0^s \int_{-\infty}^{\infty} \psi(Q)_x Q_{xt}\,dx\,d\tau \le K_3 + \eta\hat{\mathcal{C}} \sum_{j=0}^{2} K_j \\ &\quad +2\eta\Lambda \int_{-\infty}^{\infty} Q_0'^2(x)\,dx + \eta\bar{\mathcal{C}} \int_0^s \int_{-\infty}^{\infty} Q_{xx}^2\,dx\,d\tau \end{aligned} \tag{4.3.44}$$

We now observe that

$$2\psi(Q)_x Q_{xt} = (\psi'(Q)Q_x^2)_t - \psi''(Q)Q_t Q_x^2$$

so that

$$\begin{aligned} &2\alpha \int_0^s \int_{-\infty}^{\infty} \psi(Q)_x Q_{xt}\,dx\,d\tau \\ &\quad = 2\alpha\left\{\int_{-\infty}^{\infty} \psi'(Q)Q_x^2\,dx - \int_{-\infty}^{\infty} \psi'(Q_0(x))Q_0'^2(x)\,dx\right\} \\ &\quad -2\alpha \int_0^s \int_{-\infty}^{\infty} \psi''(Q)Q_t Q_x^2\,dx\,d\tau \end{aligned} \tag{4.3.45}$$

Employing (4.3.45) in (4.3.44), using the fact that $\psi'(Q(x,s)) \ge \psi_0$, $\forall\, x \in (-\infty,\infty)$, as well as the definition of Λ, then leads us to the estimate

$$\int_{-\infty}^{\infty} (Q_{xt}^2(x,s) + f_0 Q_{xx}^2(x,s))\,dx + (\alpha+\psi_0)\int_0^s \int_{-\infty}^{\infty} Q_{xt}^2\,dx\,d\tau \tag{4.3.46}$$

$$+2\alpha\psi_0 Q_x^2(x,s)\,dx \le \Big[K_3 + \eta\hat{\mathcal{C}}\sum_{j=0}^{2} K_j$$

$$+2\eta\Lambda\int_{-\infty}^{\infty} Q_0'^2(x)\,dx + 2\alpha\int_{-\infty}^{\infty}\psi'(Q_0(x))Q_0'^2(x)\,dx\Big]$$

$$+\eta\bar{\mathcal{C}}\int_0^s\int_{-\infty}^{\infty} Q_{xx}^2\,dx\,d\tau + 2\alpha\eta\Lambda\int_0^s\int_{-\infty}^{\infty} Q_x^2\,dx\,d\tau$$

If we use, at this point, the bound for $\int_0^s\int_{-\infty}^{\infty} Q_x^2\,dx\,d\tau$ which is implied by Lemma 4.4, i.e. (4.3.21), in conjunction with the fact that

$$\int_{-\infty}^{\infty} Q_0'^2(x)\,dx \le \frac{1}{f_0}K_2,$$

and we then drop the integral $\int_{-\infty}^{\infty} Q_x^2(x,s)\,dx$ on the left-hand side of (4.3.46), we find that for some $\tilde{\mathcal{C}} > 0$

$$\int_{-\infty}^{\infty}(Q_{xt}^2(x,s) + f_0 Q_{xx}^2(x,s))\,dx + (\alpha+\psi_0)\int_0^s\int_{-\infty}^{\infty} Q_{xt}^2\,dx\,d\tau \qquad (4.3.47)$$

$$\le \eta\tilde{\mathcal{C}}\sum_{j=0}^{2} K_j + K_3 + 2\alpha K_4 + \eta\bar{\mathcal{C}}\int_0^s\int_{-\infty}^{\infty} Q_{xx}^2\,dx\,d\tau$$

where K_4 is given by (4.3.16e).

It is now essential that we be able to estimate the integral $\int_0^s\int_{-\infty}^{\infty} Q_{xx}^2\,dx\,d\tau$; in order to initiate this process, we will first write (4.3.1) in the form

$$\mathcal{F}'(Q)Q_{xx} + \mathcal{F}''(Q)Q_x^2$$

$$= Q_{tt} + \psi'(Q)Q_t + \alpha Q_t + \alpha\psi(Q),$$

then multiply through by $Q_{xx}(x,\tau)$, $0 < \tau < s$, and integrate over $[0,s)\times(-\infty,\infty)$. Carrying out the aforementioned procedure, and applying the arithmetic-geometric mean inequality to the resulting equation, yields, for any $\lambda_1 > 1/2f_0$, the estimate

$$\left(f_0 - \frac{1}{2\lambda_1}\right)\int_0^s\int_{-\infty}^{\infty} Q_{xx}\,dx\,d\tau \le \eta\Lambda\int_0^s\int_{-\infty}^{\infty} Q_x^2\,dx\,d\tau \qquad (4.3.48)$$

$$+\int_0^s\int_{-\infty}^{\infty}\tilde{\psi}'(Q)Q_tQ_{xx}\,dx\,d\tau + \alpha\int_0^s\int_{-\infty}^{\infty}\psi(Q)Q_{xx}\,dx\,d\tau$$

$$+\frac{1}{2}\lambda_1\int_0^s\int_{-\infty}^{\infty} Q_{tt}^2\,dx\,d\tau$$

where $\tilde{\psi}(Q) = \psi(Q) + \alpha Q$. In view of our local existence result, if we integrate by parts in both x and t we achieve the identity

$$\int_0^s \int_{-\infty}^{\infty} \tilde{\psi}'(Q) Q_t Q_{xx}\, dx\, d\tau = -\frac{1}{2}\int_0^s \int_{-\infty}^{\infty} \tilde{\psi}''(Q) Q_t Q_x^2\, dx\, d\tau$$
$$-\frac{1}{2}\int_0^s \int_{-\infty}^{\infty} \left[\tilde{\psi}'(Q) Q_x^2\right]_t\, dx\, d\tau$$

and its use in (4.3.48) now produces the estimate

$$\frac{1}{2}(\alpha + \psi_0)\int_{-\infty}^{\infty} Q_x^2(x,s)\, dx + \left(f_0 - \frac{1}{2\lambda_1}\right)\int_0^s \int_{-\infty}^{\infty} Q_{xx}\, dx\, d\tau \tag{4.3.49}$$
$$\leq \frac{1}{2}\int_{-\infty}^{\infty} \tilde{\psi}'(Q_0(x)) Q_0'^2(x)\, dx$$
$$+\frac{3}{2}\eta\Lambda \int_0^s \int_{-\infty}^{\infty} Q_x^2\, dx\, d\tau + \alpha \int_0^s \int_{-\infty}^{\infty} \psi(Q) Q_{xx}\, dx\, d\tau$$
$$+\frac{1}{2}\gamma_1 \int_0^s \int_{-\infty}^{\infty} Q_{tt}^2\, dx\, d\tau$$

In obtaining (4.3.49), we have also used the fact that for $t < s$, $\tilde{\psi}'(Q(x,t)) \geq \psi_0 + \alpha$, $x \in (-\infty, \infty)$. Using the local existence theorem, again, we have

$$\int_0^s \int_{-\infty}^{\infty} \psi(Q) Q_{xx}\, dx\, d\tau = -\int_0^s \int_{-\infty}^{\infty} \psi'(Q) Q_x^2\, dx\, d\tau$$
$$\leq \Lambda \int_0^s \int_{-\infty}^{\infty} Q_x^2\, dx\, d\tau$$

and by employing this estimate in (4.3.49), and dropping the integral $\int_{-\infty}^{\infty} Q_x^2\, dx$, we find that

$$\left(f_0 - \frac{1}{2\lambda_1}\right)\int_0^s \int_{-\infty}^{\infty} Q_{xx}\, dx\, d\tau \leq \frac{1}{2}\int_{-\infty}^{\infty} (\psi'(Q_0(x) + \alpha)) Q_0'^2(x)\, dx \tag{4.3.50}$$
$$+\left(\frac{3}{2}\eta + \alpha\right)\Lambda \int_0^s \int_{-\infty}^{\infty} Q_x^2\, dx\, d\tau + \frac{1}{2}\lambda_1 \int_0^s \int_{-\infty}^{\infty} Q_{tt}^2\, dx\, d\tau$$

By turning again to the result of Lemma 4.4 we see that (4.3.50) implies that $\exists \mathcal{C}_1, \mathcal{C}_2 > 0$ such that

$$\int_0^s \int_{-\infty}^{\infty} Q_{xx}^2\, dx\, d\tau \leq \mathcal{C}_1 \sum_{j=0}^{4} K_j + \mathcal{C}_2 \int_0^s \int_{-\infty}^{\infty} Q_{tt}^2\, dx\, d\tau \tag{4.3.51}$$

so that it is clear that an estimate for $\int_0^s \int_{-\infty}^{\infty} Q_{xx}^2\, dx\, d\tau$ cannot be obtained without first deriving an appropriate bound for $\int_0^s \int_{-\infty}^{\infty} Q_{tt}^2\, dx\, d\tau$.

To obtain the bound for $\int_0^s \int_{-\infty}^{\infty} Q_{tt}^2\,dx\,d\tau$ which is needed in (4.3.51), we differentiate (4.3.1) through with respect to t, multiply the resulting equation by $2Q_{tt}$ and integrate over $[0,s)\times(-\infty,\infty)$; we find, for some $\mathcal{C}>0$, that the following estimate holds:

$$\begin{aligned}
&\int_{-\infty}^{\infty} Q_{tt}^2(x,s)\,dx + (\alpha+\psi_0)\int_0^s \int_{-\infty}^{\infty} Q_{tt}^2\,dx\,d\tau \\
&\quad +2\alpha\int_0^s \int_{-\infty}^{\infty} \psi(Q)_t Q_{tt}\,dx\,d\tau \\
&\qquad \leq \int_{-\infty}^{\infty} Q_{tt}^2(x,0)\,dx + 2\int_0^s \int_{-\infty}^{\infty} \mathcal{F}'(Q)Q_{xxt}Q_{tt}\,dx\,d\tau \\
&\quad +\eta\mathcal{C}\int_0^s \int_{-\infty}^{\infty} (Q_t^2+Q_x^2+Q_{tt}^2+Q_{xt}^2+Q_{xx}^2)\,dx\,d\tau
\end{aligned} \tag{4.3.52}$$

Therefore, choosing η sufficiently small, we have

$$\begin{aligned}
&\int_{-\infty}^{\infty} Q_{tt}^2(x,s)\,dx + (\alpha+\frac{1}{2}\psi_0)\int_0^s \int_{-\infty}^{\infty} Q_{tt}^2\,dx\,d\tau \\
&\quad +2\alpha\int_0^s \int_{-\infty}^{\infty} \psi(Q)_t Q_{tt}\,dx\,d\tau \\
&\qquad \leq \int_{-\infty}^{\infty} Q_{tt}^2(x,0)\,dx + 2\int_0^s \int_{-\infty}^{\infty} \mathcal{F}'(Q)Q_{xxt}Q_{tt}\,dx\,d\tau \\
&\quad +\eta\mathcal{C}\int_0^s \int_{-\infty}^{\infty} (Q_t^2+Q_x^2+Q_{xt}^2+Q_{xx}^2)\,dx\,d\tau
\end{aligned} \tag{4.3.53}$$

From the identity

$$2\psi(Q)_t Q_{tt} = \psi'(Q)(Q_t^2)_t + (\psi'(Q)Q_t^2)_t - \psi''(Q)Q_t^3$$

we have, at once,

$$\begin{aligned}
2\alpha\int_0^s \int_{-\infty}^{\infty} \psi(Q)_t Q_{tt}\,dx\,d\tau &= \alpha\int_{-\infty}^{\infty} \psi'(Q(x,s))Q_t^2(x,s)\,dx \\
&\quad -\alpha\int_{-\infty}^{\infty} \psi'(Q_0(x))Q_1^2(x)\,dx - \alpha\int_0^s \int_{-\infty}^{\infty} \psi''(Q)Q_t^3\,dx\,d\tau
\end{aligned}$$

We now make use of this last identity in (4.3.53), and upon dropping the expression $\alpha\int_{-\infty}^{\infty} \psi'(Q)Q_t^2\,dx$, and absorbing the term

$$\alpha\int_0^s \int_{-\infty}^{\infty} \psi''(Q)Q_t^3\,dx\,d\tau \leq \alpha\eta\Lambda\int_0^s \int_{-\infty}^{\infty} Q_t^2\,dx\,d\tau$$

we find that for some $\hat{\mathcal{C}} > 0$

$$\int_{-\infty}^{\infty} Q_{tt}^2(x,s)\,dx + (\alpha + \frac{1}{2}\psi_0)\int_0^s \int_{-\infty}^{\infty} Q_{tt}^2\,dx\,d\tau \tag{4.3.54}$$

$$\leq \int_{-\infty}^{\infty} (Q_{tt}^2(x,0) + \alpha\psi'(Q_0(x))Q_1^2(x))\,dx$$

$$+\eta\hat{\mathcal{C}}\int_0^s \int_{-\infty}^{\infty} (Q_t^2 + Q_x^2 + Q_{xt}^2 + Q_{xx}^2)\,dx\,d\tau$$

$$+2\int_0^s \int_{-\infty}^{\infty} \mathcal{F}'(Q)Q_{xxt}Q_{tt}\,dx\,d\tau$$

A series of integrations by parts of the last integral in (4.3.54) leads to the estimate

$$2\int_0^s \int_{-\infty}^{\infty} \mathcal{F}'(Q)Q_{xxt}Q_{tt}\,dx\,d\tau \leq -\int_{-\infty}^{\infty} \mathcal{F}'(Q(x,s))Q_{xt}^2(x,s)\,dx$$

$$+\int_{-\infty}^{\infty} \mathcal{F}'(Q_0(x)){Q_1'}^2(x)\,dx + \eta\tilde{\mathcal{C}}\int_0^s \int_{-\infty}^{\infty} (Q_{tt}^2 + Q_{xt}^2)\,dx\,d\tau$$

whose use in (4.3.54) now produces, for some $\mathcal{C}^* > 0$,

$$\int_{-\infty}^{\infty} (Q_{tt}^2(x,s) + f_0 Q_{xt}^2(x,s))\,dx + (\alpha + \frac{1}{2}\psi_0)\int_0^s \int_{-\infty}^{\infty} Q_{tt}^2\,dx\,d\tau \tag{4.3.55}$$

$$\leq \int_{-\infty}^{\infty} (Q_{tt}^2(x,0) + \alpha\psi'(Q_0(x))Q_1^2(x) + \mathcal{F}'(Q_0(x)){Q_1'}^2(x))\,dx$$

$$+\eta\mathcal{C}^*\int_0^s \int_{-\infty}^{\infty} (Q_t^2 + Q_x^2 + Q_{xt}^2 + Q_{xx}^2 + Q_{tt}^2)\,dx\,d\tau$$

We now choose η to be sufficiently small and use, once more, (4.3.21), so as to obtain from (4.3.55) the estimate

$$\int_{-\infty}^{\infty} (Q_{tt}^2(x,s) + f_0 Q_{xt}^2(x,s))\,dx + (\alpha + \frac{1}{4}\psi_0)\int_0^s \int_{-\infty}^{\infty} Q_{tt}^2\,dx\,d\tau \tag{4.3.56}$$

$$\leq \alpha K_4 + K_5 + \eta\mathcal{C}^*\sum_{j=0}^{2} K_j + \eta\mathcal{C}^*\int_0^s \int_{-\infty}^{\infty} Q_{xt}^2\,dx\,d\tau$$

$$+\eta\mathcal{C}^*\int_0^s \int_{-\infty}^{\infty} Q_{xx}^2\,dx\,d\tau$$

where K_5 is defined by (4.3.16f). We now proceed as follows: we use the bound given by (4.3.51) to estimate the last integral on the right-hand side of (4.3.56) and then choose η to be so small that we can absorb the expression which results, namely, $\eta\mathcal{C}_2\mathcal{C}^*\int_0^s \int_{-\infty}^{\infty} Q_{tt}^2\,dx\,d\tau$ into the similar expression on the left-hand side of (4.3.56);

the result of all this is an estimate of the form

$$\int_{-\infty}^{\infty} (Q_{tt}^2(x,s) + f_0 Q_{xt}^2(x,s))\,dx + (\alpha + \frac{1}{8}\psi_0)\int_0^s \int_{-\infty}^{\infty} Q_{tt}^2\,dx\,d\tau \tag{4.3.57}$$

$$\leq \mathcal{C}_* \sum_{j=0}^{5} K_j + \eta \mathcal{C}^* \int_0^s \int_{-\infty}^{\infty} Q_{xt}^2\,dx\,d\tau$$

which is valid for some $\mathcal{C}_*, \mathcal{C}^* > 0$. We now return to (4.3.47) and bound the integral $\int_0^s \int_{-\infty}^{\infty} Q_{xx}^2\,dx\,d\tau$ using (4.3.51); there results, for some $\bar{\mathcal{C}}_1, \bar{\mathcal{C}}_2 > 0$ the estimate

$$\int_{-\infty}^{\infty} (Q_{xt}^2(x,s) + f_0 Q_{xx}^2(x,s))\,dx + (\alpha + \psi_0)\int_0^s \int_{-\infty}^{\infty} Q_{xt}^2\,dx\,d\tau \tag{4.3.58}$$

$$\leq \bar{\mathcal{C}}_1 \sum_{j=0}^{4} K_j + \eta \bar{\mathcal{C}}_2 \int_0^s \int_{-\infty}^{\infty} Q_{tt}^2\,dx\,d\tau$$

Also, (4.3.57) implies that for some $\bar{\mathcal{C}}_*, \bar{\mathcal{C}}^* > 0$

$$\int_0^s \int_{-\infty}^{\infty} Q_{tt}^2\,dx\,d\tau \leq \bar{\mathcal{C}}_* \sum_{j=0}^{5} K_j + \eta \bar{\mathcal{C}}^* \int_0^s \int_{-\infty}^{\infty} Q_{xt}^2\,dx\,d\tau \tag{4.3.59}$$

If we employ (4.3.59) in (4.3.58), and choose η sufficiently small, then we clearly achieve an estimate of the type

$$\int_{-\infty}^{\infty} (Q_{xt}^2(x,s) + f_0 Q_{xx}^2(x,s))\,dx + (\alpha + \frac{1}{2}\psi_0)\int_0^s \int_{-\infty}^{\infty} Q_{xt}^2\,dx\,d\tau \tag{4.3.60}$$

$$\leq \mathcal{C} \sum_{j=0}^{5} K_j$$

for some $\mathcal{C} > 0$. As a consequence of (4.3.60) and (4.3.59) we also have, for some $\mathcal{C}_1 > 0$,

$$\int_0^s \int_{-\infty}^{\infty} Q_{tt}^2\,dx\,d\tau \leq \mathcal{C}_1 \sum_{j=0}^{5} K_j \tag{4.3.61}$$

so that (4.3.51) now produces

$$\int_0^s \int_{-\infty}^{\infty} Q_{xx}^2\,dx\,d\tau \leq \mathcal{C}_2 \sum_{j=0}^{5} K_j \tag{4.3.62}$$

Finally, combining (4.3.56) with (4.3.60) and (4.3.62) we obtain, for some $\mathcal{C}_3 > 0$, the bound

$$\int_{-\infty}^{\infty} Q_{tt}^2(x,s)\,dx \leq \mathcal{C}_3 \sum_{j=0}^{5} K_j \tag{4.3.63}$$

The required estimate, i.e., (4.3.38) now follows as a direct consequence of (4.3.60)-(4.3.63) and, therefore, Lemma 4.5 has been established. □

Our third set of estimates closely parallel those obtained for the third-order derivative terms in the analysis of the wave-dielectric interaction problem presented in § 3.5; in other words, at this stage those terms in (4.3.1) which correspond to $\alpha \neq 0$ contribute little, in terms of difficulty, once we begin to repeatedly differentiate the equation to obtain new estimates. Because of the similarity to the analysis described in § 3.5, our proof of the following result will be presented with considerable brevity:

Lemma 4.6 Under the same conditions as those which apply in Lemmas 4.4 and 4.5, for any $t < s$, and η sufficiently small, $\exists C' > 0$ such that

$$\begin{aligned} &\int_{-\infty}^{\infty} (Q_{ttt}^2 + Q_{xxt}^2 + Q_{xxx}^2 + Q_{xtt}^2)(x,s)\,dx \\ &+ \int_0^s \int_{-\infty}^{\infty} (Q_{ttt}^2 + Q_{xxt}^2 + Q_{xxx}^2 + Q_{xtt}^2)\,dx\,d\tau \\ &\leq C' \sum_{j=0}^{8} K_j \end{aligned} \tag{4.3.64}$$

Proof: We begin by noting that, as in § 3.5, the local solution on $[0, t_\infty)$ is not smooth enough to justify formal continuation of the procedure we have followed to this point because this procedure requires the existence, for $0 \leq t < T_\infty$, of various fourth-order derivatives of $Q(x,t)$; these fourth-order derivatives are not guaranteed to exist classically, as functions, by the local existence theorem. However, as we noted for the wave-dielectric interaction problem in § 3.5, the formal estimates which we will obtain below may be derived rigorously by using forward differencing operators.

The first sets of estimates to be obtained here may be formally deduced from (4.3.1) by first differentiating the equation with respect to x and squaring both sides of the resulting equation, and then by differentiating (4.3.1) with respect to t and squaring the result of that calculation; integrating each of the results obtained in this fashion over $(-\infty, \infty)$ and, then, over $[0, s) \times (-\infty, \infty)$, yields estimates of the form

$$\begin{aligned} \int_{-\infty}^{\infty} Q_{xxx}^2(x,s)\,dx \;\leq\; & \int_{-\infty}^{\infty} Q_{xxt}^2(x,s)\,dx \\ & + \; C_1 \int_{-\infty}^{\infty} (Q_{xt}^2(x,s) + Q_x^2(x,s))\,dx \end{aligned} \tag{4.3.65}$$

$$\int_0^s \int_{-\infty}^{\infty} Q_{xxx}^2\,dx\,d\tau \le \int_0^s \int_{-\infty}^{\infty} Q_{xtt}^2\,dx\,d\tau + \mathcal{C}_1 \int_0^s \int_{-\infty}^{\infty} (Q_{xt}^2 + Q_x^2)\,dx\,d\tau \tag{4.3.66}$$

$$\int_{-\infty}^{\infty} Q_{ttt}^2(x,s)\,dx \le \mathcal{C}_2 \int_{-\infty}^{\infty} (Q_t^2 + Q_{xt}^2 + Q_{tt}^2 + Q_{xxt}^2)(x,s)\,dx \tag{4.3.67}$$

and

$$\int_0^s \int_{-\infty}^{\infty} Q_{ttt}^2\,dx\,d\tau \le \mathcal{C}_2 \int_0^s \int_{-\infty}^{\infty} (Q_t^2 + Q_{xt}^2 + Q_{tt}^2 + Q_{xxt}^2)\,dx\,d\tau \tag{4.3.68}$$

for some $\mathcal{C}_1$, $\mathcal{C}_2 > 0$. Next, we differentiate (4.3.1) twice in succession with respect to x, multiply the resulting equation by $2Q_{xxt}$, and integrate over $[0,s) \times (-\infty,\infty)$; after a lengthy analysis, involving several integrations by parts, we obtain, for η sufficiently small, a bound of the form

$$\begin{aligned}\int_{-\infty}^{\infty} (Q_{xxt}^2(x,s) + f_0 Q_{xxx}^2(x,s))\,dx & \\ + \frac{1}{4}\psi_0 \int_0^s \int_{-\infty}^{\infty} Q_{xxt}^2\,dx\,d\tau & \\ \le \mathcal{C}_3 \left(\sum_{j=0}^{6} K_j\right) + \mathcal{C}_4 \int_{-\infty}^{\infty} (Q_x^2 + Q_{xx}^2 + Q_{xxx}^2)(x,0)\,dx & \\ + \eta \mathcal{C}_5 \int_0^s \int_{-\infty}^{\infty} Q_{xxx}^2\,dx\,d\tau &\end{aligned} \tag{4.3.69}$$

for some $\mathcal{C}_3$, $\mathcal{C}_4$, $\mathcal{C}_5 > 0$, where K_6 is given by (4.3.16g), and where we have also made use of the estimates of Lemmas 4.4 and 4.5. We now require a bound for the integral $\int_0^s \int_{-\infty}^{\infty} Q_{xtt}^2\,dx\,d\tau$ in (4.3.66); to this end, (4.3.1) is first differentiated with respect to x and, then, with respect to t and the equation which results is multiplied by $2Q_{xxt}$ and integrated over $[0,s) \times (-\infty,\infty)$, yielding, for sufficiently small η, an estimate of the form

$$\begin{aligned}\int_{-\infty}^{\infty} (Q_{xtt}^2(x,s) + f_0 Q_{xxt}^2(x,s))\,dx & \\ + \frac{1}{16}\psi_0 \int_0^s \int_{-\infty}^{\infty} Q_{xtt}^2\,dx\,d\tau \le \mathcal{C}_6 \sum_{j=0}^{7} K_j & \\ + \eta \mathcal{C}_7 \int_0^s \int_{-\infty}^{\infty} Q_{xxt}^2\,dx\,d\tau &\end{aligned} \tag{4.3.70}$$

for some $\mathcal{C}_6, \mathcal{C}_7 > 0$ where K_7 is given by (4.3.16h). We now combine the estimates (4.3.66), (4.3.69) and (4.3.70) by substituting the bound for the integral

$\int_0^s \int_{-\infty}^{\infty} Q_{xxx}^2 \, dx \, d\tau$, as given by (4.3.66), into (4.3.69) and then adding the estimate which results to (4.3.70); there results, for η chosen sufficiently small, and some $C > 0$, a bound of the form

$$\int_{-\infty}^{\infty} (Q_{xtt}^2 + Q_{xxt}^2 + Q_{xxx}^2)(x,s)\, dx \qquad (4.3.71)$$

$$+ \int_0^s \int_{-\infty}^{\infty} (Q_{xtt}^2 + Q_{xxt}^2)\, dx\, d\tau \le C \sum_{j=0}^{8} K_j$$

with K_8 as given by (4.3.16i). In view of (4.3.66) and (4.3.71), for η sufficiently small, $\exists \bar{C} > 0$ such that

$$\int_0^s \int_{-\infty}^{\infty} Q_{xxx}^2 \, dx\, d\tau \le \bar{C} \sum_{j=0}^{8} K_j \qquad (4.3.72)$$

while by (4.3.67), (4.3.68), and (4.3.71), for η sufficiently small, $\exists C^{\#} > 0$ such that

$$\int_{-\infty}^{\infty} Q_{ttt}^2(x,s)\, dx + \int_0^s \int_{-\infty}^{\infty} Q_{ttt}^2 \, dx\, d\tau \le C^{\#} \sum_{j=0}^{8} K_j \qquad (4.3.73)$$

The estimate required in order to establish Lemma 4.6, i.e., (4.3.64), is now a direct consequence of (4.3.71)-(4.3.73). □

We are now in a position to state and prove the main result of this section, namely,

Theorem 4.3 [15] If hypotheses (i)-(iii), i.e., (4.3.3a)-(4.3.3c), apply and the initial data Q_0, Q_1 are periodic and satisfy the smoothness hypotheses (4.3.12), with $\|Q_0\|_{H^3}^2 + \|Q_1\|_{H^2}^2$ sufficiently small, then there exists a unique solution $Q \in C^2(R^1 \times [0,\infty))$ of (4.3.1), (4.3.12) such that, for some $C > 0$,

$$\mathcal{E}_0(t) \le \mathcal{E}(t) \le C\left(\|Q_0\|_{H^3}^2 + \|Q_1\|_{H^2}^2\right) \qquad (4.3.74)$$

holds for all $t \in [0,\infty)$, where

$$\mathcal{E}_0(t) = \int_{-\infty}^{\infty} \{Q^2 + Q_t^2 + Q_x^2 + Q_{tt}^2 + Q_{xt}^2 \qquad (4.3.75a)$$

$$+ Q_{xx}^2 + Q_{ttt}^2 + Q_{xxt}^2 + Q_{xtt}^2 + Q_{xxx}^2\}(x,t)\, dx$$

and

$$\mathcal{E}(t) = \mathcal{E}_0(t) + \int_0^t \int_{-\infty}^{\infty} \{Q^2 + Q_t^2 + Q_x^2 + Q_{tt}^2 + Q_{xt}^2 \qquad (4.3.75b)$$

$$+ Q_{xx}^2 + Q_{ttt}^2 + Q_{xxt}^2 + Q_{xtt}^2 + Q_{xxx}^2\}(x,\tau)\, dx\, d\tau$$

In particular, it follows that

$$Q,\ Q_t,\ Q_x,\ Q_{tt},\ Q_{xt},\ Q_{xx},\ Q_{ttt},\ Q_{xxt},\ Q_{xtt} \text{ and } Q_{xxx} \in L^\infty([0,\infty);L^2(R^1)) \tag{4.3.76a}$$

and

$$Q,\ Q_t,\ Q_x,\ Q_{tt},\ Q_{xt},\ Q_{xx},\ Q_{ttt},\ Q_{xxt},\ Q_{xtt} \text{ and } Q_{xxx} \in L^2([0,\infty);L^2(R^1)) \tag{4.3.76b}$$

Remarks: The conclusion presented in (4.3.76b), i.e., that $Q \in L^2([0,\infty);L^2(R^1))$ has no counterpart in the analysis presented in § 3.5 for the wave-dielectric interaction problem.

Proof: (Theorem 4.3) As a consequence of Lemmas 4.4, 4.5, and 4.6, and the definition (4.3.75a), (4.3.75b) of $\mathcal{E}(t)$ we see, at once, that for η sufficiently small, $\exists \bar{\mathcal{C}} > 0$ such that

$$\mathcal{E}(s) \le \bar{\mathcal{C}} \sum_{j=0}^{8} K_j \tag{4.3.77}$$

From the structure of the K_j, (4.3.16a)-(4.3.16i), it follows that $\exists \mathcal{C}' > 0$ such that

$$\sum_{j=0}^{8} K_j \le \mathcal{C}'\mathcal{E}_0(0) \tag{4.3.78}$$

with $\mathcal{E}_0(0)$ as determined by (4.3.75a). Combining (4.3.77) and (4.3.78) we deduce, therefore, the existence of $\tilde{\mathcal{C}} > 0$ such that

$$\mathcal{E}(s) \le \tilde{\mathcal{C}}\mathcal{E}_0(0) \tag{4.3.79}$$

for η chosen sufficiently small. Combining (4.3.79) with (4.3.17), we see that $\exists \mathcal{C} > 0$ such that

$$\mathcal{E}(s) \le \mathcal{C}\left(\|Q_0\|^2_{H^3} + \|Q_1\|^2_{H^2}\right) \tag{4.3.80}$$

However, for the initial data taken so small that

$$\|Q_0\|^2_{H^3} + \|Q_1\|^2_{H^2} \le \eta^2/2^k, \qquad k \ge 2 \tag{4.3.81}$$

it is clear that we may choose k so large that

$$\mathcal{E}(s) \le \eta/2 \tag{4.3.82}$$

(in fact, by virtue of (4.3.80), we need only choose $k \geq \ln(4C)/\ln 2$). From (4.3.82), the definition of $\mathcal{E}(s)$, and the Sobolev embedding theorem, we deduce that (4.3.20) holds on some interval $[s, s_1)$, $s_1 > s$. Continuing in this fashion, we generate a sequence $\{s_k\}$, $s_{k+1} > s_k$, $s_k \to \infty$, as $k \to \infty$, such that (4.3.80) holds with C independent of s_k; that the $\{s_k\}$ cannot have an accumulation point at some finite value $t^* < \infty$ follows from an elementary analysis entirely similar to the one presented, e.g., in Slemrod [162]. Therefore,

$$\lim_{t \to T_\infty^-} \mathcal{E}_0(t) < \infty \tag{4.3.83}$$

in which case the locally defined solution of (4.3.1), (4.3.12) may be continued uniquely for all $t > 0$ and (4.3.74) holds on $[0, \infty)$. □

Corollary 4.2 [15] Under the same conditions as those which apply in Theorem 4.3, the uniquely defined, global C^2 solution of (4.3.1), (4.3.12) has the following asymptotic behavior as $t \to \infty$:

(i) Q, Q_t, Q_x, Q_{tt}, Q_{xt}, $Q_{xx} \to 0$, uniformly on R^1, as $t \to +\infty$ (4.3.84a)

(ii) $Q(\cdot,t)$, $Q_t(\cdot,t)$, $Q_x(\cdot,t)$, $Q_{tt}(\cdot,t)$, $Q_{xt}(\cdot,t)$, and $Q_{xx}(\cdot,t) \to 0$, in $L^2(R^1)$, as $t \to +\infty$. (4.3.84b)

Chapter 5

DISTRIBUTED PARAMETER NONLINEAR TRANSMISSION LINES II: Existence of Weak Solutions

5.0 Introduction

In the last chapter we introduced the concept of the distributed parameter nonlinear transmission line, allowing in the line for a voltage-dependent capacitance as well as a leakage conductance per unit length of the line; we established, for the Cauchy problem associated with the basic inhomogeneous, quasilinear, hyperbolic system which governs the evolution of the current and charge distribution in the line, i.e., the system (4.1.8), that, for initial data which is sufficiently small, smooth C^1 (in x, t) solutions will exist for all $t > 0$, while for initial data which is sufficiently large, in a sense which is made precise in § 4.1, (smooth) C^1 solutions must break down in finite time. Once smooth solutions fail to exist, however, for arbitrarily large initial data, there remains the important question concerning the existence of weak solutions, for the Cauchy problem associated with (4.1.8), which are globally defined for all $t > 0$. In this chapter we will address this issue of the global, in time, existence of weak solutions for the Cauchy problem governing the evolution of the current and charge distribution in a nonlinear, distributed parameter transmission line. Because of the similarity of our basic system (4.1.8), in that special case when the resistance $R = 0$, to the system governing the wave-dielectric interaction problem in one-space dimension, and because our results in this chapter do not depend, in any intrinsic manner, on R being different from zero, the existence result which we obtain in § 5.7 will also apply, with some obvious minor changes, to the wave-dielectric interaction problem.

We begin our work in this chapter by discussing, in § 5.1, the precise problems which underlie proving the existence of global, in time, weak solutions for the Cauchy problem associated with the Burgers' equation; in the course of doing this we present Hopf's original solution of this problem [69] and introduce the idea of obtaining such weak solutions as the limit of a suitable sequence of solutions of an associated (parabolically) regularized initial-value problem. We also discuss, in § 5.1, some of the pitfalls that one can run into in the realm of weak convergence of functions of weakly convergent sequences. To deal, in a precise manner, with the central issue of weak convergence, the concept of the Young measure is introduced in § 5.2 in two different, albeit, related ways; the importance of being able to prove, in practical situations, that the Young measure which is generated by a sequence of solutions to a parabolically regularized hyperbolic equation (or system) is, in fact, a Dirac measure, is then illustrated by using the Burgers' equation as an example once again. In § 5.3 we bring in the technique of compensated compactness, as originated by Murat [120], [121], [119], [122] and Tartar [171], [169], [168], [170], and use it, in conjunction with the concept of the Young measure, to establish the existence of global weak solutions for the Cauchy problem for the Burgers' equation. Then, in § 5.4, we present a summary of DiPerna's work [48], [47] which is directed at employing the ideas of Young measure and compensated compactness to prove the global (in time) existence of weak solutions for Cauchy problems associated with systems of homogeneous, strictly hyperbolic, genuinely nonlinear conservation laws; we also indicate under what circumstances the requirement, e.g., of genuine nonlinearity of each of the characteristic fields can be relaxed.

In § 5.5 we return to the nonlinear, distributed parameter, transmission line problem to explain why a straightforward parabolic regularization of each of the equations in the system (4.1.8) will not, in general, suffice for the purpose of proving the existence of global weak solutions; the parabolic regularization of (4.1.8) which turns out to be appropriate for our purposes is then introduced in § 5.6 through the simple device of perturbing the original constitutive relation between the voltage and the charge distribution. For the regularized (parabolic) problem proposed in § 5.6 we then have two theorems, one of which yields, for the regularization parameter chosen sufficiently small, a unique (and sufficiently regular) solution of the associated

Cauchy problem, while the second (theorem) establishes the existence of those L^∞ bounds, independent of the regularization parameter, that suffice to produce convergent subsequences of solutions of the regularized system for the transmission line; from such subsequences of solutions we construct the corresponding Young measure. Finally, in § 5.7, we prove that the Young measure constructed from a convergent subsequence of solutions of the regularized transmission line equations is, in fact, a Dirac mass; we are then able to employ the technique of compensated compactness and DiPerna's results, as presented in § 5.4, to establish the global, in time, existence of weak solutions for the Cauchy problem associated with the nonlinear transmission line equations.

5.1 Weak Solutions of the Burgers' Equation I

In § 2.2 we considered the problem of shock formation in solutions of a single hyperbolic equation in one space dimension of the form

$$u_t(x,t) + f(u(x,t))_x = 0, \qquad -\infty < x < \infty,\ t > 0, \tag{5.1.1}$$

using the Burgers' equation (2.2.1), in which $f(u) = \frac{u^2}{2}$, to show that C^1 solutions must, in general, break down in finite time even for initial data which are smooth and compactly supported. Then, in § 2.3 a definition, i.e., (2.3.16) was given of what we mean by a weak solution of (5.1.1) with associated initial data $u(x,0) = u_0(x)$, $-\infty < x < \infty$, $u_0(\cdot)$ a bounded measurable function and, indeed, the definition given in § 2.3 extends beyond (5.1.1) to systems of equations for which $\boldsymbol{u} \in R^n$ and $\boldsymbol{f} : R^n \to R^n$, $n \geq 1$. The finite time breakdown result for solutions of the initial-value problem associated with (5.1.1) was extended, in § 2.5, to 2×2 hyperbolic systems of the form (2.4.1), i.e.,

$$\begin{cases} u_t + f(u,v)_x = 0 \\ v_t + g(u,v)_x = 0 \end{cases} \tag{5.1.2}$$

by working with the Riemann invariants associated with the system and employing the analysis of Lax [99]. For the single equation (5.1.1), as well as the system (5.1.2), we require that the genuine nonlinearity condition hold.

Once it has been established, e.g., that classical solutions of (5.1.1) break down in finite time, i.e., develop shock discontinuities (and we define according to, say, (2.3.16), the notion of a weak solution for which jump discontinuities, consistent with the Rankine Hugoniot condition, occur across smooth curves $x = X(t)$ in the (x,t) plane) it is natural to inquire as to whether one can prove that such weak solutions persist globally in time. For the Burgers' equation, i.e., (5.1.1) with $f(u) = u^2/2$, it would seem that a rigorous proof of the existence of weak solutions for the initial-value problem appears, for the first time, in the paper of E. Hopf [69]. Hopf considers the initial-value problem for the heat equation

$$\begin{cases} \theta_t = \mu\theta_{xx}, \ -\infty < x < \infty, \ t > 0, \ \mu > 0 \\ \theta(x,0) = \theta_0(x), \ -\infty < x < \infty \end{cases} \tag{5.1.3}$$

which has the well-known solution (e.g. [59])

$$\theta(x,t) = \frac{1}{\sqrt{4\pi\mu t}} \int_{-\infty}^{\infty} \exp\left[\frac{-(x-y)^2}{4\mu t}\right] \theta_0(y)\, dy \tag{5.1.4}$$

Setting $u(x,t) = -2\mu\left(\dfrac{\theta_x(x,t)}{\theta(x,t)}\right)$, Hopf shows that $u(x,t)$ satisfies

$$u_t + uu_x = \mu u_{xx}, \quad -\infty < x < \infty, \ t > 0 \tag{5.1.5}$$

$$u(x,0) = -2\mu\left(\frac{\theta_0'(x)}{\theta_0(x)}\right), \quad -\infty < x < \infty \tag{5.1.6}$$

i.e., the Burgers' equation with a viscous damping term and an initial condition that is tied, of course, to the transformation cited above, which is now known as the Hopf-Cole transform. Using (5.1.4), and the Cole-Hopf transform, we find that the solution of the initial-value problem (5.1.5), (5.1.6) is given by the expression

$$u(x,t) = \frac{\displaystyle\int_{-\infty}^{\infty} \exp\left[\frac{-(x-y)^2}{4\mu t}\right]\left(\frac{x-y}{t}\right)\exp\left(-\frac{1}{2\mu}\int_{-\infty}^{y} u_0(\zeta)\, d\zeta\right) dy}{\displaystyle\int_{-\infty}^{\infty} \exp\left[\frac{-(x-y)^2}{4\mu t}\right]\exp\left(-\frac{1}{2\mu}\int_{-\infty}^{y} u_0(\zeta)\, d\zeta\right) dy} \tag{5.1.7}$$

Through a very careful analysis, Hopf [69] shows that one can pass to the limit in (5.1.7), as $\mu \to 0^+$, and obtain from (5.1.7) a function $\bar{u}(x,t)$ which is a weak solution, as per our definition (2.3.16), of the initial-value problem

$$\begin{cases} u_t + uu_x = 0, \quad -\infty < x < \infty, \ t > 0 \\ u(x,0) = u_0(x), \quad -\infty < x < \infty \end{cases} \tag{5.1.8}$$

Unfortunately, the technique used by Hopf [69] for the initial-value problem (5.1.8) can, in no way, be carried over to the study of global existence of weak solutions for the initial-value problems associated with 2×2 systems of the form (5.1.2). However, there is still the hope that by perturbing (5.1.2) into a parabolic system, with the addition of suitable viscous terms governed by a parameter (or parameters) akin to μ in (5.1.5), one might obtain the existence of a unique global weak solution for the initial-value problem associated with (5.1.2) by studying the limit of the solution of the regularized system as the parameter (or parameters) approach zero; that such a goal can be achieved for systems such as (5.1.2)—and, in fact, even for the system (4.1.8) governing the nonlinear distributed parameter transmission line—is due in large measure to the pioneering work of Murat [120], [121], [119], [122], Tartar [171], [169], [168], [170], and DiPerna [48], [47], [49] on Young measures and the compensated compactness technique and benefited greatly from earlier work of Morrey [117] and Lax [102]. A very nice compact version of the relation between the concept of the Young measure and weak convergence may be found in the recent work of J. Ball [7], while surveys of the problems of weak convergence for nonlinear partial differential equations, in general, are available in the monographs by Evans [54] and Dacorogna [39], both of which contain more extensive sets of recent references than it is necessary for us to offer here. An excellent up-to-date survey of the problems involved with studying the limit behavior of solutions to regularized hyperbolic systems may be found in Chen [32]; in particular, Chen [32] pays special attention to recent results for hyperbolic systems which are not genuinely nonlinear, i.e., possess one or more linearly degenerate characteristic fields.

In order to better understand what one is up against in trying to take the limit of the solutions of the (viscous) perturbed equation (5.1.5), without the aid of the explicit representation (5.1.7), consider the following argument: it is well known that, as a consequence of the maximum principle (e.g., John [59]), for any $\mu > 0$ the unique classical solution of the initial-value problem for (5.1.5), with initial function $u_0(\cdot) \in L^\infty(R^1)$, satisfies

$$\|u^\mu\|_{L^\infty(R^1 \times [0,\infty))} \leq \mathcal{C} \tag{5.1.9}$$

where $\mathcal{C}$ is independent of μ (and where we are now denoting the solutions of (5.1.5) by a superscript μ). As a consequence of (5.1.9) it follows that there exists a subsequence of $\{u^\mu\}$, which we still denote as $\{u^\mu\}$, such that as $\mu \to 0^+$,

$$u^\mu \overset{*}{\rightharpoonup} \bar{u} \text{ in } L^\infty(R^1 \times [0,\infty)) \tag{5.1.10}$$

where $\overset{*}{\rightharpoonup}$ denotes weak $*$ convergence in L^∞, i.e., weak convergence in the dual space $[L^\infty(R^1 \times [0,\infty))]^* = L^1(R^1 \times [0,\infty))$. Now, let $\phi \in C_0^\infty(R^1 \times [0,\infty))$ be any test function. We multiply (5.1.5) by ϕ, then integrate over $R^1 \times [0,\infty)$, and integrate by parts in both the spatial and temporal variables so as to obtain

$$\begin{aligned} -\int_0^\infty \int_{-\infty}^\infty \phi_t u^\mu \, dx\, dt \quad - \quad \frac{1}{2}\int_0^\infty \int_{-\infty}^\infty \phi_x (u^\mu)^2 \, dx\, dt \\ = \quad \mu \int_0^\infty \int_{-\infty}^\infty \phi_{xx} u^\mu \, dx\, dt \end{aligned} \tag{5.1.11}$$

Because of (5.1.9), and the continuity of $f(u) = \frac{1}{2}u^2$, it follows (possibly for some refined subsequence $\{u^\mu\}$) that as $\mu \to 0^+$

$$(u^\mu)^2 \overset{*}{\rightharpoonup} \bar{w} \quad \text{in } L^\infty(R^1 \times [0,\infty)) \tag{5.1.12}$$

Going to the limit in (5.1.11) as $\mu \to 0^+$, we obtain, therefore, as a consequence of (5.1.9), (5.1.10), (5.1.12) and the definition of weak $*$ convergence

$$-\int_0^\infty \int_{-\infty}^\infty \phi_t \bar{u} \, dx\, dt - \frac{1}{2}\int_0^\infty \int_{-\infty}^\infty \phi_x \bar{w} \, dx\, dt = 0 \tag{5.1.13}$$

Suppose, now, that we could somehow show that

$$\text{weak}^* \lim (u^\mu)^2 = (\text{weak}^* \lim u^\mu)^2 \tag{5.1.14}$$

i.e., that $\bar{w} = \bar{u}^2$ (a.e., of course, in x and t); then from (5.1.13) we would have

$$\int_0^\infty \int_{-\infty}^\infty \phi_t \bar{u} \, dx\, dt + \frac{1}{2}\int_0^\infty \int_{-\infty}^\infty \phi_x \bar{u}^2 \, dx\, dt = 0 \tag{5.1.15}$$

$\forall \phi \in C_0^\infty(R^1 \times (0,\infty))$ and, by definition, it would follow that

$$\bar{u}_t + \bar{u}\bar{u}_x = 0, \qquad -\infty < x < \infty, \quad t > 0 \tag{5.1.16}$$

in the sense of distributions. The problem, of course, is with verifying (5.1.14), or, in general, that for a nonlinear function f (which is, e.g., continuous), and a weak $*$ convergent sequence $\{v^\varepsilon\}$,

$$\text{weak}^* \lim f(v^\varepsilon) = f(\text{weak}^* \lim v^\varepsilon) \tag{5.1.17}$$

In order to see that one can usually expect problems with verifying a condition such as (5.1.17), we may cite the following simple example which is given in Slemrod [164]: Let $f(x)$ be periodic and in $L^2[0,2\pi]$; for ease of exposition we may even assume that f is continuous on $[0,2\pi]$. Then f is determined, uniquely, by its Fourier representation

$$f(x) = \frac{a_0}{2} + \sum_{j=1}^{\infty}(a_j \cos jx + b_j \sin jx) \tag{5.1.18}$$

where the a_j, $j = 0, 1, \ldots$ and the b_j, $j = 1, 2, \ldots$ are the classical Fourier coefficients of f (see, e.g., Weinberger [181]). Replacing $x \to nx$, n an integer, we obtain from (5.1.18) the representation

$$f_n(x) = \frac{a_0}{2} + \sum_{j=1}^{\infty}(a_j \cos jnx + b_j \sin jnx) \tag{5.1.19}$$

where $f_n(x) = f(nx)$. For any $\phi \in L^2[0,2\pi]$, therefore,

$$\int_0^{2\pi} \phi(x) f_n(x)\, dx = \frac{a_0}{2}\int_0^{2\pi} \phi(x) f_n(x)\, dx + \sum_{j=1}^{\infty}(a_j c_{jn} + b_j d_{jn}) \tag{5.1.20}$$

where

$$\begin{cases} c_{jn} = \displaystyle\int_0^{2\pi} \phi(x) \cos jnx\, dx \\ d_{jn} = \displaystyle\int_0^{2\pi} \phi(x) \sin jnx\, dx \end{cases} \tag{5.1.21}$$

Using the Cauchy-Schwarz inequality we have

$$\left|\sum_{j=1}^{\infty} a_j c_{jn}\right| \leq \left[\sum_{j=1}^{\infty} a_j^2\right]^{1/2} \left[\sum_{j=1}^{\infty} c_{jn}^2\right]^{1/2} \tag{5.1.22}$$

However, $\sum_{j=1}^{\infty} c_{jn}^2 < \sum_{k=n}^{\infty} c_k^2 \to 0$, as $n \to \infty$, and, therefore,

$$\begin{cases} \displaystyle\sum_{j=1}^{\infty} a_j c_{jn} \to 0, \text{ as } n \to \infty \\ \displaystyle\sum_{j=1}^{\infty} b_j c_{jn} \to 0, \text{ as } n \to \infty \end{cases} \tag{5.1.23}$$

Combining (5.1.23) with (5.1.20), we easily find that

$$\lim_{n\to\infty} \int_0^{2\pi} \phi(x) f_n(x)\, dx = \int_0^{2\pi} \phi(x) \left(\frac{a_0}{2}\right)\, dx, \tag{5.1.24}$$

i.e., that as $n \to \infty$

$$f_n(\cdot) \rightharpoonup \bar{f}(\cdot) = \frac{1}{2}a_0 = \frac{1}{2\pi}\int_0^{2\pi} f(x)\, dx \tag{5.1.25}$$

with $\rightharpoonup$ denoting weak convergence in $L^2[0, 2\pi]$. We are now in a position to show, with a very simple example, what can go wrong with (5.1.17), indeed, for the case of interest with respect to the Burgers' equation, i.e., $f(x) = x^2$. Consider $f(x) = \sin x$ so that $f_n(x) = \sin nx$. By virtue of (5.1.25) we have

$$f_n(\cdot) \rightharpoonup \frac{1}{2\pi} \int_0^{2\pi} \sin x \, dx = 0, \text{ as } n \to \infty \tag{5.1.26}$$

However, for $g(x) = f^2(x) = \sin^2 x$, $g_n(x) = g(nx)$ satisfies (again, by (5.1.25))

$$g_n(\cdot) \rightharpoonup \frac{1}{2\pi} \int_0^{2\pi} \sin^2 x \, dx = \frac{1}{2}, \text{ as } n \to \infty \tag{5.1.27}$$

and, clearly,

$$\underset{n\to\infty}{\text{weak} \lim} [f_n(\cdot)]^2 \neq [\underset{n\to\infty}{\text{weak} \lim} f_n(\cdot)]^2 \tag{5.1.28}$$

In order to overcome the difficulty involved with passing from (5.1.13) to (5.1.15), in the case of the Burgers' equation, we will proceed in the next section with showing how the concept of the Young measure can be used to characterize weak limits of sequences (and functions of sequences); the concept of compensated compactness will be introduced in § 5.3 when we, once again, return to the problem of extracting the existence of a global weak solution to the initial-value problem for the Burgers' equation by passing to the limit in (5.1.15) as $\mu \to 0^+$.

5.2 The Concept of the Young Measure

There are several (essentially equivalent) ways of developing the concept of the Young measure and showing how it can be used to characterize weak convergence; we will present two such approaches here, the first of which follows the presentation in Slemrod [164]. Let $\Omega \subseteq R^m$ and $K \subseteq R^n$ be bounded open sets and let $\boldsymbol{u}_n : \Omega \to K$ be a measurable sequence; in practice the sequence $\boldsymbol{u}_n$ would be generated by some parabolic regularization of a hyperbolic system of equations. We introduce a sequence $\{m_n\}$ of Radon measures (see, e.g., Royden [144]) on $C_0(\Omega \times R^n)$ as follows: for any $\phi \in C_0(\Omega \times R^n)$

$$\langle m_n, \phi(\boldsymbol{x}, \boldsymbol{\lambda}) \rangle = \int_\Omega \phi(\boldsymbol{x}, \boldsymbol{u}_n(\boldsymbol{x})) \, d\boldsymbol{x} \tag{5.2.1}$$

As $\boldsymbol{u}_k(\boldsymbol{x}) \subseteq K$, $\forall\, k$, the norms $\|\boldsymbol{u}_k\|_{L^\infty(R^n)}$ are uniformly bounded in k; therefore, the sequence $\{m_n\}$ has a weakly convergent subsequence, also denoted by $\{m_n\}$, such that

$$\langle m_n, \phi \rangle \longrightarrow \langle m, \phi \rangle, \qquad \forall \phi \in C_0(\Omega \times R) \tag{5.2.2}$$

and this serves to define the measure m, which is clearly a linear functional on $C_0(\Omega \times R^n)$. We now seek to deduce a representation for m. Consider the set of ϕ's in $C_0(\Omega \times R^n)$ of the form

$$\phi(\boldsymbol{x}, \boldsymbol{\lambda}) = f(\boldsymbol{\lambda})\Phi(\boldsymbol{x}), \qquad \boldsymbol{x} \in \Omega, \ \boldsymbol{\lambda} \in R^n \tag{5.2.3}$$

with f continuous on R^n and Φ continuous on Ω. For any fixed f,

$$\langle m, f(\boldsymbol{\lambda})\Phi(\boldsymbol{x}) \rangle = \lim_{n \to \infty} \int_\Omega \Phi(\boldsymbol{x}) f(\boldsymbol{u}_n(\boldsymbol{x}))\, d\boldsymbol{x} \tag{5.2.4}$$

is a bounded linear functional on $C_0(\Omega)$. By virtue of the Riesz representation theorem, there exists a measure $\omega(d\boldsymbol{x})$ such that

$$\langle m, f(\boldsymbol{\lambda})\Phi(\boldsymbol{x}) \rangle = \int_\Omega \Phi(\boldsymbol{x})\omega(d\boldsymbol{x}) \tag{5.2.5}$$

We claim that the measure ω is absolutely continuous with respect to Lebesgue measure on Ω. In order to see this we let Φ_0 be the characteristic function of any set of Lebesgue measure zero in Ω; then

$$\langle m_n, f(\boldsymbol{\lambda})\Phi_0(\boldsymbol{x}) \rangle = \int_\Omega \Phi_0(\boldsymbol{x}) f(\boldsymbol{u}_n(\boldsymbol{x}))\, d\boldsymbol{x} = 0 \tag{5.2.6}$$

However, as $n \to \infty$,

$$\langle m_n, f(\boldsymbol{\lambda})\Phi_0(\boldsymbol{x}) \rangle \to \langle m, f(\boldsymbol{\lambda})\Phi_0(\boldsymbol{x}) \rangle \tag{5.2.7}$$

Combining (5.2.6) and (5.2.7) with (5.2.5) we have

$$\int_\Omega \Phi(\boldsymbol{x})\omega(d\boldsymbol{x}) = 0 \tag{5.2.8}$$

thus establishing the absolute continuity of the measure $\omega(d\boldsymbol{x})$ with respect to Lebesgue measure $d\boldsymbol{x}$. Because $\omega(d\boldsymbol{x})$ is absolutely continuous with respect to $d\boldsymbol{x}$, the Radon-Nikodym theorem guarantees the existence of a density $\omega_f(\boldsymbol{x})$ such that

$$\omega(d\boldsymbol{x}) = \omega_f(\boldsymbol{x})\, d\boldsymbol{x} \tag{5.2.9}$$

As,

$$\begin{aligned}
&\langle m, (c_1 f + c_2 g)(\boldsymbol{\lambda})\Phi(\boldsymbol{x})\rangle \\
&\quad = c_1\langle m, f(\boldsymbol{\lambda})\Psi(\boldsymbol{x})\rangle + c_2\langle m, g(\boldsymbol{\lambda})\Psi(\boldsymbol{x})\rangle \\
&\quad = c_1 \int_\Omega \Psi(\boldsymbol{x})\omega_f(\boldsymbol{x})\, d\boldsymbol{x} + c_2 \int_\Omega \Psi(\boldsymbol{x})\omega_g(\boldsymbol{x})\, d\boldsymbol{x} \\
&\quad = \int_\Omega \Psi(\boldsymbol{x})(c_1\omega_f(\boldsymbol{x}) + c_2\omega_g(\boldsymbol{x}))\, d\boldsymbol{x} \\
&\quad = \int_\Omega \Psi(\boldsymbol{x})\omega_{c_1 f + c_2 g}(\boldsymbol{x})\, d\boldsymbol{x}
\end{aligned}$$

we have, $\forall\, \Psi \in \mathcal{C}_0(\Omega)$

$$\int_\Omega \Psi(\boldsymbol{x})[c_1\omega_f + c_2\omega_g - \omega_{c_1 f + c_2 g}](\boldsymbol{x})\, d\boldsymbol{x} = 0 \tag{5.2.10}$$

so that $c_1\omega_f + c_2\omega_g = \omega_{c_1 f + c_2 g}$, a.e., in Ω; in other words, the mapping $f \to \omega_f(\boldsymbol{x})$ is, a.e. on Ω, a linear functional, which is easily seen to be bounded. Using, once again, the Riesz representation theorem, we deduce the existence of a (representing) measure $\nu_{\boldsymbol{x}}(\boldsymbol{\lambda})$ with the property that

$$\omega_f(\boldsymbol{x}) = \int_{R^n} f(\boldsymbol{\lambda})\, d\nu_{\boldsymbol{x}}(\boldsymbol{\lambda}) \tag{5.2.11}$$

Combining (5.2.5), (5.2.9), and (5.2.11), we have

$$\langle m, f(\boldsymbol{\lambda})\Phi(\boldsymbol{x})\rangle = \lim_{n\to\infty} \int_\Omega \Phi(\boldsymbol{x}) f(\boldsymbol{u}_n(\boldsymbol{x}))\, d\boldsymbol{x} \tag{5.2.12}$$

$$= \int_\Omega \Phi(\boldsymbol{x}) \int_{R^n} f(\boldsymbol{\lambda})\, d\nu_{\boldsymbol{x}}(\boldsymbol{\lambda})\, d\boldsymbol{x} \tag{5.2.13}$$

Having considered the action of m on $\phi \in \mathcal{C}_0(\Omega \times R^n)$ of the form (5.2.3), consider now ϕ's of the form

$$\phi(\boldsymbol{x}, \boldsymbol{\lambda}) = \sum_{i=1}^{n} c_i f_i(\boldsymbol{\lambda})\Psi_i(\boldsymbol{x}) \tag{5.2.14}$$

Then,

$$\begin{aligned}
\langle m, \phi(\boldsymbol{x}, \boldsymbol{\lambda})\rangle &= \sum_{i=1}^{n} c_i \langle m, f_i(\boldsymbol{\lambda})\Phi_i(\boldsymbol{x})\rangle \\
&= \sum_{i=1}^{n} c_i \int_\Omega \Phi_i(\boldsymbol{x}) \int_{R^n} f_i(\boldsymbol{\lambda})\, d\nu_{\boldsymbol{x}}(\boldsymbol{\lambda})\, d\boldsymbol{x}
\end{aligned}$$

or

$$\langle m, \phi(\boldsymbol{x}, \boldsymbol{\lambda})\rangle = \int_\Omega \int_{R^n} \phi(\boldsymbol{x}, \boldsymbol{\lambda})\, d\nu_{\boldsymbol{x}}(\boldsymbol{\lambda})\, d\boldsymbol{x} \tag{5.2.15}$$

Furthermore, as any continuous $\phi \in \mathcal{C}_0(\Omega \times R^n)$ may be approximated by finite sums of the form (5.2.14), by continuity (5.2.15) extends to all $\phi \in \mathcal{C}_0(\Omega \times R^n)$ and we have the fundamental representation

$$m = \int_\Omega \nu_{\boldsymbol{x}}(\boldsymbol{\lambda})\, d\boldsymbol{x} \tag{5.2.16}$$

for the measure m in terms of the representing measure $\nu_{\boldsymbol{x}}$—the Young measure generated by the sequence $\boldsymbol{u}_n$.

Using the fundamental representation formula (5.2.16), we now want to characterize weak limits; once again we let $f : K \subseteq R^n \to R^1$ be continuous and $\boldsymbol{u}_n : \Omega \subseteq R^m \to K$ be a measurable sequence so that the norms $\|\boldsymbol{u}_k\|_{L^\infty(R^n)}$ are uniformly bounded independent of k. Thus, there exists a subsequence of $\{\boldsymbol{u}_k\}$, which we also denote as $\{\boldsymbol{u}_k\}$, such that

$$\boldsymbol{u}_k \rightharpoonup \bar{\boldsymbol{u}}, \text{ weak* in } L^\infty, \text{ as } k \to \infty \tag{5.2.17}$$

Because f is continuous on K, we also have, for some c independent of k,

$$\|f(\boldsymbol{u}_k(\cdot))\|_{L^\infty(R^n)} \leq c \tag{5.2.18}$$

and, therefore, we may choose, if necessary, yet another subsequence of the subsequence $\{\boldsymbol{u}_k\}$, again denoted by $\{\boldsymbol{u}_k\}$, such that

$$f(\boldsymbol{u}_k) \rightharpoonup \bar{f}, \text{ weak* in } L^\infty, \text{ as } k \to \infty \tag{5.2.19}$$

Now, let $\Phi \in \mathcal{C}_0(\Omega)$; then we know that the Radon measures m_n, defined by (5.2.1) for any $\phi \in \mathcal{C}_0(\Omega \times R^n)$, satisfy

$$\begin{aligned} \langle m_k, \Phi f \rangle &= \int_\Omega \Phi(\boldsymbol{x}) f(\boldsymbol{u}_k(\boldsymbol{x}))\, d\boldsymbol{x} \\ &\quad \to \langle m, \Phi f \rangle \\ &= \int_\Omega \int_{R^n} \Phi(\boldsymbol{x}) f(\boldsymbol{\lambda})\, d\nu_{\boldsymbol{x}}(\boldsymbol{\lambda})\, d\boldsymbol{x}, \text{ as } k \to \infty \end{aligned} \tag{5.2.20}$$

But, as (5.2.19) holds, we also have

$$\langle m_k, \Phi f \rangle = \int_\Omega \Phi(\boldsymbol{x}) \bar{f}(\boldsymbol{x})\, d\boldsymbol{x}, \text{ as } k \to \infty \tag{5.2.21}$$

Combining (5.2.20) and (5.2.21), we see that

$$\int_\Omega \Phi(\boldsymbol{x}) \bar{f}(\boldsymbol{x})\, d\boldsymbol{x} = \int_\Omega \int_{R^n} \Phi(\boldsymbol{x}) f(\boldsymbol{\lambda})\, d\nu_{\boldsymbol{x}}(\boldsymbol{\lambda})\, d\boldsymbol{x} \tag{5.2.22}$$

Therefore, as Φ is any element in $C_0(\Omega)$,

$$\bar{f}(\boldsymbol{x}) = \langle \nu_{\boldsymbol{x}}, f \rangle = \int_{R^n} f(\boldsymbol{\lambda})\, d\nu_{\boldsymbol{x}}(\boldsymbol{\lambda})\, d\boldsymbol{x}, \text{ a.e. in } \Omega \tag{5.2.23}$$

and (5.2.23) is the fundamental representation formula for weak limits in terms of the (representing) Young measure $\nu_{\boldsymbol{x}}(\boldsymbol{\lambda})$. We note that as $\bar{f} \equiv 1$, when $f \equiv 1$, it follows from (5.2.23) that

$$\int_{R^n} d\nu_{\boldsymbol{x}}(\boldsymbol{\lambda}) = 1$$

so that, inasmuch as $\nu_{\boldsymbol{x}}$ is a positive measure, it is, in fact, a probability measure.

Now, suppose that one could show that $\nu_{\boldsymbol{x}}(\boldsymbol{\lambda}) = \delta(\boldsymbol{\lambda} - \bar{\boldsymbol{u}}(\boldsymbol{x}))$, i.e., that the (representing) Young measure was the Dirac mass concentrated at the weak* limit $\bar{u}$ of the sequence $\{\boldsymbol{u}_n\}$; it would then follow, directly from (5.2.23), that

$$\bar{f}(\boldsymbol{x}) = \langle \delta(\boldsymbol{\lambda} - \bar{\boldsymbol{u}}(\boldsymbol{x})), f(\boldsymbol{\lambda}) \rangle = f(\bar{\boldsymbol{u}}(\boldsymbol{x})) \tag{5.2.24}$$

so that

$$\text{weak}^* \lim f(\boldsymbol{u}_k(\cdot)) = f(\text{weak}^* \lim \boldsymbol{u}_k(\cdot)) \tag{5.2.25}$$

It should now be clear, based on the analysis in § 5.1, that the key issue in the analysis of regularized conservation laws such as (5.1.5) is the ability to prove that the (representing) Young measure determined by a sequence $\{\boldsymbol{u}^\mu\}$ of solutions is the Dirac mass $\delta(\boldsymbol{\lambda} - \bar{\boldsymbol{u}}(\boldsymbol{x}))$, where $\bar{\boldsymbol{u}}(\cdot) = \text{weak}^* \lim\{\boldsymbol{u}^\mu(\cdot)\}$, as $\mu \to 0^+$; in the next section we will indicate, precisely, how this can be done for the parabolic regularization (5.1.5) of the Burgers' equation. Prior to proceeding, however, we present, below, an alternative formulation of the relationship between the Young measure and weak convergence which is closer to that used in recent work (e.g., Ball [7], Roytburd and Slemrod [141], Bellout, Bloom, and Nečas [12], [14]); for ease of exposition we will work in one-space dimension, i.e., $n = 2$. Thus, let $\{u^\mu\}$ be a sequence of functions on $R^1 \times [0,T]$ such that $\|u^\mu\|_{L^\infty(R^1\times[0,T])} \le c$, with c independent of $\mu > 0$, and extract a subsequence, also denoted by $\{u^\mu\}$, such that $u^\mu \rightharpoonup \bar{u}$, weak* in $L^\infty(R^1 \times [0,T])$, as $\mu \to 0^+$. We also assume that f is a continuous function so that $\|f(u^\mu)\|_{L^\infty(R^1\times[0,T])} \le c'$, with c' independent of μ, so that extracting, if necessary, yet another subsequence $\{u^\mu\}$ we have $f(u^\mu) \rightharpoonup \bar{f}$, weak* in $L^\infty(R^1 \times [0,T])$, as $\mu \to 0^+$; this last result can be expressed as

$$\int_0^T \int_{-\infty}^{\infty} \phi(x,t) f(u^\mu(x,t))\, dx\, dt \to \int_0^T \int_{-\infty}^{\infty} \phi(x,t) \bar{f}(x,t)\, dx\, dt \tag{5.2.26}$$

as $\mu \to 0^+$, $\forall\, \phi \in L^1(R^1 \times [0,t])$, or as

$$\int_0^T \int_{-\infty}^{\infty} \phi(x,t) \langle \delta_{u^\mu(x,t)}, f\rangle \, dx\, dt \tag{5.2.27}$$

$$\to \int_0^T \int_{-\infty}^{\infty} \phi(x,t) \bar{f}(x,t) \, dx\, dt$$

as $\mu \to 0^+$, $\forall\, \phi \in L^1(R^1 \times [0,t])$, where

$$f(u^\mu(x,t)) = \langle \delta_{u^\mu(x,t)}, f\rangle = \int_{-\infty}^{\infty} f(\lambda)\delta(\lambda - u^\mu(x,t))\, d\lambda \tag{5.2.28}$$

Now, consider the sequence $\langle \delta_{u^\mu(x,t)}, f\rangle$; clearly, by (5.2.28)

$$|\langle \delta_{u^\mu}, f\rangle| \le \bar{c} \|f\|_{L^\infty(R^1)} \tag{5.2.29}$$

with $\bar{c}$ independent of μ, i.e., as f is any continuous function, $\{\delta_{u^\mu}\}$ is a uniformly bounded sequence living in the dual space $C(K)$ of continuous functions defined on a compact subset $K \subseteq R^1$. Because $\{\delta_{u^\mu}\}$ is uniformly bounded in $\mathcal{C}(K)^*$, by the weak* compactness of the unit ball in $\mathcal{C}(K)^*$, there exists a subsequence of $\{\delta_{u^\mu}\}$, which we also denote as $\{\delta_{u^\mu}\}$, such that as $\mu \to 0^+$

$$\delta_{u^\mu(x,t)} \rightharpoonup \nu_{x,t} \begin{cases} \text{weak}^* \text{ in } \mathcal{C}(K)^* \\ \text{a.e. in } R^2 \end{cases} \tag{5.2.30}$$

or

$$\langle \delta_{u^\mu(x,t)}, f\rangle \to \langle \nu_{x,t}, f\rangle, \quad \text{as } \mu \to 0^+, \tag{5.2.31}$$

$\forall\, f \in \mathcal{C}(K)$, where supp $\nu_{x,t} \subset K$. Therefore,

$$\int_0^T \int_{-\infty}^{\infty} \phi(x,t) \langle \delta_{u^\mu}, f\rangle \, dx\, dt \;\to\; \int_0^T \int_{-\infty}^{\infty} \phi(x,t) \langle \nu_{x,t}, f\rangle \, dx\, dt \tag{5.2.32}$$

$$= \int_0^T \int_{-\infty}^{\infty} \phi(x,t) \bar{f}(x,t) \, dx\, dt$$

as $\mu \to 0^+$, by virtue of (5.2.27), and we again recover the representation (5.2.23), albeit for $\boldsymbol{x} \in R^2$, i.e.,

$$\bar{f}(x,t) = \langle \nu_{x,t}, f\rangle, \quad \text{a.e. on } R^1 \times [0,T] \tag{5.2.33}$$

The approach to the Young measure delineated above follows the presentation in Roytburd and Slemrod [141] and is similar, in spirit, to the somewhat deeper analysis presented by Ball [7].

5.3 Weak Solutions of the Burgers' Equation II

In this section, following the analysis in Tartar [169] and Slemrod [164], we demonstrate how we can pass to the limit, as $\mu \to 0^+$, in the regularized initial-value problem:

$$\begin{cases} u_t^\mu + u^\mu u_x^\mu = \mu u_{xx}^\mu, & x \in R^1,\, t > 0 \\ u^\mu(x,0) = u_0(x), & x \in R^1 \end{cases} \tag{5.3.1}$$

in order to obtain the weak solution of

$$\begin{cases} u_t + u u_x = 0, & x \in R^1,\, t > 0 \\ u(x,0) = u_0(x), & x \in R^1 \end{cases} \tag{5.3.2}$$

From our work in § 5.1 we know that, in essence, we must show that for $u^\mu \rightharpoonup \bar{u}$, weak* in $L^\infty(R^1 \times [0,T])$, $T > 0$, we also have that $\frac{1}{2}(u^\mu)^2 \rightharpoonup \frac{1}{2}\bar{u}^2$, weak* in $L^\infty(R^1 \times [0,T])$. Inasmuch as the sequence $\{u^\mu\}$ generates a Young measure $\nu_{x,t}$, and, with $f(\lambda) = \frac{1}{2}\lambda^2$, $f(u^\mu(x,t)) \to \langle \nu_{x,t}, f \rangle$, a.e. on R^2, as $\mu \to 0^+$, the actual task is to prove that for the associated Young measure we have $\nu_{x,t} = \delta(\lambda - \bar{u}(x,t))$.

Thus, let $\eta(u)$ be any C^2 convex function and define $q(u)$ to be any real-valued function with the property that

$$q'(u) = u\eta'(u) \tag{5.3.3}$$

A pair $(\eta(u), q(u))$ satisfying (5.3.3), with η a C^2 convex function, will be termed a *flux-entropy pair*. If we multiply the regularized (parabolic) equation for u^μ in (5.3.1) through by $\eta'(u^\mu)$, and use the obvious relation

$$(u\eta'(u))u_x = q(u)_x \tag{5.3.4}$$

we readily obtain

$$\begin{aligned} \eta(u^\mu)_t + q(u^\mu)_x &= \mu\eta'(u^\mu)u_{xx}^\mu \\ &= \mu\left[\eta(u^\mu)_{xx} - \eta''(u^\mu)(u_x^\mu)^2\right] \end{aligned} \tag{5.3.5}$$

Suppose, now, that for *any* entropy-flux pair $(\eta(u), q(u))$ we could show that

$$\eta(u^\mu)_t + q(u^\mu)_x \in \text{ a compact subset of } H_{\text{loc}}^{-1} \tag{5.3.6}$$

with u^μ the unique solution of the initial-value problem (5.3.1) for $\mu > 0$; then if (η_1, q_1), (η_2, q_2) were any two entropy-flux pairs we would have

$$\text{div}_{t,x}(\eta_1(u^\mu),\ q_1(u^\mu)) \in \text{ a compact subset of } H^{-1}_{\text{loc}} \tag{5.3.7}$$

and

$$\text{curl}_{t,x}(q_2(u^\mu), -\eta_2(u^\mu)) \in \text{ compact subset of } H^{-1}_{\text{loc}} \tag{5.3.8}$$

For the initial-value problem (5.3.1) we know that we may select a subsequence, also denoted by $\{u^\mu\}$, such that $u^\mu \rightharpoonup \bar{u}$, weak* in $L^\infty(R^1 \times [0,T])$ and as η_i, q_i, $i = 1, 2$, are continuous, further subsequences, if necessary, such that, as $\mu \to 0^+$,

$$\eta_i(u^\mu) \rightharpoonup \bar{\eta}_i,\ i = 1, 2, \text{ weak* in } L^\infty(R^1 \times [0,T]) \tag{5.3.9a}$$

$$q_i(u^\mu) \rightharpoonup \bar{q}_i,\ i = 1, 2, \text{ weak* in } L^\infty(R^1 \times [0,T]) \tag{5.3.9b}$$

Thus the sequences

$$\boldsymbol{v}_1^\mu = (\eta_1(u^\mu), q_1(u^\mu)) \tag{5.3.10a}$$

$$\boldsymbol{v}_2^\mu = (q_2(u^\mu), -\eta_2(u^\mu)) \tag{5.3.10b}$$

are weak* L^∞ convergent, as $\mu \to 0^+$, to, respectively,

$$\begin{cases} \bar{\boldsymbol{v}}_1 = (\bar{\eta}_1, \bar{q}_1) \\ \bar{\boldsymbol{v}}_2 = (\bar{q}_2, -\bar{\eta}_2) \end{cases} \tag{5.3.11}$$

and, moreover, satisfy (i.e., (5.3.7), (5.3.8))

$$\left.\begin{array}{l} \text{div}_{t,x}\, \boldsymbol{v}_1^\mu \\ \text{curl}_{t,x} \boldsymbol{v}_2^\mu \end{array}\right\} \in \text{ a compact subset of } H^{-1}_{\text{loc}} \tag{5.3.12}$$

The conditions delineated in (5.3.12) are precisely those required in order to apply the *Div-Curl Lemma* of Murat [120] (see also, e.g., Tartar [169], Dacorogna [39], or Antman [2]) so as to conclude that

$$\boldsymbol{v}_1^\mu \cdot \boldsymbol{v}_2^\mu \text{ is } \textit{weakly continuous}, \tag{5.3.13}$$

i.e., as $\mu \to 0^+$ we have, in the sense of distributions, $\boldsymbol{v}_1^\mu \cdot \boldsymbol{v}_2^\mu \to \bar{\boldsymbol{v}}_1 \cdot \bar{\boldsymbol{v}}_2$, or

$$\eta_1(u^\mu)q_2(u^\mu) - q_1(u^\mu)\eta_2(u^\mu) \to \bar{\eta}_1\bar{q}_2 - \bar{q}_1\bar{\eta}_2 \tag{5.3.14}$$

in the sense of distributions, as $\mu \to 0^+$. Because

$$\begin{cases} \bar{\eta}_i = \langle \eta_i(\lambda), \nu_{x,t} \rangle, & i = 1,2 \\ \bar{q}_i = \langle q_i(\lambda), \nu_{x,t} \rangle, & i = 1,2 \end{cases} \tag{5.3.15}$$

while, as $\mu \to 0^+$,

$$\begin{aligned} &\eta_1(u^\mu)q_2(u^\mu) - q_1(u^\mu)\eta_2(u^\mu) \\ &\qquad \rightharpoonup \langle \eta_1(\lambda)q_2(\lambda) - q_1(\lambda)\eta_2(\lambda), \nu_{x,t} \rangle \end{aligned} \tag{5.3.16}$$

weak* in L^∞, we deduce that

$$\begin{aligned} &\langle \eta_1(\lambda)q_2(\lambda) - q_1(\lambda)\eta_2(\lambda), \nu_{x,t} \rangle \\ &\qquad = \langle \eta_1(\lambda), \nu_{x,t} \rangle \langle q_2(\lambda), \nu_{x,t} \rangle - \langle q_1(\lambda), \nu_{x,t} \rangle \langle \eta_2(\lambda), \nu_{x,t} \rangle \end{aligned} \tag{5.3.17}$$

This last relation holds for all η_1, η_2 which are C^2 and convex; however, by continuity in the η_i, $i = 1, 2$, it also holds, as has been noted by Lax [101] for η_1, η_2 which are the limits of convex η_1, η_2. We now set $f(\lambda) = \frac{1}{2}\lambda^2$, define the pairs

$$\begin{cases} \eta_1(\lambda) = \begin{cases} \lambda, & \lambda \geq 0 \\ 0, & \lambda < 0 \end{cases} \\ q_1(\lambda) = \begin{cases} f(\lambda), & \lambda \geq 0 \\ 0, & \lambda < 0 \end{cases} \end{cases} \tag{5.3.18}$$

and

$$\begin{cases} \eta_2(\lambda) = |\lambda - \bar{u}| \\ q_2(\lambda) = \frac{1}{2}|f(\bar{u}) - f(\lambda)| \end{cases} \tag{5.3.19}$$

and claim the following

Lemma 5.1 If $\bar{q}_1 = \lim\limits_{\mu \to 0^+} q_1(u^\mu)$ in L^∞ (weak*) with $q_1(\cdot)$ as given by (5.3.18) and u^μ, for $\mu > 0$, the unique solution of (5.3.1), then

$$\left(\bar{q}_1 - \frac{\bar{u}^2}{2}\right) \langle \nu_{x,t}, |\lambda - \bar{u}| \rangle = 0 \tag{5.3.20}$$

where $\bar{u}(x,t)$ is the weak* L^∞ limit of $u^\mu(x,t)$, as $\mu \to 0^+$, a.e. on R^2, while $\nu_{x,t}$ is the Young measure generated by an appropriate subsequence $\{u^\mu\}$.

Proof: Using the pairs (η_i, q_i) given by (5.3.18) and (5.3.19), and substituting in (5.3.17), we deduce the relation

$$\langle\lambda|f(\bar{u})-f(\lambda)|,\nu_{x,t}\rangle-\langle f(\lambda)|\lambda-\bar{u}|,\nu_{x,t}\rangle \tag{5.3.21}$$
$$=\langle\lambda,\nu_{x,t}\rangle\langle|f(\bar{u})-f(\lambda)|,\nu_{x,t}\rangle-\langle f(\lambda),\nu_{x,t}\rangle\langle|\lambda-\bar{u}|,\nu_{x,t}\rangle$$

However, $\langle\lambda,\nu_{x,t}\rangle=\bar{u}(x,t)$, while $\langle f(\lambda),\nu_{x,t}\rangle=\bar{q}_1(x,t)$, a.e. on R^2, so from (5.3.21) we obtain

$$\langle(\lambda-\bar{u})|f(\bar{u})-f(\lambda)|,\nu_{x,t}\rangle-\langle(f(\lambda)-\bar{q}_1)|\lambda-\bar{u}|,\nu_{x,t}\rangle=0$$

or

$$\langle(\lambda-\bar{u})|f(\bar{u})-f(\lambda)|-(f(\lambda)-\bar{q}_1)|\lambda-\bar{u}|,\nu_{x,t}\rangle=0 \tag{5.3.22}$$

Consider first the case in which $\lambda\geq\bar{u}$; then

$$\begin{aligned}&(\lambda-\bar{u})|f(\bar{u})-f(\lambda)|-(f(\lambda)-\bar{q}_1)|\lambda-\bar{u}|\\&=(\lambda-\bar{u})(f(\lambda)-f(\bar{u}))-(f(\lambda)-\bar{q}_1)(\lambda-\bar{u})\\&=(\lambda-\bar{u})(\bar{q}_1-f(\bar{u}))\\&=|\lambda-\bar{u}|(\bar{q}_1-f(\bar{u}))\end{aligned}$$

A similar calculation shows that we also have

$$(\lambda-\bar{u})|f(\bar{u})-f(\lambda)|-(f(\lambda)-\bar{q}_1)|\lambda-\bar{u}|=|\lambda-\bar{u}|(\bar{q}_1-f(\bar{u})) \tag{5.3.23}$$

for the case in which $\lambda\leq\bar{u}$. Using (5.3.23) in (5.3.22), we are led immediately to (5.3.20). □

From the relation (5.3.20) it follows, of course, that either $\bar{q}_1=\frac{1}{2}\bar{u}^2$, or that

$$\langle\nu_{x,t},|\lambda-\bar{u}|\rangle=\int_{-\infty}^{\infty}|\lambda-\bar{u}(x,t)|\,d\nu_{x,t}(\lambda)=0, \tag{5.3.24}$$

in which case, as $\nu_{x,t}$ is a positive measure, $\nu_{x,t}(\lambda)=\delta(\lambda-\bar{u}(x,t))$ and

$$\begin{aligned}\bar{q}_1&=\int_{-\infty}^{\infty}q_1(\lambda)\delta(\lambda-\bar{u}(x,t))\,d\lambda\\&=\frac{1}{2}\int_{-\infty}^{\infty}\lambda^2\delta(\lambda-\bar{u}(x,t))\,d\lambda\\&=\frac{1}{2}\bar{u}^2(x,t),\ \text{a.e. on } R^2\end{aligned} \tag{5.3.25}$$

Thus, in either case, (5.1.14) holds and, when combined with the analysis in § 5.1, Lemma 5.1 yields the following

Theorem 5.1 If for any entropy-flux pair $(\eta(u), q(u))$, (5.3.6) is valid, where u^μ is the unique solution of the initial-value problem (5.3.1), for each $\mu > 0$, then as $\mu \to 0^+$, $u^\mu \rightharpoonup \bar{u}$, weak* in $L^\infty(R^1 \times [0,T])$, $T > 0$, where $\bar{u}$ satisfies (5.3.2) in the sense of distributions.

Remarks:

Actually we will have $u^\mu \to \bar{u}$ pointwise, a.e., for some subsequence $\{u^\mu\}$.

It remains, therefore, for us to verify that (5.3.6) is satisfied, or, equivalently, by virtue of (5.3.5), that for each $\mu > 0$

$$\mu\left[\eta(u^\mu)_{xx} - \eta''(u^\mu)(u_x^\mu)^2\right] \in \text{a compact subset of } H_{\text{loc}}^{-1} \tag{5.3.26}$$

for any convex C^2 function η, with u^μ, for each $\mu > 0$, the unique solution of the (regularized) parabolic initial-value problem (5.3.1); to this end we integrate (5.3.5) over $R^1 \times [0,T]$ so as to obtain

$$\int_{-\infty}^{\infty} \eta(u^\mu(x,T))\,dx + \mu \int_0^T \int_{-\infty}^{\infty} \eta''(u^\mu(x,t))(u_x^\mu(x,t))^2\,dx\,dt \tag{5.3.27}$$
$$= \int_{-\infty}^{\infty} \eta(u_0(x))\,dx$$

As we may assume, without loss of generality, that $\eta \geq 0$, it follows from (5.3.27) that

$$\mu\eta''(u^\mu)(u_x^\mu)^2 \in \text{a bounded subset of } L^1(R^1 \times [0,T]) \tag{5.3.28}$$

which is independent of μ. From the strict convexity of η it is an immediate consequence of (5.3.28) that

$$\mu(u_x^\mu)^2 \in \text{a bounded subset of } L^1(R^1 \times [0,T]) \tag{5.3.29}$$

which is independent of μ. Therefore, as we have, for some $C > 0$, independent of μ,

$$\int_0^T \int_{-\infty}^{\infty} (\mu\eta(u^\mu)_x)^2\,dx\,dt$$
$$= \int_0^T \int_{-\infty}^{\infty} \mu^2\eta'^2(u^\mu)(u_x^\mu)^2\,dx\,dt$$
$$\leq \mu C \int_0^T \int_{-\infty}^{\infty} \mu(u_x^\mu)^2\,dx\,dt$$

it clearly follows that

$$\mu\eta(u^\mu)_x \to 0, \text{ in } L^2(R^1 \times [0,T]) \text{ as } \mu \to 0^+ \tag{5.3.30}$$

from which it is immediate that, also,

$$\{\mu\eta(u^\mu)_x\} \subset \text{a compact subset of } L^2(R^1 \times [0,T]) \tag{5.3.31}$$

Finally, as a consequence of (5.3.31) we have

$$\{\mu\eta(u^\mu)_{xx}\} \subset \text{a compact subset of } H^{-1}(R^1 \times [0,T]) \tag{5.3.32}$$

We now rewrite (5.3.5) as

$$\begin{aligned} \eta_t + q_x &= \left.\mu\eta(u^\mu)_{xx}\right\} \begin{array}{l}\text{in a compact}\\ \text{subset of } H^{-1}\end{array} \\ &\quad + \left.(-\mu)\eta''(u^\mu)(u^\mu_x)^2\right\} \begin{array}{l}\text{in a bounded}\\ \text{subset of } L^1\end{array} \end{aligned} \tag{5.3.33}$$

As a consequence of one version of the Murat Lemma (e.g., [120], [169]) it now follows from (5.3.33) that (5.3.6) is valid for any entropy-flux pair (η, q), with u^μ, for $\mu > 0$, the unique solution of the initial-value problem (5.3.1), provided we also have

$$\eta(u^\mu)_t + q(u^\mu)_x \in \text{a bounded set in } W^{-1,\infty} \tag{5.3.34}$$

However, (5.3.34) follows directly from the fact that η (and q) are C^2 functions and u^μ is, for each $\mu > 0$, in a bounded set of $L^\infty(R^1 \times [0,T])$, which is independent of μ. By combining the analysis given above with Theorem 5.1, we see that we have established, via the method of compensated compactness (and through the characterization of weak convergence which is given by the Young measure) the result originally deduced by Hopf through a careful analysis of the limit of the expression (5.1.7) as $\mu \to 0^+$, namely,

Corollary 5.1 If u^μ, for each $\mu > 0$, is the unique solution of the initial-value problem (5.3.1), then as $\mu \to 0^+$, $u^\mu \rightharpoonup \bar{u}$, weak* in $L^\infty(R^1 \times [0,T])$, $T > 0$, where $\bar{u}$ satisfies (5.3.2) in the distribution sense. Also $u^\mu \to \bar{u}$ pointwise, a.e., for some subsequence u^μ.

5.4 Systems of Conservation Laws and DiPerna's Theorem

As we have already indicated, the primary goal of our work in this chapter is to prove that, under appropriate conditions, the initial-value problem associated with the system (4.1.8), governing the evolution of the current and charge distribution in the distributed parameter, nonlinear transmission line, possesses a global (in time) weak solution. In this section we will review those ideas of DiPerna [48], [47] which are connected with employing the concept of the Young measure, and the technique of compensated compactness, in order to establish the existence of global (in time) weak solutions for initial-value problems associated with strictly hyperbolic, genuinely nonlinear, homogeneous systems of conservation laws; then, in § 5.5–5.7, we will show that DiPerna's work can be extended so as to encompass, also, a nonhomogeneous hyperbolic system of the type associated with the nonlinear transmission problem. Among other extensions of DiPerna's original work in this area we may mention the recent work on homogeneous hyperbolic systems which either lack strict hyperbolicity or fail to have characteristic fields which are all genuinely nonlinear (e.g., Chen [32], [33], Serre [153], [154], [152], Rascle and Serre [133], and Weinan and Kohn [180]), the new proof of DiPerna's theorem which can be found in Morawetz [116], and the work of Nohel, Rogers, and Tzavaras [127] on weak solutions for an integrodifferential system arising in nonlinear viscoelasticity. Pertinent extensions of the work of Tartar on the Burgers' equation, which has been described in § 5.3, would include, e.g., the recent work [167] of Szepessy on measure-valued solutions to initial-boundary value problems for scalar conservation laws, the studies of Schonbek [149], [148] that are concerned with the convergence of solutions to nonlinear dispersive equations such as the Korteveg de Vries equation, and the work of Dafermos [44] on global L^∞ solutions for a single conservation law with memory.

The starting point for our description of the basic result of DiPerna [131], [27] is the system

$$\boldsymbol{u}_t + \boldsymbol{f}(\boldsymbol{u})_x = \boldsymbol{0}; \quad x \in R^1,\, t > 0 \tag{5.4.1}$$

$\boldsymbol{u} \in R^m$, $\boldsymbol{f} : R^m \to R^m$, which we assume to be strictly hyperbolic, i.e., at each $\boldsymbol{p} \in S \subseteq R^m$, S an open connected set, $\nabla \boldsymbol{f}$ has m real distinct eigenvalues $\lambda_i(\boldsymbol{p})$, $i =$

$1, \ldots, m$, which we order as $\lambda_1(\boldsymbol{p}) < \cdots < \lambda_m(\boldsymbol{p})$, with corresponding right eigenvectors $\boldsymbol{r}_i(\boldsymbol{p})$, $i = 1, \ldots, m$. We also assume that the system (5.4.1) is genuinely nonlinear which means, of course, that $(\nabla\lambda_i \cdot \boldsymbol{r}_i)(\boldsymbol{p}) \neq 0$ for each $\boldsymbol{p} \in S$ and each i, $i = 1, \ldots, m$.

Remark: We note that the system (5.4.1) always possesses oscillating solutions, i.e., if $\boldsymbol{p}, \boldsymbol{q} \in R^m$ are two states which can be connected by a shock, so that the Rankine-Hugoniot condition

$$\boldsymbol{f}(\boldsymbol{p}) - \boldsymbol{f}(\boldsymbol{q}) = \sigma(\boldsymbol{p} - \boldsymbol{q}), \ \sigma \in R, \ \boldsymbol{p} \neq \boldsymbol{q}$$

is satisfied, we may define

$$\boldsymbol{u}^k(x,t) = \boldsymbol{p} + \boldsymbol{q}\chi^k(x - \sigma t) \tag{5.4.2}$$

with χ^k a characteristic function, for each integer k, and $\{\boldsymbol{u}^k\}$ will be a sequence of functions which satisfy the system (5.4.1) and oscillate between $\boldsymbol{p}$ and $\boldsymbol{q}$. We also know that the set of all $\boldsymbol{q}$ which satisfy the Rankine-Hugoniot conditions, as cited above, must, in some neighborhood of $\boldsymbol{p}$, lie along m distinct curves in the directions of the eigenvectors $\boldsymbol{r}_i(\boldsymbol{p})$, $i = 1, \ldots, m$; this (standard) result, which is a consequence of the implicit function theorem, may be found, e.g., in the paper of Lax [100] or the text by Smoller [165]. Henceforth, however, we will restrict our attention only to solutions of (5.4.1) which satisfy a set of entropy conditions that are described below.

As was the case with the Burgers' equation, we consider a parabolic regularization of (5.4.1), i.e., the system

$$\boldsymbol{u}_t^{\varepsilon} + \boldsymbol{f}(\boldsymbol{u}^{\varepsilon})_x = \varepsilon \boldsymbol{u}_{xx}^{\varepsilon}; \quad x \in R^1, \, t > 0 \tag{5.4.3}$$

We assume that there exists a sequence $\{\boldsymbol{u}^{\varepsilon_k}\}$ of solutions to the initial-value problem (i.e., $\boldsymbol{u}^{\varepsilon_k}(x,0) = \boldsymbol{u}_0(x)$, $x \in R^1$, $\forall\, \varepsilon_k > 0$) for (5.4.3) with the property that

$$\boldsymbol{u}^{\varepsilon_k} \text{ is bounded in } (L^\infty)^m \tag{5.4.4}$$

and

$$\sqrt{\varepsilon_k}\,\boldsymbol{u}_x^{\varepsilon_k} \text{ is bounded in } (L^2)^m \tag{5.4.5}$$

for each $\varepsilon_k > 0$ independently of ε_k, where $\varepsilon_k \to 0$, as $k \to \infty$; the properties (5.4.4), (5.4.5) are the natural generalizations of the condition that the solutions u^μ of the initial-value problem (5.3.1) satisfy

$$\|u^\mu\|_{L^\infty(R^1\times[0,T])} \leq c, \text{ with } c \text{ independent of } \mu$$

and must be verified for each individual system (5.4.1), or each particular class of systems. In analogy with (5.3.3), we now define an entropy η and an associated flux q, for our system (5.4.1), to be a pair of real-valued C^2 functions on R^m which satisfy

$$\nabla\eta(\boldsymbol{p})\cdot\nabla\boldsymbol{f}(\boldsymbol{p}) = \nabla q(\boldsymbol{p}),\ \forall\boldsymbol{p}\in\mathcal{D}(\boldsymbol{f}), \tag{5.4.6}$$

$\mathcal{D}(\boldsymbol{f})$ denoting, of course, the domain of the vector-valued function $\boldsymbol{f}$. If we now take the inner product, in R^m, of both sides of (5.4.3) with $\nabla\eta(\boldsymbol{u}^\varepsilon)$ we obtain the equation

$$\begin{aligned}\eta(\boldsymbol{u}^\varepsilon)_t + q(\boldsymbol{u}^\varepsilon)_x - \varepsilon\eta(\boldsymbol{u}^\varepsilon)_{xx}&\\ +\varepsilon\nabla^2\eta(\boldsymbol{u}^\varepsilon)[\boldsymbol{u}_x^\varepsilon,\boldsymbol{u}_x^\varepsilon] = 0&\end{aligned} \tag{5.4.7}$$

where $\nabla^2\eta$ denotes the Hessian matrix associated with the entropy function $\eta(\cdot)$, while

$$\nabla^2\eta(\boldsymbol{u}^\varepsilon)[\boldsymbol{u}_x^\varepsilon,\boldsymbol{u}_x^\varepsilon] = (\boldsymbol{u}_x^\varepsilon)^t\cdot\nabla^2\eta(\boldsymbol{u}^\varepsilon)\cdot\boldsymbol{u}_x^\varepsilon \tag{5.4.8}$$

the superscript "t" in (5.4.8) denoting the transpose of the indicated vector. If, for the solution of (5.4.7), we could show that $\boldsymbol{u}^{\varepsilon_k} \to \bar{\boldsymbol{u}}$ (strong convergence, a.e.), as $k\to\infty$, it would then follow that $\bar{\boldsymbol{u}}$ would satisfy the entropy inequality

$$\eta(\bar{\boldsymbol{u}})_t + q(\bar{\boldsymbol{u}})_x \leq 0 \tag{5.4.9}$$

Because of (5.4.4)—assuming that we have established such a result for the initial-value problem associated with (5.4.3)—$\eta(\boldsymbol{u}^{\varepsilon_k})$ and $q(\boldsymbol{u}^{\varepsilon_k})$ are both bounded in $(L^\infty)^m$, independently of $\varepsilon_k > 0$. In a manner similar to the discussion in § 5.3, we may show that

$$\varepsilon_k\eta(\boldsymbol{u}^{\varepsilon_k})_{xx} \to 0, \text{ in } H^1_{\text{loc}} \tag{5.4.10}$$

In fact, we may write that

$$\varepsilon_k\eta(\boldsymbol{u}^{\varepsilon_k})_{xx} = \sqrt{\varepsilon_k}\left(\sqrt{\varepsilon_k}\nabla\eta(\boldsymbol{u}^{\varepsilon_k})\boldsymbol{u}_x^{\varepsilon_k}\right)_x$$

and $\sqrt{\varepsilon_k}\,\nabla\eta(\boldsymbol{u}^{\varepsilon_k})\boldsymbol{u}_x^{\varepsilon_k}$ is bounded in L^2, independently of $\varepsilon_k > 0$, by virtue of the assumption (5.4.5). Also, for $\eta \in C^2$, the term in (5.4.7) for $\varepsilon = \varepsilon_k$, i.e., (5.4.8), is easily seen to be bounded in L^1, while by virtue of (5.4.4), and the smoothness of η, q, the sum $\eta(\boldsymbol{u}^{\varepsilon_k})_t + q(\boldsymbol{u}^{\varepsilon_k})_x$ is bounded in $W^{-1,\infty}$. All of the requisite conditions are, therefore, in place which enable us to apply the Murat lemma ([120], [169]) so as to be able to conclude that

$$\eta(\boldsymbol{u}^{\varepsilon_k})_t + q(\boldsymbol{u}^{\varepsilon_k})_x \in \text{ a compact subset of } H^{-1}_{\text{loc}} \tag{5.4.11}$$

independent of $\varepsilon_k > 0$. Using (5.4.11), and the (hopefully) now familiar construction of the Young measure generated by the sequence of solutions $\boldsymbol{u}^{\varepsilon_k}$, of the initial-value problem associated with the system (5.4.3), we may apply the Div-Curl Lemma of Murat [120] so as to arrive at the relation

$$\langle \nu, \eta_1 q_2 - \eta_2 q_1 \rangle = \langle \nu, \eta_1 \rangle \langle \nu, q_2 \rangle - \langle \nu, \eta_2 \rangle \langle \nu, q_1 \rangle \tag{5.4.12}$$

which will be valid for any two entropy-flux pairs (η_1, q_1), (η_2, q_2) satisfying (5.4.6). In (5.4.12), as in the remainder of this section, we refrain from subscripting the Young measure, i.e., $\nu = \nu_{x,t}$.

In order to show now that solutions of the initial-value problem for (5.4.3), with $\varepsilon = \varepsilon_k$, converge weak* in L^∞, to a weak solution $\bar{\boldsymbol{u}}$ of the initial-value problem for (5.4.1), as $k \to \infty$ (which will, then, be the unique weak solution satisfying the entropy inequality (5.4.9)) we must, of course, prove that the Young measure ν, as restricted by (5.4.12), is a Dirac measure; this turns out to be a far more formidable problem than it was for the case of the Burgers' equation (where the entire issue turned, so to speak, on the judicious choice of the two entropy-flux pairs given by (5.3.18), (5.3.19)) and the essence of the solution of the problem lies with DiPerna's recognition of the significance of the special families of entropies constructed by Lax [13]. Thus, we consider, once again, the equations (5.4.6) for the entropy-flux pairs (η, q), which we now rewrite in terms of the basis of eigenvectors $\boldsymbol{r}_i(\boldsymbol{p})$, $i = 1, \ldots, m$, of $\nabla \boldsymbol{f}(\boldsymbol{p})$, i.e.,

$$\nabla q(\boldsymbol{p}) \cdot \boldsymbol{r}_i(\boldsymbol{p}) = \lambda_i(\boldsymbol{p}) \nabla \eta(\boldsymbol{p}) \cdot \boldsymbol{r}_i(\boldsymbol{p}) \tag{5.4.13}$$

for $i = 1, \ldots, m$. Taking $m = 2$ in (5.4.1) (i.e., we focus our attention on a 2×2 system of conservation laws) we change to Riemann-invariants coordinates w_i where, of course, w_i is an i-Riemann invariant if $\nabla w_i(\boldsymbol{p}) \cdot r_i(\boldsymbol{p}) = 0$, $\forall \boldsymbol{p} \in S$. Assuming that the mapping from coordinates $(u_1, u_2) \to (w_1, w_2)$ is nonsingular, the system of equations (5.4.6) may be rewritten in the form

$$\frac{\partial q}{\partial w_1} = \lambda_2 \frac{\partial \eta}{\partial w_1}, \quad \frac{\partial q}{\partial w_2} = \lambda_1 \frac{\partial \eta}{\partial w_2} \tag{5.4.14}$$

where $\lambda_1 < \lambda_2$, as well as η, q, are now functions of the Riemann-invariants w_1, w_2. Following Lax [13] we seek solutions $\eta_k(w_1, w_2)$ and $q_k(w_1, w_2)$ of (5.4.14) which, for each integer k, have the asymptotic form

$$\eta_k(w_1, w_2) = e^{kw_1}\left[A_0 + \frac{A_1}{k} + \mathcal{O}\left(\frac{1}{k^2}\right)\right] \tag{5.4.15}$$

and

$$q_k(w_1, w_2) = e^{kw_1}\left[B_0 + \frac{B_1}{k} + \mathcal{O}\left(\frac{1}{k^2}\right)\right] \tag{5.4.16}$$

as $k \to \pm\infty$, where $A_j, B_j, j = 0, 1$ are smooth functions of w_1, w_2 which do not depend on k. It can be shown [44] that solutions of (5.4.14) which behave, for $|k|$ large, exist in the form delineated in (5.4.15), (5.4.16), provided the following conditions are satisfied

$$\begin{cases} A_0 > 0 \\ B_0 = \lambda_2 A_0 \\ \dfrac{\partial B_0}{\partial w_2} = \lambda_1 \dfrac{\partial A_0}{\partial w_2} \\ B_1 + \dfrac{\partial B_0}{\partial w_1} = \lambda_2 \left(A_1 + \dfrac{\partial A_0}{\partial w_1}\right) \end{cases} \tag{5.4.17}$$

The first condition in (5.4.17) is a direct consequence of assuming that η_k, as given by (5.4.15), is a convex function on R^2, while the remaining three conditions result from substituting, for fixed k, the expressions (5.4.15), (5.4.16) into the relations (5.4.14) and equating the coefficients of powers of $1/k$. From (5.4.17) it also follows that

$$B_1 - \lambda_2 A_1 = -\frac{\partial \lambda_2}{\partial w_1} A_0 \tag{5.4.18}$$

so that $\operatorname{sgn}(B_1 - \lambda_2 A_1)$ is opposite to $\operatorname{sgn}\left(\dfrac{\partial \lambda_2}{\partial w_1}\right)$. The functions η_k, q_k represent exact progressing wave solutions of the system (5.4.14), and a similar asymptotic family, with e^{kw_2} replacing e^{kw_1}, also exists if we permute the roles of the eigenvalues λ_1, λ_2.

As we have already stated, suppose that we can use the Riemann invariants w_1, w_2 to effect a global change of coordinates on the domain S and let the Young measure

ν, which is generated by the solutions of the initial-value problems associated with the system (5.4.3), for each $\varepsilon = \varepsilon_k > 0$, satisfy (5.4.12) for each set of entropy-flux pairs $(\eta_i(w_1, w_2), q_i(w_1, w_2))$, $i = 1, 2$, satisfying (5.4.14). Also, let $\mathcal{R}$ be the smallest rectangle of the form

$$\mathcal{R} = \{(w_1, w_2) \Big| w_j^- \leq w_j \leq w_j^+, \, j = 1, 2\}, \tag{5.4.19}$$

with sides parallel to the w_1, w_2 axes, which contains the support of the Young measure ν; then DiPerna's result may be stated as follows

Theorem 5.2 ([48]): For the Young measure ν generated by the solutions of the initial-value problems associated with the regularized parabolic system (5.4.3) we have meas(supp ν) = 0, i.e., ν is a Dirac measure.

Proof: The essence of the proof is to show that $w_1^- = w_1^+$ in (5.4.19); the result then follows from our assumption that $\mathcal{R}$ is the smallest rectangle, with sides parallel to the w_1, w_2 axes, which contains the support of ν. To accomplish the task at hand, we will show that if one assumes that $w_1^- < w_1^+$, then it must follow that each closed vertical side of $\mathcal{R}$ contains a point where $\dfrac{\partial \lambda_2}{\partial w_1} = 0$ which, of course (see, e.g., the discussion immediately preceding Theorem 2.1) would violate our assumption that the characteristic fields λ_1, λ_2 are genuinely nonlinear.

We assume, therefore, that $w_1^- < w_1^+$ and begin by defining (boundary) probability measures $\mu^{\pm}$ as follows: for $|k|$ sufficiently large, the probability measure μ_k is given by

$$\langle \mu_k, f \rangle \equiv \frac{\langle \nu, \eta_k f \rangle}{\langle \nu, \eta_k \rangle} \tag{5.4.20}$$

with η_k as given in (5.4.15). In using, now, the pairs (η_k, q_k) defined by (5.4.15), (5.4.16) we will assume that the A_i, B_i, $i = 0, 1$ satisfy the compatibility conditions (5.4.17) and, thus, (5.4.18) as well. Because

$$\eta_k e^{-kw_1} = A_0 + \mathcal{O}(\frac{1}{k}) > 0, \quad |k| \, \text{large} \tag{5.4.21}$$

(5.4.20) makes sense and defines, for each k with $|k|$ sufficiently large, a probability measure with support in the rectangle $\mathcal{R}$. Using the definition (5.4.20) we extract weak* convergent subsequences $\{\mu_{k_n}\}$, $\{\mu_{\bar{k}_n}\}$ such that

$$\begin{cases} \mu_{k_n} \overset{*}{\rightharpoonup} \mu^+ \\ \mu_{\bar{k}_n} \overset{*}{\rightharpoonup} \mu^- \end{cases} \quad \text{as } k_n, \, \bar{k}_n \to \infty \tag{5.4.22}$$

In all that now follows in the proof we will write k in place of k_n so that, e.g., $\mu_k \overset{*}{\rightharpoonup} \mu^\pm$ as $k \to \pm\infty$. Because the factor e^{kw_1} in (5.4.15) gives most of the weight in η_k to a neighborhood of either w_1^- or w_1^+, as $k \to \infty$, and $\mathcal{R}$ is, by hypothesis, the smallest rectangle containing supp ν, it follows that

$$\begin{aligned} \text{supp}\mu^- &\subset \mathcal{R} \cap \{w_1 = w_1^-\} \\ \text{supp}\mu^+ &\subset \mathcal{R} \cap \{w_1 = w_1^+\} \end{aligned} \tag{5.4.23}$$

We now set

$$\lambda_2^\pm = \langle \mu^\pm, \lambda_2 \rangle \tag{5.4.24}$$

and claim that for each entropy-flux pair (η, q)

$$\langle \nu, q - \lambda_2^\pm \eta \rangle = \langle \mu^\pm, q - \lambda_2 \eta \rangle \tag{5.4.25}$$

To prove the validity of (5.4.25) we set

$$\begin{cases} (\eta_1, q_1) = (\eta, q) \\ (\eta_2, q_2) = (\eta_k, q_k) \end{cases} \tag{5.4.26}$$

in (5.4.12) and we obtain

$$\frac{\langle \nu, \eta q_k - \eta_k q \rangle}{\langle \nu, \eta_k \rangle} = \frac{\langle \nu, \eta \rangle \langle \nu, q_k \rangle}{\langle \nu, \eta_k \rangle} - \langle \nu, q \rangle \tag{5.4.27}$$

However,

$$\frac{\langle \nu, q_k \rangle}{\langle \nu, \eta_k \rangle} = \frac{\langle \nu, \lambda_2 \eta_k \rangle}{\langle \nu, \eta_k \rangle} + \frac{\langle \nu, \eta_k \cdot \mathcal{O}(1/k) \rangle}{\langle \nu, \eta_k \rangle} \tag{5.4.28}$$

so as

$$\begin{aligned} q_k &= e^{kw_1} \left[\lambda_2 A_0 + \frac{B_1}{k} + \mathcal{O}(1/k^2) \right] \\ &= \lambda_2 e^{kw_1} \left[A_0 + \left(\frac{B_1}{\lambda_2} \right) \frac{1}{k} + \mathcal{O}(1/k^2) \right], \end{aligned} \tag{5.4.29}$$

by virtue of the compatability conditions (5.4.17), we have

$$\frac{\langle \nu, q_k \rangle}{\langle \nu, \eta_k \rangle} = \langle \mu_k, \lambda_2 \rangle + \mathcal{O}(1/k) \to \lambda_2^\pm, \text{ as } k \to \pm\infty \tag{5.4.30}$$

where we have used (5.4.20) and (5.4.24). In a similar manner we find that

$$\frac{\langle \nu, \eta q_k \rangle}{\langle \nu, \eta_k \rangle} \to \langle \mu^\pm, \lambda_2 \eta \rangle, \text{ as } k \to \pm\infty \tag{5.4.31}$$

Using (5.4.30) and (5.4.31) in (5.4.27), and letting $k \to \pm\infty$, we easily obtain the relation

$$\langle \mu^{\pm}, \lambda_2 \eta \rangle - \langle \mu^{\pm}, q \rangle = -\langle \nu, q \rangle + \lambda_2^{\pm} \langle \nu, \eta \rangle \tag{5.4.32}$$

which is identical to (5.4.25).

Having established the relation (5.4.25), we now claim that $\lambda_2^- = \lambda_2^+$ with $\lambda_2^{\pm}$ given by (5.4.24). We again begin with the restriction on ν that is given by (5.4.12) taking, this time,

$$\begin{cases} (\eta_1, q_1) = (\eta_k, q_k) \\ (\eta_2, q_2) = (\eta_{-k}, q_{-k}) \end{cases} \tag{5.4.33}$$

and dividing the resulting equation through by the product $\langle \nu, \eta_k \rangle \langle \nu, \eta_{-k} \rangle$; the result of all of this is the equation

$$\frac{\langle \nu, \eta_k q_{-k} - \eta_{-k} q_k \rangle}{\langle \nu, \eta_k \rangle \langle \nu, \eta_{-k} \rangle} = \frac{\langle \nu, q_{-k} \rangle}{\langle \nu, \eta_{-k} \rangle} - \frac{\langle \nu, q_k \rangle}{\langle \nu, \eta_k \rangle} \tag{5.4.34}$$

However, as

$$\frac{\langle \nu, q_{\pm k} \rangle}{\langle \nu, \eta_{\pm k} \rangle} = \langle \mu_{\pm k}, \lambda_2 \rangle + \mathcal{O}(1/k) \to \lambda_2^{\pm}, \text{ as } k \to \infty, \tag{5.4.35}$$

for the right-hand side of (5.4.34) we have

$$\frac{\langle \nu, q_{-k} \rangle}{\langle \nu, \eta_{-k} \rangle} - \frac{\langle \nu, q_k \rangle}{\langle \nu, \eta_k \rangle} \to \lambda_2^- - \lambda_2^+, \text{ as } k \to \infty \tag{5.4.36}$$

However, $\eta_k q_{-k} - \eta_{-k} q_k$ is, by virtue of (5.4.15), (5.4.16), $\mathcal{O}(1/k)$ while $\langle \nu, \eta_k \rangle \langle \nu, \eta_{-k} \rangle \to \infty$ faster than $e^{k(w_1^+ - w_1^-)}$, because of the definition of $\mathcal{R}$. Therefore

$$\frac{\langle \nu, \eta_k q_{-k} - \eta_{-k} q_k \rangle}{\langle \nu, \eta_k \rangle \langle \nu, \eta_{-k} \rangle} \to 0, \text{ as } k \to \infty \tag{5.4.37}$$

Combining (5.4.34) with (5.4.36) and (5.4.37) we easily determine that $\lambda_2^- = \lambda_2^+$.

Finally, we claim that

$$\langle \mu^{\pm}, A_0 \frac{\partial \lambda_2}{\partial w_1} \rangle = 0 \tag{5.4.38}$$

To establish the validity of (5.4.38) we first combine (5.4.25) with the fact that $\lambda_2^- = \lambda_2^+$ and we find that, for every entropy-flux pair (η, q)

$$\begin{aligned} \langle \mu^+, q - \lambda_2 \eta \rangle &= \langle \nu, q - \lambda_2^+ \eta \rangle \\ &= \langle \nu, q - \lambda_2^- \eta \rangle \\ &= \langle \mu^-, q - \lambda_2 \eta \rangle \end{aligned} \tag{5.4.39}$$

Therefore, with $(\eta, q) = (\eta_k, q_k)$,

$$\langle \mu^+, q_k - \lambda_2 \eta_k \rangle = \langle \mu^-, q_k - \lambda_2 \eta_k \rangle \tag{5.4.40}$$

However,

$$\langle \mu^+, q_k - \lambda_2 \eta_k \rangle = e^{kw_1^+} \langle \mu^+, \frac{B_1 - \lambda_2 A_1}{k} \rangle + \mathcal{O}(1/k^2) \tag{5.4.41}$$

$$\langle \mu^-, q_k - \lambda_2 \eta_k \rangle = e^{kw_1^-} \langle \mu^-, \frac{B_1 - \lambda_2 A_1}{k} \rangle + \mathcal{O}(1/k^2) \tag{5.4.42}$$

where we have used (5.4.15), (5.4.16), as well as the result (5.4.23) on supp $\mu^\pm$. Combining (5.4.40) with (5.4.41), (5.4.42), and (now) using the assumption that $w_1^- < w_1^+$, we deduce that we must have

$$\langle \mu^\pm, B_1 - \lambda_2 A_1 \rangle = 0 \tag{5.4.43}$$

In view of (5.4.18) we are now led directly from (5.4.43) to (5.4.38). But $A_0 > 0$, and $\mu^\pm$ are both probability measures, so if $w_1^- < w_1^+$, then it follows (from (5.4.38)) that $\dfrac{\partial \lambda_2}{\partial w_1} = 0$. In view of our assumption of genuine nonlinearity, with respect to both λ_1 and λ_2, we may conclude that our assumption that $w_1^- < w_1^+$ cannot be satisfied and that, in fact, we must have $w_1^- = w_1^+$. By virtue of the definition of $\mathcal{R}$ as being the smallest rectangle of the form (5.4.19), with sides parallel to the w_1, w_2 axes, which contains the support of the Young measure ν, we may conclude that meas(supp ν) = 0 so that ν is a Dirac measure. □

As a direct consequence of DiPerna's fundamental result, i.e., Theorem 5.2, we now have a very important corollary concerning the existence of weak solutions for the system (5.4.1), which we will state only for the case $n = 2$:

Corollary 5.2 Consider the initial-value problem for a 2×2 system of conservation laws, of the form (5.4.1), which is assumed to be strictly hyperbolic and genuinely nonlinear. Then there exists a subsequence $\{\boldsymbol{u}^\varepsilon\}$ of solutions to the initial-value problem, for the parabolic regularization (5.4.3) of (5.4.1), such that we have both $\boldsymbol{u}^\varepsilon \overset{*}{\rightharpoonup} \bar{\boldsymbol{u}}$ in L^∞ and $\boldsymbol{u}^\varepsilon \to \bar{\boldsymbol{u}}$, pointwise, a.e., with $\bar{\boldsymbol{u}}$ the unique global weak solution of the initial-value problem for (5.4.1) satisfying the entropy inequality (5.4.9).

5.5 The Nonlinear Transmission Line: Some Preliminaries

In this section we turn, once again, to the initial-value problem associated with the system (4.1.8) governing the evolution of the current and charge distribution in a nonlinear, distributed parameter transmission line; we will consider only the situation in which the leakage conductance is constant with, say, $G = G_0 > 0$ and, without loss of generality, we will set the inductance $L = 1$. Using the usual subscript notation for partial differentiation our system is, therefore,

$$\begin{cases} i_t + \mathcal{V}(Q)_x = -Ri \\ Q_t + i_x = -G_0\mathcal{V}(Q) \end{cases} \tag{5.5.1}$$

The system (5.5.1) is, of course, hyperbolic provided $\mathcal{V}'(Q) > 0$, $\forall\, Q \in R^1$, which will again be the case here and in the next two sections; in fact, we will assume that for some $\beta_0 > 0$, $\beta_1 > 0$, we have

$$\mathcal{V}(\xi) = (\beta_0 + \beta_1\xi^2)\xi, \quad \forall\, \xi \in R^1 \tag{5.5.2}$$

A cubic dependence of voltage upon charge similar to that exemplified by (5.5.2) has been studied, e.g., in Landauer [97] and seems to be typical of many unbiased ferroelectric lines. The most important consequences of the constitutive assumption (5.2.2), which are pertinent to our work in the next few sections, are the following

$$\text{(i) } \mathcal{V}'(\xi) \geq \beta_0 > 0,\ \ \xi\mathcal{V}(\xi) \geq 0,\ \forall\, \xi \in R^1 \tag{5.5.3a}$$

and

$$\text{(ii) } \hat{\mathcal{V}}(\xi) \equiv \int_0^{\xi} \mathcal{V}(\rho)\, d\rho \geq 0,\ \forall\, \xi \in R^1 \tag{5.5.3b}$$

In fact, only the properties (i), (ii) of $\mathcal{V}(\cdot)$ delineated above will be relevant to what follows and not the particular form, e.g. (5.5.2), of $\mathcal{V}(\cdot)$.

Associated with the system (5.5.1) is initial data of the form

$$i(x,0) = i_0(x),\ \ Q(x,0) = Q_0(x),\ \ x \in R^1 \tag{5.5.4}$$

which, we assume, satisfy

$$i_0(\cdot) \in W^{1,2}(-\infty,\infty), \ Q_0(\cdot) \in W^{2,2}(-\infty,\infty) \tag{5.5.5}$$

Thus, we are considering the problem here as being posed for an infinitely long, distributed parameter, nonlinear transmission line, but our results will apply, of course, equally well to a finite transmission line with periodic initial current and charge distributions. A sketch of the situation we have in mind is depicted in Figure 5.1 below:

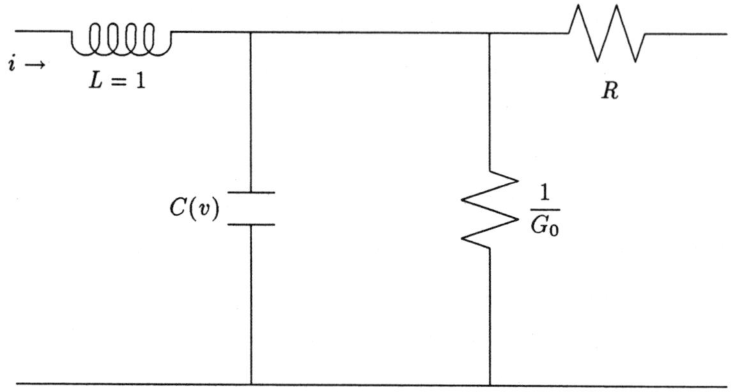

Figure 5.1

To this point it has been demonstrated, for the initial-value problem (5.5.1), (5.5.4), or for the equivalent problem on a finite line with periodic initial data, that global C^1 solutions $(i(x,t), Q(x,t))$ exist for all $t > 0$ for initial data which are sufficiently small, while shock discontinuities must develop in the line if the initial data are sufficiently large; these results form, in essence, the content of Chapter 4, where the notions of "sufficiently small" (respectively, "sufficiently large") initial data have been made precise. Our aim in the next three sections is to establish the existence of global weak solutions for the system (5.5.1) with associated arbitrarily large data $i_0(\cdot)$, $Q_0(\cdot)$, i.e., solutions having the property

$$(i,Q) \in L^\infty(R^1 \times [0,\infty)) \times L^\infty(R^1 \times [0,\infty)) \tag{5.5.6}$$

It may be easily confirmed that any effort to establish (5.5.6) through the use of classical compactness arguments, such as those used for the superconducting nonlinear dielectric problem treated in § 3.6, is doomed to failure; however, the problem that we wish to resolve here is susceptible to treatment by using the concepts of compensated compactness and the Young measure which we introduced in § 5.2–§ 5.4 for both the single conservation law and for homogeneous 2×2 systems of hyperbolic conservation laws which are genuinely nonlinear. Key elements in the analysis to be presented will be an appropriate regularization of the basic inhomogeneous system (5.5.1) that results from introducing a modification of the standard dependence of voltage upon charge in the line, a demonstration that the Div-Curl lemma applies, even though the resulting parabolic system possesses coefficients, other than those associated with the diffusive terms, which are dependent upon the regularization parameter, and, at a critical junction, an appeal to the work of DiPerna [48], [47], which has been described in the last section. For the most part, our work here will follow the recent treatment by Bellout, Nečas, and the author [55]; however, we would be remiss if we did not mention the independent work of Feireisl [57], [56] on the global existence of time-periodic solutions to nonautonomous quasilinear telegraph equations, with large data, which bears many similarities to the work in [55].

As we have already seen in the previous section, the work of DiPerna [48] guarantees the global existence of L^∞ solutions for the system (5.5.1), with $R = G_0 = 0$, even under those circumstances where $\mathcal{V}(\cdot)$ has a single inflection point (as it does if we take $\mathcal{V}(\cdot)$ to be of the form (5.5.2)). In [48] it is claimed (without proof) that the results obtained there for homogeneous hyperbolic systems, in one-spatial dimension, extend virtually without modification to nonhomogeneous systems of the form

$$\begin{cases} u_t - v_x + f(u,v) = 0 \\ v_t - \sigma(u)_x + g(u,v) = 0 \end{cases} \tag{5.5.7}$$

as long as f and g are smooth functions of their arguments. In a similar vein it is stated in [40] that, based upon DiPerna's work in [48], any sequence of solutions $\{(u^\varepsilon, v^\varepsilon\}$, for the regularization of (5.5.7) which is given by

$$\begin{cases} u_t^\varepsilon - v_x^\varepsilon + f(u^\varepsilon, v^\varepsilon) = \varepsilon u_{xx}^\varepsilon \\ v_t^\varepsilon - \sigma(u^\varepsilon)_x + g(u^\varepsilon, v^\varepsilon) = \varepsilon v_{xx}^\varepsilon \end{cases} \tag{5.5.8}$$

which converges in L^∞ weak*, coverges a.e. to a weak solution of (5.5.7). However, smoothness of f and g alone does not appear to be sufficient to substantiate such results and further restrictions on these functions would seem to be necessary; in order to make this point, let us assume that there exists a sufficiently smooth solution of the system (5.5.8), with initial data $u^\varepsilon(x,0) = u_0(x)$, $v^\varepsilon(x,0) = v_0(x)$, $x \in R^1$, on $R^1 \times [0,T]$. If we multiply the first equation in (5.5.8) by $\sigma(u^\varepsilon)$, the second by v^ε, integrate over $R^1 \times [0,T]$, and then integrate by parts, we obtain the *a priori* estimate

$$\begin{aligned} &\frac{1}{2}\int_{-\infty}^{\infty}(v^\varepsilon(x,t))^2\,dx + \int_{-\infty}^{\infty}\Sigma(u^\varepsilon(x,t))\,dx \\ &+\int_0^T\int_{-\infty}^{\infty}[f(u^\varepsilon,v^\varepsilon)\sigma(u^\varepsilon)+g(u^\varepsilon,v^\varepsilon)v^\varepsilon]\,dx\,dt \\ &+\varepsilon\int_0^T\int_{-\infty}^{\infty}\left[\sigma'(u^\varepsilon)(u_x^\varepsilon)^2+(v_x^\varepsilon)^2\right]\,dx\,dt \\ &=\int_{-\infty}^{\infty}v_0^2(x)\,dx+\int_{-\infty}^{\infty}\Sigma(u_0(x))\,dx \end{aligned} \tag{5.5.9}$$

where $\Sigma(u) = \int_0^u \sigma(\lambda)\,d\lambda$. If, for example, we take $\sigma(u) = (\sigma_0 + \sigma_1 u^2)u$ so that $\sigma(\cdot)$ has a single inflection point at 0, $\sigma'(u) \geq \sigma_0 > 0$, $\forall\, u \in R^1$, and $\Sigma(u) \geq 0$, $\forall\, u \in R^1$, we obtain from (5.5.9) the bound

$$\begin{aligned} &\int_0^T\int_{-\infty}^{\infty}(f(u^\varepsilon,v^\varepsilon)\sigma(u^\varepsilon)+g(u^\varepsilon,v^\varepsilon)v^\varepsilon)\,dx\,dt \\ &+\varepsilon\int_0^T\int_{-\infty}^{\infty}\left[\sigma_0(u_x^\varepsilon)^2+(v_x^\varepsilon)^2\right]\,dx\,dt \\ &\leq\frac{1}{2}\int_{-\infty}^{\infty}v_0^2(x)\,dx+\int_{-\infty}^{\infty}\Sigma(u_0(x))\,dx \end{aligned} \tag{5.5.10}$$

Now, in order to carry forth with the extension of the argument developed in DiPerna [48] to the system (5.5.7), we would first define (compare with (5.4.6)) entropy-flux pairs $\eta = \eta(u,v)$, $q = q(u,v)$ which satisfy

$$\begin{cases} \eta_u + q_v = 0 \\ \sigma'(u)\eta_v + q_u = 0 \end{cases} \tag{5.5.11}$$

with $\eta, q \in C^2$. Employing (5.5.8) we now easily compute that

$$\begin{aligned} &\eta(u^\varepsilon,v^\varepsilon)_t + q(u^\varepsilon,v^\varepsilon)_x \\ &= \epsilon(\eta_u u_{xx}^\varepsilon + \eta_v v_{xx}^\varepsilon) - \eta_u f(u^\varepsilon,v^\varepsilon) - \eta_v g(u^\varepsilon,v^\varepsilon) \end{aligned} \tag{5.5.12}$$

Critical to the augmentation of the compensated compactness argument now is the verification that the Div-Curl lemma is applicable, i.e., that

$$\eta_t + q_x \in \text{ a compact subset of } H_{\text{loc}}^{-1}(R^1 \times [0,T]) \tag{5.5.13}$$

Now if, in (5.5.10), we knew that $\forall u, v \in R^1$

$$f(u,v)\sigma(u) + g(u,v)v \geq 0 \tag{5.5.14}$$

it would then follow that for some $C > 0$, independent of ε

$$\epsilon \int_0^T \int_{-\infty}^{\infty} ((u_x^\varepsilon)^2 + (v_x^\varepsilon)^2)\, dx\, dt \leq C \tag{5.5.15}$$

which, in turn, would be sufficient to allow us to decompose the first expression on the right-hand side of (5.5.12) into one which is in a compact subset of $H_{\text{loc}}^{-1}(R^1 \times [0,T])$ plus one which is bounded in $L^1(R^1 \times [0,T])$. Also, for $\{(u^\varepsilon, v^\varepsilon)\}$ bounded in L^∞, and f, g both smooth, we would have $\eta_u f(u^\varepsilon, v^\varepsilon) + \eta_v g(u^\varepsilon, v^\varepsilon)$ bounded in L^∞ and, thus, also in L^1_{loc}. The Murat lemma [120] would then tell us, as $\eta_t + q_x \in W^{-1,\infty}$, that the right-hand side of (5.5.12) does, indeed, lie in a compact subset of $H_{\text{loc}}^{-1}(R^1 \times [0,T])$. Central to the entire argument delineated above, however, is the hypothesis (5.5.14), for without it the essential estimate (5.5.15) does not follow from the *a priori* bound (5.5.10). The bound cited in (5.5.15) is, indeed, realized in one example mentioned in DiPerna [48], namely, the case of one-dimensional nonlinear elasticity with frictional damping for which

$$\begin{cases} g(u,v) = \alpha v, & \alpha > 0 \\ f(u,v) = 0 \end{cases} \tag{5.5.16}$$

For the special case of (5.5.7) given by (5.5.16), it follows that (5.5.15) is satisfied irrespective of the particular form of $\sigma(\cdot)$; for more general situations it would appear that if one wishes to use the parabolic regularization of (5.5.7) which is given in (5.5.8), then f and g must not only be sufficiently smooth, but must also satisfy (5.5.14).

For the nonlinear transmission line problem, i.e., (5.5.1), it will be true that (5.5.15) is also satisfied, although the situation is somewhat more complicated. Indeed, the corresponding $\sigma(\cdot)$ of (5.5.7) in our case will not be the voltage function $\mathcal{V}(\cdot)$ which appears in (5.5.1) but, rather, a function $\bar{\mathcal{V}}_\varepsilon(\cdot)$ which depends on the regularization parameter ε; that this occurs is due in part to the fact that we wish to develop a physically justifiable regularization of the original system (5.5.1) and that cannot be accomplished by simply adding on diffusion terms of the form $\epsilon i_{xx}, \epsilon Q_{xx}$ to the (respective) right-hand sides of the equations in (5.5.1). It will then turn out that a bound, independent of $\varepsilon > 0$, will be valid for the appropriate expression which corresponds to the term on the left-hand side of (5.5.15), albeit for ε sufficiently small. Our problem is also complicated by the fact that because the entropy-flux pairs (η, q) must be independent of ε, an additional term, which is dependent on the regularization parameter, will appear on the right-hand side of the equation equivalent to (5.5.12); this term will be seen to have the form

$$\gamma^{\varepsilon}(x,t) = a(\varepsilon)u_x^{\varepsilon} + b(\varepsilon)v_x^{\varepsilon} \tag{5.5.17}$$

for particular functions $a(\cdot)$, $b(\cdot)$ whose forms will be delineated in § 5.7. It will then be incumbent upon us to show that

$$\|\gamma^{\varepsilon}(\cdot,\cdot)\|_{L^2(R^1\times[0,T])} \to 0, \text{ as } \varepsilon \to 0^+, \tag{5.5.18}$$

so that the Murat lemma can be applied in order to conclude that (5.5.13) is valid; this, in turn, will depend on having $a(\varepsilon)$ and $b(\varepsilon)$ tend to zero, in an appropriate manner, as $\varepsilon \to 0^+$.

Remarks: The condition delineated in [40], in connection with the system (5.5.7), namely,

$$(\mathrm{sgn} u)\sqrt{\sigma'(u)}f(u,v) + (\mathrm{sgn} v)g(u,v) \geq 0, \tag{5.5.19}$$

suffices, as a consequence of the analysis in Chueh, Conley, and Smoller [34], to ensure the existence of uniform L^∞ bounds for solutions of the Cauchy problem associated with the parabolic regularization (5.5.8), when $\sigma(\cdot)$ has an inflection point at 0, is convex on $(0,\infty)$, and concave on $(-\infty,0)$; however, (5.5.19) does not, in general, imply that (5.5.14) is valid. In fact, as we shall see in the next section, even the proof of the existence of a sufficiently smooth solution to the initial-value problem for a (parabolically) regularized system such as (5.5.8) seems to require a restriction similar to (5.5.14) and simple smoothness of both f and g does not appear to suffice here either.

We will close this section by stating the following:

Definition 5.1 By a global weak solution of (5.5.1), (5.5.4), on $R^1 \times [0,\infty)$, we will understand a pair $(\bar{i}(x,t), \bar{Q}(x,t)) \in L^\infty(R^1 \times [0,\infty)) \times L^\infty(R^1 \times [0,\infty))$ such that

$\forall\, \phi, \psi \in C_0^\infty(R^1 \times [0,\infty))$ we have

$$\int_0^\infty \int_{-\infty}^\infty \phi_t(x,t)\bar{i}(x,t)\,dx\,dt + \int_0^\infty \int_{-\infty}^\infty \phi_x(x,t)\mathcal{V}(\bar{Q}(x,t))\,dx\,dt \tag{5.5.20}$$
$$= R\int_0^\infty \phi(x,t)\bar{i}(x,t)\,dx\,dt - \int_{-\infty}^\infty \phi(x,0)i_0(x)\,dx$$

and

$$\int_0^\infty \int_{-\infty}^\infty \psi_t(x,t)\bar{Q}(x,t)\,dx\,dt + \int_0^\infty \int_{-\infty}^\infty \psi_x(x,t)\bar{i}(x,t)\,dx\,dt \tag{5.5.21}$$
$$= G_0\int_0^\infty \int_{-\infty}^\infty \psi(x,t)\mathcal{V}(\bar{Q}(x,t))\,dx\,dt - \int_{-\infty}^\infty \psi(x,0)Q_0(x)\,dx$$

5.6 The Regularized Transmission Line: Existence, Uniqueness, and Uniform Boundedness of Solutions

In order to deduce what will turn out to be an effective regularization of the distributed parameter, nonlinear transmission line equations (5.5.1), we replace the constitutive relation $v = \mathcal{V}(Q)$ by

$$v = \mathcal{V}(Q) + \varepsilon_1 i_x + \varepsilon_2 Q_{xx}; \quad \varepsilon_1,\, \varepsilon_2 > 0 \tag{5.6.1}$$

Using (5.6.1) we easily obtain, from Ohm's and Kirchhoff's laws, the system

$$\begin{cases} i_t + \mathcal{V}(Q)_x = -Ri - \varepsilon_1 i_{xx} - \varepsilon_2 Q_{xxx} \\ Q_t + \delta i_x = G_0\mathcal{V}(Q) - \varepsilon_2 G_0 Q_{xx} \end{cases} \tag{5.6.2}$$

where

$$\delta = 1 + \varepsilon_1 G_0 \tag{5.6.3}$$

We now introduce the new dependent variable

$$z(x,t) = i(x,t) - cQ_x(x,t), \tag{5.6.4}$$

with c a constant which will be determined below, and (5.6.2) becomes

$$z_t + \bar{\mathcal{V}}(Q)_x = (c\delta - \varepsilon_1)z_{xx} - Rz + p(c)Q_{xxx} \tag{5.6.5a}$$
$$Q_t + \delta z_x = -(c\delta + \varepsilon_2 G_0)Q_{xx} - G_0\mathcal{V}(Q) \tag{5.6.5b}$$

where

$$p(c) = \delta c^2 + (\varepsilon_2 G_0 - \varepsilon_1) - \varepsilon_2 \tag{5.6.6}$$

and

$$\bar{\mathcal{V}}(\xi) = (1 - cG_0)\mathcal{V}(\xi) + cR\xi, \ \forall \xi \in R^1 \tag{5.6.7}$$

It is desirable to require that the coefficients of the diffusive terms in (5.6.5a), (5.6.5b) be equal, in which case we must take

$$c = \frac{\epsilon_1 - \epsilon_2 G_0}{2\delta} \tag{5.6.8}$$

It is also desirable to enforce the vanishing of the coefficient of the dispersive term in (5.6.5a), i.e., to require that $p(c) = 0$; by virtue of (5.6.6) this means that c should assume the value

$$c = \frac{\epsilon_1 - \epsilon_2 G_0 \pm \sqrt{(\epsilon_1 - \epsilon_2 G_0)^2 + 4\delta\varepsilon_2}}{2\delta} \tag{5.6.9}$$

If we compare (5.6.8) with (5.6.9), we readily determine that in order for both results to hold, we must have

$$\epsilon_1 = -\epsilon_2 G_0 \ \pm 2\sqrt{-\epsilon_2} \tag{5.6.10}$$

so that, in (5.6.1), we want $\epsilon_2 < 0$. We also note that we can achieve

$$\epsilon_1 = |\epsilon_2| G_0 - 2\sqrt{|\epsilon_2|} < 0 \tag{5.6.11}$$

if $G_0 < 2/\sqrt{|\varepsilon_2|}$; however, as we will, eventually, be letting $|\epsilon_2| \to 0^+$ in (5.6.2), the condition that ϵ_1, as given by (5.6.11), be negative will be satisfied for $|\varepsilon_2|$ sufficiently small, irrespective of the value of G_0. We now define the regularization parameter by

$$\epsilon = \sqrt{|\varepsilon_2|} \tag{5.6.12}$$

Then the following results are easily verified:

$$\epsilon_1 = \epsilon^2 G_0 - 2\epsilon < 0 \ (\epsilon \text{ sufficiently small}) \tag{5.6.13a}$$

$$\epsilon_2 = -\epsilon^2 \tag{5.6.13b}$$

$$c\delta - \epsilon_1 = \frac{1}{2}(\epsilon_1 + |\epsilon_2| G_0) - \epsilon_1 = \epsilon \tag{5.6.13c}$$

$$c = c(\epsilon) = \frac{\epsilon^2 G_0 - \epsilon}{1 - |\epsilon^2 G_0 - 2\epsilon| G_0} \to 0, \text{ as } \epsilon \to 0^+ \tag{5.6.13d}$$

$$\delta = \delta(\varepsilon) = 1 - |\epsilon^2 G_0 - 2\epsilon| G_0 \to 1, \text{ as } \epsilon \to 0^+ \tag{5.6.13e}$$

Therefore, for $\epsilon > 0$ chosen arbitrarily, and ϵ_1, ϵ_2 defined as in (5.6.13a), (5.6.13b), we have $p(c(\varepsilon)) = 0$ and (5.6.5a), (5.6.5b) reduce to

$$z_t + \bar{\mathcal{V}}_\varepsilon(Q)_x = \epsilon z_{xx} - Rz \tag{5.6.14a}$$

$$Q_t + \delta(\varepsilon) z_x = \epsilon Q_{xx} - G_0 \mathcal{V}(Q) \tag{5.6.14b}$$

where $z(x,t)$ and $\delta(\varepsilon)$ are given, respectively, by (5.6.4), (5.6.13d), and (5.6.13e) while, $\forall\, \xi \in R^1$,

$$\bar{\mathcal{V}}_\varepsilon(\xi) = (1 - c(\varepsilon)G_0)\mathcal{V}(\xi) + c(\varepsilon)R\xi \tag{5.6.15}$$

Associated with the new system (5.6.14a), (5.6.14b) we also have the initial data

$$\begin{cases} z^\varepsilon(x,0) = i_0(x) - c(\varepsilon)Q_0'(x), \; x \in R^1 \\ Q^\varepsilon(x,0) = Q_0(x), \; x \in R^1 \end{cases} \tag{5.6.16}$$

where, henceforth, we will denote solutions of the system (5.6.14a), (5.6.14b), for $\varepsilon > 0$, by $(z^\varepsilon, Q^\varepsilon)$. We note, in passing, that

$$\bar{\mathcal{V}}_\varepsilon'(\xi) = (1 - c(\varepsilon)G_0)\mathcal{V}'(\xi) + c(\varepsilon)R > 0, \quad \forall\, \xi \in R^1 \tag{5.6.17}$$

for $\varepsilon > 0$ chosen sufficiently small; this is, of course, a consequence of (5.5.3a) and (5.6.13d). We are now in a position to state and prove the fundamental existence and uniqueness result for the initial-value problem consisting of (5.6.14a), (5.6.14b), (5.6.15), and (5.6.16):

Theorem 5.3 ([55]): For $\xi \in R^1$, let $\mathcal{V}(\xi)$ be given by (5.5.2) and $\bar{\mathcal{V}}_\varepsilon(\xi)$ by (5.6.15), where $c(\varepsilon)$ is defined as per (5.6.13d). Let $\delta(\varepsilon)$ be defined as in (5.6.13e). If $i_0(\cdot) \in W^{1,2}(-\infty,\infty)$ and $Q_0(\cdot) \in W^{2,2}(-\infty,\infty)$, then for $\varepsilon > 0$ sufficiently small, there exists a unique solution $(z^\varepsilon, Q^\varepsilon)$ of (5.6.14a), (5.6.14b), (5.6.16) on $R^1 \times [0,T]$, for any $T < \infty$, which satisfies:

(i) $z^\varepsilon, Q^\varepsilon \in L^\infty([0,T]; W^{1,2}(-\infty,\infty)) \cap L^2([0,T]; W^{1,2}(-\infty,\infty))$ (5.6.18a)

(ii) $z^\varepsilon, Q^\varepsilon \in L^\infty([0,T]; L^2(-\infty,\infty))$ (5.6.18b)

(iii) $\dfrac{\partial z^\varepsilon}{\partial t}, \dfrac{\partial Q^\varepsilon}{\partial t} \in L^2((-\infty,\infty) \times [0,T])$ (5.6.18c)

Proof: We begin by modifying the constitutive function $\mathcal{V}(\cdot)$ by setting, for $\lambda > 0$, $\xi \in R^1$

$$\mathcal{V}_\lambda(\xi) = \frac{\mathcal{V}(\xi)}{1+\lambda\xi^2} \tag{5.6.19}$$

in which case, for $\xi \in R^1$, $\bar{\mathcal{V}}_\varepsilon(\xi)$ is modified to

$$\bar{\mathcal{V}}_{\varepsilon,\lambda}(\xi) = (1-c(\varepsilon))\mathcal{V}_\lambda(\xi) + c(\varepsilon)R\xi \tag{5.6.20}$$

We note that by virtue of (5.5.2) and (5.6.19) the new constitutive function $\mathcal{V}_\lambda(\xi)$ exhibits linear growth in ξ for all $\lambda > 0$. We now denote by $(S_{\varepsilon,\lambda})$ the system (5.6.14a), (5.6.14b) with $\mathcal{V}(\cdot)$ replaced by $\mathcal{V}_\lambda(\cdot)$, i.e.,

$$\begin{cases} z_t + \bar{\mathcal{V}}_{\varepsilon,\lambda}(Q)_x = \epsilon z_{xx} - Rz \\ Q_t + \delta(\varepsilon) z_x = \epsilon Q_{xx} - G_0\mathcal{V}_\lambda(Q) \end{cases} \tag{5.6.21}$$

The first step in the proof of the theorem will be to show that there exist global solutions $(z^{\varepsilon,\lambda}, Q^{\varepsilon,\lambda})$ of (5.6.21), for fixed $\varepsilon > 0$, $\lambda > 0$, which satisfy the initial conditions

$$\begin{cases} z^{\varepsilon,\lambda}(x,0) = z_0(x) \equiv i_0(x) - c(\varepsilon)Q_0'(x), \quad x \in R^1 \\ Q^{\varepsilon,\lambda}(x,0) = Q_0(x), \quad x \in R^1 \end{cases} \tag{5.6.22}$$

with $i_0(\cdot) \in W^{1,2}(-\infty,\infty)$ and $Q_0(\cdot) \in W^{2,2}(-\infty,\infty)$; for this purpose we introduce an orthonormal basis $\{w_j(x)\}$ in $W^{1,2}(-\infty,\infty)$ and study the Galerkin approximations

$$\begin{cases} z_n^{\varepsilon,\lambda}(x,t) = \sum_{j=1}^{n} a_j(\varepsilon,\lambda;t) w_j(x) \\ Q_n^{\varepsilon,\lambda}(x,t) = \sum_{j=1}^{n} b_j(\varepsilon,\lambda;t) w_j(x) \end{cases} \tag{5.6.23}$$

In the standard fashion we first substitute the Galerkin approximations (5.6.23) into the system (5.6.21) so as to obtain, for fixed $\varepsilon > 0$, $\lambda > 0$, a system of first-order ordinary differential equations for the coefficients $a_j(\varepsilon,\lambda;t)$ and $b_j(\varepsilon,\lambda;t)$, in (5.6.23); this, in turn, yields an (approximate) solution for the system (5.6.21), subject to the initial conditions (5.6.22), on some time interval $[0,t_0]$. If $t_0 < \infty$, then we can extend this local (approximate) solution to all of $[0,\infty)$ as follows: for any $t < t_0$ we multiply the first equation in (5.6.21) by $z_n^{\varepsilon,\lambda}$ and the second by $Q_n^{\varepsilon,\lambda}$, and then integrate the resulting equations over $R^1 \times [0,t]$ so as to obtain

$$\frac{1}{2}\int_{-\infty}^{\infty} (z_n^{\varepsilon,\lambda}(x,t))^2\,dx - \frac{1}{2}\int_{-\infty}^{\infty} z_0^2(x)\,dx \tag{5.6.24a}$$

$$-\int_0^t \int_{-\infty}^{\infty} \bar{\mathcal{V}}_{\varepsilon,\lambda}(Q_n^{\varepsilon,\lambda}(x,\tau))(\partial_x z_n^{\varepsilon,\lambda}(x,\tau))\,dx\,d\tau$$

$$+\varepsilon \int_0^t \int_{-\infty}^{\infty} (\partial_x z_n^{\varepsilon,\lambda}(x,\tau))^2\,dx\,d\tau + R\int_0^t \int_{-\infty}^{\infty} (z_n^{\varepsilon,\lambda}(x,t))^2\,dx\,d\tau = 0$$

and

$$\frac{1}{2}\int_{-\infty}^{\infty} (Q_n^{\varepsilon,\lambda}(x,t))^2\,dx - \frac{1}{2}\int_{-\infty}^{\infty} Q_0^2(x)\,dx \tag{5.6.24b}$$

$$+\delta(\varepsilon)\int_0^t \int_{-\infty}^{\infty} (\partial_x z_n^{\varepsilon,\lambda}(x,\tau))Q_n^{\varepsilon,\lambda}(x,\tau)\,dx\,d\tau$$

$$+\varepsilon \int_0^t \int_{-\infty}^{\infty} (\partial_x Q_n^{\varepsilon,\lambda}(x,\tau))^2\,dx\,d\tau$$

$$+G_0 \int_0^t \int_{-\infty}^{\infty} \mathcal{V}_\lambda(Q_n^{\varepsilon,\lambda}(x,t))Q_n^{\varepsilon,\lambda}(x,t)\,dx\,d\tau = 0$$

If we set

$$A_0 = \frac{1}{2}\int_{-\infty}^{\infty} (z_0^2(x) + Q_0^2(x))\,dx \tag{5.6.25}$$

and take note of the fact that $\forall\,\xi \in R^1$, and all $\lambda > 0$, $\mathcal{V}_\lambda(\xi)\cdot\xi \geq 0$, then we obtain, upon adding (5.6.24a) and (5.6.24b), the following estimate

$$\frac{1}{2}\int_{-\infty}^{\infty} \left[(z_n^{\varepsilon,\lambda}(x,t))^2 + (Q_n^{\varepsilon,\lambda}(x,t))^2\right]\,dx \tag{5.6.26}$$

$$+\varepsilon \int_0^t \int_{-\infty}^{\infty} \left\{(\partial_x z_n^{\varepsilon,\lambda}(x,\tau))^2 + (\partial_x Q_n^{\varepsilon,\lambda}(x,\tau))^2\right\}\,dx\,d\tau$$

$$+R\int_0^t \int_{-\infty}^{\infty} (z_n^{\varepsilon,\lambda}(x,t))^2\,dx\,d\tau \leq A_0$$

$$+\int_0^t \int_{-\infty}^{\infty} \bar{\mathcal{V}}_{\varepsilon,\lambda}(Q_n^{\varepsilon,\lambda}(x,\tau))(\partial_x z_n^{\varepsilon,\lambda}(x,\tau))\,dx\,d\tau$$

$$+\delta(\varepsilon)\int_0^t \int_{-\infty}^{\infty} \left|Q_n^{\varepsilon,\lambda}(x,\tau)(\partial_x z_n^{\varepsilon,\lambda}(x,\tau))\right|\,dx\,d\tau$$

However, for arbitrary $\gamma_1, \gamma_2 > 0$ we may estimate the two integrals on the right-hand side of (5.6.26) as follows:

$$\int_0^t \int_{-\infty}^{\infty} \bar{\mathcal{V}}_{\varepsilon,\lambda}(Q_n^{\varepsilon,\lambda}(x,\tau))(\partial_x z_n^{\varepsilon,\lambda}(x,\tau))\,dx\,d\tau \tag{5.6.27}$$

$$\leq \frac{1}{2}\gamma_1 \int_0^t \int_{-\infty}^{\infty} \bar{\mathcal{V}}_{\varepsilon,\lambda}^2(Q_n^{\varepsilon,\lambda}(x,\tau))\,dx\,d\tau$$

$$+\frac{1}{2\gamma_1}\int_0^t \int_{-\infty}^{\infty} (\partial_x z_n^{\varepsilon,\lambda}(x,\tau))^2\,dx\,d\tau$$

$$\leq k_{\varepsilon,\lambda} \int_0^t \int_{-\infty}^{\infty} (Q_n^{\varepsilon,\lambda}(x,\tau))^2 \, dx \, d\tau$$
$$+\frac{1}{2\gamma_1} \int_0^t \int_{-\infty}^{\infty} (\partial_x z_n^{\varepsilon,\lambda}(x,\tau))^2 \, dx \, d\tau$$

for some constant $k_{\varepsilon,\lambda} > 0$, where we have used the linear growth of $\bar{\mathcal{V}}_{\varepsilon,\lambda}(\cdot)$, for fixed $\varepsilon > 0$, $\lambda > 0$, and

$$\int_0^t \int_{-\infty}^{\infty} \left| Q_n^{\varepsilon,\lambda}(x,\tau)(\partial_x z_n^{\varepsilon,\lambda}(x,\tau)) \right| \, dx \, d\tau \tag{5.6.28}$$
$$\leq \frac{1}{2}\gamma_2 \int_0^t \int_{-\infty}^{\infty} (Q_n^{\varepsilon,\lambda}(x,\tau))^2 \, dx \, d\tau$$
$$+\frac{1}{2\gamma_2} \int_0^t \int_{-\infty}^{\infty} (\partial_x z_n^{\varepsilon,\lambda}(x,\tau))^2 \, dx \, d\tau$$

If we now choose γ_1, $\gamma_2 > 0$ so large that

$$\epsilon - \frac{1}{2}\left(\frac{1}{\gamma_1} + \frac{1}{\gamma_2}\right) \geq \frac{1}{2}\varepsilon \tag{5.6.29}$$

and set

$$m_{\varepsilon,\lambda} = k_{\varepsilon,\lambda} + \frac{1}{2}\gamma_2 \tag{5.6.30}$$

then from (5.6.26)-(5.6.28) we obtain the estimate

$$\frac{1}{2}\int_{-\infty}^{\infty} \left[(z_n^{\varepsilon,\lambda}(x,t))^2 + (Q_n^{\varepsilon,\lambda}(x,t))^2 \right] dx \tag{5.6.31}$$
$$+\epsilon \int_0^t \int_{-\infty}^{\infty} \left\{ \frac{1}{2}(\partial_x z_n^{\varepsilon,\lambda}(x,\tau))^2 + (\partial_x Q_n^{\varepsilon,\lambda}(x,\tau))^2 \right\} dx \, d\tau$$
$$+R \int_0^t \int_{-\infty}^{\infty} (z_n^{\varepsilon,\lambda}(x,\tau))^2 \, dx \, d\tau$$
$$\leq A_0 + m_{\varepsilon,\lambda} \int_0^t \int_{-\infty}^{\infty} (Q_n^{\varepsilon,\lambda}(x,\tau))^2 \, dx \, d\tau$$

From the estimate (5.6.31) we obtain, directly, that

$$H(t) - l_{\varepsilon,\lambda} \int_0^t H(\tau) \, d\tau \leq H_0, \quad t < t_0 \tag{5.6.32}$$

with $l_{\varepsilon,\lambda} = 2m_{\varepsilon,\lambda}$, $H_0 = 2A_0$, and

$$H(t) = \int_{-\infty}^{\infty} (Q_n^{\varepsilon,\lambda}(x,t))^2 \, dx \tag{5.6.33}$$

Applying Gronwall's inequality to (5.6.32) now yields the estimates

$$\int_{-\infty}^{\infty} (Q_n^{\varepsilon,\lambda}(x,t))^2 \, dx \leq H_0 \exp(l_{\varepsilon,\lambda} t), \tag{5.6.34a}$$

and

$$\int_0^t \int_{-\infty}^{\infty} (Q_n^{\varepsilon,\lambda}(x,\tau))^2 \, dx \, d\tau \leq \frac{H_0}{l_{\varepsilon,\lambda}} \left\{ \exp(l_{\varepsilon,\lambda} t) - 1 \right\}, \tag{5.6.34b}$$

for $t < t_0$, so that both $\lim\limits_{t \to t_0^-} \int_{-\infty}^{\infty} (Q_n^{\varepsilon,\lambda}(x,t))^2 \, dx$ and $\lim\limits_{t \to t_0^-} \int_0^t \int_{-\infty}^{\infty} (Q_n^{\varepsilon,\lambda}(x,\tau))^2 \, dx \, d\tau$ are bounded, for arbitrary $\varepsilon > 0$, with the bounds being independent of n. From (5.6.31) and (5.6.34a), (5.6.34b), it now follows that the integrals listed below are also bounded, as $t \to t_0^-$, independently of n:

$$\begin{cases} \int_{-\infty}^{\infty} (z_n^{\varepsilon,\lambda}(x,t))^2 \, dx, & \int_0^t \int_{-\infty}^{\infty} (z_n^{\varepsilon,\lambda}(x,\tau))^2 \, dx \, d\tau \\ \int_0^t \int_{-\infty}^{\infty} (\partial_x z_n^{\varepsilon,\lambda}(x,\tau))^2 \, dx \, d\tau, & \int_0^t \int_{-\infty}^{\infty} (\partial_x Q_n^{\varepsilon,\lambda}(x,\tau))^2 \, dx \, d\tau \end{cases} \tag{5.6.35}$$

Our next step is to multiply, for $t < t_0$, the first equation in (5.6.21) by $\partial_t z_n^{\varepsilon,\lambda}$, and the second equation in (5.6.21) by $\partial_t Q_n^{\varepsilon,\lambda}$; then integrating the results over $R^1 \times [0,t]$ we obtain

$$\begin{aligned} &\int_0^t \int_{-\infty}^{\infty} (\partial_t z_n^{\varepsilon,\lambda}(x,\tau))^2 \, dx \, d\tau + \frac{\varepsilon}{2} \int_{-\infty}^{\infty} (\partial_x z_n^{\varepsilon,\lambda}(x,t))^2 \, dx \\ &\quad + \frac{1}{2} R \int_{-\infty}^{\infty} (z_n^{\varepsilon,\lambda}(x,t))^2 \, dx + \int_0^t \int_{-\infty}^{\infty} (\partial_x \bar{\mathcal{V}}_{\varepsilon,\lambda}(Q_n^{\varepsilon,\lambda}(x,\tau))) \partial_t z_n^{\varepsilon,\lambda}(x,\tau) \, dx \, d\tau = A_1^\varepsilon \end{aligned} \tag{5.6.36}$$

with

$$A_1^\varepsilon = \frac{\varepsilon}{2} \int_{-\infty}^{\infty} (z_0'(x))^2 \, dx + \frac{R}{2} \int_{-\infty}^{\infty} z_0^2(x) \, dx \tag{5.6.37}$$

and

$$\begin{aligned} &\int_0^t \int_{-\infty}^{\infty} (\partial_t Q_n^{\varepsilon,\lambda}(x,\tau))^2 \, dx \, d\tau + \frac{\varepsilon}{2} \int_{-\infty}^{\infty} (\partial_x Q_n^{\varepsilon,\lambda}(x,t))^2 \, dx \\ &\quad + \delta(\varepsilon) \int_0^t \int_{-\infty}^{\infty} (\partial_t Q_n^{\varepsilon,\lambda}(x,\tau)) \partial_x z_n^{\varepsilon,\lambda}(x,\tau) \, dx \, d\tau \\ &\quad + G_0 \int_0^t \int_{-\infty}^{\infty} \mathcal{V}_\lambda(Q_n^{\varepsilon,\lambda}(x,\tau)) \partial_t Q_n^{\varepsilon,\lambda}(x,\tau) \, dx \, d\tau = A_2^\varepsilon \end{aligned} \tag{5.6.38}$$

with

$$A_2^\varepsilon = \frac{\varepsilon}{2} \int_{-\infty}^{\infty} (Q_0'(x))^2 \, dx \tag{5.6.39}$$

However, as $\bar{\mathcal{V}}_{\varepsilon,\lambda}'(\xi)$ is bounded, for $\xi \in R^1$, for any $\varepsilon > 0$, $\lambda > 0$, there exists $n_{\varepsilon,\lambda} > 0$ such that for any $\gamma_3 > 0$,

$$\int_0^t \int_{-\infty}^{\infty} \left| (\partial_x \bar{\mathcal{V}}_{\varepsilon,\lambda}(Q_n^{\varepsilon,\lambda}(x,\tau))) \partial_t z_n^{\varepsilon,\lambda}(x,\tau) \right| \, dx \, d\tau \tag{5.6.40}$$

$$\leq \frac{1}{2}\gamma_3 \int_0^t \int_{-\infty}^{\infty} \bar{\mathcal{V}}'^2_{\varepsilon,\lambda}(Q_n^{\varepsilon,\lambda}(x,\tau))(\partial_x Q_n^{\varepsilon,\lambda}(x,\tau))^2\, dx\, d\tau$$
$$+\frac{1}{2\gamma_3}\int_0^t \int_{-\infty}^{\infty} (\partial_t z_n^{\varepsilon,\lambda}(x,\tau))^2\, dx\, d\tau$$
$$\leq n_{\varepsilon,\lambda}\int_0^t \int_{-\infty}^{\infty} (\partial_x Q_n^{\varepsilon,\lambda}(x,\tau))^2\, dx\, d\tau$$
$$+\frac{1}{2\gamma_3}\int_0^t \int_{-\infty}^{\infty} (\partial_t z_n^{\varepsilon,\lambda}(x,\tau))^2\, dx\, d\tau$$

Therefore, choosing γ_3 so large that $1-\frac{1}{2\gamma_3}\geq\frac{1}{2}$, we obtain from (5.6.36) the estimate

$$\frac{1}{2}\int_0^t \int_{-\infty}^{\infty} (\partial_t z_n^{\varepsilon,\lambda}(x,\tau))^2\, dx\, d\tau + \frac{\varepsilon}{2}\int_{-\infty}^{\infty} (\partial_x z_n^{\varepsilon,\lambda}(x,t))^2\, dx \tag{5.6.41}$$
$$+\frac{1}{2}R\int_{-\infty}^{\infty} (z_n^{\varepsilon,\lambda}(x,t))^2\, dx \leq A_1^\varepsilon + n_{\varepsilon,\lambda}\int_0^t \int_{-\infty}^{\infty} (\partial_x Q_n^{\varepsilon,\lambda}(x,t))^2\, dx\, d\tau$$

Also, as $\int_0^t \int_{-\infty}^{\infty} (\partial_x Q_n^{\varepsilon,\lambda}(x,\tau))^2\, dx\, d\tau$ is bounded, for fixed $\varepsilon, \lambda > 0$, independently of n, as $t \to t_0^-$, it follows now from (5.6.41) that the same conclusion is valid for the integrals

$$\begin{cases} \int_{-\infty}^{\infty} (\partial_x z_n^{\varepsilon,\lambda}(x,t))^2\, dx \\ \int_0^t \int_{-\infty}^{\infty} (\partial_t z_n^{\varepsilon,\lambda}(x,\tau))^2\, dx\, d\tau \end{cases} \tag{5.6.42}$$

Finally, we turn to (5.6.38); with $\hat{\mathcal{V}}_\lambda(\xi) = \int_0^\xi \mathcal{V}_\lambda(\rho)\, d\rho$, $\forall\, \xi \in R^1$, we first note that

$$\int_0^t \int_{-\infty}^{\infty} \mathcal{V}_\lambda(Q_n^{\varepsilon,\lambda}(x,\tau))\partial_t Q_n^{\varepsilon,\lambda}(x,\tau)\, dx\, d\tau$$
$$= \int_{-\infty}^{\infty} \left[\hat{\mathcal{V}}_\lambda(Q_n^{\varepsilon,\lambda}(x,t)) - \hat{\mathcal{V}}_\lambda(Q_0(x))\right] dx$$

Therefore, because $\hat{\mathcal{V}}_\lambda(\xi) \geq 0$, $\forall\, \xi \in R^1$, we obtain from (5.6.38), for some $\gamma_4 > 0$, an estimate of the form

$$\left(1 - \frac{\delta(\varepsilon)}{2\gamma_4}\right)\int_0^t \int_{-\infty}^{\infty} (\partial_t Q_n^{\varepsilon,\lambda}(x,\tau))^2\, dx\, d\tau \tag{5.6.43}$$
$$+\frac{\varepsilon}{2}\int_{-\infty}^{\infty} (\partial_x Q_n^{\varepsilon,\lambda}(x,t))^2\, dx \leq A_3^\varepsilon$$
$$+\frac{\gamma_4\delta(\varepsilon)}{2}\int_0^t \int_{-\infty}^{\infty} (\partial_x z_n^{\varepsilon,\lambda}(x,\tau))^2\, dx\, d\tau$$

where we have set

$$A_3^\varepsilon = \frac{\varepsilon}{2}\int_{-\infty}^{\infty} Q_0'^2(x)\, dx + G_0\int_{-\infty}^{\infty} \hat{\mathcal{V}}_\lambda(Q_0(x))\, dx \tag{5.6.44}$$

Choosing γ_4 so large that $2\gamma_4 - \delta(\varepsilon) > 0$, and using the fact that $\int_0^t \int_{-\infty}^{\infty} (\partial_x z_n^{\varepsilon,\lambda}(x,\tau))^2\, dx\, d\tau$ is bounded, independent of n, for fixed $\varepsilon, \lambda > 0$, as $t \to t_0^-$, we now find that as a consequence of (5.6.43) the following two integrals are also bounded, as $t \to t_0^-$, for fixed $\varepsilon, \lambda > 0$, independent of n:

$$\begin{cases} \int_0^t \int_{-\infty}^{\infty} (\partial_t Q_n^{\varepsilon,\lambda}(x,\tau))^2\, dx\, d\tau \\ \int_{-\infty}^{\infty} (\partial_x Q_n^{\varepsilon,\lambda}(x,t))^2\, dx \end{cases} \tag{5.6.45}$$

Combining our results we see that we can achieve an estimate of the form

$$\begin{aligned} &\int_{-\infty}^{\infty} \left[(z_n^{\varepsilon,\lambda})^2 + (Q_n^{\varepsilon,\lambda})^2 + (\partial_x z_n^{\varepsilon,\lambda})^2 + (\partial_x Q_n^{\varepsilon,\lambda})^2 \right] dx \\ &\quad + \int_0^t \int_{-\infty}^{\infty} \left[(z_n^{\varepsilon,\lambda})^2 + (Q_n^{\varepsilon,\lambda})^2 + (\partial_x z_n^{\varepsilon,\lambda})^2 + (\partial_x Q_n^{\varepsilon,\lambda})^2 + (\partial_t z_n^{\varepsilon,\lambda})^2 + (\partial_t Q_n^{\varepsilon,\lambda})^2 \right] dx\, d\tau \\ &\qquad \le \Gamma(\varepsilon, \lambda; t) \end{aligned} \tag{5.6.46}$$

for $t < t_0^-$, where $\Gamma(\varepsilon, \lambda; t)$ is independent of n and satisfies, for each fixed pair $\varepsilon, \lambda > 0$, $\lim_{t \to t_0^-} \Gamma(\varepsilon, \lambda; t) < \infty$. As a direct consequence of (5.6.46), we can conclude that the local solution $(z_n^{\varepsilon,\lambda}, Q_n^{\varepsilon,\lambda})$ on $[0, t_0)$ may be continued, in t, to all of $[0, \infty)$; also, as $\Gamma(\varepsilon, \lambda; t)$ in (5.6.46) is independent of n, we may let $n \to \infty$ and, in this manner, we obtain for each fixed $\lambda > 0$ a global solution $(z^{\varepsilon,\lambda}, Q^{\varepsilon,\lambda})$ of the Cauchy problem (5.6.21), (5.6.22). It also follows from (5.6.46) that, for each fixed $T > 0$, the functions $(z^{\varepsilon,\lambda}, Q^{\varepsilon,\lambda})$ which are obtained as the limits of the Galerkin approximants $(z_n^{\varepsilon,\lambda}, Q_n^{\varepsilon,\lambda})$ satisfy, for each fixed pair $\varepsilon, \lambda > 0$, the regularity which is given in (i)–(iii) of the theorem. In fact, it can also be proven that $(z^{\varepsilon,\lambda}, Q^{\varepsilon,\lambda}) \in C(\bar{S}_T) \times C(\bar{S}_T)$ and that $Q^{\varepsilon,\lambda} \in L^4(S_T) \cap L^6(S_T)$, where $S_T = R^1 \times [0, T)$, $T < \infty$, but these results will not be needed for what follows.

We now want to pass to the limit in the solutions $(z^{\varepsilon,\lambda}, Q^{\varepsilon,\lambda})$ as $\lambda \to 0^+$ but, because of the λ dependence of $\Gamma(\varepsilon, \lambda; t)$, we cannot accomplish this by appealing to (5.6.46). By tracing backwards the steps which led to (5.6.46), it is not difficult to see that the problem here lies with the estimates (5.6.34a), (5.6.34b); this difficulty may, however, be circumvented. Denoting, as above, the global (in time) solutions of (5.6.21), (5.6.22) by $z^{\varepsilon,\lambda}$, $Q^{\varepsilon,\lambda}$, we multiply the first equation in (5.6.21) by $\delta(\varepsilon) q^{\varepsilon,\lambda}$, the second by $\bar{V}_{\varepsilon,\lambda}(Q^{\varepsilon,\lambda})$, integrate over $R^1 \times [0, T]$, and then add the resulting equations; there results the equation

$$
\begin{aligned}
&\frac{\delta(\varepsilon)}{2}\int_{-\infty}^{\infty}(z^{\varepsilon,\lambda}(x,T))^2\,dx+\int_{-\infty}^{\infty}\hat{\mathcal{V}}_\lambda(Q^{\varepsilon,\lambda}(x,T))\,dx \qquad (5.6.47)\\
&+\varepsilon\delta(\varepsilon)\int_0^T\int_{-\infty}^{\infty}(z_x^{\varepsilon,\lambda}(x,t))^2\,dx\,dt\\
&+\varepsilon\int_0^T\int_{-\infty}^{\infty}\bar{\mathcal{V}}'_{\varepsilon,\lambda}(Q^{\varepsilon,\lambda}(x,t))(Q_x^{\varepsilon,\lambda}(x,t))^2\,dx\,dt\\
&+R\delta(\varepsilon)\int_0^T\int_{-\infty}^{\infty}(z^{\varepsilon,\lambda}(x,t))^2\,dx\,dt\\
&+G_0\int_0^T\int_{-\infty}^{\infty}\mathcal{V}_\lambda(Q^{\varepsilon,\lambda}(x,t))\bar{\mathcal{V}}_{\varepsilon,\lambda}(Q^{\varepsilon,\lambda})\,dx\,dt\\
&=\frac{1}{2}\delta(\varepsilon)\int_{-\infty}^{\infty}(z^{\varepsilon,\lambda}(x,0))^2\,dx+\int_{-\infty}^{\infty}\hat{\mathcal{V}}_\lambda(Q^{\varepsilon,\lambda}(x,0))\,dx
\end{aligned}
$$

However,

$$
\begin{aligned}
&\int_0^T\int_{-\infty}^{\infty}\mathcal{V}_\lambda(Q^{\varepsilon,\lambda}(x,t))\bar{\mathcal{V}}_{\varepsilon,\lambda}(Q^{\varepsilon,\lambda}(x,t))\,dx\,dt \qquad (5.6.48)\\
&=(1-c(\varepsilon)G_0)\int_0^T\int_{-\infty}^{\infty}\mathcal{V}_\lambda^2(Q^{\varepsilon}(x,t))\,dx\,dt\\
&+Rc(\varepsilon)\int_0^T\int_{-\infty}^{\infty}Q^{\varepsilon,\lambda}(x,t)\mathcal{V}_\lambda(Q^{\varepsilon,\lambda}(x,t))\,dx\,dt\\
&\geq(1-c(\varepsilon)G_0)\int_0^T\int_{-\infty}^{\infty}\mathcal{V}_\lambda^2(Q^{\varepsilon,\lambda}(x,t))\,dx\,dt
\end{aligned}
$$

as $\xi\mathcal{V}_\lambda(\xi)\geq 0$, $\forall\xi\in R^1$. Using (5.6.48) in (5.6.47), we now obtain the estimate

$$
\begin{aligned}
&\int_{-\infty}^{\infty}(z^{\varepsilon,\lambda}(x,T))^2\,dx+\frac{\varepsilon}{2}\int_0^T\int_{-\infty}^{\infty}(z_x^{\varepsilon,\lambda}(x,t))^2\,dx\,dt \qquad (5.6.49)\\
&+\varepsilon\int_0^T\int_{-\infty}^{\infty}\bar{\mathcal{V}}'_{\varepsilon,\lambda}(Q^{\varepsilon,\lambda}(x,t))(Q_x^{\varepsilon,\lambda}(x,t))^2\,dx\,dt\\
&+\frac{R}{2}\int_0^T\int_{-\infty}^{\infty}(z^{\varepsilon,\lambda}(x,t))^2\,dx\,dt+\int_{-\infty}^{\infty}\hat{\mathcal{V}}_\lambda(Q^{\varepsilon,\lambda}(x,t))\,dx\\
&+G_0(1-c(\varepsilon)G_0)\int_0^T\int_{-\infty}^{\infty}\mathcal{V}_\lambda^2(Q^{\varepsilon}(x,t))\,dx\,dt\\
&\leq\frac{1}{2}\int_{-\infty}^{\infty}z_0^2(x)\,dx+\int_{-\infty}^{\infty}\hat{\mathcal{V}}_\lambda(Q_0(x))\,dx
\end{aligned}
$$

We now make the following observations with regard to the estimate (5.6.49):

(a) The results of Chueh, Conley, and Smoller [34] on invariant regions for parabolic systems may be applied to the system (5.6.21) so as to yield the conclusion that

$$\|(z^{\varepsilon,\lambda}, Q^{\varepsilon,\lambda}\|_{(L^\infty(R^1\times[0,\infty)))^2} \le k(\varepsilon) \tag{5.6.50}$$

for each $\varepsilon > 0$, with $k(\varepsilon) > 0$ independent of $\lambda > 0$.

(b) For any real valued Q defined on $R^1 \times [0,\infty)$ such that $\|Q\|_{L^\infty(R^1\times[0,\infty))} \le k$

$$\begin{aligned} \hat{\mathcal{V}}_\lambda(Q) &= \int_0^Q \left(\frac{\beta_0+\beta_1\xi^2}{1+\lambda\xi^2}\right)\xi\,d\xi \qquad (5.6.51)\\ &= \left[\frac{\beta_0+\beta_1\bar{\xi}^2}{1+\lambda\bar{\xi}^2}\right]\frac{Q^2}{2} \qquad (0\le\bar{\xi}\le Q)\\ &\ge \frac{1}{2}\left[\frac{\beta_0}{1+\frac{1}{2}k^2}\right]Q^2 \end{aligned}$$

and, in particular, by virtue of (5.6.50), (5.6.51) applies with $Q^{\varepsilon,\lambda}$, $k(\varepsilon)$ replacing Q, k.

(c) By virtue of (5.5.2), for both $\varepsilon > 0$ and $\lambda > 0$ sufficiently small,

$$\bar{\mathcal{V}}'_{\varepsilon,\lambda}(Q^{\varepsilon,\lambda}) = (1-c(\varepsilon))\mathcal{V}'_\lambda(Q^{\varepsilon,\lambda}) + c(\varepsilon)RQ^{\varepsilon,\lambda} \ge \frac{1}{2}\beta_0 \tag{5.6.52}$$

as $\|Q^{\varepsilon,\lambda}\|_{L^\infty(R^1\times[0,\infty))} \le k(\varepsilon)$.

(d) For λ sufficiently small and any $\varepsilon > 0$,

$$\begin{aligned} \mathcal{V}^2_\lambda(Q^{\varepsilon,\lambda}) &= \left[\frac{\beta_0+\beta_1(Q^{\varepsilon,\lambda})^2}{1+\lambda(Q^{\varepsilon,\lambda})^2}\right](Q^{\varepsilon,\lambda})^2 \qquad (5.6.53)\\ &\ge \left[\frac{\beta_0}{1+\frac{1}{2}k^2(\varepsilon)}\right](Q^{\varepsilon,\lambda})^2 \end{aligned}$$

Employing (5.6.51)-(5.6.53) in (5.6.49) we obtain, for $\varepsilon > 0, \lambda > 0$ sufficiently small, the estimate

$$\begin{aligned} &\int_{-\infty}^{\infty}(z^{\varepsilon,\lambda}(x,T))^2\,dx + \frac{\varepsilon}{2}\int_0^T\int_{-\infty}^{\infty}(z_x^{\varepsilon,\lambda}(x,t))^2\,dx\,dt \qquad (5.6.54)\\ &+\frac{\varepsilon\beta_0}{2}\int_0^t\int_{-\infty}^{\infty}(Q_x^{\varepsilon,\lambda}(x,t))^2\,dx\,dt + \frac{R}{2}\int_0^T\int_{-\infty}^{\infty}(z^{\varepsilon,\lambda}(x,t))^2\,dx\,dt\\ &+f(\varepsilon)\int_{-\infty}^{\infty}(Q^{\varepsilon,\lambda}(x,t))^2\,dx + g(\varepsilon)\int_0^T\int_{-\infty}^{\infty}(Q^{\varepsilon,\lambda}(x,t))^2\,dx\,dt\\ &\le \frac{1}{2}\int_{-\infty}^{\infty} z_0^2(x)\,dx + \int_{-\infty}^{\infty}\hat{\mathcal{V}}_\lambda(Q_0(x))\,dx \end{aligned}$$

with

$$f(\varepsilon) = \frac{1}{2}\left[\frac{\beta_0}{1+\frac{1}{2}k^2(\varepsilon)}\right] \tag{5.6.55a}$$

$$g(\varepsilon) = 2G_0(1 - c(\varepsilon)G_0)f(\varepsilon) \tag{5.6.55b}$$

However, in a manner similar to (5.6.1),

$$\hat{\mathcal{V}}_\lambda(Q_0) \le \left[\frac{\beta_0 + \beta_1\|Q_0\|^2_{L^\infty(R^1)}}{2}\right] Q_0^2 \tag{5.6.56}$$

so that $\forall\, \lambda > 0$

$$\int_{-\infty}^{\infty} \hat{\mathcal{V}}_\lambda(Q_0(x))\, dx \le \frac{1}{2}\left(\beta_0 + \beta_1\|Q_0\|^2_{L^\infty(R^1)}\right)\int_{-\infty}^{\infty} Q_0^2(x)\, dx \tag{5.6.57}$$

As a consequence of (5.6.57), it is now clear that the right-hand side of the estimate (5.6.54) is bounded by a nonnegative functional of the initial data which is independent of $\lambda > 0$.

We now return to the estimate (5.6.31) which also holds, of course, in the limit as $n \to \infty$, i.e., for $z^{\varepsilon,\lambda}$, $Q^{\varepsilon,\lambda}$ replacing $z_n^{\varepsilon,\lambda}$, $Q_n^{\varepsilon,\lambda}$. In (5.6.31) we now have on the right-hand side, in the limit as $n \to \infty$, the expression

$$A_0 + m_{\varepsilon,\lambda}\int_0^t \int_{-\infty}^{\infty} (Q^{\varepsilon,\lambda}(x,\tau))^2\, dx\, d\tau \tag{5.6.58}$$

but now we see that by virtue of the estimate (5.6.54), the integral in (5.6.58) is bounded, independent of λ. Also, for the coefficient $m_{\varepsilon,\lambda}$ in (5.6.58), we recall that $m_{\varepsilon,\lambda} = \frac{1}{2} + k_{\varepsilon,\lambda}$ (i.e., (5.6.30)) where $k_{\varepsilon,\lambda}$ enters the picture via the bound

$$\bar{\mathcal{V}}^2_{\varepsilon,\lambda}(Q^{\varepsilon,\lambda}) \le k_{\varepsilon,\lambda}(Q^{\varepsilon,\lambda})^2 \tag{5.6.59}$$

However, we can now do better than (5.6.59). In fact, in view of the L^∞ bound for $Q^{\varepsilon,\lambda}$, (5.6.19), and (5.6.20), we may achieve, in lieu of (5.6.59), an estimate of the form

$$\bar{\mathcal{V}}^2_{\varepsilon,\lambda}(Q^{\varepsilon,\lambda}) \le \tilde{k}_\varepsilon(Q^{\varepsilon,\lambda})^2 \tag{5.6.60}$$

with $\tilde{k}_\varepsilon$ independent of $\lambda > 0$. In light of the discussion above, it is clear that (5.6.58), which appears on the right-hand side of the estimate (5.6.31), in the limit as $n \to \infty$, may be replaced, for $\varepsilon > 0$ sufficiently small, by a nonnegative functional of the initial data which is independent of λ. Therefore, if we go to the limit in (5.6.31) as $n \to \infty$, we obtain bounds for the following integrals which are independent of λ when ε is sufficiently small:

$$\left\{ \begin{array}{c} \int_{-\infty}^{\infty} (z^{\varepsilon,\lambda})^2\,dx, \int_{-\infty}^{\infty} (Q^{\varepsilon,\lambda})^2\,dx \\ \int_0^t \int_{-\infty}^{\infty} (Q_x^{\varepsilon,\lambda})^2\,dx\,d\tau, \int_0^t \int_{-\infty}^{\infty} (z_x^{\varepsilon,\lambda})^2\,dx\,d\tau, \int_0^t \int_{-\infty}^{\infty} (z^{\varepsilon,\lambda})^2\,dx\,d\tau \end{array} \right. \tag{5.6.61}$$

Using an argument which is entirely analogous to the one above, it is easy to show that the constant $n_{\varepsilon,\lambda}$, which appears in the *a priori* estimate (5.6.41), may be replaced by a nonnegative constant $\tilde{n}_\varepsilon$ which is independent of $\lambda > 0$. Going to the limit, therefore, in (5.6.41) as $n \to \infty$ we now obtain bounds for the integrals

$$\int_{-\infty}^{\infty} (z_x^{\varepsilon,\lambda})^2\,dx, \quad \int_0^t \int_{-\infty}^{\infty} (z_t^{\varepsilon,\lambda})^2\,dx\,d\tau \tag{5.6.62}$$

which are independent of λ for ε sufficiently small. Similar bounds for

$$\int_{-\infty}^{\infty} (Q_x^{\varepsilon,\lambda})^2\,dx, \quad \int_0^t \int_{-\infty}^{\infty} (Q_t^{\varepsilon,\lambda})^2\,dx\,d\tau \tag{5.6.63}$$

follow directly from (5.6.43), and the definition of $\mathcal{V}_\lambda(\cdot)$, if we let $n \to \infty$ in (5.6.43). Of course, bounds which are independent of λ for many of the integrals cited in (5.6.62)-(5.6.63) also follow directly from the estimate (5.6.54).

We may now pass to the limit as $\lambda \to 0^+$ so as to obtain from the solution $(z^{\varepsilon,\lambda}, Q^{\varepsilon,\lambda})$ of the Cauchy problem for (5.6.21), a unique solution $(z^\varepsilon, Q^\varepsilon)$ of the Cauchy problem associated with (5.6.14a), (5.6.14b) on $R^1 \times [0,T]$, for any $T < \infty$, when $\varepsilon > 0$ is sufficiently small; the proof of Theorem 5.3 is now complete. □

As a direct consequence of the work of Chueh, Conley, and Smoller [34], which we have already cited in the proof of Theorem 5.3, we have the following

Theorem 5.4 Under the same conditions as those which prevail in Theorem 5.3 (in particular, the hypothesis that $\varepsilon > 0$ be sufficiently small), the unique solution $(z^\varepsilon, Q^\varepsilon)$ of the Cauchy problem (5.6.14a), (5.6.14b), (5.6.16), on $R^1 \times [0,\infty)$, satisfies, for some $k > 0$ which is independent of ε, the L^∞ bound

$$\|(z^\varepsilon, Q^\varepsilon)\|_{(L^\infty(R^1\times[0,\infty)))^2} \leq k \tag{5.6.64}$$

Remarks: L^∞ estimates of the type (5.6.64) for systems similar to (5.6.14a), (5.6.14b) are also discussion in Dafermos [40] and Feireisl [57], e.g., Proposition 2 of [57].

Now, as a consequence of the L^∞ bound of Theorem 5.4, it follows, in the usual fashion, that for some sequence $\{\varepsilon_n\}$, $\varepsilon_n \to 0^+$, as $n \to \infty$,

$$\begin{cases} z^{\varepsilon_n}(x,t) \rightharpoonup z^*(x,t) \\ Q^{\varepsilon_n}(x,t) \rightharpoonup Q^*(x,t) \end{cases} \quad \text{in } L^\infty, \text{ weak}^* \tag{5.6.65}$$

and, as $\mathcal{V}(\cdot)$ is continuous, we also have (perhaps for some subsequence which we also label as $\{\varepsilon_n\}$)

$$\mathcal{V}(Q^{\varepsilon_n}(x,t)) \rightharpoonup \mathcal{V}^*(x,t), \text{ in } L^\infty, \text{ weak}^* \tag{5.6.66}$$

Owing to the fact that $c(\varepsilon) \to 0$, as $\varepsilon \to 0^+$, i.e., (5.6.13d), we have, additionally, that

$$\begin{aligned} \bar{\mathcal{V}}_{\varepsilon_n}(Q^{\varepsilon_n}(x,t)) &= (1 - c(\varepsilon_n)G_0)\mathcal{V}(Q^{\varepsilon_n}(x,t)) \\ &\quad + c(\varepsilon_n)RQ^{\varepsilon_n}(x,t) \\ &\rightharpoonup \mathcal{V}^*(x,t), \text{ in } L^\infty, \text{ weak}^*. \end{aligned} \tag{5.6.67}$$

Suppose that $\phi, \psi \in C_0^\infty(R^1 \times [0,\infty))$. For $\varepsilon_n > 0$ chosen so small that $c(\varepsilon_n) < 1/G_0$, and $\delta(\varepsilon_n) > 0$, let $(z^{\varepsilon_n}, Q^{\varepsilon_n})$ be the unique solution of the Cauchy problem (5.6.14a), (5.6.14b), (5.6.16), with $\varepsilon = \varepsilon_n$, whose existence is guaranteed by Theorem 5.3. Multiplying (5.6.14a) by ϕ and (5.6.14b) by ψ, and integrating over $R^1 \times [0,\infty)$, we find that

$$\begin{aligned} &\int_0^\infty \int_{-\infty}^\infty \phi_t(x,t) z^{\varepsilon_n}(x,t)\,dx\,dt \\ &\qquad + \int_0^\infty \int_{-\infty}^\infty \phi_x(x,t)\bar{\mathcal{V}}_{\varepsilon_n}(Q^{\varepsilon_n}(x,t))\,dx\,dt \\ &= -\varepsilon_n \int_0^\infty \int_{-\infty}^\infty \phi_{xx}(x,t) z^{\varepsilon_n}(x,t)\,dx\,dt \\ &\quad + R\int_0^\infty \int_{-\infty}^\infty \phi(x,t) z^{\varepsilon_n}(x,t)\,dx\,dt - \int_{-\infty}^\infty \phi(x,0) z^{\varepsilon_n}(x,0)\,dx \end{aligned} \tag{5.6.68}$$

and

$$\int_0^\infty \int_{-\infty}^\infty \psi_t(x,t) Q^{\varepsilon_n}(x,t)\,dx\,dt \tag{5.6.69}$$

$$+\delta(\varepsilon_n)\int_0^\infty\int_{-\infty}^\infty \psi_x(x,t)z^{\varepsilon_n}(x,t)\,dx\,dt$$

$$= -\varepsilon_n\int_0^\infty\int_{-\infty}^\infty \psi_{xx}(x,t)Q^{\varepsilon_n}(x,t)\,dx\,dt$$

$$+G_0\int_0^\infty\int_{-\infty}^\infty \psi(x,t)\mathcal{V}(Q^{\varepsilon_n}(x,t))\,dx\,dt - \int_{-\infty}^\infty \psi(x,0)Q^{\varepsilon_n}(x,0)\,dx$$

Using (5.6.13c), (5.6.16), (5.6.65)-(5.6.67), and taking note of the fact that as $\varepsilon_n \to 0$,

$$i^{\varepsilon_n}(x,t) = z^{\varepsilon_n}(x,t) - c(\varepsilon_n)Q_x^{\varepsilon_n}(x,t) \tag{5.6.70}$$

$$\rightharpoonup z^*(x,t) \equiv i^*(x,t),\ \text{in } L^\infty,\ \text{weak}^*$$

we obtain from (5.6.68), (5.6.69) the equations

$$\int_0^\infty\int_{-\infty}^\infty \phi_t(x,t)i^*(x,t)\,dx\,dt + \int_0^\infty\int_{-\infty}^\infty \phi_x(x,t)\mathcal{V}^*(x,t)\,dx\,dt \tag{5.6.71}$$

$$= R\int_0^\infty\int_{-\infty}^\infty \phi(x,t)i^*(x,t)\,dx\,dt - \int_{-\infty}^\infty \phi(x,0)i_0(x)\,dx$$

and

$$\int_0^\infty\int_{-\infty}^\infty \psi_t(x,t)Q^*(x,t)\,dx\,dt + \int_0^\infty\int_{-\infty}^\infty \psi_x(x,t)i^*(x,t)\,dx\,dt \tag{5.6.72}$$

$$= G_0\int_0^\infty\int_{-\infty}^\infty \psi(x,t)\mathcal{V}^*(x,t)\,dx\,dt - \int_{-\infty}^\infty \psi(x,0)Q_0(x)\,dx$$

By comparing (5.6.71), (5.6.72) with (5.5.20), (5.5.21), we see that the pair $(i^*(x,t), Q^*(x,t))$ will be a global weak solution of the Cauchy problem (5.5.1), (5.5.4) if we can prove that

$$\mathcal{V}^*(x,t) = \mathcal{V}(Q^*(x,t)),\ \text{a.e. on } R^1\times[0,\infty) \tag{5.6.73}$$

In other words, we must establish that for $Q^{\varepsilon_n}(x,t) \rightharpoonup Q^*(x,t)$ in L^∞, weak*, a.e. on $R^1\times[0,\infty)$, we have $\mathcal{V}(Q^{\varepsilon_n}(x,t)) \rightharpoonup \mathcal{V}(Q^*(x,t))$, in L^∞, weak*, a.e. on $R^1\times[0,\infty)$.

5.7 Existence of Global Weak Solutions for the Transmission Line

In the previous section we showed that there exists a sequence $\{\boldsymbol{u}^{\varepsilon_n}\}$, $\boldsymbol{u}^{\varepsilon_n} = (z^{\varepsilon_n}, Q^{\varepsilon_n})$ such that

$$\boldsymbol{u}^{\varepsilon_n} \rightharpoonup (z^*, Q^*),\ \text{in } L^\infty,\ \text{weak}^* \tag{5.7.1}$$

For $\lambda_1, \lambda_2 \in R^1$ we now set

$$\tilde{\mathcal{V}}(\lambda_1, \lambda_2) = \mathcal{V}(\lambda_2) \tag{5.7.2}$$

It then follows from our work in § 5.2 (on the Young measure) that the sequence $\{\boldsymbol{u}^{\varepsilon_n}\}$ generates a probability measure $\nu_{x,t}(\lambda_1, \lambda_2)$, the Young measure associated with $\{\boldsymbol{u}^{\varepsilon_n}\}$ such that, in the sense of L^∞ weak* convergence,

$$\tilde{\mathcal{V}}(z^{\varepsilon_n}(x,t), Q^{\varepsilon_n}(x,t)) \rightharpoonup \langle \tilde{\mathcal{V}}(\lambda_1, \lambda_2), \nu_{x,t}(\lambda_1, \lambda_2)\rangle \tag{5.7.3}$$

a.e. for $(x,t) \in R^1 \times [0, \infty)$, where for any continuous function f on $R^1 \times [0, \infty)$

$$\langle f(\lambda_1, \lambda_2), \nu_{x,t}(\lambda_1, \lambda_2)\rangle = \int_{-\infty}^{\infty}\int_{-\infty}^{\infty} f(\lambda_1, \lambda_2)\, d\nu_{x,t}(\lambda_1, \lambda_2) \tag{5.7.4}$$

What we intend to do now, in this section, is to prove that the Div-Curl lemma of Murat [120] may be applied, in the present situation, for appropriately chosen entropy-flux pairs, and that the arguments of DiPerna, as described in § 5.4, may be employed, essentially without change, in order to show that for the Young measure $\nu_{x,t}(\lambda_1, \lambda_2)$, generated by $\{\boldsymbol{u}^{\varepsilon_n}\}$, we have, a.e. in x, t on $fR^1 \times [0, \infty)$

$$\nu_{x,t}(\lambda_1, \lambda_2) = \delta(\lambda_1 - z^*(x,t), \lambda_2 - Q^*(x,t)) \tag{5.7.5}$$

Once (5.7.5) has been established, it will then follow from (5.7.3), and the definition of $\tilde{\mathcal{V}}(\lambda_1, \lambda_2)$, that (5.6.73) holds and, therefore, that the pair $(i^*(x,t), Q^*(x,t))$ is a global weak solution of the Cauchy problem (5.5.1), (5.5.4). We now state

Theorem 5.5 Let $(z^{\varepsilon_n}, Q^{\varepsilon_n}) \equiv \boldsymbol{u}^{\varepsilon_n}$, $\varepsilon_n > 0$, $\varepsilon_n \to 0^+$ as $n \to \infty$, be a sequence of solutions of the Cauchy problem (5.6.14a), (5.6.14b), (5.6.16), with $\varepsilon = \varepsilon_n$, on $R^1 \times [0, \infty)$, such that (5.6.65) holds for some z^*, $Q^* \in L^\infty(R^1 \times [0, \infty))$; then the Young measure $\nu_{x,t}(\lambda_1, \lambda_2)$ generated by $\{\boldsymbol{u}^{\varepsilon_n}\}$ is a Dirac mass concentrated at (z^*, Q^*).

Proof: We begin by defining suitable entropy-flux pairs $(\eta(z,Q), q(z,Q))$ for the system

$$\begin{cases} z_t + \mathcal{V}(Q)_x = 0 \\ Q_t + z_x = 0 \end{cases} \tag{5.7.6}$$

Such a pair (η, q) will consist of C^2 functions which satisfy the system of equations

$$\begin{cases} q_Q - \mathcal{V}'(Q)\eta_z = 0 \\ q_z - \eta_Q = 0 \end{cases} \tag{5.7.7}$$

so that the entropy $\eta(z,Q)$ satisfies the linear wave equation

$$\eta_{QQ} = \mathcal{V}'(Q)\eta_{zz} \tag{5.7.8}$$

For $\varepsilon > 0$ chosen sufficiently small, let $(z^\varepsilon, Q^\varepsilon)$ be the unique solution of (5.6.14a), (5.6.14b), (5.6.16) whose existence is guaranteed by Theorem 5.3; an elementary calculation shows that

$$\begin{aligned}&\frac{\partial}{\partial t}\eta(z^\varepsilon(x,t),Q^\varepsilon(x,t)) + \frac{\partial}{\partial x}q(z^\varepsilon(x,t),Q^\varepsilon(x,t)) \\ &\quad = \{-\eta_z\,[\mathcal{V}'(Q^\varepsilon)(1-c(\varepsilon)G_0) + c(\varepsilon)R] + q_Q\}\,Q^\varepsilon_x \\ &\qquad + \{(h(\varepsilon)-1)\eta_Q + q_z\}\,z^\varepsilon_x \\ &\quad + \varepsilon\eta_z z^\varepsilon_{xx} + \varepsilon\eta_Q Q^\varepsilon_{xx} - R\eta_z z^\varepsilon - G_0\eta_Q\mathcal{V}(Q^\varepsilon)\end{aligned} \tag{5.7.9}$$

with $h(\varepsilon) = 1 - \delta(\varepsilon)$. Employing (5.7.7), i.e., the definition of the entropy-flux pair (η, q), (5.7.9) reduces to the statement that for each $\varepsilon > 0$ we have

$$\begin{aligned}&\eta(z^\varepsilon,Q^\varepsilon)_t + q(z^\varepsilon,Q^\varepsilon)_x \\ &\quad = (\Gamma^\varepsilon_1)_x + \Gamma^\varepsilon_2 + \Gamma^\varepsilon_3 - R\eta_z z^\varepsilon - G_0\eta_Q\mathcal{V}(Q^\varepsilon)\end{aligned} \tag{5.7.10}$$

with Γ^ε_i, $i = 1,2,3$ defined as follows:

$$\Gamma^\varepsilon_1(x,t) = \varepsilon\,[\eta_z z^\varepsilon_x + \eta_Q Q^\varepsilon_x] \tag{5.7.11a}$$

$$\Gamma^\varepsilon_2(x,t) = -\varepsilon\left[\eta_{zz}(z^\varepsilon_x)^2 + 2\eta_{zQ} z^\varepsilon_x Q^\varepsilon_x + \eta_{QQ}(Q^\varepsilon_x)^2\right] \tag{5.7.11b}$$

$$\Gamma^\varepsilon_3(x,t) = c(\varepsilon)\eta_z\,[G_0\mathcal{V}'(Q^\varepsilon) - RQ^\varepsilon_x + h(\varepsilon)\eta_Q z^\varepsilon_x] \tag{5.7.11c}$$

Our goal now is, of course, to show that

$$\eta(z^\varepsilon,Q^\varepsilon)_t + q(z^\varepsilon,Q^\varepsilon)_x \in \text{ a compact subset of } H^{-1}_{\text{loc}}(R^1\times[0,T]) \tag{5.7.12}$$

for any $T < \infty$; to this end we will derive some suitable *a priori* estimates. With $\varepsilon > 0$ chosen sufficiently small, we begin by taking our first equation, (5.6.14a), and multiplying it by $\delta(\varepsilon)z^\varepsilon(x,t)$ and our second equation, (5.6.14b), and multiplying it by

$\bar{\mathcal{V}}_\varepsilon(Q^\varepsilon(x,t))$; we then integrate each of the resulting equations over $(-\infty,\infty)\times[0,T]$, and add, so as to produce

$$\begin{aligned}
&\frac{1}{2}\delta(\varepsilon)\int_{-\infty}^{\infty}(z^\varepsilon(x,T))^2\,dx+\int_{-\infty}^{\infty}\hat{\mathcal{V}}(Q^\varepsilon(x,T))\,dx \\
&\quad+\varepsilon\delta(\varepsilon)\int_0^T\int_{-\infty}^{\infty}(z_x^\varepsilon(x,t))^2\,dx\,dt+\varepsilon\int_0^T\int_{-\infty}^{\infty}\bar{\mathcal{V}}_\varepsilon'(Q^\varepsilon)(Q_x^\varepsilon)^2\,dx\,dt \\
&\quad+R\delta(\varepsilon)\int_0^T\int_{-\infty}^{\infty}(z^\varepsilon(x,t))^2\,dx\,dt+G_0\int_0^T\int_{-\infty}^{\infty}\mathcal{V}(Q^\varepsilon)\bar{\mathcal{V}}_\varepsilon(Q^\varepsilon)\,dx\,dt \\
&\quad=\frac{1}{2}\delta(\varepsilon)\int_{-\infty}^{\infty}(z^\varepsilon(x,0))^2\,dx+\int_{-\infty}^{\infty}\hat{\mathcal{V}}(Q_0(x))\,dx
\end{aligned}\tag{5.7.13}$$

with $\hat{\mathcal{V}}(\xi)=\int_0^\xi\mathcal{V}(\lambda)\,d\lambda\geq 0$, $\forall\xi\in R^1$, by virtue of (5.5.3b). As in the proof of Theorem 5.3, we now avail ourselves of the fact that

$$\begin{aligned}
&\int_0^T\int_{-\infty}^{\infty}\mathcal{V}(Q^\varepsilon)\bar{\mathcal{V}}_\varepsilon(Q^\varepsilon)\,dx\,dt \\
&\quad=(1-c(\varepsilon)G_0)\int_0^T\int_{-\infty}^{\infty}\mathcal{V}^2(Q^\varepsilon)\,dx\,dt+Rc(\varepsilon)\int_0^T\int_{-\infty}^{\infty}Q^\varepsilon\mathcal{V}(Q^\varepsilon)\,dx\,dt \\
&\quad\geq(1-c(\varepsilon)G_0)\int_0^T\int_{-\infty}^{\infty}\mathcal{V}^2(Q^\varepsilon)\,dx\,dt
\end{aligned}\tag{5.7.14}$$

because $\xi\mathcal{V}(\xi)\geq 0$, $\forall\xi\in R^1$, and $1-c(\varepsilon)G_0>0$, for $\varepsilon>0$ sufficiently small. We now recall that $\delta(\varepsilon)=1-|\varepsilon^2G_0-2\varepsilon|G_0$ so that for $\varepsilon>0$ sufficiently small we have $\frac{1}{2}<\delta(\varepsilon)<1$; using this fact, as well as the lower bound of (5.7.14), in (5.7.13), we are led to the following estimate for $\varepsilon>0$ sufficiently small:

$$\begin{aligned}
&\int_{-\infty}^{\infty}(z^\varepsilon(x,T))^2\,dx+\frac{\varepsilon}{2}\int_0^T\int_{-\infty}^{\infty}(z_x^\varepsilon(x,t))^2\,dx\,dt \\
&\quad+\varepsilon\int_0^T\int_{-\infty}^{\infty}\bar{\mathcal{V}}_\varepsilon'(Q^\varepsilon)(Q_x^\varepsilon)^2\,dx\,dt+\frac{R}{2}\int_0^T\int_{-\infty}^{\infty}(z^\varepsilon(x,t))^2\,dx\,dt \\
&\quad+G_0(1-c(\varepsilon)G_0)\int_0^T\int_{-\infty}^{\infty}\mathcal{V}^2(Q^\varepsilon)\,dx\,dt \\
&\quad\leq\frac{1}{2}\int_{-\infty}^{\infty}(z^\varepsilon(x,0))^2\,dx+\int_{-\infty}^{\infty}\hat{\mathcal{V}}(Q_0(x))\,dx
\end{aligned}\tag{5.7.15}$$

where $z^\varepsilon(x,0)$, $\varepsilon>0$, $x\in R^1$, is given by (5.6.16). Using the definition (5.7.11a), we now compute that for some $C_1>0$, independent of ε,

$$\begin{aligned} \|\Gamma_1^\varepsilon(\cdot,\cdot)\|_{L^2(R^1\times[0,T])} \\ &= \varepsilon^2 \int_0^T \int_{-\infty}^{\infty} \left[\eta_z z_x^\varepsilon + \eta_Q Q_x^\varepsilon\right]^2 \, dx\, dt \\ &\le C_1\varepsilon^2 \int_0^T \int_{-\infty}^{\infty} \left[(z_x^\varepsilon)^2 + (Q_x^\varepsilon)^2\right] \, dx\, dt \\ &\longrightarrow 0 \end{aligned} \tag{5.7.16}$$

as $\varepsilon \to 0^+$, because the last integral on the right-hand side of (5.7.16) is bounded, independent of ε, as a consequence of (5.7.15); in applying (5.7.15), we have used (5.5.2), as well as the fact that $\forall\, \xi \in R^1$, $\bar{\mathcal{V}}_\varepsilon'(\xi) \ge \frac{1}{2}\beta_0$, for $\varepsilon > 0$ sufficiently small. Of course, in deriving the estimate displayed in (5.7.16), we have made use of Theorem 5.4 and our hypothesis that $\eta \in C^2$. Because

$$\|\Gamma_1^\varepsilon(\cdot,\cdot)\|_{L^2(R^1\times[0,T])} \to 0, \text{ as } \varepsilon \to 0^+,$$

for any $T < \infty$, it follows that for $\varepsilon \to 0^+$,

$$(\Gamma_1^\varepsilon)_x \to 0, \text{ in } H^{-1}_{\text{loc}}(R^1 \times [0,T]),\, T < \infty$$

so that for $\varepsilon > 0$ sufficiently small

$$(\Gamma_1^\varepsilon)_x \in \left\{ \begin{array}{c} \text{a compact subset} \\ \text{of } H^{-1}_{\text{loc}}(R^1 \times [0,T]) \end{array} \right\} \tag{5.7.17}$$

Also, in view of (5.7.11b)

$$\begin{aligned} \|\Gamma_2^\varepsilon(\cdot,\cdot)\|_{L^1(R^1\times[0,T])} \\ &\le \varepsilon \int_0^T \int_{-\infty}^{\infty} \left|\eta_{zz}(z_x^\varepsilon)^2 + 2\eta_{zQ} z_x^\varepsilon Q_x^\varepsilon + \eta_{QQ}(Q_x^\varepsilon)^2\right| \, dx\, dt \\ &\le C_2\varepsilon \int_0^T \int_{-\infty}^{\infty} \left[(z_x^\varepsilon)^2 + (Q_x^\varepsilon)^2\right] \, dx\, dt, \end{aligned} \tag{5.7.18}$$

for some $C_2 > 0$, independent of ε. For ε sufficiently small, therefore,

$$\Gamma_2^\varepsilon \in \left\{ \begin{array}{c} \text{a bounded subset} \\ \text{of } L^1(R^1 \times [0,T]) \end{array} \right\} \tag{5.7.19}$$

for any $T < \infty$. We now note that, in view of the definitions of $c(\varepsilon)$ and $h(\varepsilon)$,

$$h^2(\varepsilon) \equiv |\varepsilon^2 G_0 - 3\varepsilon|^2 = 4\varepsilon^2 \left(1 - \frac{\varepsilon}{2} G_0\right)^2 < 4\varepsilon^2 \tag{5.7.20a}$$

and

$$c^2(\varepsilon) = \left[\frac{(\varepsilon^2 G_0 - \varepsilon)}{1 - |\varepsilon^2 G_0 - 2\varepsilon|G_0}\right]^2 < (1 - \varepsilon G_0)^2 < \varepsilon^2 \tag{5.7.20b}$$

for $\varepsilon > 0$ sufficiently small. Therefore, by (5.7.11c), for some $C_3 > 0$, independent of ε,

$$\begin{aligned} &\|\Gamma_3^\varepsilon(\cdot,\cdot)\|^2_{L^1(R^1\times[0,T])} \\ &\leq C_3\varepsilon \int_0^T \int_{-\infty}^{\infty} \left[(z_x^\varepsilon)^2 + (Q_x^\varepsilon)^2\right] \, dx\, dt, \\ &\to 0, \text{ as } \varepsilon \to 0^+ \end{aligned} \tag{5.7.21}$$

and, for $\varepsilon > 0$ sufficiently small,

$$\Gamma_3^\varepsilon \in \left\{ \begin{array}{c} \text{a bounded subset} \\ \text{of } L^2(R^1 \times [0,T]) \end{array} \right\} \tag{5.7.22}$$

for any $T < \infty$. From (5.7.22) we may conclude that

$$\Gamma_3^\varepsilon \in \left\{ \begin{array}{c} \text{a compact subset} \\ \text{of } H^{-1}_{\text{loc}}(R^1 \times [0,T]) \end{array} \right\} \tag{5.7.23}$$

for any $T < \infty$, if $\varepsilon > 0$ is sufficiently small. Finally,

$$\Gamma_4^\varepsilon(x,t) \equiv R\eta_z z^\varepsilon(x,t) + G_0\eta_z \mathcal{V}(Q^\varepsilon) \tag{5.7.24}$$

lies, as a consequence of the estimate (5.7.15), in a bounded subset of $L^2(R^1 \times [0,T])$, for any $T < \infty$, from which we also conclude that

$$\Gamma_4^\varepsilon \in \left\{ \begin{array}{c} \text{a compact subset} \\ \text{of } H^{-1}_{\text{loc}}(R^1 \times [0,T]) \end{array} \right\} \tag{5.7.25}$$

We now return to (5.7.10) and set

$$\Gamma^\varepsilon(x,t) = (\Gamma_1^\varepsilon(x,t))_x + \Gamma_2^\varepsilon(x,t) + \Gamma_3^\varepsilon(x,t) + \Gamma_4^\varepsilon(x,t) \tag{5.7.26}$$

so that

$$\eta(z^\varepsilon, Q^\varepsilon)_t + q(z^\varepsilon, Q^\varepsilon)_x = \Gamma^\varepsilon(x,t) \tag{5.7.27}$$

As a consequence of Theorems 5.3 and 5.4, we know that $\Gamma^\varepsilon \in W^{-1,\infty}(R^1 \times [0,T])$ for any $T < \infty$. Also, for $\varepsilon > 0$ sufficiently small, $\Gamma^\varepsilon(x,t)$, as given by (5.7.26), is the sum of four terms, three of which we have shown lie in compact subsets of $H^{-1}_{\text{loc}}(R^1 \times [0,T])$, for any $T < \infty$, while the fourth belongs to a bounded subset of $L^1(R^1 \times [0,T])$. By the Murat lemma [120] it follows that Γ^ε and, hence, $\eta_t + q_x$, lie in a compact subset of $H^{-1}_{\text{loc}}(R^1 \times [0,T])$ for any $T < \infty$, if $\varepsilon > 0$ is sufficiently small.

Now, suppose that (η_1, q_1) and (η_2, q_2) are any two entropy-flux pairs associated with the system (5.7.6). As

$$(z^{\varepsilon_n}, Q^{\varepsilon_n}) \rightharpoonup (z^*, Q^*), \text{ in } L^\infty, \text{ weak}^*,$$

by considering appropriate subsequences, which we also label as $\{\varepsilon_n\}$,

$$\begin{cases} \eta_1(z^{\varepsilon_n}, Q^{\varepsilon_n}) \rightharpoonup \eta_1^*, \text{ in } L^\infty, \text{ weak}^*, \\ q_1(z^{\varepsilon_n}, Q^{\varepsilon_n}) \rightharpoonup q_1^*, \text{ in } L^\infty, \text{ weak}^*, \\ \eta_2(z^{\varepsilon_n}, Q^{\varepsilon_n}) \rightharpoonup \eta_2^*, \text{ in } L^\infty, \text{ weak}^*, \\ q_2(z^{\varepsilon_n}, Q^{\varepsilon_n}) \rightharpoonup q_2^*, \text{ in } L^\infty, \text{ weak}^*, \end{cases} \tag{5.7.28}$$

where, as a consequence of our work in § 5.2

$$\begin{cases} \eta_1^* = \langle \eta_1(\lambda_1, \lambda_2), \nu_{x,t}(\lambda_1, \lambda_2) \rangle \\ q_1^* = \langle q_1(\lambda_1, \lambda_2), \nu_{x,t}(\lambda_1, \lambda_2) \rangle \\ \eta_2^* = \langle \eta_2(\lambda_1, \lambda_2), \nu_{x,t}(\lambda_1, \lambda_2) \rangle \\ q_2^* = \langle q_2(\lambda_1, \lambda_2), \nu_{x,t}(\lambda_1, \lambda_2) \rangle \end{cases} \tag{5.7.29}$$

and the representations (5.7.29) hold a.e. in $R^1 \times [0, \infty)$, with $\nu_{x,t}(\lambda_1, \lambda_2)$ the Young measure generated by $\{\boldsymbol{u}^{\varepsilon_n}\} = \{(z^{\varepsilon_n}, Q^{\varepsilon_n})\}$. However, for any entropy-flux pair (η, q), as defined by (5.7.7), if n is sufficiently large, then $\eta(z^{\varepsilon_n}, Q^{\varepsilon_n})_t + q(z^{\varepsilon_n}, Q^{\varepsilon_n})_x$ lies in a compact subset of $H^{-1}_{\text{loc}}(R^1 \times [0,T])$, for any $T < \infty$, and we may apply the Div-Curl lemma [171] so as to conclude that

$$(\eta_1(z^{\varepsilon_n}, Q^{\varepsilon_n}), q_1(z^{\varepsilon_n}, Q^{\varepsilon_n})) \cdot (q_2(z^{\varepsilon_n}, Q^{\varepsilon_n}), -\eta_2(z^{\varepsilon_n}, Q^{\varepsilon_n}))$$

is weakly continuous. In other words, in the sense of L^∞ weak* convergence we have

$$\eta_1(z^{\varepsilon_n}, Q^{\varepsilon_n}) q_2(z^{\varepsilon_n}, Q^{\varepsilon_n}) - \eta_2(z^{\varepsilon_n}, Q^{\varepsilon_n}) q_1(z^{\varepsilon_n}, Q^{\varepsilon_n}) \tag{5.7.30}$$

$$\rightharpoonup \eta_1^* q_2^* - \eta_2^* q_1^*, \quad \text{as } n \to \infty$$

Combining (5.7.29) with (5.7.30) and using the fundamental representation

$$\eta_1(z^{\varepsilon_n}, Q^{\varepsilon_n})q_2(z^{\varepsilon_n}, Q^{\varepsilon_n}) - \eta_2(z^{\varepsilon_n}, Q^{\varepsilon_n})q_1(z^{\varepsilon_n}, Q^{\varepsilon_n}) \tag{5.7.31}$$

$$\rightharpoonup \langle \eta_1(\lambda_1, \lambda_2)q_2(\lambda_1, \lambda_2) - \eta_2(\lambda_1, \lambda_2)q_1(\lambda_1, \lambda_2), \nu_{x,t}(\lambda_1, \lambda_2)\rangle$$

in L^∞, weak*, we find that for all entropy-flux pairs (η_1, q_1), (η_2, q_2) satisfying (5.7.6) we have the identity

$$\langle \eta_1(\lambda_1, \lambda_2)q_2(\lambda_1, \lambda_2) - \eta_2(\lambda_1, \lambda_2)q_1(\lambda_1, \lambda_2), \nu_{x,t}(\lambda_1, \lambda_2)\rangle \tag{5.7.32}$$

$$= \langle \eta_1(\lambda_1, \lambda_2), \nu_{x,t}(\lambda_1, \lambda_2)\rangle\langle q_2(\lambda_1, \lambda_2), \nu_{x,t}(\lambda_1, \lambda_2)\rangle$$

$$-\langle \eta_2(\lambda_1, \lambda_2), \nu_{x,t}(\lambda_1, \lambda_2)\rangle\langle q_1(\lambda_1, \lambda_2), \nu_{x,t}(\lambda_1, \lambda_2)\rangle$$

a.e. in $R^1 \times [0, T]$, for any $T < \infty$. The crucial observation now is that beginning with the identity (5.7.32), and using the fact that our entropy-flux pairs (η, q) satisfy a hyperbolic system, i.e. (5.7.6), which is *completely analogous* to the one employed by DiPerna (i.e., compare with (5.4.6)), the arguments of § 5.4 can be used to yield the conclusion that for the Young measure $\nu_{x,t}(\lambda_1, \lambda_2)$, generated by $\{\boldsymbol{u}^{\varepsilon_n}\} = \{(z^{\varepsilon_n}, Q^{\varepsilon_n})\}$, (5.7.5) is valid. □

In view of (5.7.5), (5.7.2), and (5.6.66), it follows that

$$\mathcal{V}(Q^{\varepsilon_n}) = \tilde{\mathcal{V}}(z^{\varepsilon_n}, Q^{\varepsilon_n}) \tag{5.7.33}$$

$$\rightharpoonup \langle \tilde{\mathcal{V}}(\lambda_1, \lambda_2), \delta(\lambda_1 - z^*, \lambda_2 - Q^*)\rangle = \tilde{\mathcal{V}}(z^*, Q^*) = \mathcal{V}(Q^*)$$

in the sense of L^∞ weak* convergence and we have, therefore, established the following result:

Theorem 5.6 The L^∞ weak* convergent sequence $\{(z^{\varepsilon_n}, Q^{\varepsilon_n})\}$, where $(z^{\varepsilon_n}, Q^{\varepsilon_n})$ is the unique solution of the Cauchy problem (5.6.14a), (5.6.14b), (5.6.16) for $\varepsilon = \varepsilon_n$, converges to a weak solution (i^*, Q^*) of the Cauchy problem (5.5.1), (5.5.4), pointwise a.e. on $R^1 \times [0, T]$, for any $T < \infty$, where $i^*(x, t) \equiv z^*(x, t)$.

Remarks: Although the system (5.7.6) satisfied by our entropy-flux pairs (η, q) is analogous to the one appearing in the construction used by DiPerna [48], as a consequence of our constitutive hypothesis (5.5.2) we have $\mathcal{V}''(0) = 0$; however,

$\mathcal{V}''(\xi) \neq 0$, $\forall \xi \neq 0$ and following the work of Chen [33], Serre [153] and Rascle and Serre [133], the reduction of $\nu_{x,t}$, in this case, to a Dirac measure still applies as $\mathcal{V}''(\cdot)$ has, at most, one zero. As we have noted before, we may dispense with the specific form (5.5.2) chosen for the voltage $\mathcal{V}(\cdot)$ and require, instead, that $\mathcal{V}(\cdot)$ respect the conditions $\mathcal{V}'(\xi) \geq \beta_0 > 0$, $\forall \xi \in R^1$, $\xi\mathcal{V}(\xi) \geq 0$, $\forall \xi \in R^1$, and $\int_0^\xi \mathcal{V}(\lambda)\,d\lambda \geq 0$, $\forall \xi \in R^1$. Another interesting extension of Theorem 5.6 would involve removing the restriction that $G = G_0 =$ const. and to allow, as we did in Chapter 4, for a voltage-dependent leakage conductance.

Chapter 6

SOME NONLOCAL ELECTROMAGNETIC PROBLEMS

6.0 Introduction

Our work in this last chapter differs somewhat from the material presented in Chapters 2 through 5, both in terms of flavor and content. In the last four chapters, we have been concerned with two basic problems, namely, the propagation of electromagnetic waves in nonlinear dielectric media and the evolution of the current and charge distribution in a distributed parameter, nonlinear transmission line; both of these basic problems were posed either in terms of systems of first order, inhomogeneous quasilinear equations or in terms of nonlinearly damped, second order, nonlinear wave equations. Throughout Chapters 2–5 our main interest was either in the question of shock formation or in the complementary problems of existence of both classical and weak solutions (globally, in time).

In this chapter we will be concerned with equilibrium problems for electromagnetic media (often, deformable) which are subject to a self-interaction arising from the Biot-Savart law. We make no effort here to treat problems for deformable (say, elastic) electromagnetic media in general; such an effort would require a separate treatise and there is, already, a very extensive literature in this area. In particular, we may refer the interested reader to the following: the foundation papers of Eringen [52], Jordan and Eringen [86] and Toupin [173], [172], the basic survey paper of Pao [129], the edited collections of papers by Maugin [113] and Parkus [130], and the books of Brown [31], Hutter and Van de Ven [72], and Moon [115]. Much of what we do present here has not appeared before in book form and has as its source three distinct sets of work, namely, the work of Wolfe [184] on (nonself-interacting) elastic wires in magnetic fields, the unpublished results of the author [21] on the equilibrium states of self-interacting current bearing elastic wires, and the work of Rogers [138] and Antman and Rogers [4] on some general classes of non-local, steady-state, electromagnetic problems for both rigid and deformable media in which the Biot-Savart law, as in the analysis in [21], plays a prominent role.

Although no self-interaction via Biot-Savart is present in Wolfe's original work [184], or in the complementary papers by Wolfe [183] and Seidman and Wolfe [185] on the equilibrium states of elastic conducting rods in a magnetic field, we present it in § 6.1 as a convenient means by which to introduce the more general model of the self-interacting, nonlinearly elastic, current bearing wire in § 6.2. In § 6.3 we show, through use of a simple semi-inverse method, that straight equilibrium states exist for the self-interacting model, although the demonstration is by no means as trivial as it is for the nonself-interacting model of § 6.1; moreover, these straight states exist for all values of the parameter $\lambda = IB$, where I is the current in the wire and B the (constant) strength of the magnetic induction field in which the wire is placed. Next, in § 6.4, we are able to demonstrate the existence of plane circular equilibrium states for the nonlinearly elastic, self-interacting wire and to delineate conditions on λ under which such states exist. While helical equilibrium states exist, as shown in [184], for the nonself-interacting wire, the analysis in § 6.5 points quite strongly to the fact that such states, in all likelihood, do not exist in the self-interacting model, although, in the final analysis, this thesis in § 6.5 still has the status of a conjecture. In a similar fashion, in § 6.6, we conjecture, for the self-interacting, current bearing wire, the existence of a class of *wiggly* plane circular equilibrium states which perturb, in a precise manner, the circular equilibrium states of § 6.4. Finally, in § 6.6, we present some of the work of Rogers [138] and Antman and Rogers [4] on equilibrium problems for electromagnetic media, both rigid as well as nonlinearly elastic, in which the Biot-Savart law plays a critical role. We begin with a description of the results for (in general) three-dimensional rigid conductors and then proceed to generalize these results to the deformable case, ignoring in both cases the thermal effects which are considered in the work in [4]; in both cases one has to deal with nonlocal problems which lead to systems of partial-functional differential equations. We conclude with a discussion of semi-inverse problems (suitable for applications to problems involving wires or tubes) which are posed in terms of a set of cylindrical coordinates and which lead to a system of ordinary-functional differential equations; thus, the work in this last case is somewhat closer in spirit to the problem of the self-interacting, current bearing wire presented in § 6.2–§ 6.5.

The work reported in § 6.2–§ 6.5 is based on previously unpublished notes of the author [21] and resulted from research the author conducted during the spring of 1981 while he was on leave at the University of Maryland; much of what we present here benefited greatly from extensive discussions, during that period, with my colleague Stuart Antman.

6.1 Equilibrium States of an Elastic Conductor in a Magnetic Field: Non-Self Interaction Case

In order to put into perspective the problem of determining the set of equilibrium states of a nonlinearly elastic, current bearing wire, which is placed in an ambient magnetic field, and is subject to a self-interactive force arising from an application of the Biot-Savart law, we first present a synopsis of some work of Wolfe [184] on the simpler problem in which the thickness of the wire and the self-interactive force acting at points of the wire are ignored.

In the problem studied in [184], the wire is assumed to be perfectly flexible and is suspended between fixed supports; the distance between the supports is assumed to be greater than the natural length of the wire, so that the wire will always be in tension, and the magnetic field is assumed to be constant and directed parallel to the line between the supports. Wolfe [184] shows that this problem can be solved exactly and that the set of equilibrium solutions exhibits the classical bifurcation phenomena, with bifurcation parameter $\lambda = IB$, I the current in the wire and B the strength of the magnetic field; in particular, for all values of $\lambda > 0$ there exists a trivial solution in which the wire remains straight but, at the eigenvalues of the problem obtained by linearizing the equilibrium equations about this trivial solution, bifurcation occurs and we obtain branches of nontrivial solutions. To be more specific, we consider a nonlinearly elastic conducting wire with natural length 1 and we identify each material point of the wire by its coordinate $s \in [0,1]$; we let $\boldsymbol{r}(s)$ denote the deformed position in Euclidean 3-space $\mathbb{R}^3$ of the material point s and, then $\nu(s) = \|\boldsymbol{r}'(s)\|$ is the *stretch* at s. Thus, by definition, we have that $\boldsymbol{e}(s) = \boldsymbol{r}'(s)/\nu(s)$ is a unit tangent vector to the wire at s. The wire is stretched between fixed supports at $\mathbf{0}$ and $b\boldsymbol{k}$, $b > 1$, so that $\boldsymbol{r}(0) = \mathbf{0}$, while $\boldsymbol{r}(1) = b\boldsymbol{k}$. We assume that the wire carries a current I and that a constant magnetic field of the form $\boldsymbol{B} = B\boldsymbol{k}$ is present; for such a scenario the force on the wire is given by

$$\begin{aligned} \boldsymbol{f}(s) &= I\boldsymbol{r}'(s) \times B\boldsymbol{k} \\ &= I\nu\boldsymbol{e}'(s) \times B\boldsymbol{k} \end{aligned} \qquad (6.1.1)$$

The situation described above is depicted below in Figure 6.1:

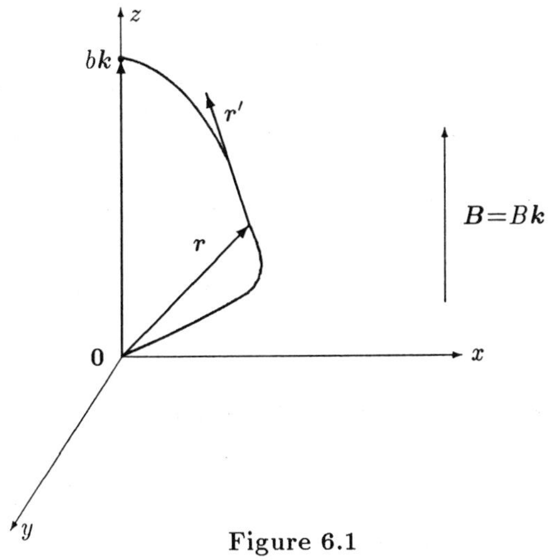

Figure 6.1

If $\boldsymbol{n}(s)$ denotes the resultant contact force exerted by the material of $(s,1]$ on the material of $[0,s]$, then the equilibrium equation for the wire is

$$\boldsymbol{n}'(s) + \boldsymbol{f}(s) = 0, \quad 0 \le s \le 1 \qquad (6.1.2)$$

Also, as the wire is assumed to be perfectly flexible, $\boldsymbol{n}(s) = N(s)\boldsymbol{e}(s)$, where $N(s)$ is the *tension* in the wire at s; in addition, the wire will be taken to be *nonlinearly elastic* which means, in this context, that there exists a continuously differentiable function $\eta(\nu, s)$, $\nu \in (0, \infty)$, $s \in (0, 1]$ with

$$\begin{cases} \eta_\nu(\nu, s) > 0; \quad \eta(1, s) = 0; \quad \eta(\nu, s) \to \infty, \text{ as } \nu \to \infty \\ \qquad\qquad \text{and} \quad \eta(\nu, s) \to -\infty, \text{ as } \nu \to 0 \end{cases} \tag{6.1.3}$$

such that $N(s) = \eta(\nu(s), s)$. The boundary-value problem for the wire then consists of finding classical solutions to

$$(N(s)\boldsymbol{e}(s))' + \lambda\nu(s)\boldsymbol{e}(s) \times \boldsymbol{k} = 0, \; 0 < s < 1 \tag{6.1.4a}$$

and

$$\boldsymbol{r}(0) = \boldsymbol{0}, \; \boldsymbol{r}(1) = b\boldsymbol{k}, \; b > 1 \tag{6.1.4b}$$

where $\lambda = IB$ and $\boldsymbol{e}(s) = \boldsymbol{r}'(s)/\|\boldsymbol{r}'(s)\|$.

We will only outline, for (6.1.4a), (6.1.4b), the procedure which is followed in [184] for the case of a homogeneous wire, i.e., for the case in which $\eta(\nu, s)$ is independent of s so that $N(s) = \eta(\nu(s))$. We begin by writing (6.1.4a) in the form

$$\boldsymbol{n}'(s) + IB\boldsymbol{r}'(s) \times \boldsymbol{k} = \boldsymbol{0}, \quad 0 < s < 1, \tag{6.1.5}$$

from which it follows, by taking the dot product with $\boldsymbol{n}(s) = \eta(\nu, s)\boldsymbol{e}(s)$ that, for $0 < s < 1$,

$$\begin{aligned} \boldsymbol{n}(s) \cdot \boldsymbol{n}'(s) &= -\boldsymbol{n}'(s) \cdot (\nu(s)IB\boldsymbol{e}(s) \times \boldsymbol{k}) \\ &= -\nu(s)\eta(\nu(s))IB\boldsymbol{e}(s) \cdot (\boldsymbol{e}(s) \times \boldsymbol{k}) \\ &= 0 \end{aligned} \tag{6.1.6}$$

Therefore,

$$\frac{d}{ds}(\|\boldsymbol{n}(s)\|^2) = \frac{d}{ds}(\eta^2(\nu(s))\boldsymbol{e}(s) \cdot \boldsymbol{e}(s)) = 0 \tag{6.1.7}$$

or, as $\|\boldsymbol{e}(s)\| = 1$,

$$N(s) = \eta(\nu(s)) = \text{ const.}, \quad 0 \le s \le 1 \tag{6.1.8}$$

The equilibrium equation (6.1.4a) then reduces to

$$N(s)\boldsymbol{e}'(s) + \lambda\nu(s)(\boldsymbol{e}(s) \times \boldsymbol{k}) = \boldsymbol{0}, \quad 0 < s < 1 \tag{6.1.9}$$

As a consequence of (6.1.8) we also note that $\nu(s)$ is constant along the wire and that we must have $\nu \geq b > 1$ so that $N > 0$. If we now take the dot product of (6.1.9) with $\boldsymbol{k}$ we clearly obtain $\eta(\nu)\boldsymbol{e}'(s) \cdot \boldsymbol{k} = 0$ or

$$\boldsymbol{e}(s) \cdot \boldsymbol{k} = \text{ const. } \equiv \cos\psi \tag{6.1.10}$$

where ψ is just the angle between $\boldsymbol{e}(s)$ and $\boldsymbol{k}$. In the trivial case we would have $\cos\psi = 1$, in which case $\boldsymbol{e}(s) = \boldsymbol{k}$, $0 \leq s \leq 1$, so that with $\boldsymbol{r}'(s) = \nu\boldsymbol{e}(s) = \nu\boldsymbol{k}$, we find for all $\lambda = IB$ the trivial solution $(\nu \equiv b)$

$$\boldsymbol{r}(s) = bs\boldsymbol{k}, \quad 0 \leq s \leq 1, \tag{6.1.11}$$

representing a perfectly straight wire stretched between $\mathbf{0}$ and $b\boldsymbol{k}$. On the other hand, if $|\cos\psi| < 1$, then we may represent $\boldsymbol{e}(s)$ in the form

$$\boldsymbol{e}(s) = \cos\theta\sin\psi\boldsymbol{i} + \sin\theta\sin\psi\boldsymbol{j} + \cos\psi\boldsymbol{k} \tag{6.1.12}$$

If we now substitute the representation (6.1.12) into (6.1.9), and then take the dot product of the resulting equation with $\cos\theta\boldsymbol{j} - \sin\theta\boldsymbol{j}$, we find that

$$\theta'(s) = \nu\lambda/N(s) \tag{6.1.13}$$

Next, we note that, by virtue of the second boundary condition in (6.1.4b), and the fact that $\boldsymbol{r}'(s) = \nu\boldsymbol{e}(s)$, we have

$$\nu\int_0^1 \boldsymbol{e}(s)\,ds = b\boldsymbol{k} \tag{6.1.14}$$

If, therefore, we substitute the representation (6.1.12) into (6.1.14), and use the fact that $\psi =$ const., we easily obtain the relations

$$\begin{cases} \displaystyle\int_0^1 \cos\theta(s)\,ds = 0 \\ \displaystyle\int_0^1 \sin\theta(s)\,ds = 0 \\ \nu\cos\psi = b \end{cases} \tag{6.1.15}$$

However, in view of (6.1.13),

$$\theta(s) = \frac{\nu\lambda}{\eta(\nu)} \cdot s + \theta_0, \quad 0 \leq s \leq 1 \tag{6.1.16}$$

for some $\theta_0 = \text{const}$. Combining (6.1.15) with (6.1.16), we are led to the conclusions

$$\sin\left(\frac{\nu\lambda}{\eta(\nu)} + \theta_0\right) = \sin\theta_0, \qquad \cos\left(\frac{\nu\lambda}{\eta(\nu)} + \theta_0\right) = \cos\theta_0 \tag{6.1.17}$$

from which it follows that $\dfrac{\nu\lambda}{\eta(\nu)} = 2\pi m$, $m = 1, 2, 3, \ldots$ Finally, we substitute (6.1.16) back into the representation (6.1.12), use the last relation in (6.1.5), as well as the results of (6.1.17), and then integrate with respect to s so as to obtain the following:

Theorem 6.1 (Wolfe, [184]) Suppose that $N = \eta(\nu)$ and let $\lambda > 0$ and $\nu > b$ satisfy $\dfrac{\nu\lambda}{\eta(\nu)} = 2\pi m$ for some positive integer m. Then the boundary-value problem (6.1.4a), (6.1.4b) has solutions of the form

$$\begin{aligned} \boldsymbol{r}_{\theta_0}(s) = {} & \frac{\eta(\nu)}{\lambda}\left\{\sin\left(\theta_0 + \frac{\nu\lambda}{\eta(\nu)}\right)s - \sin\theta_0\right\}\sin\psi\boldsymbol{i} \\ & + \frac{\eta(\nu)}{\lambda}\left\{\cos\theta_0 - \cos\left(\theta_0 + \frac{\nu\lambda}{\eta(\nu)}\right)s\right\}\sin\psi\boldsymbol{j} + \nu(\cos\psi)s\boldsymbol{k}, \\ & 0 \le s \le 1, \text{ where } \psi = \cos^{-1}(b/\nu). \end{aligned} \tag{6.1.18}$$

Thus, for each pair (ν, λ) satisfying the relation $\dfrac{\nu\lambda}{\eta(\nu)} = 2\pi m$, for some positive integer m, we obtain a one-parameter family of solutions (6.1.18) of the boundary-value problem (6.1.4a), (6.1.4b); the members of this one parameter family are helices, as depicted in Figure 6.2 below:

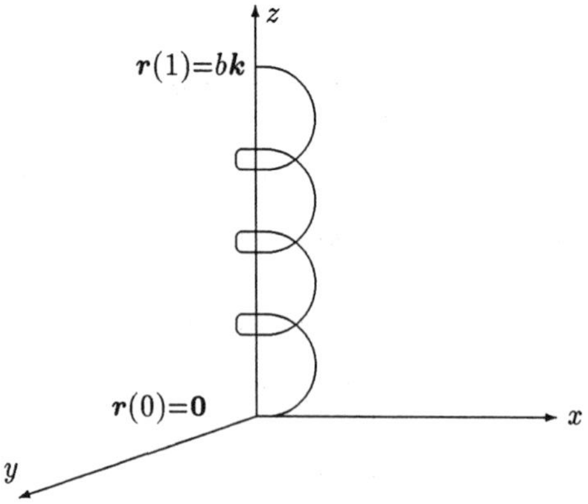

Figure 6.2

As pointed out in [184], the essential reason for the multiplicity of solutions (6.1.18) is the fact that the boundary-value problem (6.1.4a), (6.1.4b) is invariant with respect to rotations about the z axis. It is also demonstrated in [184] that the set of solutions (6.1.18) exhibits the classical bifurcation phenomenon; to verify the validity of such a claim one begins by linearizing the equilibrium equation (6.1.4a) about the trivial solution given by (6.1.11) (i.e., about the straight state joining $\mathbf{0}$ to $b\boldsymbol{k}$) so as to obtain the set of equations

$$\begin{cases} \frac{\eta(b)}{b}x''(s) + \lambda y'(s) = 0, \\ \frac{\eta(b)}{b}y''(s) - \lambda x'(s) = 0, \\ \eta_\nu(b)z''(s) = 0, \end{cases} \tag{6.1.19}$$

for $0 < s < 1$, with associated boundary conditions

$$\begin{cases} x(0) = y(0) = z(0) = 0 \\ x(1) = y(1) = z(1) = 0 \end{cases} \tag{6.1.20}$$

Elementary computations yield the result that (6.1.19), (6.1.20) possesses eigenvalues

$$\lambda_k = 2\pi k\frac{\eta(b)}{b} \tag{6.1.21}$$

with associated eigenfunctions

$$\begin{cases} x_k(s) = C_1(1 - \cos 2\pi ks) + C_2 \sin 2\pi ks \\ y_k(s) = -C_1 \sin 2\pi ks + C_2(1 - \cos 2\pi ks) \\ z_k(s) = 0 \end{cases} \tag{6.1.22}$$

The eigenvalues λ_k, as given by (6.1.22) are, of course, the bifurcation points, and the branch of solutions (of the form (6.1.18)) on which $\frac{\nu\lambda}{\eta(\nu)} = 2\pi k$ holds, for some positive integer k, bifurcates off from the trivial (straight state) solution (6.1.11) for $\lambda = \lambda_k$. As noted in [184], the number of solution branches present, for a given λ, depends on the behavior of the function $\eta(\nu)/\nu$; if $\frac{d}{d\nu}(\eta(\nu)/\nu) > 0$, then the bifurcation diagram is as depicted in Figure 6.3:

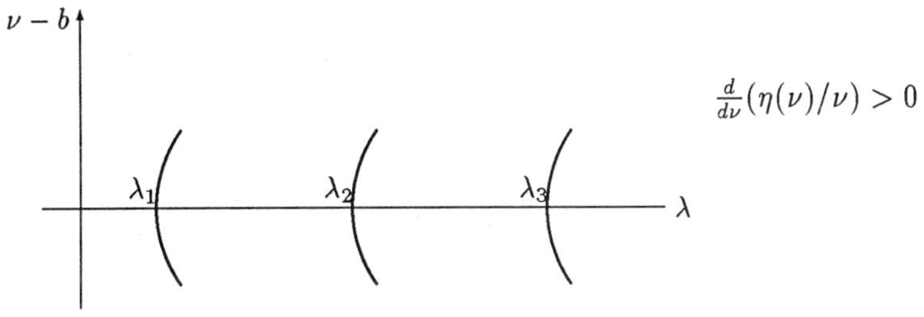

Figure 6.3

Before turning from the work in [184] to our description of self-interacting current bearing wires, it is worth noting that the equilibrium equation (6.1.4a) for the nonself-interacting case is just the Euler-Lagrange equation for the rotationally invariant potential energy functional

$$\Psi = \int_0^1 \left[\Lambda(\nu, s) + \frac{\lambda}{2}(\boldsymbol{r} \times \boldsymbol{k}) \cdot \boldsymbol{r}' \right] ds \tag{6.1.23}$$

where

$$\Lambda(\nu, s) = \int_1^{\nu} \eta(\xi, s)\, d\xi \tag{6.1.24}$$

For the case of a homogeneous wire, we have, of course, $\eta_s(\xi, s) = 0$, so that $\Lambda = \Lambda(\nu)$. Also, given the equilibrium solutions (helices) (6.1.18), it is a relatively easy matter to compute that on the k-th branch the potential energy is given by

$$\Psi = \Lambda(\nu) - \frac{\lambda}{2\pi k}(\sin^2 \psi)\nu^2 \tag{6.1.25}$$

which, in view of the relations $\dfrac{\nu\lambda}{\eta(\nu)} = 2\pi k$ and $\psi = \cos^{-1}(b/\nu)$ may be expressed as

$$\Psi = \Lambda(\nu) - \frac{1}{2}\nu\eta(\nu) + \frac{1}{2}b^2\frac{\eta(\nu)}{\nu} \tag{6.1.26}$$

It is also true that (6.1.26) holds on the (trivial) branch corresponding to the (straight state) solution (6.1.11), i.e., on the branch where $\nu = b$. We also have the following result of Wolfe [184]:

Theorem 6.2 If $\dfrac{d}{d\nu}(\eta(\nu)/\nu) > 0$ then, for a given $\lambda > 2\pi\dfrac{\eta(b)}{b}$, among all solutions of the boundary-value problem (6.1.4a), (6.1.4b), the potential energy Ψ is minimized at solutions of the form (6.1.18) with $\dfrac{\nu}{\eta(\nu)} = \dfrac{2\pi}{\lambda}$, i.e., at solutions which lie on the first solution branch in Figure 6.3.

Proof: We note that for a given λ, the stretch ν is largest on the branch corresponding to $\lambda = \lambda_1$; to demonstrate that Ψ is minimized along this branch of solutions, it suffices to show that for $\nu > b$ we have $\dfrac{d}{d\nu}\Psi(\nu) < 0$ with $\Psi(\nu)$ as expressed by (6.1.26). However,

$$\frac{d}{d\nu}\Psi(\nu) = \frac{1}{2}(b^2 - \nu^2)\frac{d}{d\nu}\left(\frac{\eta(\nu)}{\nu}\right) < 0, \tag{6.1.27}$$

for $\nu > b$, if $\dfrac{d}{d\nu}\left(\dfrac{\eta(\nu)}{\nu}\right) > 0$. □

While the analysis in [57] continues with a discussion of the equilibrium case when $\eta_s(\nu, s) \neq 0$, as well as with a study of the dynamical problem governing the motion of the elastic conductor in a magnetic field, when $\eta_s(\nu, s) = 0$, we will not pursue these issues here; we do note, however, that the discussion of linearized stability in [184] tends to support the idea that the first branch ($\lambda = \lambda_1$) of solutions of the form (6.1.18) is stable while the succeeding branches are not, i.e., the set of solutions (6.1.18), with $\frac{\nu\lambda}{\eta(\nu)} = 2\pi k$, $k = 1, 2, 3 \ldots$ form, for $k = 1$, a stable invariant manifold for the dynamical problem given by

$$\rho \boldsymbol{r}_{tt} = (\eta(\nu(s))\boldsymbol{e}(s))' + IB\nu(s)\boldsymbol{e}(s) \times \boldsymbol{k} \tag{6.1.28}$$

where the density $\rho > 0$ is constant. We also make note of the fact that the analysis presented in [184], which we have described above, has been extended by Wolfe [183], and Seidman and Wolfe [185], to cover the description of the equilibrium states of nonlinearly elastic conducting rods in a magnetic field when the rod is modelled by the special Cosserat theory as described, e.g., in [3].

The work described above assumes that we are dealing with an idealized, thin, nonlinearly elastic wire which is subjected to no self-interaction force; this hypothesis will be weakened in the next section.

6.2 The Self-Interacting Current Bearing Wire

Although we assumed, in § 6.1, that the current bearing, nonlinearly elastic wire had, in essence, zero thickness, in reality any such wire will have some thickness and current flowing in one part of the wire will create a magnetic field which will then exert a force at other points of the wire; this self-interaction, which is determined by the classical Biot-Savart law (see (1.2.25), (1.2.29), and (1.2.30)), and its effect on the nature of the equilibrium states of a flexible, current bearing wire, does not seem to have been considered before in the literature. To set the notation we refer to Figure 6.4.

The figure depicts a segment of the wire, which is assumed to have a circular cross-section: $\boldsymbol{r}(s)$ is the deformed position of the center line of the wire (the curve $\mathcal{C}$) at the material point s, $\boldsymbol{p}(\xi, \eta, s)$ is the position vector to a point in the cross-sectional cut through the wire at s, orthogonal to C, ξ and η are coordinates in that cross-section (of constant area A) with respect to an orthogonal pair of basis vectors $\boldsymbol{d}_1(s), \boldsymbol{d}_2(s)$ in the plane at s; thus, $\boldsymbol{p}(\xi, \eta, s) = \boldsymbol{r}(s) + \xi \boldsymbol{d}_1(s) + \eta \boldsymbol{d}_2(s)$. If $\bar{\boldsymbol{p}}(\bar{\xi}, \bar{\eta}, \bar{\sigma})$ is the

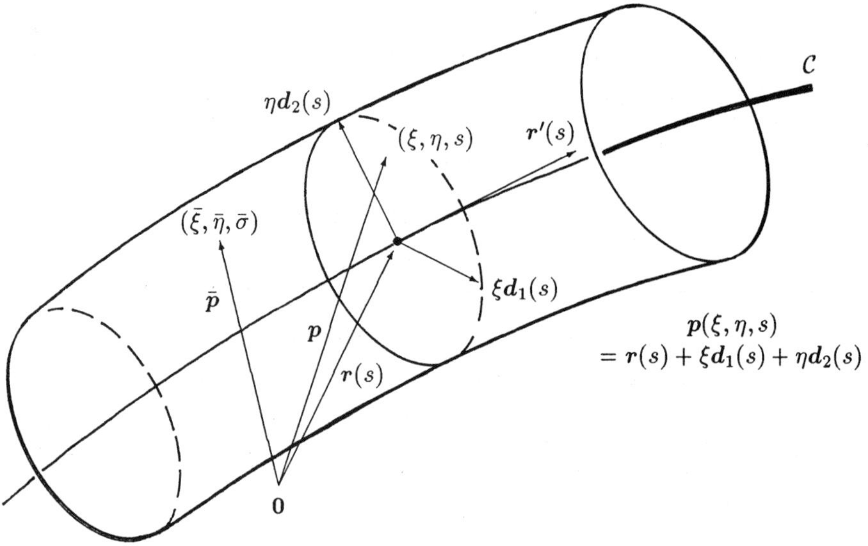

Figure 6.4

position vector from the origin to another point with coordinates $(\bar{\xi}, \bar{\nu}, \bar{\sigma})$ in the wire, i.e., in the cross-section at $s = \bar{\sigma}$, then the magnetic field experienced at (ξ, η, s) due to the flow of current past all other points $(\bar{\xi}, \bar{\eta}, \bar{\sigma})$ in the volume element V containing (ξ, η, s) is given, according to the Biot-Savart law, by

$$\boldsymbol{b}^*(\xi, \eta, s) = -\iiint_V \frac{\boldsymbol{p}(\xi, \eta, s) - \bar{\boldsymbol{p}}(\bar{\xi}, \bar{\eta}, \bar{\sigma})}{\|\boldsymbol{p} - \bar{\boldsymbol{p}}\|^3} \times \left(\frac{I}{A}\right) \boldsymbol{r}'(\bar{\sigma})\, d\bar{\xi}\, d\bar{\eta}\, d\bar{\sigma} \tag{6.2.1}$$

and the self-interaction force acting on the cross-section of wire at the material point s on C is then given by

$$\boldsymbol{f}^*(s) = \left(\frac{I}{A}\right) \boldsymbol{r}'(s) \times \iint_A \boldsymbol{b}^*(\xi, \eta, s)\, d\xi\, d\eta \tag{6.2.2}$$

To $\boldsymbol{f}^*$ we must add the force acting at s due to the ambient magnetic field $\boldsymbol{B} = B\boldsymbol{k}$, thus obtaining the vector equilibrium equation

$$\boldsymbol{n}'(s) + \left(\frac{I}{A}\right)^2 \boldsymbol{r}'(s) \times \iint_A \Bigg(\iiint_V \frac{\boldsymbol{p}(\xi, \eta, s) - \bar{\boldsymbol{p}}(\bar{\xi}, \bar{\eta}, \bar{\sigma})}{\|\boldsymbol{p} - \bar{\boldsymbol{p}}\|^3} \tag{6.2.3}$$

$$\times \boldsymbol{r}'(\bar{\sigma})\, d\bar{\xi}\, d\bar{\eta}\, d\bar{\sigma} \Bigg) d\xi\, d\eta + IB\boldsymbol{r}'(s) \times \boldsymbol{k} = \boldsymbol{0}$$

For the form of $\boldsymbol{n}(s)$ (which again denotes the resultant contact force exerted by the material of, say, $(s, 1]$, in the wire, on the material of $[0, s]$) we have, as in § 6.1, $\boldsymbol{n}(s) = N(s)\boldsymbol{e}(s)$, with $\boldsymbol{e}(s) = \boldsymbol{r}'(s)/\nu(s)$ and $\nu(s) = \|\boldsymbol{r}'(s)\|$; also we again assume that $N(s)$ has the form $N(s) = \mu(\nu(s))$ so that

$$\boldsymbol{n}(s) = \mu(\nu(s))\boldsymbol{e}(s) \equiv \frac{\mu(\nu(s))}{\nu(s)}\boldsymbol{r}'(s) \tag{6.2.4}$$

Therefore, if we take the dot product of (6.2.3) with $\boldsymbol{n}(s)$ we find, just as in § 6.1, that $\boldsymbol{n}'(s) \cdot \boldsymbol{n}(s) = 0$, in which case we have, once more, the condition that $\nu(s) = \text{const}$. In the following section we will describe those calculations which show that we still have the existence of *straight* equilibrium states even for the equilibrium equations (6.2.3) possessing the self-interaction term; similar results will be established in § 6.4 with regard to the existence of plane, circular equilibrium states.

6.3 Self-Interacting Wires: Existence of Straight Equilibrium States

Although no longer quite as trivial as was the situation in § 6.1 for the perfectly flexible, (idealized) zero thickness wire, we can still exhibit the existence of straight equilibrium states for the equations (6.2.3). We refer to the sketch in Figure 6.5:

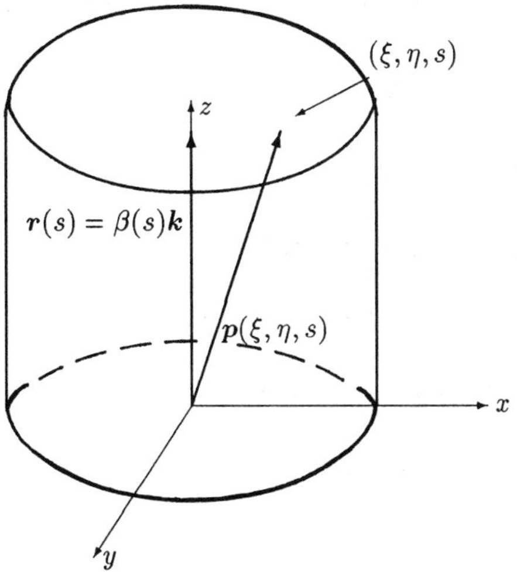

Figure 6.5

We begin by looking for solutions of the form

$$\boldsymbol{p}(\xi,\eta,s)=\xi\boldsymbol{i}+\eta\boldsymbol{j}+\beta(s)\boldsymbol{k} \tag{6.3.1}$$

so that

$$\boldsymbol{r}(s)=\beta(s)\boldsymbol{k} \tag{6.3.2}$$

In (6.3.1) we may assume that $\xi^2+\eta^2\le 1$ without loss of generality; we also note (see Figure 6.4) that we have taken, $\forall\, s$, $\boldsymbol{d}_1(s)=\boldsymbol{i}$, $\boldsymbol{d}_2(s)=\boldsymbol{j}$. In Figure 6.5 (and (6.3.1)), $\boldsymbol{p}(\xi,\eta,s)$ is, of course, the position vector to a point in the cross-sectional cut through the wire at s, orthogonal to the center line $\mathcal{C}$, which is coincident with the z-axis. As in Figure 6.4, we also take

$$\boldsymbol{p}(\bar{\xi},\bar{\eta},\bar{\sigma})=\bar{\xi}\boldsymbol{i}+\bar{\eta}\boldsymbol{j}+\beta(\bar{\sigma})\boldsymbol{k} \tag{6.3.3}$$

to be the position vector from the origin to another point with coordinates $(\bar{\xi},\bar{\eta},\bar{\sigma})$ in the cross-section at $s=\bar{\sigma}$, where $\bar{\xi}^2+\bar{\eta}^2\le 1$. We now have the following

Lemma 6.1 If $\boldsymbol{p}$, $\bar{\boldsymbol{p}}$ are given, respectively, by (6.3.1), (6.3.3), then

$$\boldsymbol{r}'(s)\times((\boldsymbol{p}-\bar{\boldsymbol{p}})\times\boldsymbol{r}'(\bar{\sigma}))=\beta(s)\beta'(\bar{\sigma})[(\xi-\bar{\xi})\boldsymbol{i}+(\eta-\bar{\eta})\boldsymbol{j}] \tag{6.3.4a}$$

and

$$|\boldsymbol{p}-\bar{\boldsymbol{p}}|^3=[(\xi-\bar{\xi})^2+(\eta-\bar{\eta})^2+(\beta(s)-\beta(\bar{\sigma}))^2]^{3/2} \tag{6.3.4b}$$

Proof: By direct calculation we have

$$\begin{aligned}
&\boldsymbol{r}'(s)\times(\boldsymbol{p}(\xi,\eta,s)\times\boldsymbol{r}'(\bar{\sigma})) \qquad (6.3.5)\\
&=\boldsymbol{p}(\xi,\eta,s)(\boldsymbol{r}'(s)\cdot\boldsymbol{r}'(\bar{\sigma}))-\boldsymbol{r}'(\bar{\sigma})(\boldsymbol{r}'(s)\cdot\boldsymbol{p}(\xi,\eta,s))\\
&=\boldsymbol{p}(\xi,\eta,s)\beta'(s)\beta'(\bar{\sigma})-\beta'(\bar{\sigma})\beta'(s)\beta(s)\boldsymbol{k}\\
&=\beta'(s)\beta'(\bar{\sigma})(\boldsymbol{p}(\xi,\eta,s)-\beta(s)\boldsymbol{k})\\
&=\beta'(s)\beta'(\bar{\sigma})(\xi\boldsymbol{i}+\eta\boldsymbol{j})
\end{aligned}$$

and a similar computation produces

$$\boldsymbol{r}'(s)\times(\bar{\boldsymbol{p}}(\bar{\xi},\bar{\eta},\bar{\sigma})\times\boldsymbol{r}'(\bar{\sigma}))=\beta'(s)\beta'(\bar{\sigma})(\bar{\xi}\boldsymbol{i}+\bar{\eta}\boldsymbol{j}) \tag{6.3.6}$$

so that (6.3.4a) follows by combining (6.3.5) with (6.3.6). Next we compute that

$$\begin{cases} \boldsymbol{p}\cdot\boldsymbol{p} = \xi^2 + \eta^2 + \beta^2(s) \\ \bar{\boldsymbol{p}}\cdot\bar{\boldsymbol{p}} = \bar{\xi}^2 + \bar{\eta}^2 + \beta^2(\bar{\sigma}) \\ \boldsymbol{p}\cdot\bar{\boldsymbol{p}} = \xi\bar{\xi} + \eta\bar{\eta} + \beta(s)\beta(\bar{\sigma}) \end{cases} \tag{6.3.7}$$

so that

$$\boldsymbol{p}\cdot\boldsymbol{p} - 2\boldsymbol{p}\cdot\bar{\boldsymbol{p}} + \bar{\boldsymbol{p}}\cdot\bar{\boldsymbol{p}} = \beta^2(s) + \beta^2(\bar{\sigma}) \tag{6.3.8}$$

$$-2\beta(s)\beta(\bar{\sigma}) + \xi^2 + \bar{\xi}^2 + \eta^2 + \bar{\eta}^2 - 2\xi\bar{\xi} - 2\eta\bar{\eta}$$

However,

$$|\boldsymbol{p} - \bar{\boldsymbol{p}}|^3 = (\boldsymbol{p}\cdot\boldsymbol{p} - 2\boldsymbol{p}\cdot\bar{\boldsymbol{p}} + \bar{\boldsymbol{p}}\cdot\bar{\boldsymbol{p}})^{3/2} \tag{6.3.9}$$

and, therefore, (6.3.4b) follows by combining (6.3.8) with (6.3.9). □

As a consequence of (6.2.4), (6.3.2), and the fact that the stretch $\nu(s) = \text{const.}$, we easily find that

$$\begin{cases} \boldsymbol{n}(s) = \dfrac{\mu(\nu)}{\nu}\beta'(s)\boldsymbol{k} \\ \boldsymbol{n}'(s) = \dfrac{\mu(\nu)}{\nu}\beta''(s)\boldsymbol{k} \end{cases} \tag{6.3.10}$$

while

$$IB\boldsymbol{r}'(s) \times \boldsymbol{k} = IB\beta'(s)\boldsymbol{k} \times \boldsymbol{k} = \boldsymbol{0} \tag{6.3.11}$$

Combining (6.3.10), (6.3.11), with Lemma 6.1 and the general form (6.2.3) of the equilibrium equations, we easily determine that a solution of (6.2.3) of the form (6.3.1) exists iff

$$\frac{\mu(\nu)}{\nu}\beta''(s)\boldsymbol{k} \tag{6.3.12}$$

$$+ \left(\frac{I}{A}\right)^2 \int_A\!\!\int \left(\iiint_V \frac{\beta'(s)\beta'(\bar{\sigma})[(\xi - \bar{\xi})\boldsymbol{i} + (\eta - \bar{\eta})\boldsymbol{j}]\, d\bar{\xi}\, d\bar{\eta}\, d\bar{\sigma}}{\left[(\xi - \bar{\xi})^2 + (\eta - \bar{\eta})^2 + (\beta(s) - \beta(\bar{\sigma})^2\right]^{3/2}} \right) d\xi\, d\eta = \boldsymbol{0}$$

Now, by virtue of the (assumed) circular geometry of the cross-sections of the wire, and the fact that the numerator of the integrand of the integral in (6.3.12) is skew-symmetric with respect to the transformation $\xi \to \bar{\xi}$, $\eta \to \bar{\eta}$, while the denominator is, obviously, invariant with respect to the same transformation, the integral in (6.3.12) vanishes identically and the equilibrium equations reduce to the statement that

$$\frac{\mu(\nu)}{\nu}\beta''(s)\boldsymbol{k} = \boldsymbol{0} \tag{6.3.13}$$

From (6.3.13), (6.3.1) it then follows that we have established the following

Theorem 6.3 For all values of the parameter $\lambda = IB$ there exist straight (equilibrium) solutions of the system (6.2.3) of the form

$$\begin{cases} \boldsymbol{p}(\xi, \eta, s) = \xi\boldsymbol{i} + \eta\boldsymbol{j} + (\alpha_0 + \alpha_1 s)\boldsymbol{k} \\ \boldsymbol{r}(s) = (\alpha_0 + \alpha_1 s)\boldsymbol{k} \end{cases} \tag{6.3.14}$$

where α_0, α_1 are constants determined by the boundary conditions, i.e., the prescribed values of $\boldsymbol{r}(0)$ and $\boldsymbol{r}(1)$.

6.4 Self-Interacting Wire: Existence of Plane Circular States

The demonstration that there exist other, nontrivial, equilibrium states for the self-interacting, current bearing wire, which is governed by the system (6.2.3), other than the straight states (6.3.14) is far from simple; one such set of states, whose existence has been established, are plane circular states of the form

$$\boldsymbol{p}(\xi, \eta, s) = \boldsymbol{r}(s) + \xi\boldsymbol{d}_1(s) + \eta\boldsymbol{d}_2(s) \tag{6.4.1}$$

with

$$\boldsymbol{r}(s) = \alpha(\cos\omega s\boldsymbol{i} + \sin\omega s\boldsymbol{j}) \tag{6.4.2}$$

where α, ω are parameters to be determined below. It is not difficult to show that for the orthogonal pair of basis vectors $\boldsymbol{d}_1(s)$, $\boldsymbol{d}_2(s)$, in the plane at s, we may choose

$$\boldsymbol{d}_1(s) = \frac{\boldsymbol{r}''(s)}{|\boldsymbol{r}''(s)|} = -\frac{1}{\alpha}\boldsymbol{r}(s) \tag{6.4.3a}$$

and

$$\begin{aligned} \boldsymbol{d}_2(s) &= \frac{\boldsymbol{r}'(s)}{|\boldsymbol{r}'(s)|} \times \boldsymbol{d}_1(s) \\ &= (-\sin\omega s\boldsymbol{i} + \cos\omega s\boldsymbol{j}) \times \left[-\frac{1}{\alpha}\boldsymbol{r}(s)\right] \\ &= \boldsymbol{k} \end{aligned} \tag{6.4.3b}$$

where, of course, we have used (6.4.2). Therefore, solutions of (6.2.3) are being sought which have the form

$$\boldsymbol{p}(\xi, \eta, s) = \left(1 - \frac{\xi}{\alpha}\right) \boldsymbol{r}(s) + \eta \boldsymbol{k} \tag{6.4.4}$$

with $\boldsymbol{r}(s)$ given by (6.4.2); it is a straightforward matter to establish the following

Lemma 6.2 If $\boldsymbol{p}$ is given by (6.4.4), $\boldsymbol{r}(s)$ by (6.4.2) and

$$\bar{\boldsymbol{p}}(\bar{\xi}, \bar{\eta}, \bar{\sigma}) = \left(1 - \frac{\bar{\xi}}{\alpha}\right) \boldsymbol{r}(\bar{\sigma}) + \bar{\eta} \boldsymbol{k} \tag{6.4.5}$$

then

$$\begin{aligned} &\boldsymbol{r}'(s) \times [(\boldsymbol{p} - \bar{\boldsymbol{p}}) \times \boldsymbol{r}'(\bar{\sigma})] \\ &\quad = K_1(s, \bar{\sigma})(\boldsymbol{p} - \bar{\boldsymbol{p}}) + (1 - \bar{\xi}/\alpha) K_2(s, \bar{\sigma}) \boldsymbol{r}'(\bar{\sigma}) \end{aligned} \tag{6.4.6}$$

with

$$\begin{cases} K_1(s, \bar{\sigma}) = \alpha^2 \omega^2 \cos \omega(s - \bar{\sigma}) \\ K_2(s, \bar{\sigma}) = -\alpha^2 \omega^2 \sin \omega(s - \bar{\sigma}) \end{cases} \tag{6.4.7}$$

and

$$|\boldsymbol{p} - \bar{\boldsymbol{p}}|^3 = \left[(\xi - \bar{\xi})^2 + (\eta - \bar{\eta})^2 + 2(\alpha - \xi)(\alpha - \bar{\xi}) K_3(s, \bar{\sigma})\right]^{3/2} \tag{6.4.8}$$

with

$$K_3(s, \bar{\sigma}) = 1 - \cos \omega(s - \bar{\sigma}) \tag{6.4.9}$$

Proof: As $\boldsymbol{r}'(s) = -\alpha\omega(\sin \omega \boldsymbol{i} - \cos \omega \boldsymbol{j})$, we begin by noting that

$$\boldsymbol{r}(s) \cdot \boldsymbol{k} = \boldsymbol{r}'(s) \cdot \boldsymbol{k} = 0 \tag{6.4.10a}$$

$$\begin{aligned} \boldsymbol{r}'(s) \cdot \boldsymbol{p}(\xi, \eta, s) &= (1 - \xi/\alpha) \boldsymbol{r}'(s) \cdot \boldsymbol{r}(s) \\ &\quad + \eta \boldsymbol{r}'(s) \cdot \boldsymbol{k} = 0 \end{aligned} \tag{6.4.10b}$$

and, in a similar fashion

$$\boldsymbol{r}'(\bar{\sigma}) \cdot \bar{\boldsymbol{p}}(\bar{\xi}, \bar{\eta}, \bar{\sigma}) = 0 \tag{6.4.10c}$$

For the numerator of the integrand in the self-interaction term appearing in the equilibrium equations (6.2.3) we have

$$\begin{aligned}
&\boldsymbol{r}'(s) \times [(\boldsymbol{p} - \bar{\boldsymbol{p}}) \times \boldsymbol{r}'(\bar{\sigma})] \\
&= \boldsymbol{r}'(s) \cdot \boldsymbol{r}'(\bar{\sigma})(\boldsymbol{p} - \bar{\boldsymbol{p}}) - \boldsymbol{r}'(\bar{\sigma})[\boldsymbol{r}'(s) \cdot (\boldsymbol{p} - \bar{\boldsymbol{p}})] \\
&= \boldsymbol{r}'(s) \cdot \boldsymbol{r}'(\bar{\sigma})(\boldsymbol{p} - \bar{\boldsymbol{p}}) + \boldsymbol{r}'(\bar{\sigma})[\boldsymbol{r}'(s) \cdot \bar{\boldsymbol{p}}(\bar{\xi}, \bar{\eta}, \bar{\sigma})]
\end{aligned} \tag{6.4.11}$$

where we have used (6.4.10b). However,

$$\boldsymbol{r}'(s) \cdot \boldsymbol{r}'(\bar{\sigma}) = \alpha^2\omega^2 \cos\omega(s - \bar{\sigma}) \equiv K_1(s, \bar{\sigma}) \tag{6.4.12}$$

and

$$\boldsymbol{p} - \bar{\boldsymbol{p}} = (1 - \xi/\alpha)\boldsymbol{r}(s) - (1 - \bar{\xi}/\alpha)\boldsymbol{r}(\bar{\sigma}) + (\eta - \bar{\eta})\boldsymbol{k} \tag{6.4.13}$$

while

$$\begin{aligned}
\boldsymbol{r}'(s) \cdot \bar{\boldsymbol{p}}(\bar{\xi}, \bar{\eta}, \bar{\sigma}) &= \boldsymbol{r}'(s) \cdot [(1 - \bar{\xi}/\alpha)\boldsymbol{r}(\bar{\sigma}) + \bar{\eta}\boldsymbol{k}] \\
&= (1 - \bar{\xi}/\alpha)\boldsymbol{r}'(s) \cdot \boldsymbol{r}(\bar{\sigma}) \\
&= -\alpha^2\omega(1 - \bar{\xi}/\alpha)\sin\omega(s - \bar{\sigma}) \\
&\equiv (1 - \bar{\xi}/\alpha)K_2(s, \bar{\sigma})
\end{aligned} \tag{6.4.14}$$

where $K_1(s, \bar{\sigma})$ and $K_2(s, \bar{\sigma})$ are given by (6.4.7). The relation (6.4.6) now follows if we combine (6.4.11)–(6.4.14). For our next set of calculations we recall the identity (6.3.9) and, based on the forms (6.4.2), (6.4.4) of $\boldsymbol{r}(s)$ and $\boldsymbol{p}(\xi, \eta, s)$, respectively, we compute that

$$\begin{cases}
\boldsymbol{p} \cdot \boldsymbol{p} = \alpha^2(1 - \xi/\alpha)^2 + \eta^2 \\
\bar{\boldsymbol{p}} \cdot \bar{\boldsymbol{p}} = \alpha^2(1 - \bar{\xi}/\alpha)^2 + \bar{\eta}^2 \\
\boldsymbol{p} \cdot \bar{\boldsymbol{p}} = (1 - \xi/\alpha)(1 - \bar{\xi}/\alpha)\boldsymbol{r}(s) \cdot \boldsymbol{r}(\bar{\sigma}) + \eta\bar{\eta}
\end{cases} \tag{6.4.15}$$

However,

$$\boldsymbol{r}(s) \cdot \boldsymbol{r}(\bar{\sigma}) = \alpha^2 \cos\omega(s - \bar{\sigma}) \equiv \frac{1}{\omega^2}K_1(s, \bar{\sigma}) \tag{6.4.16}$$

Therefore,

$$\begin{aligned}
|\boldsymbol{p} - \bar{\boldsymbol{p}}|^2 = \alpha^2 \left[(1 - \xi/\alpha)^2 + (1 - \bar{\xi}/\alpha)^2\right] + \eta^2 + \bar{\eta}^2 \\
- 2\eta\bar{\eta} - \frac{2}{\omega^2}(1 - \xi/\alpha)(1 - \bar{\xi}/\alpha)K_1(s, \bar{\sigma})
\end{aligned} \tag{6.4.17}$$

in which case

$$|\boldsymbol{p}-\bar{\boldsymbol{p}}|^3 = \left[(\alpha-\xi)^2+(\alpha-\bar{\xi})^2+(\eta-\bar{\eta})^2-\frac{2}{\alpha^2\omega^2}(\alpha-\xi)(\alpha-\bar{\xi})K_1(s,\bar{\sigma})\right]^{3/2} \tag{6.4.18}$$

Using the definition (6.4.7) of $K_1(s,\bar{\sigma})$, we may rewrite (6.4.18) in the form

$$\begin{aligned} |\boldsymbol{p}-\bar{\boldsymbol{p}}|^3 &= \left[(\alpha-\xi)^2+(\alpha-\bar{\xi})^2+(\eta-\bar{\eta})^2-2(\alpha-\xi)(\alpha-\bar{\xi})\cos\omega(s-\bar{\sigma})\right]^{3/2} \\ &= \left[(\eta-\bar{\eta})^2+(\xi-\bar{\xi})^2+2(\alpha-\xi)(\alpha-\bar{\xi})(1-\cos\omega(s-\bar{\sigma}))\right]^{3/2} \end{aligned} \tag{6.4.19}$$

which is, of course, equivalent to (6.4.8) in view of the definition, (6.4.9), of $K_3(s,\bar{\sigma})$. □

Inasmuch as $\boldsymbol{n}(s)=\frac{\mu(\nu)}{\nu}\boldsymbol{r}'(s)$, with $\boldsymbol{r}(s)$ given by (6.4.2), we easily find that

$$\boldsymbol{n}'(s)=\frac{\mu(\nu)}{\nu}\boldsymbol{r}''(s)\equiv-\frac{\mu(\nu)}{\nu}\omega^2\boldsymbol{r}(s) \tag{6.4.20}$$

while

$$\begin{aligned} IB\boldsymbol{r}'(s)\times\boldsymbol{k} &= -\alpha\omega IB[\sin\omega s\boldsymbol{i}-\cos\omega s\boldsymbol{j}]\times\boldsymbol{k} \\ &= \omega IB\boldsymbol{r}(s). \end{aligned} \tag{6.4.21}$$

Using (6.4.20), (6.4.21), and the results of Lemma 6.2 (i.e., (6.4.6) and (6.4.8)) we now determine that there exists an (equilibrium) solution of the system of equations (6.2.3) for the nonlinearly elastic, current bearing, self-interacting wire, which has the form (6.4.4), with $\boldsymbol{r}(s)$ given by (6.4.2), if and only if for some values of the parameters $\lambda = IB$, ω, and α

$$\begin{aligned} &\left(-\frac{\mu(\nu)}{\nu}\omega^2+\omega IB\right)\boldsymbol{r}(s) \\ &+\left(\frac{I}{A}\right)^2\iint_A\left(\iiint_V\frac{(K_1(s,\bar{\sigma})(\boldsymbol{p}-\bar{\boldsymbol{p}})+(1-\bar{\xi}/\alpha)K_2(s,\bar{\sigma})\boldsymbol{r}'(\bar{\sigma}))\,d\bar{\xi}\,d\bar{\eta}\,d\bar{\sigma}}{[(\eta-\bar{\eta})^2+(\xi-\bar{\xi})^2+2(\alpha-\xi)(\alpha-\bar{\xi})K_3(s,\bar{\sigma})]^{3/2}}\right)d\xi\,d\eta \\ &=0 \end{aligned} \tag{6.4.22}$$

If we recall (6.4.13) and set

$$\begin{aligned} &\Gamma_\alpha(\xi,\eta,s;\bar{\xi},\bar{\eta},\bar{\sigma}) = \\ &\qquad (\eta-\bar{\eta})^2+(\xi-\bar{\xi})^2+2(\alpha-\xi)(\alpha-\bar{\xi})K_3(s,\bar{\sigma}) \end{aligned} \tag{6.4.23}$$

it is a relatively trivial exercise to show that (6.4.22) can be rewritten in the form

$$\left[-\frac{\mu(\nu)}{\nu}+\omega IB+\left(\frac{I}{A}\right)^2\iint_A\left(\iiint_V\frac{(1-\xi/\alpha)K_1(s,\bar\sigma)}{\Gamma_\alpha^{3/2}}\,d\bar\xi\,d\bar\eta\,d\bar\sigma\right)d\xi\,d\eta\right]\boldsymbol{r}(s)$$

$$=\left(\frac{I}{A}\right)^2\iint_A\left(\iiint_V\frac{(1-\bar\xi/\alpha)[K_1(s,\bar\sigma)\boldsymbol{r}(\bar\sigma)-K_2(s,\bar\sigma)\boldsymbol{r}'(\bar\sigma)]}{\Gamma_\alpha^{3/2}}\,d\bar\xi\,d\bar\eta\,d\bar\sigma\right)d\xi\,d\eta$$

$$-\left(\frac{I}{A}\right)^2\iint_A\left(\iiint_V\frac{(\eta-\bar\eta)K_1(s,\bar\sigma)\boldsymbol{k}}{\Gamma_\alpha^{3/2}}\,d\bar\xi\,d\bar\eta\,d\bar\sigma\right)d\xi\,d\eta \qquad (6.4.24)$$

However, for the last integral in (6.4.24) we have

$$\iint_A\left(\iiint_V\frac{(\eta-\bar\eta)K_1(s,\bar\sigma)}{\Gamma_\alpha^{3/2}}\,d\bar\xi\,d\bar\eta\,d\bar\sigma\right)d\xi\,d\eta=0 \qquad (6.4.25)$$

as a direct consequence of the circular geometry of the cross-sections of the wire, the skew symmetry of the integrand, with respect to $\eta\to\bar\eta$, and the invariance of Γ_α, as given by (6.4.23), with respect to $\eta\to\bar\eta$. Our equilibrium equations have now been reduced to the system

$$\left[-\frac{\mu(\nu)}{\nu}+\omega IB+\left(\frac{I}{A}\right)^2\iint_A\left(\iiint_V\frac{(1-\xi/\alpha)K_1(s,\bar\sigma)}{\Gamma_\alpha^{3/2}}\,d\bar\xi\,d\bar\eta\,d\bar\sigma\right)d\xi\,d\eta\right]\boldsymbol{r}(s)$$

$$=\left(\frac{I}{A}\right)^2\iint_A\left(\iiint_V\frac{(1-\bar\xi/\alpha)K_1(s,\bar\sigma)\boldsymbol{r}(\bar\sigma)}{\Gamma_\alpha^{3/2}}\,d\bar\xi\,d\bar\eta\,d\bar\sigma\right)d\xi\,d\eta \qquad (6.4.26)$$

$$-\left(\frac{I}{A}\right)^2\iint_A\left(\iiint_V\frac{(1-\bar\xi/\alpha)K_2(s,\bar\sigma)\boldsymbol{r}'(\bar\sigma)}{\Gamma_\alpha^{3/2}}\,d\bar\xi\,d\bar\eta\,d\bar\sigma\right)d\xi\,d\eta$$

As $\boldsymbol{r}(s)\cdot\boldsymbol{r}'(s)=0$, for any s, if we take the dot product on both sides of (6.4.26) with $\boldsymbol{r}'(s)$, we determine that we must have

$$\iint_A\left(\iiint_V\frac{(1-\bar\xi/\alpha)[K_1(s,\bar\sigma)\boldsymbol{r}(\bar\sigma)\cdot\boldsymbol{r}'(s)]}{\Gamma_\alpha^{3/2}}\,d\bar\xi\,d\bar\eta\,d\bar\sigma\right)d\xi\,d\eta \qquad (6.4.27)$$

$$=\iint_A\left(\iiint_V\frac{(1-\bar\xi/\alpha)K_2(s,\bar\sigma)\boldsymbol{r}'(\bar\sigma)\cdot\boldsymbol{r}'(s)]}{\Gamma_\alpha^{3/2}}\,d\bar\xi\,d\bar\eta\,d\bar\sigma\right)d\xi\,d\eta$$

However,

$$\left\{\begin{array}{l}\boldsymbol{r}'(s)\cdot\boldsymbol{r}'(\bar\sigma)=K_1(s,\bar\sigma)\\ \boldsymbol{r}'(s)\cdot\boldsymbol{r}(\bar\sigma)=K_2(s,\bar\sigma)\end{array}\right. \qquad (6.4.28)$$

so that (6.4.27) is satisfied.

Having shown that (6.4.26) is satisfied, for any values of $\lambda = IB$, ω, and α, if we project onto the direction of $\boldsymbol{r}'$ at s, we now want to project (6.4.26) onto the direction of $\boldsymbol{r}$ at s (it is clear that (6.4.26) is trivially satisfied if we project it onto the direction of $\boldsymbol{k}$). Thus, taking the dot product of the vector equation (6.4.26) with $\boldsymbol{r}(s)$, and using the fact that $\|\boldsymbol{r}(s)\| = \alpha$, we obtain the equation

$$\alpha^2 \left[-\frac{\mu(\nu)}{\nu}\omega^2 + \omega IB + \left(\frac{I}{A}\right)^2 \iint_A \left(\iiint_V \frac{(1-\bar{\xi}/\alpha)K_1(s,\bar{\sigma})}{\Gamma_\alpha^{3/2}} \, d\bar{\xi}\, d\bar{\eta}\, d\bar{\sigma} \right) d\xi\, d\eta \right] \tag{6.4.29}$$
$$= \left(\frac{I}{A}\right)^2 \iint_A \left(\iiint_V \left\{ \frac{(1-\bar{\xi}/\alpha)}{\Gamma_\alpha^{3/2}} \right\} \left[\frac{K_1^2(s,\bar{\sigma})}{\omega^2} + K_2^2(s,\bar{\sigma}) \right] d\bar{\xi}\, d\bar{\eta}\, d\bar{\sigma} \right) d\xi\, d\eta$$

In arriving at (6.4.29) we have also used the fact that

$$\begin{cases} \boldsymbol{r}'(\bar{\sigma}) \cdot \boldsymbol{r}(s) = K_2(\bar{\sigma}, s) = -K_2(s, \bar{\sigma}) \\ \boldsymbol{r}(\bar{\sigma}) \cdot \boldsymbol{r}(s) = K_1(\bar{\sigma}, s) = K_1(s, \bar{\sigma}) \end{cases} \tag{6.4.30}$$

However, as a consequence of the definitions of $K_1(s,\bar{\sigma})$ and $K_2(s,\bar{\sigma})$, i.e., (6.4.7), we find that

$$\frac{K_1^2(s,\bar{\sigma})}{\omega^2} + K_2^2(s,\bar{\sigma}) = \alpha^4\omega^2 \tag{6.4.31}$$

which reduces (6.4.29) to

$$\frac{1}{\alpha^2}\left(\frac{IB}{\omega} - \frac{\mu(\nu)}{\nu}\right) = G_1(s) - \frac{1}{\alpha^2\omega^2}G_2(s) \tag{6.4.32}$$

where

$$G_1(s) = \left(\frac{I}{A}\right)^2 \iint_A \left(\iiint_V \frac{(1-\bar{\xi}/\alpha)}{\Gamma_\alpha^{3/2}} \, d\bar{\xi}\, d\bar{\eta}\, d\bar{\sigma} \right) d\xi\, d\eta \tag{6.4.33a}$$

and

$$G_2(s) = \left(\frac{I}{A}\right)^2 \iint_A \left(\iiint_V \frac{(1-\bar{\xi}/\alpha)}{\Gamma_\alpha^{3/2}} K_1(s,\bar{\sigma})\, d\bar{\xi}\, d\bar{\eta}\, d\bar{\sigma} \right) d\xi\, d\eta \tag{6.4.33b}$$

We now examine the structure of the right-hand side of (6.4.32); by virtue of (6.4.33a), (6.4.33b), we obviously have

$$G_1(s) - \frac{1}{\alpha^2\omega^2}G_2(s) \tag{6.4.34}$$
$$= \left(\frac{I}{A}\right)^2 \iint_A \left(\iiint_V \left\{ \frac{(1-\bar{\xi}/\alpha)}{\Gamma_\alpha^{3/2}} \right\} \left[1 - \frac{K_1(s,\bar{\sigma})}{\alpha^2\omega^2} \right] d\bar{\xi}\, d\bar{\eta}\, d\bar{\sigma} \right) d\xi\, d\eta$$

while, in view of (6.4.7)

$$\begin{aligned} 1-\frac{K_1(s,\bar{\sigma})}{\alpha^2\omega^2} &= 1-\cos\omega(s-\bar{\sigma}) \\ &\equiv K_3(s,\bar{\sigma}) \\ &= 2\sin^2\frac{\omega}{2}(s-\bar{\sigma}) \end{aligned} \tag{6.4.35}$$

Combining (6.4.32) with (6.4.34), and (6.4.35), we now record that the projection of the equilibrium equation on the direction of $\boldsymbol{r}$ at s has been reduced to the statement that

$$\begin{aligned} &\frac{IB}{\omega}-\frac{\mu(\nu)}{\nu} \\ &= 2\alpha^2\left(\frac{I}{A}\right)^2\iint\limits_A\left(\iiint\limits_V\frac{(1-\bar{\xi}/\alpha)}{\Gamma_\alpha^{3/2}}\sin^2\frac{\omega}{2}(s-\bar{\sigma})\,d\bar{\xi}\,d\bar{\eta}\,d\bar{\sigma}\right)d\xi\,d\eta \end{aligned} \tag{6.4.36}$$

However, from (6.4.2), we have

$$\nu=\|\boldsymbol{r}'(s)\|=\alpha\omega \tag{6.4.37}$$

and, therefore, in place of (6.4.36) we have

$$\frac{IB}{\omega}-\frac{\mu(\alpha\omega)}{\alpha\omega}=2\alpha^2G(s) \tag{6.4.38}$$

with

$$G(s)=\left(\frac{I}{A}\right)^2\iint\limits_A\left(\iiint\limits_V\frac{(1-\bar{\xi}/\alpha)}{\Gamma_\alpha^{3/2}}\sin^2\frac{\omega}{2}(s-\bar{\sigma})\,d\bar{\xi}\,d\bar{\eta}\,d\bar{\sigma}\right)d\xi\,d\eta \tag{6.4.39}$$

If we set

$$\Lambda_\alpha(\xi,\eta,s;\bar{\xi},\bar{\eta},\bar{\sigma})=\sin^2\frac{\omega}{2}(s-\bar{\sigma})\Gamma_\alpha^{-3/2}(\xi,\eta,s;\bar{\xi},\bar{\eta},\bar{\sigma})$$

so that

$$\begin{aligned} &\Lambda_\alpha(\xi,\eta,s;\bar{\xi},\bar{\eta},\bar{\sigma}) \\ &= \sin^2\frac{\omega}{2}(s-\bar{\sigma})\left[(\eta-\bar{\eta})^2+(\xi-\bar{\xi})^2+4(\alpha-\xi)(\alpha-\bar{\xi})\sin^2\frac{\omega}{2}(s-\bar{\sigma})\right]^{-3/2} \end{aligned} \tag{6.4.40}$$

then a lengthy, but straightforward, computation shows that $\dfrac{\partial\Lambda_\alpha}{\partial s}$ is an odd function of $\tau=s-\bar{\sigma}$; therefore, by (6.4.39), (6.4.40),

$$G'(s)=\left(\frac{I}{A}\right)^2\iint\limits_A\left(\iiint\limits_V(1-\bar{\xi}/\alpha)\frac{\partial\Lambda_\alpha}{\partial s}\,d\bar{\xi}\,d\bar{\eta}\,d\bar{\sigma}\right)d\xi\,d\eta=0 \tag{6.4.41}$$

where we have again used the circular symmetry of the cross-sections of the wire as defined by (6.4.2). From (6.4.41) it follows, directly, that for some function G_0 defined on $R^+ \times R^+ \times R$ we have $G(s) = G_0(\alpha, \omega, I)$. Combining all of our results to this point we find that we are able to state the following

Theorem 6.4 There exist plane circular (equilibrium) solutions of the system (6.2.3) which have the form

$$\boldsymbol{p}(\xi, \eta, s) = \alpha\left(1 - \frac{\xi}{\alpha}\right)(\cos\omega s\boldsymbol{i} + \sin\omega s\boldsymbol{j}) + \eta\boldsymbol{k} \tag{6.4.42}$$

provided $\lambda = IB$, α, and ω satisfy

$$\lambda = \frac{\mu(\alpha\omega)}{\alpha} + 2\alpha^2 G_0(\alpha, \omega, I) \tag{6.4.43}$$

where $G_0(\alpha, \omega, I)$ is given by (6.4.39), (6.4.23), and (6.4.9).

6.5 Helical States of Self-Interacting Wires

While the existence of helical equilibrium states for thin, nonlinearly elastic wires placed in an ambient magnetic field has been established in § 6.1, the existence of such helical states for current bearing wires subject to a self-interaction force is not clear (although we will conjecture in this section that such states probably do not exist). In looking for helical equilibrium states we seek solutions of the vector equation (6.2.3) of the form

$$\boldsymbol{p}(\xi, \eta, s) = \boldsymbol{r}(s) + \xi\boldsymbol{d}_1(s) + \eta\boldsymbol{d}_2(s)$$

with

$$\begin{cases} \boldsymbol{r}(s) = \boldsymbol{R}(s) + \beta(s)\boldsymbol{k} \\ \boldsymbol{R}(s) = \alpha(\cos\omega s\boldsymbol{i} + \sin\omega s\boldsymbol{j}) \end{cases} \tag{6.5.1}$$

We easily compute that

$$\boldsymbol{r}'(s) = -\alpha\omega(\sin\omega s\boldsymbol{i} - \cos\omega s\boldsymbol{j}) + \beta'(s)\boldsymbol{k} \tag{6.5.2}$$

so that

$$\begin{cases} \|\boldsymbol{r}(s)\| = \sqrt{\alpha^2 + \beta^2(s)} \\ \|\boldsymbol{r}'(s)\| = \sqrt{\alpha^2\omega^2 + \beta'^2(s)} \end{cases} \tag{6.5.3}$$

Also,

$$\boldsymbol{d}_1(s) = \frac{\boldsymbol{r}''(s)}{\|\boldsymbol{r}''(s)\|} = \frac{-\omega^2\boldsymbol{R}(s) + \beta''(s)\boldsymbol{k}}{\sqrt{\alpha^2\omega^4 + \beta''^2(s)}} \tag{6.5.4a}$$

and

$$\boldsymbol{d}_2(s) = \frac{\boldsymbol{r}'(s)}{\|\boldsymbol{r}'(s)\|} \times \boldsymbol{d}_1(s) = \frac{\boldsymbol{r}'(s) \times \boldsymbol{d}_1(s)}{\sqrt{\alpha^2\omega^2 + \beta'^2(s)}} \tag{6.5.4b}$$

In exactly the same fashion as in § 6.1–§ 6.4 we determine that $\boldsymbol{n}(s) = \dfrac{\mu(\nu(s))}{\nu(s)}\boldsymbol{r}'(s)$ must satisfy $\dfrac{d}{ds}\|\boldsymbol{n}(s)\| = 0$ so that $\|\boldsymbol{r}'(s)\| = \nu(s) = \text{const}$. From (6.5.3) it then follows that $\beta(s) = \beta_1 s + \beta_0$ with

$$\nu^2 - \alpha^2\omega^2 = \beta_1^2 \tag{6.5.5}$$

As $\beta''(s) = 0$ we easily find that $\boldsymbol{d}_1(s)$ in (6.5.4a) reduces to

$$\boldsymbol{d}_1(s) = \frac{-\omega^2\boldsymbol{R}(s)}{\sqrt{\alpha^2\omega^4}} \equiv -\frac{1}{\alpha}\boldsymbol{R}(s) \tag{6.5.6a}$$

while (6.5.4b) becomes

$$\begin{aligned} \boldsymbol{d}_2(s) &= \frac{\boldsymbol{r}'(s)}{\sqrt{\alpha^2\omega^2 + \beta_1^2}} \times \boldsymbol{d}_1(s) \\ &= \frac{-1}{\alpha\sqrt{\alpha^2\omega^2 + \beta_1^2}}[(\boldsymbol{R}'(s) + \beta_1\boldsymbol{k}) \times \boldsymbol{R}(s)] \end{aligned}$$

But

$$\boldsymbol{R}'(s) \times \boldsymbol{R}(s) = -\alpha^2\omega\boldsymbol{k}$$

as a consequence of (6.5.1) and, therefore,

$$\boldsymbol{d}_2(s) = \frac{\alpha\omega}{\sqrt{\alpha^2\omega^2 + \beta_1^2}}\boldsymbol{k} + \frac{\beta_1\boldsymbol{R}(s) \times \boldsymbol{k}}{\alpha\sqrt{\alpha^2\omega^2 + \beta_1^2}} \tag{6.5.6b}$$

We note, in passing, that the helical state defined by (6.5.1), with $\beta(s) = \beta_1 s + \beta_0$, reduces to a plane circular state iff $\beta_1 = 0$, in which case, by virtue of (6.5.6b), $\boldsymbol{d}_2(s) = \boldsymbol{k}$. Without loss of generality, however, we may set $\beta_0 = 0$. We also introduce the notation

$$\begin{cases} \lambda_1(\alpha, \omega, \beta_1) = \alpha\omega/(\alpha^2\omega^2 + \beta_1^2)^{1/2} \\ \lambda_2(\alpha, \omega, \beta_1) = \beta_1/\alpha(\alpha^2\omega^2 + \beta_1^2)^{1/2} \end{cases} \tag{6.5.7}$$

so that $\lambda_1 \left(\frac{\beta_1}{\alpha^2\omega}\right) = \lambda_2$. With the notation as in (6.5.7) we now have

$$\begin{cases} \boldsymbol{d}_1(s) = -\frac{1}{\alpha}\boldsymbol{R}(s) \\ \boldsymbol{d}_2(s) = \lambda_1\boldsymbol{k} + \lambda_2\boldsymbol{R}(s) \times \boldsymbol{k} \end{cases} \tag{6.5.8}$$

and

$$\boldsymbol{p}(\xi, \eta, s) = \left(1 - \frac{\xi}{\alpha}\right)\boldsymbol{R}(s) + (\beta_1 s + \lambda_1\eta)\boldsymbol{k} + \lambda_2\eta\boldsymbol{R}(s) \times \boldsymbol{k} \tag{6.5.9}$$

Because the calculations which lead (in this particular situation) to those expressions which determine the self-interaction term in the equilibrium equation (6.2.3) are more involved than those which we encountered in § 6.3 and § 6.4, we will spread these computations out over the course of two lemmas, the first of which is

Lemma 6.3 For $\boldsymbol{p}(\xi, \eta, s)$ having the form (6.5.9), where $\boldsymbol{R}(s)$ is given as in (6.5.1), we have

$$\begin{aligned} \boldsymbol{r}'(s) \times (\boldsymbol{p} - \bar{\boldsymbol{p}}) \times \boldsymbol{r}'(\bar{\sigma}) &= \tilde{K}_1(s - \bar{\sigma})(\boldsymbol{p} - \bar{\boldsymbol{p}}) \\ &- \beta_1^2(s - \bar{\sigma})\boldsymbol{r}'(\bar{\sigma}) + \left(1 - \frac{\bar{\xi}}{\alpha}\right) K_2(s, \bar{\sigma})\boldsymbol{r}'(\bar{\sigma}) \\ &+ \lambda_1\beta_1\eta(1 - \cos\omega(s - \bar{\sigma}))\boldsymbol{r}'(\bar{\sigma}), \end{aligned} \tag{6.5.10}$$

where

$$\tilde{K}_1(s, \bar{\sigma}) = \alpha^2\omega^2[\cos\omega(s - \bar{\sigma})] + \beta_1^2 \tag{6.5.11}$$

while $K_2(s, \bar{\sigma})$ is given by (6.4.7).

Remarks: It is easy to check that for $\beta_1 = 0$ the expression (6.5.10) reduces to the corresponding relation (6.4.6) associated with the plane circular equilibrium states of § 6.4.

Proof: (Lemma 6.3) We begin, once again, with the identity

$$\begin{aligned} &\boldsymbol{r}'(s) \times \boldsymbol{p}(\xi, \eta, s) \times \boldsymbol{r}'(\bar{\sigma}) \\ &= \boldsymbol{p}(\xi, \eta, s)(\boldsymbol{r}'(s) \cdot \boldsymbol{r}'(\bar{\sigma})) - \boldsymbol{r}'(\bar{\sigma})(\boldsymbol{r}'(s) \cdot \boldsymbol{p}(\xi, \eta, s)) \end{aligned} \tag{6.5.12}$$

Next, we compute that

$$\boldsymbol{r}'(s) \cdot \boldsymbol{r}'(\bar{\sigma}) = (\boldsymbol{R}'(s) + \beta_1\boldsymbol{k}) \cdot (\boldsymbol{R}'(\bar{\sigma}) + \beta_1\boldsymbol{k}) \tag{6.5.13}$$

$$
\begin{aligned}
&= \boldsymbol{R}'(s) \cdot \boldsymbol{R}'(\bar{\sigma}) + \beta_1^2 \\
&= \alpha^2\omega^2[\cos\omega(s-\bar{\sigma})] + \beta_1^2 \\
&\equiv \tilde{K}_1(s;\bar{\sigma})
\end{aligned}
$$

and

$$
\begin{aligned}
\boldsymbol{r}'(s) \cdot \boldsymbol{p}(\xi,\eta,s) &= [\boldsymbol{R}'(s) + \beta_1\boldsymbol{k}] \cdot \left[\left(1-\frac{\xi}{\alpha}\right)\boldsymbol{R}(s)\right. \\
&\quad \left. +(\beta_1 s + \lambda_1\eta)\boldsymbol{k} + \lambda_2\eta\boldsymbol{R}(s) \times \boldsymbol{k}\right] \\
&= \lambda_2\eta\boldsymbol{R}'(s) \cdot (\boldsymbol{R}(s) \times \boldsymbol{k}) + \beta_1(\beta_1 s + \lambda_1\eta) \\
&= \alpha\lambda_2\eta\boldsymbol{R}'(s) \cdot [-\cos\omega s\boldsymbol{j} + \sin\omega s\boldsymbol{i}] \\
&\quad +\beta_1(\beta_1 s + \lambda_1\eta) \\
&= -\alpha^2\omega\lambda_2\eta + \beta_1(\beta_1 s + \lambda_1\eta)
\end{aligned}
$$

as $\boldsymbol{R}'(s) \cdot \boldsymbol{R}(s) = 0$. Therefore, we have

$$
\boldsymbol{r}'(s) \cdot \boldsymbol{p}(\xi,\eta,s) = \beta_1^2 s + (\lambda_1\beta_1 - \alpha^2\omega^2\lambda_2\eta) \equiv \beta_1^2 s \tag{6.5.14}
$$

because $\lambda_1\left(\dfrac{\beta_1}{\alpha^2\omega}\right) = \lambda_2$, as previously noted.

Combining (6.5.12) with (6.5.13), and (6.5.14), we find that

$$
\boldsymbol{r}'(s) \times \boldsymbol{p}(\xi,\eta,s) \times \boldsymbol{r}'(\bar{\sigma}) = \tilde{K}_1(s;\bar{\sigma})\boldsymbol{p}(\xi,\eta,s) - \beta_1^2 s\boldsymbol{r}'(\bar{\sigma}) \tag{6.5.15}
$$

In an analogous fashion,

$$
\begin{aligned}
\boldsymbol{r}'(s) \times \bar{\boldsymbol{p}}(\bar{\xi},\bar{\eta},\bar{\sigma}) \times \boldsymbol{r}'(\bar{\sigma}) &= \bar{\boldsymbol{p}}(\bar{\xi},\bar{\eta},\bar{\sigma})(\boldsymbol{r}'(s) \cdot \boldsymbol{r}'(\bar{\sigma})) \\
&\quad - \boldsymbol{r}'(\bar{\sigma}))(\boldsymbol{r}'(s) \cdot \bar{\boldsymbol{p}}(\bar{\xi},\bar{\eta},\bar{\sigma})) \\
&= \tilde{K}_1(s;\bar{\sigma})\bar{\boldsymbol{p}}(\bar{\xi},\bar{\eta},\bar{\sigma}) \\
&\quad - \boldsymbol{r}'(\bar{\sigma}))(\boldsymbol{r}'(s) \cdot \bar{\boldsymbol{p}}(\bar{\xi},\bar{\eta},\bar{\sigma}))
\end{aligned} \tag{6.5.16}
$$

However,

$$
\boldsymbol{r}'(s) \cdot \bar{\boldsymbol{p}}(\bar{\xi},\bar{\eta},\bar{\sigma}) = [\boldsymbol{R}'(s) + \beta_1\boldsymbol{k}] \cdot \left[\left(1-\frac{\bar{\xi}}{\alpha}\right)\boldsymbol{R}(\bar{\sigma})\right. \tag{6.5.17}
$$

$$
\begin{aligned}
&+(\beta_1\bar{\sigma}+\lambda_1\bar{\eta})\boldsymbol{k}+\lambda_2\bar{\eta}\boldsymbol{R}(\bar{\sigma})\times\boldsymbol{k}\Big] \\
=\;& \left(1-\frac{\xi}{\alpha}\right)\boldsymbol{R}'(s)\cdot\boldsymbol{R}(\bar{\sigma})+\lambda_2\bar{\eta}\boldsymbol{R}'(s)\cdot(\boldsymbol{R}(\bar{\sigma})\times\boldsymbol{k}) \\
&+\beta_1(\beta_1\bar{\sigma}+\lambda_1\bar{\eta})
\end{aligned}
$$

while

$$
\boldsymbol{R}'(s)\cdot\boldsymbol{R}(\bar{\sigma})=-\alpha^2\omega\sin\omega(s-\bar{\sigma}) \tag{6.5.18}
$$

and

$$
\lambda_2\bar{\eta}\boldsymbol{R}'(s)\cdot(\boldsymbol{R}(\bar{\sigma})\times\boldsymbol{k})=-\alpha^2\omega\lambda_2\bar{\eta}\cos\omega(s-\bar{\sigma}) \tag{6.5.19}
$$

Therefore, if we combine (6.5.17)-(6.5.19) we see that

$$
\begin{aligned}
\boldsymbol{r}'(s)\times\bar{\boldsymbol{p}}(\bar{\xi},\bar{\eta},\bar{\sigma}) &= \beta_1^2\bar{\sigma}-\alpha^2\omega\left(1-\frac{\bar{\xi}}{\alpha}\right)\sin\omega(s-\bar{\sigma}) \\
&\quad+\left\{\beta_1\lambda_1-\alpha^2\omega\lambda_2\cos\omega(s-\bar{\sigma})\right\}\bar{\eta} \\
&= \beta_1^2\bar{\sigma}-\alpha^2\omega\left(1-\frac{\bar{\xi}}{\alpha}\right)\sin\omega(s-\bar{\sigma}) \\
&\quad+\lambda_1\beta_1\bar{\eta}(1-\cos\omega(s-\bar{\sigma}))
\end{aligned} \tag{6.5.20}
$$

where we have, once again, used the relation between λ_1 and λ_2 as defined by (6.5.7). Inserting (6.5.20) into (6.5.16) we now have

$$
\begin{aligned}
&\boldsymbol{r}'(s)\times\bar{\boldsymbol{p}}(\bar{\xi},\bar{\eta},\bar{\sigma})\times\boldsymbol{r}'(\bar{\sigma})=\tilde{K}_1(s;\bar{\sigma})\bar{\boldsymbol{p}}(\bar{\xi},\bar{\eta},\bar{\sigma}) \\
&-\boldsymbol{r}'(\bar{\sigma})[\beta_1^2\bar{\sigma}-\alpha^2\omega\left(1-\frac{\bar{\xi}}{\alpha}\right)\sin\omega(s-\bar{\sigma}) \\
&+\lambda_1\beta_1\bar{\eta}(1-\cos\omega(s-\bar{\sigma}))]
\end{aligned} \tag{6.5.21}
$$

The required result, i.e., (6.5.10) then follows by combining (6.5.15) with (6.5.21), thus establishing Lemma 6.3. □

Our next result may be stated as follows:

Lemma 6.4 Under the same conditions as those which apply in Lemma 6.3,

$$
\begin{aligned}
\|\boldsymbol{p}-\bar{\boldsymbol{p}}\|^2 &= [\beta_1(s-\bar{\sigma})+\lambda_1(\eta-\bar{\eta})]^2+\alpha^2\lambda_2(\eta^2+\bar{\eta}^2) \\
&\quad+(\xi-\bar{\xi})^2+4(\alpha-\xi)(\alpha-\bar{\xi})\sin^2\frac{\omega}{2}(s-\bar{\sigma}) \\
&\quad-\;2\alpha\omega\lambda_2[\eta(\alpha-\bar{\xi})-\bar{\eta}(\alpha-\xi)]\sin\omega(s-\bar{\sigma})
\end{aligned} \tag{6.5.22}
$$

Proof: We begin by noting that as a consequence of the definition of $\boldsymbol{R}(s)$, i.e., (6.5.1), it follows that

$$\boldsymbol{R}(s) \times \boldsymbol{k} = -\frac{1}{\omega}\boldsymbol{R}(s) \tag{6.5.23}$$

in which case

$$\begin{cases} \boldsymbol{p}(\xi,\eta,s) = \left(1 - \dfrac{\xi}{\alpha}\right)\boldsymbol{R}(s) - \dfrac{\lambda_2\eta}{\omega}\boldsymbol{R}'(s) + (\beta_1 s + \lambda_1\eta)\boldsymbol{k} \\ \bar{\boldsymbol{p}}(\bar{\xi},\bar{\eta},\bar{\sigma}) = \left(1 - \dfrac{\bar{\xi}}{\alpha}\right)\boldsymbol{R}(\bar{\sigma}) - \dfrac{\lambda_2\bar{\eta}}{\omega}\boldsymbol{R}'(\bar{\sigma}) + (\beta_1\bar{\sigma} + \lambda_1\bar{\eta})\boldsymbol{k} \end{cases} \tag{6.5.24}$$

We also note the obvious relations

$$\boldsymbol{R}(s)\cdot\boldsymbol{R}'(s) = \boldsymbol{R}(s)\cdot\boldsymbol{k} = \boldsymbol{R}'(s)\cdot\boldsymbol{k} = 0 \tag{6.5.25}$$

Using the orthogonality of $\boldsymbol{R}(s)$, $\boldsymbol{R}'(s)$, and $\boldsymbol{k}$ we easily compute, directly from (6.5.24), and the definition of $\boldsymbol{R}(s)$, that

$$\|\boldsymbol{p}(\xi,\eta,s)\|^2 = (\alpha-\xi)^2 + \alpha^2\lambda_2^2\eta^2 + (\beta_1 s + \lambda_1\eta)^2 \tag{6.5.26a}$$

and

$$\|\bar{\boldsymbol{p}}(\bar{\xi},\bar{\eta},\bar{\sigma})\|^2 = (\alpha-\bar{\xi})^2 + \alpha^2\lambda_2^2\bar{\eta}^2 + (\beta_1\bar{\sigma} + \lambda_1\bar{\eta})^2 \tag{6.5.26b}$$

while

$$\begin{aligned} \boldsymbol{p}\cdot\bar{\boldsymbol{p}} \;=\;& \left(1-\frac{\xi}{\alpha}\right)\left(1-\frac{\bar{\xi}}{\alpha}\right)\boldsymbol{R}(s)\cdot\boldsymbol{R}(\bar{\sigma}) \\ & -\frac{\lambda_2\bar{\eta}}{\omega}\left(1-\frac{\xi}{\alpha}\right)\boldsymbol{R}(s)\cdot\boldsymbol{R}'(\bar{\sigma}) - \frac{\lambda_2\eta}{\omega}\left(1-\frac{\bar{\xi}}{\alpha}\right)\boldsymbol{R}(\bar{\sigma})\cdot\boldsymbol{R}'(s) \\ & +\frac{\lambda_2^2}{\omega^2}\eta\bar{\eta}\boldsymbol{R}'(s)\cdot\boldsymbol{R}'(\bar{\sigma}) + (\beta_1 s + \lambda_1\eta)(\beta_1\bar{\sigma} + \lambda_1\bar{\eta}) \end{aligned} \tag{6.5.27}$$

Employing the identities:

$$\begin{cases} \boldsymbol{R}(s)\cdot\boldsymbol{R}(\bar{\sigma}) = \alpha^2\cos\omega(s-\bar{\sigma}) \equiv K_1(s;\bar{\sigma})/\omega^2 \\ \boldsymbol{R}'(s)\cdot\boldsymbol{R}'(\bar{\sigma}) = \alpha^2\omega^2\cos\omega(s-\bar{\sigma}) \equiv K_1(s;\bar{\sigma}) \\ \boldsymbol{R}(s)\cdot\boldsymbol{R}'(\bar{\sigma}) = \alpha^2\omega^2\sin\omega(s-\bar{\sigma}) \equiv -K_2(s;\bar{\sigma}) \\ \boldsymbol{R}(\bar{\sigma})\cdot\boldsymbol{R}'(s) = \alpha^2\omega^2\sin\omega(\bar{\sigma}-s) = K_2(s;\bar{\sigma}) \end{cases} \tag{6.5.28}$$

we easily compute that

$$\begin{aligned} \boldsymbol{p}\cdot\bar{\boldsymbol{p}} &= (\alpha-\xi)(\alpha-\bar{\xi})\cos\omega(s-\bar{\sigma}) \\ &\quad -\lambda_2\alpha\omega\bar{\eta}(\alpha-\xi)\sin\omega(s-\bar{\sigma}) \\ &\quad +\lambda_2\alpha\omega\eta(\alpha-\bar{\xi})\sin\omega(s-\bar{\sigma}) \\ &\quad +(\beta_1 s+\lambda_1\eta)(\beta_1\bar{\sigma}+\lambda_1\bar{\eta}) \end{aligned}$$

or

$$\begin{aligned} \boldsymbol{p}\cdot\bar{\boldsymbol{p}} &= (\alpha-\xi)(\alpha-\bar{\xi})\cos\omega(s-\bar{\sigma}) \\ &\quad +\lambda_2\alpha\omega[\eta(\alpha-\bar{\xi})-\bar{\eta}(\alpha-\xi)]\sin\omega(s-\bar{\sigma}) \\ &\quad +(\beta_1 s+\lambda_1\eta)(\beta_1\bar{\sigma}+\lambda_1\bar{\eta}) \end{aligned} \tag{6.5.29}$$

Combining (6.5.26a), (6.5.26b) with (6.5.29) we are led to the relations

$$\begin{aligned} \|\boldsymbol{p}-\bar{\boldsymbol{p}}\|^2 &= [\beta_1(s-\bar{\sigma})+\lambda_1(\eta-\bar{\eta})]^2+\alpha^2\lambda_2^2(\eta^2+\bar{\eta}^2) \\ &\quad +(\alpha-\xi)^2+(\alpha-\bar{\xi})^2-2(\alpha-\xi)(\alpha-\bar{\xi})\cos\omega(s-\bar{\sigma}) \\ &\quad -2\alpha\omega\lambda_2[\eta(\alpha-\bar{\xi})-\bar{\eta}(\alpha-\xi)]\sin\omega(s-\bar{\sigma}) \\ &= [\beta_1(s-\bar{\sigma})+\lambda_1(\eta-\bar{\eta})]^2+\alpha^2\lambda_2^2(\eta^2+\bar{\eta}^2) \\ &\quad +(\xi-\bar{\xi})^2+4(\alpha-\xi)(\alpha-\bar{\xi})\sin^2\frac{\omega}{2}(s-\bar{\sigma}) \\ &\quad -2\alpha\omega\lambda_2[\eta(\alpha-\bar{\xi})-\bar{\eta}(\alpha-\xi)]\sin\omega(s-\bar{\sigma}) \end{aligned} \tag{6.5.30}$$

the last of which is, of course, (6.5.22). □

Remarks: As a check on our calculations in Lemma 6.4 we note that if $\beta_1 = 0$ then, by virtue of (6.5.7), $\lambda_2 = 0$, while $\lambda_1 = 1$; in this case $\boldsymbol{p}$, as given by (6.5.9), reduces to (6.4.4), i.e., we are back in the situation of looking for plane, circular equilibrium states and (6.5.22) then reduces to (6.4.17).

As in all the earlier cases studied, we once more have $\nu(s) = \text{const.}$ and, therefore,

$$\boldsymbol{n}'(s) = \frac{\mu(\nu)}{\nu}\boldsymbol{r}''(s) = \frac{\mu(\nu)}{\nu}\boldsymbol{R}''(s)$$

because of (6.5.1) and the fact that $\beta''(s) = 0$. Using (6.5.1) again we easily see that

$$\boldsymbol{n}'(s) = -\frac{\omega^2 \mu(\nu)}{\nu} \boldsymbol{R}(s) \tag{6.5.31}$$

Also,

$$\begin{aligned} IB\boldsymbol{r}'(s) \times \boldsymbol{k} &= IB(\boldsymbol{R}'(s) + \beta_1 \boldsymbol{k}) \times \boldsymbol{k} \\ &= IB\boldsymbol{R}'(s) \times \boldsymbol{k} \\ &= -\omega IB(\boldsymbol{R}(s) \times \boldsymbol{k}) \times \boldsymbol{k} \\ &= \omega IB(\boldsymbol{R}(s) - (\boldsymbol{k} \cdot \boldsymbol{R}(s))\boldsymbol{k}) \\ &= \omega IB\boldsymbol{R}(s) \end{aligned} \tag{6.5.32}$$

where we have used (6.5.23) and (6.5.25). Thus, if helical states of the form (6.5.9), with $\boldsymbol{R}(s)$ given by (6.5.1), exist then combining the equilibrium equations (6.2.3) with Lemmas 6.3 and 6.4, and the relations (6.5.31), (6.5.32), we are led to the vector equation

$$\begin{aligned} &\left(\omega IB - \frac{\omega^2 \mu(\nu)}{\nu}\right) \boldsymbol{R}(s) \\ &+ \left(\frac{I}{A}\right)^2 \iint\limits_A \left(\iiint\limits_V \frac{\boldsymbol{r}'(s) \times (\boldsymbol{p}(\xi,\eta,s) - \bar{\boldsymbol{p}}(\bar{\xi},\bar{\eta},\bar{\sigma})) \times \boldsymbol{r}'(\bar{\sigma})}{\|\boldsymbol{p}(\xi,\eta,s) - \bar{\boldsymbol{p}}(\bar{\xi},\bar{\eta},\bar{\sigma})\|^3}\, d\bar{\xi}\, d\bar{\eta}\, d\bar{\sigma} \right) d\xi\, d\eta \\ &= 0 \end{aligned} \tag{6.5.33}$$

for $\boldsymbol{p}(\xi, \eta, s)$ of the form (6.5.9) with $\boldsymbol{R}(s)$ as given by (6.5.1). In (6.5.33) the numerator of the integrand is given by (6.5.10) while the denominator is just the right-hand side of (6.5.22) raised to the $\frac{3}{2}$ power. If we write $\boldsymbol{p}(\xi, \eta, s)$ in the form given by (6.5.24), i.e.,

$$\boldsymbol{p}(\xi,\eta,s) = (1 - \xi/\alpha)\, \boldsymbol{R}(s) - \frac{\lambda_2 \eta}{\omega} \boldsymbol{R}'(s) + (\beta_1 s + \lambda_1 \eta)\boldsymbol{k}$$

then it is clear that it is sufficient to show that the parameters α, ω, and β_1 can be chosen so that the projections of (6.5.33) onto the three mutually perpendicular directions $\boldsymbol{R}(s)$, $\boldsymbol{R}'(s)$, and $\boldsymbol{k}$, at s, are satisfied for some choice of the parameter $\lambda = IB$; as of the time of the completion of this volume, we have been unsuccessful in our efforts to accomplish this task and the available evidence would seem to point to the conjecture that pure helical equilibrium states do not exist for the nonlinearly elastic, self-interacting, current bearing wire. Suppose, for example, that we project (6.5.33) onto the direction of the unit vector $\boldsymbol{k}$. As $\boldsymbol{R}(s)\cdot\boldsymbol{k}=0$, our task is to show that for all values of $\lambda = IB$ we have, for appropriately chosen α, ω, and β_1,

$$\boldsymbol{k}\cdot\iint\limits_A\left(\iiint\limits_V \frac{\boldsymbol{r}'(s)\times(\boldsymbol{p}(\xi,\eta,s)-\bar{\boldsymbol{p}}(\bar{\xi},\bar{\eta},\bar{\sigma}))\times\boldsymbol{r}'(\bar{\sigma})}{\|\boldsymbol{p}(\xi,\eta,s)-\bar{\boldsymbol{p}}(\bar{\xi},\bar{\eta},\bar{\sigma})\|^3}\,d\bar{\xi}\,d\bar{\eta}\,d\bar{\sigma}\right)d\xi\,d\eta=0 \qquad (6.5.34)$$

However,

$$\begin{aligned}&\boldsymbol{r}'(s)\times(\boldsymbol{p}(\xi,\eta,s)-\bar{\boldsymbol{p}}(\bar{\xi},\bar{\eta},\bar{\sigma}))\times\boldsymbol{r}'(\bar{\sigma})\\&\quad=(\boldsymbol{p}(\xi,\eta,s)-\bar{\boldsymbol{p}}(\bar{\xi},\bar{\eta},\bar{\sigma}))(\boldsymbol{r}'(s)\cdot\boldsymbol{r}'(\bar{\sigma}))\\&\qquad-\boldsymbol{r}'(\bar{\sigma})(\boldsymbol{r}'(s)\cdot[\boldsymbol{p}(\xi,\eta,s)-\bar{\boldsymbol{p}}(\bar{\xi},\bar{\eta},\bar{\sigma})])\end{aligned}$$

so

$$\begin{aligned}&\boldsymbol{k}\cdot[\boldsymbol{r}'(s)\times(\boldsymbol{p}(\xi,\eta,s)-\bar{\boldsymbol{p}}(\bar{\xi},\bar{\eta},\bar{\sigma}))\times\boldsymbol{r}'(\bar{\sigma})]\\&\quad=\boldsymbol{k}\cdot(\boldsymbol{p}(\xi,\eta,s)-\bar{\boldsymbol{p}}(\bar{\xi},\bar{\eta},\bar{\sigma}))(\boldsymbol{r}'(s)\cdot\boldsymbol{r}'(\bar{\sigma}))\\&\qquad-\boldsymbol{k}\cdot\boldsymbol{r}'(\bar{\sigma})(\boldsymbol{r}'(s)\cdot[\boldsymbol{p}(\xi,\eta,s)-\bar{\boldsymbol{p}}(\bar{\xi},\bar{\eta},\bar{\sigma})])\end{aligned}$$

or, as

$$\boldsymbol{r}'(\bar{\sigma})=\boldsymbol{R}'(\bar{\sigma})+\beta_1\boldsymbol{k},$$

and

$$\begin{aligned}\boldsymbol{p}(\xi,\eta,s)-\bar{\boldsymbol{p}}(\bar{\xi},\bar{\eta},\bar{\sigma})&=(1-\xi/\alpha)\,\boldsymbol{R}(s)-\left(1-\bar{\xi}/\alpha\right)\boldsymbol{R}(\bar{\sigma})\\&\quad-\frac{\lambda_2}{\omega}(\eta\boldsymbol{R}'(s)-\bar{\eta}\boldsymbol{R}'(\bar{\sigma}))\\&\quad+[\beta_1(s-\bar{\sigma})+\lambda_1(\eta-\bar{\eta})]\boldsymbol{k},\end{aligned}$$

we can compute that

$$\begin{aligned}&\boldsymbol{k}\cdot[\boldsymbol{r}'(s)\times[\boldsymbol{p}(\xi,\eta,s)-\bar{\boldsymbol{p}}(\bar{\xi},\bar{\eta},\bar{\sigma})]\times\boldsymbol{r}'(\bar{\sigma})]\\&\quad=(\beta_1(s-\bar{\sigma})+\lambda_1(\eta-\bar{\eta}))(\boldsymbol{R}'(s)\cdot\boldsymbol{R}'(\bar{\sigma})+\beta_1^2)\\&\qquad-\beta_1(\boldsymbol{R}'(s)+\beta_1\boldsymbol{k})\cdot(\boldsymbol{p}(\xi,\eta,s)-\bar{\boldsymbol{p}}(\bar{\xi},\bar{\eta},\bar{\sigma}))\end{aligned}$$

$$
\begin{aligned}
&= (\beta_1(s-\bar{\sigma}) + \lambda_1(\eta-\bar{\eta}))(\boldsymbol{R}'(s)\cdot\boldsymbol{R}'(\bar{\sigma}) + \beta_1^2) \\
&\quad - \beta_1^2[\beta_1(s-\bar{\sigma}) + \lambda_1(\eta-\bar{\eta})] \\
&\quad + \beta_1\left(1-\bar{\xi}/\alpha\right)(\boldsymbol{R}'(s)\cdot\boldsymbol{R}(\bar{\sigma})) + \frac{\beta_1\lambda_2}{\omega}\eta\|\boldsymbol{R}'(s)\|^2 \\
&\quad - \frac{\beta_1\lambda_2}{\omega}\bar{\eta}(\boldsymbol{R}'(s)\cdot\boldsymbol{R}(\bar{\sigma})) \\
&= (\beta_1(s-\bar{\sigma}) + \lambda_1(\eta-\bar{\eta}))K_1(s,\bar{\sigma}) + \alpha^2\omega\beta_2\lambda_2\eta \\
&\quad + \beta_1\left(1-\bar{\xi}/\alpha\right)K_2(s,\bar{\sigma}) - \frac{\beta_1\lambda_2}{\omega}\bar{\eta}K_1(s,\bar{\sigma})
\end{aligned}
$$

with $K_1(s,\bar{\sigma})$ and $K_2(s,\bar{\sigma})$ as given by (6.4.7). Finally, using the relation

$$
\lambda_1\left(\frac{\beta_1}{\alpha^2\omega}\right) = \lambda_2
$$

we may put this last result into the form

$$
\begin{aligned}
&\boldsymbol{k}\cdot[\boldsymbol{r}'(s)\times[\boldsymbol{p}(\xi,\eta,s) - \bar{\boldsymbol{p}}(\bar{\xi},\bar{\eta},\bar{\sigma})]\times\boldsymbol{r}'(\bar{\sigma})] \qquad (6.5.35) \\
&= \left\{\beta_1(s-\bar{\sigma}) + \lambda_1(\eta-\bar{\eta}) - \frac{\beta_1\lambda_2}{\omega}\bar{\eta}\right\}K_1(s,\bar{\sigma}) \\
&\quad + \beta_1\left(1-\frac{\bar{\xi}}{\alpha}\right)K_2(s,\bar{\sigma}) + \lambda_1\beta_1^2
\end{aligned}
$$

The presence of $\beta_1 \neq 0$ in (6.5.35) would appear to foreclose on any possibility of using symmetry arguments to show that (6.5.34) is satisfied but such a statement has no greater status than that of a conjecture at the present time.

6.6 Perturbations of Equilibrium States of Self-Interacting Current Bearing Wires

In § 6.4 we demonstrated the existence of circular equilibrium states, i.e., solutions of (6.2.3) of the form (6.4.4), (6.4.2) with α, ω, and I subject to the constraint delineated in the algebraic condition (6.4.43). In order to show that there are other equilibrium solutions which branch off from (6.4.4) at critical values of $\lambda = IB$, it would be necessary to examine the linear boundary value problem obtained by linearizing the vector equation (6.2.3) about the solution (6.4.4); indeed, such a linearization has been attempted [5], but the results are very much inconclusive due to the serious difficulties associated with dealing effectively with the integral representing the self-interaction term.

One set of possible equilibrium solutions for (6.2.3), which represents a perturbation of the plane circular solutions of § 6.4, and whose existence has been conjectured in [21], consists of planar equilibrium states which *wiggle* about the circular states described by (6.4.4), (6.4.2). To be somewhat more precise, these *wiggly* circular states are of the form (see Figure 6.6 below)

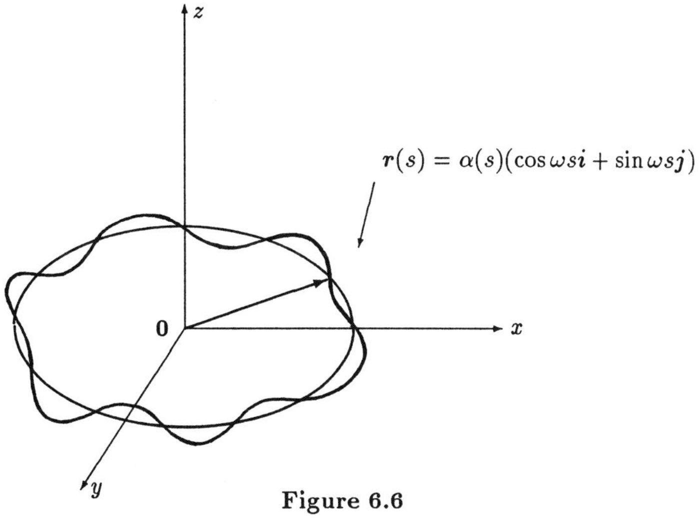

Figure 6.6

$$p(\xi, \eta, s) = r(s) + \xi d_1(s) + \eta d_2(s)$$

with

$$r(s) = \alpha(s)(\cos \omega s i + \sin \omega s j) \tag{6.6.1}$$

As we are going to deal again with deformations of the wire corresponding to constant stretch, we compute that

$$\|r'(s)\| \equiv \nu(s) = [\alpha'^2(s) + \alpha^2(s)\omega^2]^{1/2} = \text{ const.} \tag{6.6.2}$$

so that $\alpha(\cdot)$ is to satisfy the nonlinear ODE

$$\alpha'^2(s) + \omega^2 \alpha^2(s) = \nu^2 \tag{6.6.3}$$

a solution of which, for given $\nu = $ const., and any ω, is

$$\alpha(s) = \frac{\nu}{\omega} \sin \omega s \tag{6.6.4}$$

Therefore,

$$\boldsymbol{r}(s) = \frac{\nu}{\omega} \sin \omega s (\cos \omega s \boldsymbol{i} + \sin \omega s \boldsymbol{j}) \tag{6.6.5}$$

We easily compute that

$$\begin{cases} \boldsymbol{r}'(s) = \nu \cos 2\omega s \boldsymbol{i} + \nu \sin 2\omega s \boldsymbol{j} \\ \boldsymbol{r}''(s) = -2\nu\omega \sin 2\omega s \boldsymbol{i} + 2\nu\omega \cos 2\omega s \boldsymbol{j} \end{cases} \tag{6.6.6}$$

so that $\|\boldsymbol{r}''(s)\| = 2\nu\omega$. We then have

$$\boldsymbol{d}_1(s) = \frac{\boldsymbol{r}''(s)}{\|\boldsymbol{r}''(s)\|} = -\sin 2\omega s \boldsymbol{i} + \cos 2\omega s \boldsymbol{j}$$

or

$$\boldsymbol{d}_1(s) = -\frac{\boldsymbol{r}'(s)}{\nu} \times \boldsymbol{k} \tag{6.6.7}$$

while

$$\boldsymbol{d}_2(s) = \frac{\boldsymbol{r}'(s)}{\nu} \times \boldsymbol{d}_1 = \boldsymbol{k} \tag{6.6.8}$$

We make note of the fact that

$$\begin{cases} \boldsymbol{r}'(s) \cdot \boldsymbol{d}_1(s) = \boldsymbol{r}'(s) \cdot \boldsymbol{d}_2(s) = 0 \\ \boldsymbol{r}(s) \cdot \boldsymbol{d}_2(s) = \boldsymbol{d}_1(s) \cdot \boldsymbol{d}_2(s) = 0 \end{cases} \tag{6.6.9}$$

and

$$\boldsymbol{r}(s) \cdot \boldsymbol{r}'(s) = \frac{1}{2}\frac{d}{ds}\alpha^2(s) \equiv \frac{\nu^2}{2\omega} \sin 2\omega s \tag{6.6.10}$$

Using (6.6.7) and (6.6.8), we now determine that

$$\boldsymbol{p}(\xi, \eta, s) = \boldsymbol{r}(s) - \left(\frac{\xi}{\nu}\right) \boldsymbol{r}'(s) \times \boldsymbol{k} + \eta \boldsymbol{k}, \tag{6.6.11}$$

with an analogous expression for $\bar{\boldsymbol{p}}(\bar{\xi}, \bar{\eta}, \bar{\sigma})$, where $\boldsymbol{r}(s)$ is given by (6.6.5). A laborious but straightforward computation, similar to those effected in § 6.3–§ 6.5, yields

$$\boldsymbol{r}'(s) \times [(\boldsymbol{p} - \bar{\boldsymbol{p}}) \times \boldsymbol{r}'(\bar{\sigma})] \tag{6.6.12}$$

$$= \nu^2 \cos 2\omega(s - \bar{\sigma})(\boldsymbol{p} - \bar{\boldsymbol{p}}) + (\bar{\xi}\nu - \frac{\nu^2}{2\omega}) \sin 2\omega(s - \bar{\sigma}) \boldsymbol{r}'(\bar{\sigma})$$

where

$$\boldsymbol{p}-\bar{\boldsymbol{p}}=[\boldsymbol{r}(s)-\boldsymbol{r}(\bar{\sigma})]-\left[\left(\frac{\xi}{\nu}\right)\boldsymbol{r}'(s)-\left(\frac{\bar{\xi}}{\nu}\right)\boldsymbol{r}'(\bar{\sigma})\right]\times\boldsymbol{k} \tag{6.6.13}$$

In arriving at (6.6.12), we have used the following, easily verified, facts:

$$\begin{cases} \boldsymbol{r}'(s)\cdot\boldsymbol{r}'(\bar{\sigma})=\nu^2\cos 2\omega(s-\bar{\sigma})\equiv K_4(s,\bar{\sigma}) \\ \boldsymbol{r}'(s)\cdot\boldsymbol{p}(\xi,\eta,s)=\dfrac{\nu^2}{2\omega}\sin 2\omega s\equiv\boldsymbol{r}(s)\cdot\boldsymbol{r}'(s) \\ \boldsymbol{r}'(s)\cdot\bar{\boldsymbol{p}}(\bar{\xi},\bar{\eta},\bar{\sigma})=\dfrac{\nu^2}{\omega}\sin\omega\bar{\sigma}\cos\omega(\bar{\sigma}-2s)+\bar{\xi}\nu\sin 2\omega(s-\bar{\sigma}) \end{cases} \tag{6.6.14}$$

In an analogous fashion we may show, after a lengthy computation, that

$$\begin{cases} \|\boldsymbol{p}\|^2=\left(\dfrac{\nu^2}{\omega^2}-2\xi\dfrac{\nu}{\omega}\right)\sin^2\omega s+\xi^2+\eta^2 \\ \|\bar{\boldsymbol{p}}\|^2=\left(\dfrac{\nu^2}{\omega^2}-2\bar{\xi}\dfrac{\nu}{\omega}\right)\sin^2\omega\bar{\sigma}+\bar{\xi}^2+\bar{\eta}^2 \end{cases} \tag{6.6.15}$$

while

$$\begin{aligned} \boldsymbol{p}\cdot\bar{\boldsymbol{p}}&=\alpha(s)\alpha(\bar{\sigma})\cos\omega(s-\bar{\sigma})+\eta\bar{\eta} \\ &\quad+\xi\bar{\xi}\cos 2\omega(s-\bar{\sigma})+\bar{\xi}K_5(s,\bar{\sigma})+\xi K_5(\bar{\sigma},s) \end{aligned} \tag{6.6.16}$$

where $\alpha(s)$ is given by (6.6.4) and

$$K_5(s,\bar{\sigma})=\alpha(s)\sin\omega(s-2\bar{\sigma}) \tag{6.6.17}$$

Combining (6.6.15) with (6.6.16) we find, after some algebra, and with the help of standard trigonometric identities, that

$$\begin{aligned} \|\boldsymbol{p}-\bar{\boldsymbol{p}}\|^2&=(\xi-\bar{\xi})^2+(\eta-\bar{\eta})^2 \\ &\quad+\left\{\frac{\nu^2}{\omega^2}(\sin^2\omega s+\sin^2\omega\bar{\sigma})-2\alpha(s)\alpha(\bar{\sigma})\cos\omega(s-\bar{\sigma})\right\} \\ &\quad+\left\{4\xi\bar{\xi}\sin^2\frac{\omega}{2}(s-\bar{\sigma})-\frac{2\nu}{\omega}(\xi\sin^2\omega s+\bar{\xi}\sin^2\omega\bar{\sigma})\right. \\ &\qquad\left.-2(\bar{\xi}K_5(s,\bar{\sigma})+\xi K_5(\bar{\sigma},s))\right\} \end{aligned} \tag{6.6.18}$$

with $K_5(s,\bar{\sigma})$ as in (6.6.17). Employing (6.6.12) and (6.6.18), it is a simple matter to see that, for the conjectured *wiggly* perturbations of plane circular states under consideration here, the integral in the equilibrium equation (6.2.3) assumes the form

$$\iint_A \left(\iiint_V \frac{\boldsymbol{r}'(s) \times [(\boldsymbol{p}-\bar{\boldsymbol{p}}) \times \boldsymbol{r}'(\bar{\sigma})]}{\|\boldsymbol{p}-\bar{\boldsymbol{p}}\|^3} \, d\bar{\xi}\, d\bar{\eta}\, d\bar{\sigma} \right) d\xi\, d\eta \tag{6.6.19}$$

$$= \iint_A \left(\iiint_V \frac{\nu^2 \cos 2\omega(s-\bar{\sigma})}{\|\boldsymbol{p}-\bar{\boldsymbol{p}}\|^3} \cdot \Big\{ (\boldsymbol{r}(s) - \boldsymbol{r}(\bar{\sigma})) \right.$$

$$\left. + (\eta - \bar{\eta})\boldsymbol{k} - \left[\left(\frac{\xi}{\nu}\right) \boldsymbol{r}'(s) - \left(\frac{\bar{\xi}}{\nu}\right) \boldsymbol{r}'(\bar{\sigma}) \right] \times \boldsymbol{k} \Big\} \, d\bar{\xi}\, d\bar{\eta}\, d\bar{\sigma} \right) d\xi\, d\eta$$

$$+ \iint_A \left(\iiint_V \frac{\left(\bar{\xi}\nu - \frac{\nu^2}{2\omega}\right)}{\|\boldsymbol{p}-\bar{\boldsymbol{p}}\|^3} \sin 2\omega(s-\bar{\sigma}) \boldsymbol{r}'(\bar{\sigma}) \, d\bar{\xi}\, d\bar{\eta}\, d\bar{\sigma} \right) d\xi\, d\eta$$

where $\|\boldsymbol{p}-\bar{\boldsymbol{p}}\|^3$ is given by the right-hand side of (6.6.18) raised to the power 3/2 and $\boldsymbol{r}$, $\boldsymbol{r}'$ are determined as in (6.6.5), (6.6.6). However, as the right-hand side of (6.6.18) is invariant under the transformation $\eta \to \bar{\eta}$ we may again exploit symmetry, this time to show that

$$\iint_A \left(\iiint_V \frac{\nu^2 \cos 2\omega(s-\bar{\sigma})}{\|\boldsymbol{p}-\bar{\boldsymbol{p}}\|^3} (\eta - \bar{\eta}) \boldsymbol{k} \, d\bar{\xi}\, d\bar{\eta}\, d\bar{\sigma} \right) d\xi\, d\eta = \boldsymbol{0} \tag{6.6.20}$$

Also, as is easily verified,

$$\boldsymbol{n}'(s) + IB\boldsymbol{r}'(s) \times \boldsymbol{k} = \left(IB - \frac{2\omega\mu(\nu)}{\nu} \right) \boldsymbol{r}'(s) \times \boldsymbol{k} \tag{6.6.21}$$

Combining (6.6.19)–(6.6.21) we obtain the equilibrium equations (6.2.3) in the form

$$\left\{ \left(IB - \frac{2\omega\mu(\nu)}{\nu} \right) \right. \tag{6.6.22}$$

$$\left. - \left(\frac{I}{A}\right)^2 \iint_A \left(\iiint_V \frac{\xi\nu \cos 2\omega(s-\bar{\sigma})}{\|\boldsymbol{p}-\bar{\boldsymbol{p}}\|^3} \, d\bar{\xi}\, d\bar{\eta}\, d\bar{\sigma} \right) d\xi\, d\eta \right\} \boldsymbol{r}'(s) \times \boldsymbol{k}$$

$$= \left(\frac{I}{A}\right)^2 \iint_A \left(\iiint_V \left(\frac{\nu^2}{2\omega} - \bar{\xi}\nu \right) \frac{\sin 2\omega(s-\bar{\sigma}) \boldsymbol{r}'(\bar{\sigma})}{\|\boldsymbol{p}-\bar{\boldsymbol{p}}\|^3} \, d\bar{\xi}\, d\bar{\eta}\, d\bar{\sigma} \right) d\xi\, d\eta$$

$$- \left(\frac{I}{A}\right)^2 \iint_A \left(\iiint_V \frac{\nu^2 \cos 2\omega(s-\bar{\sigma})}{\|\boldsymbol{p}-\bar{\boldsymbol{p}}\|^3} (\boldsymbol{r}(s) - \boldsymbol{r}(\bar{\sigma})) \, d\bar{\xi}\, d\bar{\eta}\, d\bar{\sigma} \right) d\xi\, d\eta$$

$$- \left(\frac{I}{A}\right)^2 \iint_A \left(\iiint_V \frac{(\bar{\xi}\nu \cos 2\omega(s-\bar{\sigma}))}{\|\boldsymbol{p}-\bar{\boldsymbol{p}}\|^3} (\boldsymbol{r}'(\bar{\sigma}) \times \boldsymbol{k}) \, d\bar{\xi}\, d\bar{\eta}\, d\bar{\sigma} \right) d\xi\, d\eta$$

We want to compute the projections of the vector equation (6.6.22) onto the three linearly independent directions defined by $\boldsymbol{r}'(s)$, $\boldsymbol{r}'(s) \times \boldsymbol{k}$, and $\boldsymbol{k}$. As $\boldsymbol{r}(s) \cdot \boldsymbol{k} =$

$\boldsymbol{r}(\bar{\sigma}) \cdot \boldsymbol{k} = \boldsymbol{r}'(s) \cdot \boldsymbol{k} = \boldsymbol{r}'(\bar{\sigma}) \cdot \boldsymbol{k} = 0$, and $(\boldsymbol{r}'(s) \times \boldsymbol{k}) \cdot \boldsymbol{k} = (\boldsymbol{r}'(\bar{\sigma}) \times \boldsymbol{k}) \cdot \boldsymbol{k} = 0$, it is trivial that the relation obtained by taking the dot product of both sides of (6.6.22) with $\boldsymbol{k}$ is satisfied identically for all values of $\lambda = IB$. If we take the dot product of (6.6.22) with $\boldsymbol{r}'(s)$ then, as $\boldsymbol{r}'(s) \cdot (\boldsymbol{r}'(s) \times \boldsymbol{k}) = 0$, we must have

$$\begin{aligned} &\iint_A \left(\iiint_V \frac{(\nu^2 - \bar{\xi}\nu) \sin 2\omega(s-\bar{\sigma})}{\|\boldsymbol{p} - \bar{\boldsymbol{p}}\|^3} \boldsymbol{r}'(\bar{\sigma}) \cdot \boldsymbol{r}'(s)\, d\bar{\xi}\, d\bar{\eta}\, d\bar{\sigma} \right) d\xi\, d\eta \qquad (6.6.23) \\ &- \iint_A \left(\iiint_V \frac{\nu^2 \cos 2\omega(s-\bar{\sigma})}{\|\boldsymbol{p} - \bar{\boldsymbol{p}}\|^3} (\boldsymbol{r}(s) - \boldsymbol{r}(\bar{\sigma})) \cdot \boldsymbol{r}'(s)\, d\bar{\xi}\, d\bar{\eta}\, d\bar{\sigma} \right) d\xi\, d\eta \\ &- \iint_A \left(\iiint_V \frac{(\bar{\xi}\nu \cos 2\omega(s-\bar{\sigma}))}{\|\boldsymbol{p} - \bar{\boldsymbol{p}}\|^3} (\boldsymbol{r}'(\bar{\sigma}) \times \boldsymbol{k}) \cdot \boldsymbol{r}'(s)\, d\bar{\xi}\, d\bar{\eta}\, d\bar{\sigma} \right) d\xi\, d\eta = 0 \end{aligned}$$

and computation of the indicated inner products in (6.6.23) reduces this equation to

$$\begin{aligned} &\iint_A \left(\iiint_V \frac{\nu^2 \sin 2\omega(s-\bar{\sigma}) \cos 2\omega(s-\bar{\sigma})}{\omega \|\boldsymbol{p} - \bar{\boldsymbol{p}}\|^3} d\bar{\xi}\, d\bar{\eta}\, d\bar{\sigma} \right) d\xi\, d\eta \qquad (6.6.24) \\ &- \iint_A \left(\iiint_V \frac{\nu^4 \cos 2\omega(s-\bar{\sigma}) \sin 2\omega s}{2\omega \|\boldsymbol{p} - \bar{\boldsymbol{p}}\|^3} d\bar{\xi}\, d\bar{\eta}\, d\bar{\sigma} \right) d\xi\, d\eta \\ &+ \iint_A \left(\iiint_V \frac{\nu^4 \cos 2\omega(s-\bar{\sigma}) \sin \omega\bar{\sigma} \cos \omega(\bar{\sigma} - 2s)}{\omega \|\boldsymbol{p} - \bar{\boldsymbol{p}}\|^3} d\bar{\xi}\, d\bar{\eta}\, d\bar{\sigma} \right) d\xi\, d\eta \\ &= 0. \end{aligned}$$

However, it is easily verified that

$$2 \sin \omega\bar{\sigma} \cos \omega(\bar{\sigma} - 2s) - \sin 2\omega s = -\sin 2\omega(s - \bar{\sigma}) \qquad (6.6.25)$$

so that (6.6.24) is satisfied for all values of $\lambda = IB$. What remains, therefore, is for us to extract the dot product of (6.6.22) with $\boldsymbol{r}'(s) \times \boldsymbol{k}$; the result of this operation is the following scalar equation

$$\begin{aligned} &\left(\frac{I}{A}\right)^{-2} \left\{ \left(IB - \frac{2\omega\mu(\nu)}{\nu} \right) - \left(\frac{I}{A}\right)^2 \iint_A \left(\iiint_V \frac{\xi\nu \cos 2\omega(s-\bar{\sigma})}{\|\boldsymbol{p} - \bar{\boldsymbol{p}}\|^3} d\bar{\xi}\, d\bar{\eta}\, d\bar{\sigma} \right) d\xi\, d\eta \right\} \qquad (6.6.26) \\ &= \iint_A \left(\iiint_V \frac{\left(\frac{\nu^2}{2\omega} - \bar{\xi}\nu\right)}{\|\boldsymbol{p} - \bar{\boldsymbol{p}}\|^3} \cdot \sin^2 2\omega(s-\bar{\sigma})\, d\bar{\xi}\, d\bar{\eta}\, d\bar{\sigma} \right) d\xi\, d\eta \\ &+ \frac{\nu^2}{2\omega} \iint_A \left(\iiint_V \frac{[\cos^2 2\omega(s-\bar{\sigma}) - \cos 2\omega(s-\bar{\sigma})]}{\|\boldsymbol{p} - \bar{\boldsymbol{p}}\|^3} d\bar{\xi}\, d\bar{\eta}\, d\bar{\sigma} \right) d\xi\, d\eta \end{aligned}$$

$$- \nu \iint_A \left(\iiint_V \frac{\bar{\xi} \cos^2 2\omega(s - \bar{\sigma})}{\|\boldsymbol{p} - \bar{\boldsymbol{p}}\|^3} \, d\bar{\xi} \, d\bar{\eta} \, d\bar{\sigma} \right) \, d\xi \, d\eta$$

Using standard trigonometric identities to combine the iterated integrals on the right-hand side of (6.6.26), and noting that we also have

$$\iint_A \left(\iiint_V \frac{(\bar{\xi} - \xi) \sin^2 2\omega(s - \bar{\sigma})}{\|\boldsymbol{p} - \bar{\boldsymbol{p}}\|^3} \, d\bar{\xi} \, d\bar{\eta} \, d\bar{\sigma} \right) \, d\xi \, d\eta = 0, \tag{6.6.27}$$

by virtue of the skew-symmetry of the numerator in the integrand with respect to the transformation $\xi \to \bar{\xi}$, we obtain, as the condition to be satisfied,

$$\left(\frac{I}{A}\right)^{-2} \left\{ \left(IB - \frac{2\omega\mu(\nu)}{\nu} \right) - \left(\frac{I}{A}\right)^2 \iint_A \left(\iiint_V \frac{\xi\nu \cos 2\omega(s - \bar{\sigma})}{\|\boldsymbol{p} - \bar{\boldsymbol{p}}\|^3} \, d\bar{\xi} \, d\bar{\eta} \, d\bar{\sigma} \right) d\xi \, d\eta \right\} \tag{6.6.28}$$

$$= \iint_A \left(\iiint_V \frac{\left(\frac{\nu^2}{2\omega} - \bar{\xi}\nu \right)}{\|\boldsymbol{p} - \bar{\boldsymbol{p}}\|^3} \, d\bar{\xi} \, d\bar{\eta} \, d\bar{\sigma} \right) \, d\xi \, d\eta$$

$$- \frac{\nu^2}{2\omega} \iint_A \left(\iiint_V \frac{\cos 2\omega(s - \bar{\sigma})}{\|\boldsymbol{p} - \bar{\boldsymbol{p}}\|^3} \, d\bar{\xi} \, d\bar{\eta} \, d\bar{\sigma} \right) \, d\xi \, d\eta$$

Replacing $\bar{\xi}\nu$ by $\bar{\xi}\nu[\cos 2\omega(s - \bar{\sigma}) + 2 \sin^2 \omega(s - \bar{\sigma})]$ in (6.6.28), and noting (by skew symmetry with respect to the map $\xi \to \bar{\xi}$) that we have

$$\iint_A \left(\iiint_V \frac{(\xi - \bar{\xi}) \cos 2\omega(s - \bar{\sigma})}{\|\boldsymbol{p} - \bar{\boldsymbol{p}}\|^3} \, d\bar{\xi} \, d\bar{\eta} \, d\bar{\sigma} \right) \, d\xi \, d\eta = 0 \tag{6.6.29}$$

we are now able to show that (6.6.28) reduces to the statement that

$$\left(IB - \frac{2\omega\mu(\nu)}{\nu} \right) = \left(\frac{I}{A}\right)^2 \mathcal{J}_\nu(\omega; s) \tag{6.6.30}$$

where

$$\mathcal{J}_\nu(\omega; s) = \iint_A \left(\iiint_V \left(\frac{\nu^2}{\omega} - 2\nu\bar{\xi} \right) \frac{\sin^2 \omega(s - \bar{\sigma})}{\|\boldsymbol{p} - \bar{\boldsymbol{p}}\|^3} \, d\bar{\xi} \, d\bar{\eta} \, d\bar{\sigma} \right) \, d\xi \, d\eta \tag{6.6.31}$$

with

$$\begin{aligned} \|\boldsymbol{p} - \bar{\boldsymbol{p}}\|^3 = \Bigg[& (\xi - \bar{\xi})^2 + (\eta - \bar{\eta})^2 \\ & + \left\{ \frac{\nu^2}{\omega^2} (\sin^2 \omega s + \sin^2 \omega\bar{\sigma}) - 2\alpha(s)\alpha(\bar{\sigma}) \cos \omega(s - \bar{\sigma}) \right\} \\ & + \Bigg\{ 4\xi\bar{\xi} \sin^2 \frac{\omega}{2}(s - \bar{\sigma}) - \frac{2\nu}{\omega}(\xi \sin^2 \omega s + \bar{\xi} \sin^2 \omega\bar{\sigma}) \\ & - 2(\bar{\xi} K_5(s, \bar{\sigma}) + \xi K_5(\bar{\sigma}, s)) \Bigg\} \Bigg]^{3/2} \end{aligned} \tag{6.6.32}$$

In (6.6.32), $\alpha(s)$ is given by (6.6.4), while $K_5(s,\bar{\sigma}) = \alpha(s)\sin\omega(s-2\bar{\sigma})$. As a consequence of (6.6.30), it should be clear that the problem of determining whether or not there exist solutions of the equilibrium equation (6.2.3) of the form (6.6.11), with $\boldsymbol{r}(s)$ given by (6.6.5) (i.e., *wiggly* circular states) is equivalent to determining whether or not $\frac{\partial}{\partial s}\mathcal{J}_\nu(\omega; s) = 0$; such a result would follow if we could show, e.g., that

$$\frac{\partial}{\partial s}\left[\frac{\sin^2\omega(s-\bar{\sigma})}{\|\boldsymbol{p}-\bar{\boldsymbol{p}}\|^3}\right]$$

is an odd function of $\tau = s - \bar{\sigma}$; this problem remains open as this volume goes to press.

6.7 Nonlocal Problems for Conducting Bodies

In § 6.1 we considered the problem of determining the equilibrium states of a current bearing nonlinearly elastic wire placed in an (ambient) magnetic field. Our purpose in § 6.1 was to set the stage for the somewhat more general, nonlocal problems, discussed in § 6.2–§ 6.6, in which the deformable wire interacted with itself through a self-field arising from the Biot-Savart law (1.2.31); the work in § 6.2–§ 6.6 had as its specific goal the delineation of possible, specific equilibrium states of the wire. In recent years several authors, most notably Rogers [139], [138] and Antman and Rogers [4] have considered more general problems of electromagnetic-elastic interactions in which the Biot-Savart law plays a prominent role and leads, as in § 6.2–§ 6.6, to the phenomena of self-interaction; in this section we will briefly outline some of that work. As we have indicated in the introduction to this chapter, we have no intention of delving, in any great depth, into the general problem of the interaction of electromagnetic fields with deformable bodies; there are now many excellent surveys in this area among which we mention the following: the foundation papers of Eringen [52], Jordan and Eringen [86], and Toupin [173], [172], the review paper of Pao [129], the collections of papers edited by Maugin [113] and Parkus [130], and the books of Brown [31], Hutter and Van De Ven [72], and Moon [115]. We also make note here of the general theory of nonlocal electromagnetic solids constructed by Eringen [53].

The work in [139], [138], and [4] is, as we have mentioned, concerned with problems in classical electromagnetic theory which model nonlocal effects; as we have shown in § 6.2–§ 6.6, the Biot-Savart law (1.2.31) states that an electric current creates a magnetic field in a conducting body at points in the body distant from the source. When the current in the body is allowed to depend on the self-field associated with (1.2.31) through a generalized Ohm's law, then one has to deal with a situation in which the magnetic field, for static problems, should be given as a function of the global values of a set of electric and magnetic potentials; this problem has been examined in detail in Antman and Rogers [4] with the results summarized in Rogers [138]. In our discussion here we will, for the most part, follow the presentation in [138]; however, in order to be able to reflect somewhat more closely the connection to the work delineated in § 6.2–§ 6.6, we will also present some of the highlights of the discussion of semi-inverse problems which appears in part II of the paper of Antman and Rogers [4]. Also, as the presentation in [138] deals primarily with rigid conducting bodies, we will indicate the changes which must be incorporated so as to deal with deformable media. We mention, in passing, that the paper [138] also contains a discussion of a proposed nonlocal constitutive equation to model the interaction of particles in a ferromagnetic material. So as to keep our considerations in this section in line with the discussion in § 6.2–§ 6.6 (i.e., the role of the Biot-Savart law in generating a self-field in a current bearing, conducting body), we will forgo here any examination of the nonlocal ferromagnetic model proposed in [138]; the interested reader may consult § 5 of that paper, as well as the earlier paper of Eringen [53], and the more recent work of Rogers [137] and Rogers and Brandon [140]. Our goal here will be to briefly present the equations of steady-state electromagnetism in conducting bodies, as formulated in [138], [4], and to show how the nonlinear coupling between the electromagnetic fields yields a system of functional-differential equations for a conducting body; we also present a general existence theorem for such systems which is given in [138]. For ease of exposition, the presentation in Rogers [138] deals with rigid conducting bodies; as indicated above, we will discuss this situation first and then make note of the pertinent changes which occur if one extends the analysis to elastic conducting media.

We begin by recalling Maxwell's equations (1.3.3a)–(1.3.3d), which for the steady-state electromagnetic behavior of a rigid material body, which occupies a closed domain $\Omega \subseteq R^3$, reduce to

$$\nabla \cdot \boldsymbol{D} = \rho \tag{6.7.1a}$$

$$\nabla \cdot \boldsymbol{B} = 0 \tag{6.7.1b}$$

$$\nabla \times \boldsymbol{E} = \boldsymbol{0} \tag{6.7.1c}$$

$$\nabla \times \boldsymbol{H} = \boldsymbol{J} \tag{6.7.1d}$$

where ρ is the free charge density. Of course, (6.7.1a)–(6.7.1d) must be supplemented by a suitable set of boundary conditions and constitutive relations; for the case of dielectric media, these constitutive relations, in [138], are assumed to have the form

$$\begin{aligned} \boldsymbol{D}(\boldsymbol{x}) &= \epsilon_0 \boldsymbol{E}(\boldsymbol{x}) + \hat{\boldsymbol{P}}(\boldsymbol{E}(\boldsymbol{x}), \boldsymbol{H}(\boldsymbol{x}), \boldsymbol{x}) && (6.7.2a) \\ &\equiv \hat{\boldsymbol{D}}(\boldsymbol{E}(\boldsymbol{x}), \boldsymbol{H}(\boldsymbol{x}), \boldsymbol{x}) \\ \boldsymbol{B}(\boldsymbol{x}) &= \mu_0 \boldsymbol{E}(\boldsymbol{x}) + \hat{\boldsymbol{M}}(\boldsymbol{E}(\boldsymbol{x}), \boldsymbol{H}(\boldsymbol{x}), \boldsymbol{x}) && (6.7.2b) \\ &\equiv \hat{\boldsymbol{B}}(\boldsymbol{E}(\boldsymbol{x}), \boldsymbol{H}(\boldsymbol{x}), \boldsymbol{x}) \end{aligned}$$

where $\hat{\boldsymbol{P}}$ and $\hat{\boldsymbol{M}}$ are, of course, the polarization and magnetization vectors, respectively. As a consequence of (6.7.1c), (6.7.1d), it follows that there exist scalar fields ϕ and ψ, the electric and magnetic scalar potentials, such that

$$\begin{aligned} \boldsymbol{E} &= \nabla\phi(\boldsymbol{x}) && (6.7.3a) \\ \boldsymbol{H}(\boldsymbol{x}) &= \nabla\psi(\boldsymbol{x}) + \int_\Omega \frac{\boldsymbol{J}(\boldsymbol{y}) \times (\boldsymbol{y} - \boldsymbol{x})}{\|\boldsymbol{y} - \boldsymbol{x}\|^3}\, d\boldsymbol{y} && (6.7.3b) \end{aligned}$$

Of course, the second term on the right-hand side of (6.7.3b) is the contribution to the magnetic field (compare with the expression (1.2.31) for the magnetic induction) attributable to the Biot-Savart law; we will stay, here, with the notation in [138] writing, as indicated, $\boldsymbol{x}$, $\boldsymbol{y}$, etc. for position vectors to points in the body Ω, in lieu of $\boldsymbol{r}_1$, $\boldsymbol{r}_2$, as in (1.2.31). We begin by taking the conducting body Ω to be rigid, which has the effect of eliminating the self-effects that result when electromagnetic forces influence the deformation of the body, thus causing the domain of integration to change in (6.7.3b); the modifications necessary to deal with the more general case of a deformable body, i.e., the case discussed in Antman and Rogers [4], will be indicated below.

When the body is nonconducting, i.e., $\boldsymbol{J} \equiv \mathbf{0}$, we may easily combine (6.7.1a)–(6.7.1d), (6.7.2a), (6.7.2b), and (6.7.3a), (6.7.3b) so as to obtain the following system of second-order, quasilinear, partial differential equations in divergence form:

$$\nabla \cdot \hat{\boldsymbol{D}}(\nabla\phi(\boldsymbol{x}), \nabla\psi(\boldsymbol{x}), \boldsymbol{x}) = \rho(\boldsymbol{x}) \tag{6.7.4a}$$

$$\nabla \cdot \hat{\boldsymbol{B}}(\nabla\phi(\boldsymbol{x}), \nabla\psi(\boldsymbol{x}), \boldsymbol{x}) = 0 \tag{6.7.4b}$$

With $\boldsymbol{\nu}(\boldsymbol{x})$ the exterior unit normal to $\partial\Omega$, at $\boldsymbol{x} \in \partial\Omega$, we may consider the following boundary conditions along with (6.7.4a), (6.7.4b):

$$\phi(\boldsymbol{x}) = \tilde{\phi}(\boldsymbol{x}), \quad \text{or} \quad \boldsymbol{D}(\boldsymbol{x}) \cdot \boldsymbol{\nu}(\boldsymbol{x}) = \tilde{\delta}(\boldsymbol{x}), \quad \boldsymbol{x} \in \partial\Omega \tag{6.7.5a}$$

$$\psi(\boldsymbol{x}) = \tilde{\psi}(\boldsymbol{x}), \quad \text{or} \quad \boldsymbol{B}(\boldsymbol{x}) \cdot \boldsymbol{\nu}(\boldsymbol{x}) = \tilde{\beta}(\boldsymbol{x}), \quad \boldsymbol{x} \in \partial\Omega \tag{6.7.5b}$$

with $\tilde{\phi}$, $\tilde{\psi}$, $\tilde{\delta}$, and $\tilde{\beta}$ prescribed on $\partial\Omega$. By definition, a weak solution of the boundary value problem (6.7.4a), (6.7.4b), (6.7.5a), (6.7.5b) is a pair of scalar functions $\phi(\boldsymbol{x})$, $\psi(\boldsymbol{x})$ such that

$$\begin{aligned} &\int_\Omega [\hat{\boldsymbol{D}}(\nabla\phi(\boldsymbol{y}), \nabla\psi(\boldsymbol{y}), \boldsymbol{y}) \cdot \nabla\Phi(\boldsymbol{y}) \\ &\qquad + \hat{\boldsymbol{B}}(\nabla\phi(\boldsymbol{y}), \nabla\psi(\boldsymbol{y}), \boldsymbol{y}) \cdot \nabla\Psi(\boldsymbol{y})]\, d\boldsymbol{y} \\ &\qquad + \int_\Omega \rho(\boldsymbol{y})\Phi(\boldsymbol{y})\, d\boldsymbol{y} = \int_{\partial\Omega} [\tilde{\delta}(\boldsymbol{y})\Phi(\boldsymbol{y}) + \tilde{\beta}(\boldsymbol{y})\Psi(\boldsymbol{y})]\, dS_{\boldsymbol{y}}, \end{aligned} \tag{6.7.6}$$

for all sufficiently smooth test functions Φ, Ψ which have zero trace at all points $\boldsymbol{x} \in \partial\Omega$ at which the Dirichlet boundary conditions in (6.7.5a), (6.7.5b) are prescribed. As has been pointed out in [138], the existence and regularity theory for systems in divergence form, such as (6.7.4a), (6.7.4b), has been developed by many authors, e.g., Giaquinta [60], and is usually dependent on some kind of ellipticity condition; such a condition would follow, as Rogers [138] indicates, from a constitutive hypothesis of the form

$$\begin{aligned} &\left(\frac{\partial \hat{D}_i}{\partial E_j} U_i U_j + \frac{\partial \hat{D}_i}{\partial H_j} U_i V_j + \frac{\partial \hat{B}_i}{\partial E_j} V_i U_j \right. \\ &\qquad \left. + \frac{\partial \hat{B}_i}{\partial H_j} V_i V_j \right) > 0, \ \forall \boldsymbol{U}, \boldsymbol{V} \in R^3/\{\mathbf{0}\} \end{aligned} \tag{6.7.7}$$

where, as usual, we sum on repeated indices in (6.7.7).

For the more complicated problem of a conducting (albeit, still rigid) body, we adopt for $\boldsymbol{J}$ a nonlinear version of Ohm's law, in the same spirit as our work in Chapters 2 and 3 on wave-dielectric interactions, namely,

$$\boldsymbol{J}(\boldsymbol{x}) = \hat{\boldsymbol{J}}(\boldsymbol{E}(\boldsymbol{x}), \boldsymbol{H}(\boldsymbol{x}), \boldsymbol{x}) \tag{6.7.8}$$

and substituting (6.7.8) into the expression (6.7.3b) for the magnetic field, we obtain

$$\begin{aligned} \boldsymbol{H}(\boldsymbol{x}) - \nabla\psi(\boldsymbol{x}) &= \int_\Omega \frac{\hat{\boldsymbol{J}}(\nabla\phi(\boldsymbol{y}), \boldsymbol{H}(\boldsymbol{y}), \boldsymbol{y}) \times (\boldsymbol{y} - \boldsymbol{x})}{\|\boldsymbol{y} - \boldsymbol{x}\|^3}\, d\boldsymbol{y} \\ &\equiv \boldsymbol{K}(\nabla\phi(\cdot), H(\cdot))(\boldsymbol{x}) \end{aligned} \tag{6.7.9}$$

When $\hat{\boldsymbol{J}}$ is independent of $\boldsymbol{H}(\cdot)$, as it was in our work in Chapters 2 and 3, then (6.7.9) yields an explicit representation for $\boldsymbol{H}$ in terms of $\nabla\phi$ and $\nabla\psi$; if $\hat{\boldsymbol{J}}$ depends on $\boldsymbol{H}(\cdot)$, then the solution of (6.7.9) is to be determined by a fixed point theorem and one such result, which follows directly from the analysis in Antman and Rogers [4], is the following:

Theorem 6.5 Let $\alpha > 1$ and suppose that the domain $\Omega \subset B_\sigma$, a ball of radius $\sigma > 0$ with center at $\mathbf{0}$. Let $\nabla\psi(\cdot), \nabla\phi(\cdot) \in L^\alpha(\Omega)$ and assume that there are positive numbers $p_i > 0$, $i = 1, 2, 3$ such that $3p_3 > \alpha$, while

$$\|\hat{\boldsymbol{J}}(\nabla\phi, \nabla\psi, \boldsymbol{x})\| \le p_1(1 + \|\nabla\phi\|^{1+p_3} + \|\nabla\psi\|^{1+p_3}) \tag{6.7.10}$$

$$\|\nabla_{\boldsymbol{H}}\hat{\boldsymbol{J}}(\nabla\phi, \nabla\psi, \boldsymbol{x})\| \le p_2(1 + \|\nabla\phi\|^{1+p_3} + \|\nabla\psi\|^{1+p_3}) \tag{6.7.11}$$

Then for σ and p_2 sufficiently small, (6.7.9) has a unique solution of the form

$$\boldsymbol{H}(\boldsymbol{x}) = \nabla\psi(\boldsymbol{x}) + \hat{\boldsymbol{K}}(\nabla\phi(\cdot), \nabla\psi(\cdot))(\boldsymbol{x}) \tag{6.7.12}$$

and the mapping

$$(\nabla\phi, \nabla\psi) \in L^\alpha(\Omega) \mapsto \hat{\boldsymbol{K}}(\nabla\phi, \nabla\psi)(\cdot) \in L^\alpha(\Omega) \tag{6.7.13}$$

is both continuous and compact.

By combining (6.7.1a), (6.7.1b), (6.7.2a), (6.7.2b), (6.7.3a), and (6.7.12) we obtain the following system of second-order functional partial differential equations:

$$\nabla \cdot \hat{\boldsymbol{D}}(\nabla\phi(\boldsymbol{x}), \nabla\psi(\boldsymbol{x}) + \hat{\boldsymbol{K}}(\nabla\phi(\cdot), \nabla\psi(\cdot))(\boldsymbol{x}), \boldsymbol{x}) = \rho(\boldsymbol{x}) \tag{6.7.14a}$$

$$\nabla \cdot \hat{\boldsymbol{B}}(\nabla\phi(\boldsymbol{x}), \nabla\psi(\boldsymbol{x}) + \hat{\boldsymbol{K}}(\nabla\phi(\cdot), \nabla\psi(\cdot))(\boldsymbol{x}), \boldsymbol{x}) = 0 \tag{6.7.14b}$$

We now describe, briefly, the existence theory developed in [4], and summarized by Rogers [138], which can be applied to the system (6.7.14a), (6.7.14b); the results will first be presented in an abstract form and then applied to the system of steady-state electromagnetic equations in divergence form. We begin by assuming that $\partial\Omega$ has a locally Lipschitz continuous graph and we set

$$\bar{\boldsymbol{U}}(\boldsymbol{x}) = (u_1(\boldsymbol{x}), \dots u_m(\boldsymbol{x})), \ \boldsymbol{x} \in \Omega \tag{6.7.15}$$

with a similar definition for $\bar{\boldsymbol{V}}$. For $p \in (1, \infty)$ we define the map

$$\begin{aligned}(\bar{\boldsymbol{U}}, \nabla\bar{\boldsymbol{V}}) &\in (L^p(\Omega))^m \times (L^p(\Omega))^{3m} \\ &\mapsto \hat{\boldsymbol{k}}(\bar{\boldsymbol{U}}, \nabla\bar{\boldsymbol{V}})(\cdot) \in (L^p(\Omega))^r\end{aligned} \tag{6.7.16}$$

which takes bounded sets into bounded sets. Now, let

$$\begin{aligned}&(\boldsymbol{\xi}, \boldsymbol{\eta}, \boldsymbol{\zeta}, \boldsymbol{x}) \in R^m \times R^{3m} \times R^r \times \Omega \\ &\mapsto \left\{ \begin{array}{l} \boldsymbol{A}^i(\boldsymbol{\xi}, \boldsymbol{\eta}, \boldsymbol{\zeta}, \boldsymbol{x}) \in R^3 \\ \lambda^i(\boldsymbol{\xi}, \boldsymbol{\eta}, \boldsymbol{\zeta}, \boldsymbol{x}) \in R \end{array} \right\}, \ i = 1, \dots, m\end{aligned} \tag{6.7.17}$$

and

$$(\boldsymbol{\xi}, \boldsymbol{x}) \in R^m \times \partial\Omega \mapsto \gamma^i(\boldsymbol{\xi}, \boldsymbol{x}) \in R, \ i = 1, \dots, m \tag{6.7.18}$$

satisfy

(i) a.e. in Ω, the functions $\boldsymbol{A}^i(\cdot, \cdot, \cdot, \boldsymbol{x})$ and $\lambda^i(\cdot, \cdot, \cdot, \boldsymbol{x})$ are continuous, while $\forall\, \boldsymbol{\xi}, \boldsymbol{\eta}, \boldsymbol{\zeta}$, the functions $\boldsymbol{A}^i(\boldsymbol{\xi}, \boldsymbol{\eta}, \boldsymbol{\zeta}, \cdot)$ and $\lambda^i(\boldsymbol{\xi}, \boldsymbol{\eta}, \boldsymbol{\zeta}, \cdot)$ are measurable.

(ii) a.e. on Ω, the functions $\gamma^i(\cdot, \boldsymbol{x})$ are continuous and, $\forall\, \boldsymbol{\xi}$, the functions $\gamma^i(\boldsymbol{\xi}, \cdot)$ are measurable on $\partial\Omega$ with respect to the two-dimensional Lebesgue measure.

(iii) there exists a constant $c_1 > 0$ and a function $k_1(\cdot) \in L^q(\Omega)$, $q = p/(p-1)$, such that both

$$\begin{aligned}&\|\boldsymbol{A}^i(\boldsymbol{\xi}, \boldsymbol{\eta}, \boldsymbol{\zeta}, \boldsymbol{x})\| \ \text{and} \ |\lambda^i(\boldsymbol{\xi}, \boldsymbol{\eta}, \boldsymbol{\zeta}, \boldsymbol{x})| \\ &\leq c_1[\|\boldsymbol{\xi}\|^{p-1} + \|\boldsymbol{\eta}\|^{p-1} + \|\boldsymbol{\zeta}\|^{p-1} + k_1(\boldsymbol{x})]\end{aligned} \tag{6.7.19}$$

for $i = 1, 2, \dots, m$.

As a consequence of the Hölder inequality it then follows that the functions

$$\begin{cases} \boldsymbol{A}^i(\bar{\boldsymbol{U}}, \nabla\bar{\boldsymbol{V}}(\cdot), \hat{\boldsymbol{k}}(\bar{\boldsymbol{U}}, \nabla\bar{\boldsymbol{V}})(\cdot), \cdot) \\ \lambda^i(\bar{\boldsymbol{U}}, \nabla\bar{\boldsymbol{V}}(\cdot), \hat{\boldsymbol{k}}(\bar{\boldsymbol{U}}, \nabla\bar{\boldsymbol{V}})(\cdot), \cdot) \end{cases}$$

are in $L^q(\Omega)$, $\forall \bar{\boldsymbol{U}}, \bar{\boldsymbol{V}} \in (W^{1,p}(\Omega))^m$ and, therefore, the functional

$$\begin{aligned} a(\bar{\boldsymbol{U}}, \bar{\boldsymbol{W}}) \equiv & \int_\Omega \Big[\boldsymbol{A}^i(\bar{\boldsymbol{U}}(\boldsymbol{x}), \nabla\bar{\boldsymbol{U}}(\boldsymbol{x}), \hat{\boldsymbol{k}}(\bar{\boldsymbol{U}}, \nabla\bar{\boldsymbol{U}})(\boldsymbol{x}), \boldsymbol{x}) \cdot \nabla w_i(\boldsymbol{x}) \\ & +\lambda^i(\bar{\boldsymbol{U}}(\boldsymbol{x}), \nabla\bar{\boldsymbol{U}}(\boldsymbol{x}), \hat{\boldsymbol{k}}(\bar{\boldsymbol{U}}, \nabla\bar{\boldsymbol{U}})(\boldsymbol{x}), \boldsymbol{x}) w_i(\boldsymbol{x})\Big]\, d\boldsymbol{x} \\ & + \int_{\partial\Omega} \gamma^i(\bar{\boldsymbol{U}}(\boldsymbol{x})) w_i(\boldsymbol{x})\, dS_{\boldsymbol{x}} \end{aligned} \tag{6.7.20}$$

is well-defined, $\forall \bar{\boldsymbol{U}}, \bar{\boldsymbol{W}} \in (W^{1,p}(\Omega))^m$. We now prescribe the components $u_1, \ldots, u_m$ of $\bar{\boldsymbol{U}}$, respectively, on subsets $\partial\Omega_1, \ldots, \partial\Omega_m$ of $\partial\Omega$ and we assume that the $\partial\Omega_i$, $i = 1, \ldots, m$ are measurable. By $\mathcal{V}$ we denote that closed subspace of $(W^{1,p}(\Omega))^m$, containing $(W_0^{1,p}(\Omega))^m$, which consists of those functions $(w_1, \ldots, w_m)$ for which $w_i = 0$ on $\partial\Omega_i$, $i = 1, \ldots, m$, in the sense of vanishing trace. If $\boldsymbol{U}^*$ is a given element of $(W^{1,p}(\Omega))^m$, we will require that u_i agree with u_i^* on $\partial\Omega_i$, $i = 1, \ldots, m$, by seeking solutions $\bar{\boldsymbol{U}} \in (W^{1,p}(\Omega))^m$ which satisfy $\bar{\boldsymbol{U}} - \boldsymbol{U}^* \in \mathcal{V}$. Next, we define a map $\boldsymbol{G} : \mathcal{V} \to \mathcal{V}^*$ by

$$\langle \boldsymbol{G}(\bar{\boldsymbol{U}}), \bar{\boldsymbol{W}} \rangle \equiv a(\bar{\boldsymbol{U}}, \bar{\boldsymbol{W}}),\ \forall \bar{\boldsymbol{W}} \in \mathcal{V} \tag{6.7.21}$$

where $\langle \bar{\boldsymbol{V}}, \bar{\boldsymbol{W}} \rangle$ is the duality pairing of $\bar{\boldsymbol{V}} \in \mathcal{V}^*$ with $\bar{\boldsymbol{W}} \in \mathcal{V}$. For $\boldsymbol{A}^i$ of class C^1, and $\bar{\boldsymbol{U}}$ of class C^2 and vanishing on $\partial\Omega$, we have by (6.7.20)

$$\begin{aligned} \langle \boldsymbol{G}(\bar{\boldsymbol{U}}), \bar{\boldsymbol{W}} \rangle = & \int_\Omega [-\nabla \cdot \boldsymbol{A}^i(\bar{\boldsymbol{U}}, \nabla\bar{\boldsymbol{U}}, \hat{\boldsymbol{k}}(\bar{\boldsymbol{U}}, \nabla\bar{\boldsymbol{U}}), \boldsymbol{x}) \\ & + \lambda^i(\bar{\boldsymbol{U}}, \nabla\bar{\boldsymbol{U}}, \hat{\boldsymbol{k}}(\bar{\boldsymbol{U}}, \nabla\bar{\boldsymbol{U}}), \boldsymbol{x})]\, w_i(\boldsymbol{x})\, d\boldsymbol{x} \end{aligned} \tag{6.7.22}$$

Finally, we set

$$\boldsymbol{\eta} = (\boldsymbol{\eta}_1, \ldots, \boldsymbol{\eta}_m),\ \boldsymbol{\eta}_i \in R^3 \tag{6.7.23}$$

so that $\boldsymbol{\eta} \in R^{3m}$. Then from [138] we have, with our previously stated assumptions relative to p and $\partial\Omega$, as well as hypotheses (i)–(iii), delineated above, the following abstract result:

Theorem 6.6 Suppose that for $\bar{\boldsymbol{V}} \in \mathcal{V}$

$$\frac{a(\bar{\boldsymbol{V}}, \bar{\boldsymbol{V}})}{\|\bar{\boldsymbol{V}}\|^2_{(W^{1,p}(\Omega))^m}} \to \infty, \text{ as } \|\bar{\boldsymbol{V}}\|_{(W^{1,p}(\Omega))^m} \to \infty \tag{6.7.24}$$

and

$$[\boldsymbol{A}^i(\boldsymbol{\xi},\boldsymbol{\eta}+\boldsymbol{\lambda},\boldsymbol{\zeta},\boldsymbol{x}) - \boldsymbol{A}^i(\boldsymbol{\xi},\boldsymbol{\eta},\boldsymbol{\zeta},\boldsymbol{x})]\cdot\boldsymbol{\lambda} > 0, \tag{6.7.25}$$

$\forall\,\boldsymbol{\lambda} = (\boldsymbol{\lambda}_1,\ldots,\boldsymbol{\lambda}_m) \neq 0$, $\boldsymbol{\lambda}_i \in R^3$, $i = 1,2,\ldots,m$. For Ω bounded, let

$$\frac{\boldsymbol{A}^i(\boldsymbol{\xi},\boldsymbol{\eta},\boldsymbol{\zeta},\boldsymbol{x})\cdot\boldsymbol{\eta}_i}{[\|\boldsymbol{\eta}\| + \|\boldsymbol{\eta}\|^{p-1}]} \to \infty, \text{ as } \|\boldsymbol{\eta}\| \to \infty, \tag{6.7.26}$$

a.e. in Ω, for bounded $(\boldsymbol{\xi},\boldsymbol{\eta}) \in R^m \times R^{3m}$, while for Ω unbounded assume that

$$\boldsymbol{A}^i(\boldsymbol{\xi},\boldsymbol{\eta},\boldsymbol{\zeta},\boldsymbol{x})\cdot\boldsymbol{\eta}_i \leq c_2\|\boldsymbol{\eta}\|^p - k_2(\boldsymbol{x}) \tag{6.7.27}$$

for some constant $c_2 > 0$ and some $k_2(\cdot) \in L^1(\Omega)$. Define $\tilde{\boldsymbol{k}}$ via

$$\tilde{\boldsymbol{k}}(\bar{\boldsymbol{U}})(\cdot) = \hat{\boldsymbol{k}}(\bar{\boldsymbol{U}},\nabla\bar{\boldsymbol{U}})(\cdot), \tag{6.7.28}$$

$\forall\,\bar{\boldsymbol{U}} \in (W^{1,p}(\Omega))^m$ and let χ_S be the characteristic function of a set $S \subset \Omega$. For every subdomain S of Ω, which has compact closure in Ω, assume that the mapping

$$\bar{\boldsymbol{U}} \in (W^{1,p}(\Omega))^m \mapsto \chi_S(\cdot)\tilde{\boldsymbol{k}}(\bar{\boldsymbol{U}})(\cdot) \in (L^p(S))^r \tag{6.7.29}$$

is compact. Then $\forall\,\bar{\boldsymbol{F}} \in \mathcal{V}^*$ and $\forall\,\boldsymbol{U}^* \in (W^{1,p}(\Omega))^m$, $\exists\,\bar{\boldsymbol{U}} \in (W^{1,p}(\Omega))^m$, with $\bar{\boldsymbol{U}} - \boldsymbol{U}^* \in \mathcal{V}$, such that

$$\langle \boldsymbol{G}(\bar{\boldsymbol{U}}),\bar{\boldsymbol{V}}\rangle = \langle \bar{\boldsymbol{F}},\bar{\boldsymbol{V}}\rangle,\ \forall\,\bar{\boldsymbol{V}} \in \mathcal{V} \tag{6.7.30}$$

Remarks: The proof of Theorem 6.6 is achieved by modifying the arguments of Brezis [28], and Browder [30]; the proof makes essential use of the monotonicity with respect to the local values of the highest order derivatives (condition (6.7.25)) as well as compactness with respect to the global values of the highest order derivatives.

We now want to indicate the relationship of Theorem 6.6 to the boundary value problem associated with the second-order system of functional partial differential equations (6.7.14a), (6.7.14b). We set

$$\bar{\boldsymbol{U}} = (\phi,\psi) \tag{6.7.31a}$$

$$\tilde{\boldsymbol{k}}(\bar{\boldsymbol{U}})(\cdot) = \hat{\boldsymbol{K}}(\nabla\phi,\nabla\psi)(\cdot) \tag{6.7.31b}$$

with $\hat{\boldsymbol{K}}$ as in (6.7.12). For the variables appearing in (6.7.20) we now have

$$\boldsymbol{A}^1(\bar{\boldsymbol{U}}, \nabla\bar{\boldsymbol{U}}(\boldsymbol{x}), \tilde{\boldsymbol{k}}(\bar{\boldsymbol{U}})(\boldsymbol{x}), \boldsymbol{x}) \tag{6.7.32a}$$
$$= \hat{\boldsymbol{D}}(\nabla\phi(\boldsymbol{x}), \nabla\psi(\boldsymbol{x}) + \hat{\boldsymbol{K}}(\nabla\phi, \nabla\psi)(\boldsymbol{x}), \boldsymbol{x})$$

$$\boldsymbol{A}^2(\bar{\boldsymbol{U}}, \nabla\bar{\boldsymbol{U}}(\boldsymbol{x}), \tilde{\boldsymbol{k}}(\bar{\boldsymbol{U}})(\boldsymbol{x}), \boldsymbol{x}) \tag{6.7.32b}$$
$$= \hat{\boldsymbol{B}}(\nabla\phi(\boldsymbol{x}), \nabla\psi(\boldsymbol{x}) + \hat{\boldsymbol{K}}(\nabla\phi, \nabla\psi)(\boldsymbol{x}), \boldsymbol{x})$$

$$\lambda^1(\bar{\boldsymbol{U}}, \nabla\bar{\boldsymbol{U}}(\boldsymbol{x}), \tilde{\boldsymbol{k}}(\bar{\boldsymbol{U}})(\boldsymbol{x}), \boldsymbol{x}) = \rho(\boldsymbol{x}) \tag{6.7.32c}$$

$$\lambda^2(\bar{\boldsymbol{U}}, \nabla\bar{\boldsymbol{U}}(\boldsymbol{x}), \tilde{\boldsymbol{k}}(\bar{\boldsymbol{U}})(\boldsymbol{x}), \boldsymbol{x}) = 0 \tag{6.7.32d}$$

and

$$\gamma^1(\bar{\boldsymbol{U}}, \boldsymbol{x}) = -\tilde{\delta}(\boldsymbol{x}) \tag{6.7.32e}$$

$$\gamma^2(\bar{\boldsymbol{U}}, \boldsymbol{x}) = -\tilde{\beta}(\boldsymbol{x}) \tag{6.7.32f}$$

We identify $\bar{\boldsymbol{W}}$ with (Φ, Ψ) in (6.7.6) and note that (6.7.25) is a consequence of the ellipticity condition (6.7.7); then we have the following

Theorem 6.7 ([138]) Let $\hat{\boldsymbol{D}}$, $\hat{\boldsymbol{B}}$, $\tilde{\delta}$, and $\tilde{\beta}$ satisfy the hypotheses of Theorem 6.6, modulo the identifications prescribed by (6.7.32a)–(6.7.32f). Then (6.7.6) is satisfied $\forall\,(\Phi, \Psi) \in \mathcal{V}$.

As noted by Rogers [138], the question of regularity for the solutions of nonlocal problems such as (6.7.14a), (6.7.14b) remains an open one.

In [4], Antman and Rogers consider a more general class of functional partial differential equations which describe the steady-state behavior of solids which can sustain mechanical, electromagnetic and thermal effects; their results also include those for rigid conductors which we have described above, as well as a class of semi-inverse problems which lead to ordinary functional differential equations. As in the description of the situation for rigid conductors, given above, we identify a material body with the closure Ω of a domain in R^3 and we equip R^3 with the standard Euclidean norm. If we identify points of the body with their positions $\boldsymbol{x}$ in Ω, then by $\boldsymbol{y}(\boldsymbol{x})$ we will denote the position of $\boldsymbol{x}$ in some deformed configuration. The deformation gradient for the map $\boldsymbol{x} \mapsto \boldsymbol{y}(\boldsymbol{x})$ is given by $\boldsymbol{F} = \nabla_{\boldsymbol{x}} \boldsymbol{y}$ and, if we assume that this map preserves orientation, then $\det \boldsymbol{F} > 0$; unlike the analysis in [4] we will, for the sake of simplicity here, suppress all dependence of the relevant constitutive functions on thermodynamical quantities and focus, instead, on the mechanical-electromagnetic interaction. Then for the electric and magnetic fields we have

$$\begin{cases} \boldsymbol{E}(\boldsymbol{x}) = \hat{\boldsymbol{E}}(\boldsymbol{y}(\boldsymbol{x})) \cdot \boldsymbol{F}(\boldsymbol{x}) \\ \boldsymbol{H}(\boldsymbol{x}) = \hat{\boldsymbol{H}}(\boldsymbol{y}(\boldsymbol{x})) \cdot \boldsymbol{F}(\boldsymbol{x}) \end{cases} \tag{6.7.33}$$

with similar relations for the electric displacement field at $\boldsymbol{x}$, $\boldsymbol{D}(\boldsymbol{x})$, and the magnetic induction field $\boldsymbol{B}(\boldsymbol{x})$ at $\boldsymbol{x}$. If $\boldsymbol{T}(\boldsymbol{x})$ is the effective Cauchy stress in the material (undeformed) representation of the body, then to the steady-state Maxwell's equations (6.7.1a)–(6.7.1d) we must append the local form of the equations expressing balance of momentum and energy, namely,

$$\begin{cases} \nabla \cdot \boldsymbol{T} + \boldsymbol{f} = \boldsymbol{0} \\ \nabla \cdot \boldsymbol{q} + \boldsymbol{J} \cdot \boldsymbol{E} = 0 \end{cases} \tag{6.7.34}$$

where $\boldsymbol{f}$ is the external body force vector and $\boldsymbol{q}$ is the heat flux vector; in (6.7.34), $\boldsymbol{f}$ represents body forces of electromagnetic origin which have not been absorbed in $\boldsymbol{T}$ (which is the sum of the mechanical Cauchy stress and the Maxwell stress) and the divergence operators are taken with respect to material coordinates $\boldsymbol{x}$. For simplicity we will take $\boldsymbol{f} = \boldsymbol{0}$. The quantity $\boldsymbol{J} \cdot \boldsymbol{E}$ is known as the Joule heating and has appeared before, i.e., see (1.3.9) and the ensuing discussion in Chapter 1. As was the case with the discussion of the results in [138], the relations (6.7.3a), (6.7.3b) still apply but we will now rewrite these in the form

$$\boldsymbol{E} = \nabla \phi(\boldsymbol{x}) \tag{6.7.35a}$$

$$\boldsymbol{H}(\boldsymbol{x}) = \nabla \psi(\boldsymbol{x}) + \int_{\Omega} \frac{\boldsymbol{J}(\boldsymbol{z}) \times (\boldsymbol{z} - \boldsymbol{x})}{\|\boldsymbol{z} - \boldsymbol{x}\|^3} \, d\boldsymbol{z} \tag{6.7.35b}$$

so as not to confuse the role of $\boldsymbol{y}$ in (6.7.3b), i.e., dummy variable of integration, with its current role as the position of the material point at $\boldsymbol{x}$ after the deformation $\boldsymbol{x} \mapsto \boldsymbol{y}(\boldsymbol{x})$. For the independent constitutive variable we take the set

$$\boldsymbol{\Gamma} \equiv (\boldsymbol{F}, \boldsymbol{E}, \boldsymbol{H}) \in R^9 \times R^3 \times R^3 \tag{6.7.36}$$

and we assume for our constitutive hypotheses that

$$\boldsymbol{T}(\boldsymbol{x}) = \hat{\boldsymbol{T}}(\boldsymbol{\Gamma}(\boldsymbol{x}), \boldsymbol{x}) \tag{6.7.37}$$

with similar constitutive functions $\hat{\boldsymbol{Q}}$, $\hat{\boldsymbol{D}}$, $\hat{\boldsymbol{B}}$, and $\hat{\boldsymbol{J}}$ for the fields $\boldsymbol{q}$, $\boldsymbol{D}$, $\boldsymbol{B}$, and $\boldsymbol{J}$; all constitutive functions are assumed to be of class C^1. When $\boldsymbol{J} \neq \boldsymbol{0}$, we again have a relation of the form (6.7.9) for the magnetic field, namely,

$$\begin{aligned} \boldsymbol{H}(\boldsymbol{x}) - \nabla\psi(\boldsymbol{x}) &= \int_\Omega \frac{\hat{\boldsymbol{J}}(\boldsymbol{H}(\boldsymbol{z}), \boldsymbol{\Sigma}(\boldsymbol{z}), \boldsymbol{z}) \times (\boldsymbol{z} - \boldsymbol{x})}{\|\boldsymbol{z} - \boldsymbol{x}\|^3}\, d\boldsymbol{z} \\ &\equiv \boldsymbol{K}(H(\cdot), \Sigma(\cdot))(\boldsymbol{x}) \end{aligned} \tag{6.7.38}$$

where $\boldsymbol{\Sigma} \equiv (\boldsymbol{F}, \boldsymbol{E})$. In (6.7.9), of course, $\boldsymbol{\Sigma} = \boldsymbol{E} \equiv \nabla\phi$. Under conditions similar to those delineated in Theorem 6.5, the authors [4] propose, for the case in which $\hat{\boldsymbol{J}}$ is not independent of $\boldsymbol{H}$, growth conditions on $\hat{\boldsymbol{J}}$ and $\nabla_{\boldsymbol{H}}\hat{\boldsymbol{J}}$ which insure (as in the representation (6.7.12) for the case of a rigid, conducting body) that $\boldsymbol{H}$ can be expressed in terms of $\boldsymbol{F}$, $\nabla\phi$, and $\nabla\psi$; the relevant growth conditions are given in (6.7.39), below.

Remarks: As in Theorem 6.5, if $\Omega \subset B_\sigma$, a ball of radius $\sigma > 0$ with center at $\boldsymbol{0}$, then the arguments presented in Antman and Rogers [4] also establish the fact that the integral operator on the right-hand side of (6.7.35b) is compact as a mapping from $L^\beta(B_\sigma) \to L^\alpha(B_\sigma)$, when $\nabla\psi(\cdot), \nabla\phi(\cdot) \in L^\alpha(\Omega)$, $\alpha > 1$, and $\beta = \alpha(1+\zeta)^{-1}$, where $\zeta < \alpha/3$ serves to control the growth of $\hat{\boldsymbol{J}}$ and $\nabla_{\boldsymbol{H}}\hat{\boldsymbol{J}}$, i.e.,

$$\begin{cases} \|\hat{\boldsymbol{J}}(\boldsymbol{\Gamma}, \boldsymbol{x})\| \le \mu\left(1 + \|\boldsymbol{\Gamma}\|^{1+\zeta}\right) \\ \|\nabla_{\boldsymbol{H}}\hat{\boldsymbol{J}}(\boldsymbol{\Gamma}, \boldsymbol{x})\| \le \theta\left(1 + \|\boldsymbol{\Gamma}\|^{1+\zeta}\right) \end{cases} \tag{6.7.39}$$

for some μ, $\theta > 0$.

Assuming that $\hat{\boldsymbol{J}}$ is such that $\boldsymbol{H}$ admits a representation of the form

$$\boldsymbol{H}(\boldsymbol{x}) = \nabla\psi(\boldsymbol{x}) + \hat{\boldsymbol{K}}(\boldsymbol{\Delta})(\boldsymbol{x}), \tag{6.7.40}$$

$\boldsymbol{\Delta} = (\boldsymbol{F}, \nabla\phi, \nabla\psi)$, we may substitute from (6.7.40) into the right-hand side of (6.7.37) so as to obtain

$$\boldsymbol{T}(\boldsymbol{x}) = \hat{\boldsymbol{T}}(\boldsymbol{F}(\boldsymbol{x}), \nabla\phi(\boldsymbol{x}), \nabla\psi(\boldsymbol{x}) + \hat{\boldsymbol{K}}(\boldsymbol{\Delta})(\boldsymbol{x}), \boldsymbol{x}) \tag{6.7.41}$$

with similar relations for $\boldsymbol{q}$, $\boldsymbol{D}$, and $\boldsymbol{B}$, at $\boldsymbol{x}$; finally, (6.7.41), and the analogous relations for $\boldsymbol{q}$, $\boldsymbol{D}$, and $\boldsymbol{B}$ are inserted into the balance equations (6.7.34), and the first two of Maxwell's equations (6.7.1a), (6.7.1b), so as to produce the quasilinear system of partial functional differential equations

$$\begin{cases} \nabla \cdot \hat{\boldsymbol{T}} = \boldsymbol{0} \\ \nabla \cdot \hat{\boldsymbol{Q}} + \hat{\boldsymbol{J}} \cdot \nabla \phi = 0 \\ \nabla \cdot \hat{\boldsymbol{D}} = \rho \\ \nabla \cdot \hat{\boldsymbol{B}} = 0 \end{cases} \tag{6.7.42}$$

the arguments of $\hat{\boldsymbol{T}}$, $\hat{\boldsymbol{Q}}$, $\hat{\boldsymbol{J}}$, $\hat{\boldsymbol{D}}$, and $\hat{\boldsymbol{B}}$ being as in (6.7.41).

In [4] a weak formulation of the boundary value problem for the system (6.7.42) is given in terms of a principle of virtual work, which generalizes the virtual work formulation (6.7.6) for the rigid conducting case, and incorporates both the equations and the boundary data; we will pass over giving here a detailed technical description of the arguments employed in [4] to formulate and prove a general existence theorem for boundary value problems associated with the system (6.7.42), inasmuch as those arguments are quite similar to the ones used for the special case of the rigid conductor. A few remarks, however, are in order. First of all, a strong ellipticity condition (generalizing the condition (6.7.7) which applies in the case of a rigid conductor) and a restricted strong ellipticity condition are formulated; as indicated by the authors [4], the strong ellipticity condition is the generalization to elliptic systems in divergence form of the Legendre-Hadamard condition of the calculus of variations. Actually, (6.7.7) is a monotonicity condition, which is what the restricted strong ellipticity condition in [4] reduces to for the special case of a rigid conductor. Next, conditions are imposed, which are consistent with the strong ellipticity condition in [4] and which constrain the behavior of the constitutive functions at extreme values of their arguments; these conditions constitute a generalization of those in Antman [1] and imply, in particular, that the material response in large compression is dominated by that for purely mechanical response, i.e., that part of the stress attributable to the presence of the electromagnetic field (the Maxwell stress) does not compete effectively with the purely mechanical part of the stress (the Cauchy stress) when $\det \boldsymbol{F}$ is small. Finally, the general existence theorems which are proven in [4] are obtained for two special classes of problems which are amenable to treatment by means of relatively recent results for elliptic systems (i.e., the work of Giaquinta [60], Brezis [28], and Browder [30] we have already referenced above); the first of these classes of problems covers the situation already described here in some detail, i.e., the case of a rigid conducting material, while the second class includes problems (with no electrical conduction) for which there exists a stored energy function. For the second class of problems the reduced problem admits a variational formulation; thermal effects are ignored in this case but, unlike our description of the problem, above, are included in the treatment of rigid conducting materials. Also, the proof of the existence theorem for the second class of problems relies heavily on ideas originally formulated by J. Ball [8] to treat the problem of existence of weak solutions to boundary-value problems in nonlinear hyperelasticity.

We will close with a few remarks concerning the semi-inverse problems which are considered in [4], and which are closer in spirit to the type of problems formulated in § 6.2–§ 6.6, for the self-interacting, current bearing, nonlinear elastic wire, than those which we have discussed, to this point, in this section. Following the general notation in [4], we let $\{\boldsymbol{i}_1, \boldsymbol{i}_2, \boldsymbol{i}_3\}$ be a fixed right-handed orthonormal basis for R^3; we introduce a set $\boldsymbol{x} = (s, \theta, z)$ of cylindrical coordinates defined by

$$\boldsymbol{y} = \tilde{\boldsymbol{y}}(\boldsymbol{x}) \equiv s\boldsymbol{k}_1(\theta) + z\boldsymbol{k}_3(\theta) \tag{6.7.43}$$

where

$$\begin{cases} \boldsymbol{k}_1(\theta) = \cos\theta \boldsymbol{i}_1 + \sin\theta \boldsymbol{i}_2 \\ \boldsymbol{k}_2(\theta) = -\sin\theta \boldsymbol{i}_1 + \cos\theta \boldsymbol{i}_2 \\ \boldsymbol{k}_3(\theta) = \boldsymbol{i}_3 \end{cases} \tag{6.7.44}$$

We assume that (6.7.43) is equivalent to $\boldsymbol{x} = \tilde{\boldsymbol{x}}(\boldsymbol{y})$, with $\tilde{\boldsymbol{x}}$ the inverse of $\tilde{\boldsymbol{y}}$, and will use the notation, e.g., $\tilde{\boldsymbol{z}}(\boldsymbol{x}) = \boldsymbol{z}(\tilde{\boldsymbol{y}}(\boldsymbol{x}))$. Semi-inverse problems are considered (we will again suppress here all thermal effects for ease of exposition) in which $\tilde{\boldsymbol{z}}$, $\tilde{\varphi}$, and $\tilde{\boldsymbol{H}}$ have the form

$$\tilde{\boldsymbol{z}}(\boldsymbol{x}) = w_1(s)\boldsymbol{e}_1(\boldsymbol{x}) + [w_3(s) + \alpha_{32}\theta + \alpha_{33}z]\boldsymbol{e}_3(\boldsymbol{x}) \tag{6.7.45}$$

with

$$\begin{cases} \boldsymbol{e}_1(\boldsymbol{x}) = \cos\omega(\boldsymbol{x})\boldsymbol{i}_1 + \sin\omega(\boldsymbol{x})\boldsymbol{i}_2 \\ \boldsymbol{e}_2(\boldsymbol{x}) = -\sin\omega(\boldsymbol{x})\boldsymbol{i}_1 + \cos\omega(\boldsymbol{x})\boldsymbol{i}_2 \\ \boldsymbol{e}_3(\boldsymbol{x}) = \boldsymbol{i}_3 \end{cases} \tag{6.7.46}$$

$$\omega(\boldsymbol{x}) = w_2(s) + \alpha_{22}\theta + \alpha_{23}z \tag{6.7.47}$$

and

$$\tilde{\varphi}(\boldsymbol{x}) = w_4(s) + \alpha_{52}\theta + \alpha_{53}z \tag{6.7.48}$$

$$\tilde{\boldsymbol{H}}(\boldsymbol{x}) = H_i(s)\boldsymbol{k}_i(\theta) \tag{6.7.49}$$

Also, constitutive assumptions are made relative to $\hat{\boldsymbol{J}}$ which insure that $\tilde{\psi}(\boldsymbol{x})$ has the form

$$\tilde{\psi}(\boldsymbol{x}) = w_5(s) + \alpha_{62}\theta + \alpha_{63}z \tag{6.7.50}$$

The body Ω is taken to be

$$\Omega \equiv \tilde{\boldsymbol{y}}([a,1] \times [-\Theta,\Theta] \times [-Z,Z]) \tag{6.7.51}$$

with $0 < a < 1$, $0 < \Theta \leq \pi$, and $Z > 0$ so that Ω is a (possibly, slit) cylindrical tube if $\Theta = \pi$ and a sector of such a tube if $\Theta < \pi$. From the chain rule it follows that

$$\begin{aligned} \boldsymbol{F}(\tilde{\boldsymbol{y}}(\boldsymbol{x})) = {} & [w_1'(s)\boldsymbol{e}_1(\boldsymbol{x}) + w_1(s)w_1'(s)\boldsymbol{e}_2(\boldsymbol{x}) \\ & + w_3'(s)\boldsymbol{e}_3(\boldsymbol{x})]\boldsymbol{k}_1(\theta) \\ & + s^{-1}[\alpha_{22}w_1(s)\boldsymbol{e}_2(\boldsymbol{x}) + \alpha_{32}\boldsymbol{e}_3(\boldsymbol{x})]\boldsymbol{k}_2(\theta) \\ & + [\alpha_{23}w_1(s)\boldsymbol{e}_2(\boldsymbol{x}) + \alpha_{33}\boldsymbol{e}_3(\boldsymbol{x})]\boldsymbol{k}_3(\theta) \end{aligned} \tag{6.7.52}$$

where juxtaposition of vectors indicates the dyadic product, the dyadic product $\boldsymbol{ab}$ being the second-order tensor satisfying $(\boldsymbol{ab})\cdot\boldsymbol{c} = (\boldsymbol{b}\cdot\boldsymbol{c})\boldsymbol{a}$ for all vectors $\boldsymbol{c}$. We also have

$$\begin{aligned} \boldsymbol{E}(\tilde{\boldsymbol{y}}(\boldsymbol{x})) &= \nabla_{\boldsymbol{y}}\varphi(\tilde{\boldsymbol{y}}(\boldsymbol{x})) \\ &= w_4'(s)\boldsymbol{k}_1(\theta) + \alpha_{52}s^{-1}\boldsymbol{k}_2(\theta) + \alpha_{53}\boldsymbol{k}_3(\theta) \end{aligned} \tag{6.7.53}$$

and

$$\nabla_{\boldsymbol{y}}\psi(\tilde{\boldsymbol{y}}(\boldsymbol{x})) = w_5'(s)\boldsymbol{k}_1(\theta) + \alpha_{62}s^{-1}\boldsymbol{k}_2(\theta) + \alpha_{63}\boldsymbol{k}_3(\theta) \tag{6.7.54}$$

The requirement that $\det \boldsymbol{F} > 0$ is equivalent to the condition that, a.e.

$$(\alpha_{22}\alpha_{33} - \alpha_{23}\alpha_{32}) \left(\frac{w_1(s)}{s} \right) w_1'(s) > 0 \tag{6.7.55}$$

However, as $w_1(\cdot)$ represents radial distance, we require that $w_1(s) > 0$, $\forall\, s \in [a, 1]$, so that (6.7.55) reduces to the statement

$$(\alpha_{22}\alpha_{33} - \alpha_{23}\alpha_{32}) w_1'(s) > 0, \text{ a.e.} \tag{6.7.56}$$

For the sake of simplicity, in [4] it is, in fact, assumed that $w_1'(s) > 0$, a.e. In place of the symbol $\boldsymbol{\Gamma}$, as given by (6.7.36), we now have as a consequence of (6.7.45)–(6.7.50), (6.7.52)–(6.7.54)

$$\begin{aligned} \gamma(s) = (&w_1'(s), w_1(s), w_2'(s), w_3'(s), \alpha_{22}s^{-1}w_1(s), \\ &\alpha_{32}s^{-1}, \alpha_{23}w_1(s), \alpha_{33}, w_4'(s), \alpha_{52}s^{-1}, \alpha_{53}, \\ &H_1(s), H_2(s), H_3(s)) \end{aligned} \tag{6.7.57}$$

The constitutive functions may now be represented entirely in terms of γ and s, i.e.,

$$\begin{cases} T_{ij} = \hat{T}_{ij}(\gamma(s), s) \\ q_j = \hat{Q}_j(\gamma(s), s) \\ D_j = \hat{D}_j(\gamma(s), s) \\ B_j = \hat{B}_j(\gamma(s), s) \\ J_j = J_j(\gamma(s), s) \end{cases} \tag{6.7.58}$$

Furthermore, it can be shown that

$$\begin{cases} \hat{J}_1(\gamma(s), s) = 0 \\ s\hat{J}_3(\gamma(s), s) = [sH_2(s)]' \\ -\hat{J}_2(\gamma(s), s) = H_3'(s) \end{cases} \tag{6.7.59}$$

so that in the representation (6.7.49) for $\tilde{\boldsymbol{H}}$ we have

$$\begin{cases} H_1(s) = w_5'(s) \\ H_2(s) = s^{-1} \left[\alpha_{62} - \int_s^1 \zeta \hat{J}_3(\gamma(\zeta), \zeta)\, d\zeta \right] \\ H_3(s) = \alpha_{63} + \int_s^1 \hat{J}_2(\gamma(\zeta), \zeta)\, d\zeta \end{cases} \tag{6.7.60}$$

The condition $\hat{J}_1(\gamma(s), s) = 0$ is either satisfied identically if $\hat{J}_1 \equiv 0$ (so that the material cannot conduct electricity in the radial direction) or may be viewed as a restriction on γ; in the latter case it is assumed in [4] that the first relation in (6.7.59) can be uniquely solved for w_4' in terms of the other elements of γ. Also, by controlling the dependence of $\hat{J}_2$ and $\hat{J}_3$ on $H_2(s)$ and $H_3(s)$, it is possible to show that the last two relations in (6.7.60) can be replaced by relations of the form

$$\begin{cases} H_2(s) = s^{-1}[\alpha_{62} + \Lambda_2(\boldsymbol{w}(\cdot), \boldsymbol{\alpha}, s) \\ H_3(s) = \alpha_{63} + \Lambda_3(\boldsymbol{w}(\cdot), \boldsymbol{\alpha}, s) \end{cases} \tag{6.7.61}$$

where

$$\boldsymbol{w} = (w_1, \ldots, w_5), \qquad \boldsymbol{\alpha} = (\alpha_{22}, \ldots, \alpha_{63}) \tag{6.7.62}$$

If one assumes that there is a number $p > 1$ such that $\hat{J}_2(\gamma(\cdot), \cdot)$ and $\hat{J}_3(\gamma(\cdot), \cdot)$ are integrable on $[a, 1]$, when $\gamma \in L^p([a, 1])$, and that there is a number $\kappa(\boldsymbol{w}, \boldsymbol{\alpha})$ such that

$$\sum_{\alpha,\beta=1} \left| \frac{\partial \hat{J}_\alpha(\gamma(s), s)}{\partial H_\beta} \right| \leq \kappa(\boldsymbol{w}, \boldsymbol{\alpha}),$$

when $\gamma \in L^p([a, 1])$, then it can also be proven that the maps $\boldsymbol{w}(\cdot) \in W^{1,p}((a, 1)) \mapsto \Lambda_2(\boldsymbol{w}(\cdot), \boldsymbol{\alpha}, \cdot)$, $\Lambda_3(\boldsymbol{w}(\cdot), \boldsymbol{\alpha}, \cdot) \in C^0([a, 1])$ are compact.

If we denote by $\ell_i \boldsymbol{e}_i$ the axial vector corresponding to $\boldsymbol{L}$, where

$$\boldsymbol{T}\boldsymbol{F}^t - \boldsymbol{F}\boldsymbol{T}^t = \boldsymbol{L} \tag{6.7.63}$$

is the local form for the equation representing balance of torque, $\boldsymbol{L}$ a skew symmetric tensor depending upon the electromagnetic fields and the choice of the Maxwell stress tensor, and we assume (i) that the only body force, the Lorentz force, has been absorbed into the effective stress, and (ii) that the only heat source is due to Joule heating, then the governing equations (6.7.42) can be shown to reduce to the following system of ordinary-functional differential equations for $\boldsymbol{w}$, $\boldsymbol{\alpha}$:

$$(s\hat{\boldsymbol{\xi}})' = s\hat{\boldsymbol{\eta}} = \boldsymbol{0} \tag{6.7.64}$$

where $\hat{\boldsymbol{\xi}}$ and $\hat{\boldsymbol{\eta}}$ are defined, respectively, by

$$\hat{\boldsymbol{\xi}} = (\hat{\xi}_1, \ldots, \hat{\xi}_6), \qquad \hat{\boldsymbol{\eta}} = (\hat{\eta}_1, \hat{\eta}_2, \hat{\eta}_3, \hat{\eta}_4, \hat{\eta}_5, \hat{\eta}_6) \tag{6.7.65}$$

with

$$\begin{cases} \hat{\boldsymbol{\xi}}(\boldsymbol{w}', \boldsymbol{w}, \boldsymbol{\alpha}, \boldsymbol{w}(\cdot), s) \equiv (\hat{T}_{11}, w_1\hat{T}_{21}, \hat{T}_{31}, \hat{Q}_1, \hat{D}_1, \hat{B}_1) \\ \hat{\eta}_1(\boldsymbol{w}', \boldsymbol{w}, \boldsymbol{\alpha}, \boldsymbol{w}(\cdot), s) \equiv \hat{T}_{21}w_2' + \alpha_{22}s^{-1}\hat{T}_{22} + \alpha_{23}\hat{T}_{23} \\ \hat{\eta}_2 = \hat{\eta}_3 = \hat{\eta}_6 = 0 \\ \hat{\eta}_4(\boldsymbol{w}', \boldsymbol{w}, \boldsymbol{\alpha}, \boldsymbol{w}(\cdot), s) \equiv \alpha_{52}s^{-1}\hat{J}J_2 + \alpha_{53}\hat{J}_3 \\ \hat{\eta}_5(s) = -\rho \end{cases} \tag{6.7.66}$$

the arguments of the constitutive functions which appear on the right-hand sides in (6.7.66) being $\gamma(s), s$, with $H_i(s)$, $i = 1, 2, 3$, as in (6.7.60). For boundary conditions to go along with (6.7.64), (6.7.66), we have either

$$w_1(1) = \bar{w}_1(1) > 0 \tag{6.7.67}$$

on the cylindrical face $s = 1$ of $\partial\Omega$, with $\bar{w}_1(1)$ a given (positive) number which fixes the outer radius, or we prescribe the traction

$$\hat{\xi}_1(\boldsymbol{w}'(1), \boldsymbol{w}(1), \boldsymbol{\alpha}, \boldsymbol{w}(\cdot), 1) = \bar{\xi}_1(1) \tag{6.7.68}$$

where $\bar{\xi}_1(1)$ is a given number. The deformation is also fixed to within a rigid displacement by setting

$$w_2(1) = w_3(1) = 0 \tag{6.7.69}$$

and the data of the potentials φ and ψ are fixed by requiring that

$$w_4(1) = w_5(1) = 0 \tag{6.7.70}$$

On the cylindrical face at $s = a$, alternative boundary conditions are expressed in an analogous manner and, when both $\bar{w}_1(1)$ and $\bar{w}_1(a)$ are prescribed, we require that $\bar{w}_1(1) > \bar{w}_1(a)$, so as to be consistent with the condition that $w_1' > 0$. If the first relation in (6.7.59) is viewed as a restriction on $\hat{J}_1$, then it turns out that we are not free to prescribe both $w_4(1)$ and $w_4(a)$; for Ω an entire tube, i.e., $\Theta = \pi$, the requirement that $\boldsymbol{z}$, φ, and ψ be continuous is ensured by the conditions

$$\alpha_{22} = \pm 1, \ \alpha_{32} = 0, \ \alpha_{52} = 0, \ \alpha_{62} = 0 \tag{6.7.71}$$

For alternative conditions when $\Theta < \pi$, or $\Theta = \pi$ but the faces at $\Theta = \pm\pi$ are not identified with one another, the reader should consult the discussion in [4]. The analysis in [4] continues with a description of the consequences of the strong ellipticity condition, as it applies to the semi-inverse problems described above, proceeds with the delineation of one additional growth condition, and concludes with the proof of existence of regular solutions under various sets of conditions relative to the precise manner in which the boundary data is prescribed; for the proof of the main result, which we do not reproduce here, the interested reader should consult Theorem 12.2 of [4].

Bibliography

[1] Antman, S. S., "Regular and Singular Problems for Large Elastic Deformations of Tubes, Wedges, and Cylinders", *Arch. Rat. Mech. Anal.*, vol. 83, (1983), 1-52.

[2] Antman, S. S., "The Influence of Elasticity on Analysis: Modern Developments", *Bulletin of the Amer. Math. Soc.*, vol. 9, (1983), 267-291.

[3] Antman, S. S. and C. S. Kenney, "Large Buckled States of Nonlinearly Elastic Rods Under Torsion, Thrust, and Gravity", *Arch. Rat. Mech. Anal.*, vol. 76, (1981), 289-338.

[4] Antman, S. S. and R. C. Rogers, "Steady-State Problems of Nonlinear Electro-magneto-thermo-elasticity", *Arch. Rat. Mech. Anal.*, vol. 95, (1986), 279-323.

[5] Antman, S. S., private communication.

[6] Baldwin, G., *An Introduction to Nonlinear Optics*, (1969), Plenum Press, N.Y.

[7] Ball, J. M., "A Version of the Fundamental Theorem for Young Measures", in *P.D.E.'s and Continuum Models of Phase Transitions, Lecture Notes in Physics*, vol. 344, Rascle, M., Serre, D., and M. Slemrod, eds., 207-215, (1989), Springer-Verlag, N.Y.

[8] Ball, J. M., "Convexity Conditions and Existence Theorems in Nonlinear Elasticity", *Arch. Rat. Mech. Anal.*, vol. 63, (1977), 337-403.

[9] Ball, J. M., "Finite Time Blow-Up in Nonlinear Problems", in *Nonlinear Evolution Equations*, M. G. Crandall, ed., (1978), 189-205, Academic Press, N.Y.

[10] Ball, J. M., "Remarks on Blow-Up and Nonexistence Theorems for Nonlinear Evolution Equations", *Quart. J. Math.*, vol. 28, (1977), 473-486.

[11] Bellout, H., Bloom, F., and J. Nečas, "A Model of Wave Propagation in a Nonlinear, Superconducting Dielectric", *Differential and Integral Equations*, vol. 5, (1992), 1185-1199.

[12] Bellout, H., Bloom, F. and J. Nečas, "Bounds for the Dimensions of the Attractors of Nonlinear Bipolar Viscous Fluids", to appear in *Commun. in P.D.E.*

[13] Bellout, H., Bloom, F., and J. Nečas, "Global Existence of Weak Solutions to the Nonlinear Transmission Line Problem", *Nonlinear Analysis, T.M.A.*, vol. 17, (1991), 903-921.

[14] Bellout, H., Bloom, F. and J. Nečas, "Young Measure Solutions for Non-Newtonian Incompressible Viscous Fluids", to appear in *Commun. in P.D.E.*

[15] Bellout, H. and F. Bloom, "Existence and Asymptotic Behavior of Smooth Solutions for the Nonlinear Transmission Line Problem", *Applicable Analysis*, vol. 39, (1990), 35-57.

[16] Bellout, H. and F. Bloom, "Global Existence and Asymptotic Stability of Solutions to the Cauchy Problem for Wave Propagation in Nonlinear Dielectric Media", to appear in *Nonlinear Analysis, T.M.A.*

[17] Bergman, S., *Integral Operators in the Theory of Linear Partial Differential Equations*, Springer-Verlag, (1971), N.Y.

[18] Bloembergen, N., *Nonlinear Optics*, (1965), W. A. Benjamin, Inc., N.Y.

[19] Bloom, F., "A Condition Implying the Global Existence of $\mathbb{C}^1$ Solutions for a Class of Nonlinear Transmission Lines", *Comput. Math Applic.*, vol. 13, (1987), 861-879.

[20] Bloom, F., "Almost Global Existence in the Plane Wave-Nonlinear Dielectric Interaction Problem", *Comput. Math. Applic.*, vol. 15, (1988), 491-510.

[21] Bloom, F., "Equilibrium States of Self-Interacting Current Bearing Wires", unpublished.

[22] Bloom, F., "Formation of Shock Discontinuities for a Class of Nonlinear Transmission Lines", *Comput. Math. Applic.*, vol. 15, (1988), 459-472.

[23] Bloom, F., "Influence of Nonlinear Conduction on Shock Formation in the Intense Plane-Wave Nonlinear Dielectric Interaction Problem", in *Advances in Hyperbolic Partial Differential Equations*, vol. III, (1986), 477-489, Pergamon Press Ltd., Oxford, U.K.

[24] Bloom, F., "Nonexistence of Smooth Electromagnetic Fields in Nonlinear Dielectrics I: Infinite Cylindrical Dielectrics", *Math Modelling*, vol. 6, (1985), 125-144.

[25] Bloom, F., "Nonexistence of Smooth Electromagnetic Fields in Nonlinear Dielectrics, II: Shock Development in a Half-space", *J. Math. Anal. Applic.*, vol. 115, (1986), 245-261.

[26] Bloom, F., "Systems of Nonlinear Hyperbolic Equations Associated with Problems of Classical Electromagnetic Theory", in *Advances in Hyperbolic Partial Differential Equations*, vol. I, (1985), Pergamon Press, Oxford, U.K.

[27] Bloom, F., *Ill-Posed Problems for Integro-differential Equations in Mechanics and Electromagnetic Theory*, (1981), SIAM, Philadelphia.

[28] Brezis, H., "Équations et Inéquations Nonlinéaires dans les E'spaces Vectoriels en Dualité", Ann. Inst. Fourier, vol. 118, (1968), 115-175.

[29] Broer, L. J. F., "Wave Propagation in Nonlinear Media", *ZAMP*, vol. 16, (1965), 11-26.

[30] Browder, F. E., "Pseudo-monotone Operators and Nonlinear Elliptic Boundary Value Problems on Unbounded Domains", *Proc. Natl. Acad. Sci.*, vol. 74, (1977), 2659-2661.

[31] Brown, W. F., *Magnetoelastic Interactions*, (1966), Springer-Verlag, N.Y.

[32] Chen, G.-Q., "Limit Behaviors of Approximate Solutions to Conservation Laws", preprint.

[33] Chen, G.-Q., "Propagation and Cancellation of Oscillations in Hyperbolic Systems of Conservation Laws", preprint.

[34] Chuch, K. N., Conley, C., and J. A. Smoller, "Positively Invariant Regions for Systems of Nonlinear Diffusion Equations", *Indiana Univ. Math. J.*, vol. 26, (1977), 372-411.

[35] Coleman, B. D., Hrusa, W. J., and D. R. Owen, "Stability of Equilibrium for a Nonlinear Hyperbolic System Describing Heat Propagation by Second Sound in Solids", *Arch. Rat. Mech. Anal.*, vol. 96, (1986), 267-289.

[36] Courant, R., Friedrichs, K. P., and H. Lewy, "Über die Partiellen Differential Gleichungen der Mathematischen Physik", *Math. Ann.*, vol. 100, (1928), 32-74.

[37] Courant, R. and D. Hilbert, *Methods of Mathematical Physics*, vol. 2 (1962), Wiley-Interscience, N.Y.

[38] Cumberbatch, E., "Nonlinear Effects in Transmission Lines", *SIAM J. Appl. Math*, vol. 15, (1967), 450-463.

[39] Dacorogna, B., *Weak Continuity and Weak Lower Semicontinuity of Nonlinear Functionals, Lecture Notes in Math*, vol. 922, (1982), Springer-Verlag, N.Y.

[40] Dafermos, C., "Estimates for Conservation Laws with Little Viscosity", *SIAM J. Math Anal.*, vol. 19, (1987), 409-421.

[41] Dafermos, C., "Hyperbolic Systems of Conservation Laws", in *Systems of Nonlinear Partial Differential Equations*, J. M. Ball, ed., 25-70, (1983), D. Reidel Pub. Co., Dordrecht.

[42] Dafermos, C. M. and J. A. Nohel, "A Nonlinear Hyperbolic Volterra Equation in Viscoelasticity", *Am. J. Math, Supplement*, (1981), 87-116.

[43] Dafermos, C. M. and J. A. Nohel, "Energy Methods for Nonlinear Hyperbolic Volterra Integrodifferential Equations", *Comm. in P.D.E.*, vol. 4, (1979), 219-278.

[44] Dafermos, C., "Solutions in L^∞ for a Conservation Law with Memory", preprint.

[45] Dafermos, C., "Stabilizing Effects on Dissipation", in *Partial Differential Equations and Dynamical Systems*, W. E. Fizgibbon III, ed., (1984), 134-157, Pitman Pub. Inc., Marshfield, Mass.

[46] DeMartini, F., Townes, C. H., Gustafson, T. K., and P. L. Kelley, "Self-Steepening of Light Pulses", *Phys. Rev.*, vol. 164, (1967), 312-323.

[47] DiPerna, R., "Compensated Compactness and General Systems of Conservation Laws", *Trans. A.M.S.*, vol. 292, (1985), 383-419.

[48] DiPerna, R., "Convergence of Approximate Solutions to Conservation Laws", *Arch. Rat. Mech. Anal.*, vol. 82, (1983), 27-70.

[49] DiPerna, R., "Oscillations in Solutions to Nonlinear Differential Equations", in *Oscillation Theory, Computation, and Methods of Compensated Compactness*, Dafermos, C., Ericksen, J., and D. Kinderlehrer, eds., *IMA Volume in Math and Applications*, vol. 2, (1986), Springer-Verlag, N.Y.

[50] Donato, A. and D. Fusco, "Some Applications of the Riemann Method to Electromagnetic Wave Propagation in Nonlinear Media", *ZAMM*, vol. 60, (1980), 539-542.

[51] Douglis, A., "Some Existence Theorems for Hyperbolic Systems of Partial Differential Equations in Two Independent Variables", *Comm. Pure Appl. Math.*, vol. 5, (1952), 119-154.

[52] Eringen, A. C., "On the Foundations of Electroelastodynamics", *Int. J. Eng. Sci.*, vol. 1, (1963), 127-153.

[53] Eringen, A. C., "Theory of Nonlocal Eletromagnetic Solids", *J. Math. Phys.*, vol. 14, (1973), 733-740.

[54] Evans, L. C., *Weak Convergence Methods for Nonlinear Partial Differential Equations, C.B.M.S.*, vol. 74, (1990), Amer. Math. Soc., Providence, R.I.

[55] Feirseisl, E., "Compensated Compactness and Time-Periodic Solutions to Non-Autonomous Quasilinear Telegraph Equations", *Aplikace Matematiky*, vol. 35, (1990), 192-208.

[56] Feirseisl, E., "Time-Independent Invariant Regions for Parabolic Systems Related to One-Dimensional Nonlinear Elasticity", *Aplikace Matematiky*, vol. 35, (1990), 184-191.

[57] Feirseisl, E., "Time-Periodic Solutions to Quasilinear Telegraph Equations with Large Data", *Arch. Rat. Mech. Anal.*, vol. 112, (1990), 45-62.

[58] Friedman, A., *Partial Differential Equations*, (1969), Holt-Rinehart & Winston, N.Y.

[59] Garabedian, P., *Partial Differential Equations*, (1964), Wiley-Interscience, N.Y.

[60] Giaquinta, M., *Multiple Integrals in the Calculus of Variations and Nonlinear Elliptic Systems*, (1983), Princeton Univ. Press, Princeton, N.J.

[61] Glassey, R., "Blow-up Theorems for Nonlinear Wave Equations", *Math. Z.*, vol. 132, (1973), 183-203.

[62] Glassey, R., "Finite-Time Blow-Up for Solutions of Nonlinear Wave Equations", *Math. Z.*, vol. 177, (1981), 323-340.

[63] Grindlay, J., *An Introduction to the Phenomenological Theory of Ferroelectricity*, (1970), Pergamon Press Ltd., Oxford, U.K.

[64] Hagan, R. and J. Serrin, "Dynamic Changes of Phase in a Van der Waal's Fluid", MRC (Univ. of Wisconsin) Tech. Rep. (1984).

[65] Hagan, R. and M. Slemrod, "Viscosity-Capillarity Admissibility Criteria with Applications to Shock and Phase Transitions", *Arch. Rat. Mech. Anal.*, vol. 83, (1983), 333-361.

[66] Hale, J. K., *Ordinary Differential Equations*, (1969), Wiley-Interscience, N.Y.

[67] Hattori, H., "Breakdown of Smooth Solutions in Dissipative Nonlinear Hyperbolic Equations", *Quart. Appl. Math.*, vol. 40, (1982), 113-127.

[68] Hoff, D., "A Characterization of the Blow-Up Time for the Solution of a Conservation Law in Several Space Variables", *Comm. in P.D.E.*, vol. 7, (1982), 141-151.

[69] Hopf, E., "The Partial Differential Equation $u_t + uu_x = \mu u_{xx}$", *Comm. Pure Appl. Math.*, vol. 3, (1950), 201-230.

[70] Hrusa, W. J., "A Nonlinear Functional Differential Equation in Banach Space with Applications to Materials with Fading Memory", *Arch. Rat. Mech. Anal.*, vol. 84, (1983), 99-137.

[71] Hrusa, W. J. and J. A. Nohel, "Global Existence and Asymptotics in One-Dimensional Nonlinear Viscoelasticity", in *Proceedings, 5th Sympos. on Trends in Appl. Pure Mech., Lecture Notes in Physics*, vol. 195, (1984), 165-187, Springer-Verlag, N.Y.

[72] Hutter, K. and A. A. F. Van Deven, *Field Matter Interactions in Thermoelastic Solids, Lecture Notes in Physics*, vol. 88, (1978), Springer-Verlag, N.Y.

[73] Jackson, J. D., *Classical Electrodynamics*, 2nd edition (1975), John Wiley & Sons, N.Y.

[74] Jeffrey, A., "Wave Propagation and Electromagnetic Shock Wave Formation in Transmission Lines", *J. Math. Mech.*, vol. 15, (1966), 1-13.

[75] Jeffrey, A., "Non-Dispersive Wave Propagation in Nonlinear Dielectrics", *ZAMP*, vol. 16, (1968), 741-745.

[76] Jeffrey, A., "The Development of Jump Discontinuities in Nonlinear Hyperbolic Systems of Equations in Two Independent Variables", *Arch. Rat. Mech. Anal.*, vol. 14, (1963), 27-37.

[77] Jeffrey, A., "The Evolution of Discontinuities in Solutions of Homogeneous Nonlinear Hyperbolic Equations Having Smooth Initial Data", *J. Math. Mech.*, vol. 17, (1967), 331-352.

[78] Jeffrey, A. and T. Taniuti, *Nonlinear Wave Propagation*, (1964), Academic Press, N.Y.

[79] Jeffrey, A. and V. P. Korobeinikov, "Formation and Decay of Electromagnetic Shock Waves", *ZAMP*, vol. 20, (1969), 440-447.

[80] John, F., "Delayed Singularity Formation in Solutions of Nonlinear Wave Equations in Higher Dimensions", *Comm. Pure Appl. Math.*, vol. 29, (1976), 649-681.

[81] John, F., "Delayed Singularity Formation in Solutions of Nonlinear Wave Equations in Higher Dimensions", *Comm. in Pure and Appl. Math*, vol. 37, (1984), 433-455.

[82] John, F., "Long Time Effects of Nonlinearity in Second Order Hyperbolic Equations", *Comm. Pure Appl. Math.*, vol. 39, (1986), 139-148.

[83] John, F., "Lower Bounds for the Life Span of Solutions of Nonlinear Wave Equations in Three Dimensions", *Comm. Pure Appl. Math.*, vol. 36, (1983), 1-35.

[84] John, F. and S. Klainerman, "Almost Global Existence to Nonlinear Wave Equations in Three Space Dimensions", *Comm. Pure Appl. Math.*, vol. 37 (1984), 443-455.

[85] John, F., *Partial Differential Equations*, (1982), Springer-Verlag, N.Y.

[86] Jordan, N. F. and A. C. Eringen, "On the Static Nonlinear Theory of Electromagnetic Thermoelastic Solids, I & II", *Int. J. Eng. Sci.*, vol. 2, (1964), 59-114.

[87] Kataev, I. G., *Electromagnetic Shock Waves*, (1966), Iliffe Pub., London, U.K.

[88] Kato, T., "Quasi-linear Equations of Evolutions, with Applications to Partial Differential Equations", *Lecture Notes in Math*, vol. 448, (1975), 25-70.

[89] Kato, T., "The Cauchy Problem for Quasi-Linear Symmetric Hyperbolic Systems", *Arch. Rat. Mech. Anal.*, (1975), 181-205.

[90] Kazakia, J. and R. Venkataraman, "Propagation of Electromagnetic Waves in a Nonlinear Dielectric Slab", *ZAMP*, vol. 26, (1975), 61-76.

[91] Klainerman, S., "On Almost Global Solutions to Quasilinear Wave Equations in Three Space Dimensions", *Comm. Pure Appl. Math.*, vol. 36, (1983), 325-344.

[92] Klainerman, S. and A. Majda, "Formation of Singularities for Wave Equations Including the Nonlinear Vibrating String", *Comm. Pure Appl. Math*, vol. 23, (1980), 241-264.

[93] Knops, R. J., Levine, H. A., and L. E. Payne, "Nonexistence, Instability, and Growth Theorems for Solutions to an Abstract Nonlinear Equation with Applications to Elastodynamics", *Arch. Rat. Mech. Anal.*, vol. 55, (1974), 52-72.

[94] Korteweg, D. J., "Sur la Forme que prenuent les equations des ouvement des fluides si 'on tient comple des forces capillaries par des variations de densite", *Archives Neerlandaises des Sciences Exactes and Naturelles*, Series II, vol. 6, (1901), 1-24.

[95] Lakshmikantham, V. and S. Leela, *Differential and Integral Inequalities*, vol. I, (1969), Academic Press, N.Y.

[96] Landauer, R., "Shock Wave Structure in Nonlinear Dielectrics", *Ferroelectrics*, vol. 10, (1976), 237-240.

[97] Landauer, R., "Shock Waves in Nonlinear Transmission Lines and their Effect on Parametric Amplification", *IBM J.*, vol. 4, (1960), 391-401.

[98] Landauer, R. and S. T. Peng, "Effect of Dispersion on Steady State Electromagnetic Shock Profiles", *IBM J.*, vol. 17, (1973), 299-306.

[99] Lax, P., "Development of Singularities of Solutions of Nonlinear Hyperbolic Partial Differential Equations", *J. Math. Physics*, vol. 5, (1964), 611-613.

[100] Lax, P., "Hyperbolic Systems of Conservation Laws, II", *Comm. Pure Appl. Math*, vol. 10, (1957), 537-566.

[101] Lax, P., "Shock Waves, Increase of Entropy, and Loss of Information", in *Seminar on Nonlinear Partial Differential Equations*, Math. Sci. Res. Inst. Pub., S. S. Chern, ed., 129-173, (1984), Springer-Verlag, N.Y.

[102] Lax, P., "Shock Waves and Entropy", in *Contributions to Nonlinear Functional Analysis*, E. Zarentonello, ed., 603-634, (1971), Academic Press, N.Y.

[103] Lax, P., *Hyperbolic Systems of Conservation Laws and the Mathematical Theory of Shock Waves*, Conf. Board Math. Sci., vol. 11, (1973), SIAM Pub., Philadelphia.

[104] Levine, H. A., "Instability and Nonexistence of Global Solutions to Nonlinear Wave Equations of the Form $Pu_{tt} = -A + F(u)$", *Trans. Amer. Math. Soc.*, vol. 192, (1974), 1-21.

[105] Levine, H. A., "Some Additional Remarks on the Nonexistence of Global Solutions to Nonlinear Wave Equations", *SIAM J. Math. Anal.*, vol. 5, (1974), 138-146.

[106] Levine, H. A. and M. Protter, "The Breakdown of Solutions of Quasilinear First Order Systems of Partial Differential Equations", *Arch. Rat. Mech. Anal.*, vol. 87, (1985), 253-267.

[107] Lindquist, W. B., "The Scalar Riemann Problem in Two Spatial Dimensions: Piecewise Smoothness of Solutions and Its Breakdown", *SIAM J. Math. Anal.*, vol. 17, (1986), 1178-1197.

[108] Ludford, G. S. S., "On an Extension of Riemann's Method of Integration with Applications to One-Dimensional Gas Dynamics", *Proc. Camb. Phil. Soc.*, vol. 48, (1952), 499-510.

[109] Majda, A., *Compressible Fluid Flow and Sytems of Conservation Laws in Several Space Variables*, (1984), Springer-Verlag, N.Y.

[110] Majda, A., *The Existence of Multi-Dimensional Shock Fronts, Memoirs of the Amer. Math. Soc.*, #281, (1983).

[111] Majda, A., *The Stability of Multi-Dimensional Shock Fronts–A New Problem for Linear Hyperbolic Equations, Memoirs of the Amer. Math. Soc.*, #273, (1983).

[112] Matsumura, A., "Global Existence and Asymptotics of the Solutions of the Second Order Quasilinear Hyperbolic Equations with First Order Dissipation", *Publ. Res. Inst. Math Sci.*, Kyoto Univ., ser. A, vol. 13, (1977), 349-379.

[113] Maugin, G. A., (ed.), *The Mechanical Behavior of Electromagnetic Solid Continua*, Proc. IUTAM-IUPAP Symp., (1983), North Holland Pub., N.Y.

[114] Moloney, J. and A. C. Newell, *Nonlinear Optics*, (1991), Addison Wesley Pub., Reading, Mass.

[115] Moon, F., *Magneto-Solid Mechanics*, (1984), Wiley-Interscience, N.Y.

[116] Morawetz, C. S., "An Alternative Proof of DiPerna's Theorem", *Comm. in Pure and Appl. Math*, vol. 44, (1991), 1081-1090.

[117] Morrey, C. B., *Multiple Integrals in the Calculus of Variations*, (1966), Springer-Verlag, N.Y.

[118] Mullick, S. K., "Propagation of Signals in Nonlinear Transmission Lines", *IBM J.*, vol. 11, (1967), 558.

[119] Murat, F., "Compacité par Compensation II", in *Recent Methods in Nonlinear Analysis*, De Giorgi, Magenes, and Mosco, eds., (1979), Pitagora Editrice, Bologna, Italy.

[120] Murat, F., "Compacité par Compensation", *Annali Scuola Norm. Sup. Pisa*, vol. 5, (1978), 485-507.

[121] Murat, F., "H-Convergence", Rapport du Séminaire d'Analyse Fonctionelle et Numérique de l'Université Alger (1978).

[122] Murat, F., "L'injection du Čone Positif de H^{-1} dans $W^{-1,q}$ est Compacte pour tout $q < 2$", *J. Math Pures et. Appl.*, vol. 60, (1981), 309-322.

[123] Nečas, J., *Les Méthodes Directes en Théorie des Équations Elliptiques*, (1967), Academia, Prague.

[124] Nishida, J., "Global Smooth Solutions for the Second-Order Quasilinear Wave Equation with the First-Order Dissipation", *University of Wisconsin, MRC*, Technical Summary Report, No. 1799 (1966).

[125] Nishida, T., *Nonlinear Hyperbolic Equations and Related Topics in Fluid Dynamics*, (1978), Publications Mathematiques D'Orsay 78.02, Université de Paris-Sud.

[126] Nohel, J. A., "A Forced Quasilinear Wave Equation with Dissipation", *Proc. Equadiff. IV, Lecture Notes in Math.*, vol. 703, (1977), 318-327, Springer-Verlag, N.Y.

[127] Nohel, J. A., Rogers, R. C., and A. E. Tzavaras, "Weak Solutions for a Nonlinear System in Viscoelasticity", *Comm. in P.D.E.*, vol. 13, (1988), 97-127.

[128] Ostrovskii, L. A., "Formation and Development of Electromagnetic Shock Waves in Transmission Lines Containing an Unsaturated Ferrite", *Sov. Phys. Tech. Phys.*, vol. 8, (1964), 805-813.

[129] Pao, Y. H., "Electromagnetic Forces in Deformable Continua", in *Mechanics Today*, vol. 4, 209-305, (1978), Pergamon Press, Oxford, U.K.

[130] Parkus, H., (ed.), *Electromagnetic Interaction in Elastic Solids, C.I.S.M.*, vol. 257, (1979), Springer-Verlag, Vienna.

[131] Payne, L. E., *Improperly Posed Problems in Partial Differential Equations*, CBMS Regional Conf. Series in Appl. Math., vol. 22, (1975), SIAM, Philadelphia.

[132] Pego, R. L., "Phase Transitions in One-Dimensional Nonlinear Viscoelasticity: Admissibility and Stability", *Arch. Rat. Mech. Anal.*, 97, (1987), 353-394.

[133] Rascle, M. and D. Serre, "Compacité par Compensation et Systemes Hyperboliques de Lois de Conservation, Applications", *Compt. Rend. Acad. Sci. Paris, Serie I*, vol. 299, (1984), 673-676.

[134] Reitz, J. R., Milford, F. J., and R. W. Christy, *Foundations of Electromagnetic Theory*, 3rd edition, (1979), Addison-Wesley Pub. Co., Reading, Mass.

[135] Riley, R. B., "An Analysis of a Nonlinear Transmission Line", Stanford Solid State Electronics Laboratory Rep. 1707-1, (1961).

[136] Rogers, C., Cekirge, H., and A. Askar, "Electromagnetic Wave Propagation in Nonlinear Dielectric Media", *Acta Mechanica*, vol. 26, (1977), 59-73.

[137] Rogers, R. C., "A Nonlocal Model for the Exchange Energy in Ferromagnetic Materials", *J. of Integral Equations and Applications*, vol. 3, (1991), 85-127.

[138] Rogers, R. C., "Nonlocal Problems in Electromagnetism", in *Metastability and Incompletely Posed Problems*, Antman, S. S., Ericksen, J. L., Kinderlehrer, D., and I. Müller, eds., *IMA Volume in Math and Applications*, vol. 3, (1986), Springer-Verlag, N.Y.

[139] Rogers, R. C., "Nonlocal Variational Problems in Nonlinear Electromagnetoelastostatics", *SIAM J. Math. Anal.*, vol. 19, (1988), 1329-1347.

[140] Rogers, R. C. and D. Brandon, "On the Coercivity Paradox and Nonlocal Ferromagnetism", to appear in *Continuum Mechanics and Thermodynamics.*

[141] Roytburd, V. and M. Slemrod, "Dynamic Phase Transitions and Compensated Compactness", IMA preprint.

[142] Roytburd, V. and M. Slemrod, "Positively Invariant Regions for a Problem in Phase Transitions", *Arch. Rat. Mech. Anal.*, vol. 93, (1986), 61-79.

[143] Rozhdestvenskii, B., "Discontinuous Solutions of Hyperbolic Systems of Quasilinear Equations", *Russian Math Surveys*, vol. 15, (1960), 53-111.

[144] Rudin, W., *Real and Complex Analysis*, (1966), McGraw-Hill, N.Y.

[145] Ryzhov, O. S., "The Decay of Shock Waves in Non-Uniform Media", *P.M.T.F.*, vol. 2, (1961), 15-25.

[146] Salinger, H., "Über die Fortpflanzung von Telegraphierzeichen auf Krarupkabeln", *Arch. Elektrotech.*, vol. 12, (1923), 268-285.

[147] Schelkunoff, S. A., *Electromagnetic Waves*, (1943), Van Nostrand, N.Y.

[148] Schonbek, M., "Applications of the Theory of Compensated Compactness", in *Oscillation Theory, Computation, and Methods of Compensated Compactness*, Dafermos, C., Ericksen, J., and D. Kinderlehrer, eds., *IMA Volume in Math and Applications*, vol. 2, (1986), Springer-Verlag, N.Y.

[149] Schonbek, M., "Convergence of Solutions to Nonlinear Dispersive Equations", *Comm. in P.D.E.*, vol. 7, (1982), 959-1000.

[150] Scott, A. C., "Steady Propagation on Nonlinear Transmission Lines", *IEEE Trans. on Commun. Technology*, CT-11, (1964), 146.

[151] Scott, A. C., *Active and Nonlinear Wave Propagation in Electronics*, (1970), Wiley-Interscience, N.Y.

[152] Serre, D., "Compacité par Compensation et Systemes Hyperboliques de Lois de Conservation", *Compt. Rend. Acad. Sci. Paris, Serie I*, vol. 299, (1984), 555-558.

[153] Serre, D., "La Compacité par Compensation pour les Systèmes Hyperboliques Nonlinéaires à une Dimension d'espace", *J. Math Pure et Appl.*, vol. 65, (1987), 423-468.

[154] Serre, D., "Solutions a Variation Borneé pour Certains Systemes Hyperboliques de Lois de Conservation", *J. Diff. Eqs.*, vol. 67, (1987).

[155] Serrin, J., "Phase Transitions and Interfacial Layers for Van der Waal's Fluids", in *Recent Methods in Nonlinear Analysis and Applications*, A. Canfora, ed., 169-175, (1980), Instituto di Mathematica, Universitá di Napoli.

[156] Serrin, J., "The Form of Interfacial Surfaces in Korteweg's Theory of Phase Equilibria", *Quart. Appl. Math.*, vol. 41, (1983), 357-364.

[157] Sideris, T. C., "Formation of Singularities in Solutions to Nonlinear Hyperbolic Equations", *Arch. Rat. Mech. Anal.*, vol. 86, (1984), 369-381.

[158] Slemrod, M., "Admissibility Criteria for Propagating Phase Boundaries in a Van der Waal's Fluid", *Arch. Rat. Mech. Anal.*, vol. 81, (1983), 301-315.

[159] Slemrod, M., "An Admissibility Criterion for Fluids Exhibiting Phase Transitions", in *Systems of Nonlinear P.D.E.*, J. M. Ball, ed., (1983), D. Reidel Pub. Co., Dordrecht.

[160] Slemrod, M., "Dynamic Phase Transitions in a Van der Waal's Fluid", *J. Diff. Eqs.*, vol. 52, (1984), 1-23.

[161] Slemrod, M., "Dynamics of First Order Phase Transitions", in *Phase Transformations and Material Instabilities in Solids*, M. E. Gurtin, ed., 163-204, (1984), Academic Press, N.Y.

[162] Slemrod, M., "Global Existence, Uniqueness, and Asymptotic Stability of Classical Smooth Solutions in One-Dimensional, Nonlinear Thermoelasticity", *Arch. Rat. Mech. Anal.*, vol. 76, (1981), 97-133.

[163] Slemrod, M., "Instability of Steady Shearing Flows in a Non-linear Viscoelastic Fluid", *Arch. Rat. Mech. Anal.*, vol. 68, (1978), 211-255.

[164] Slemrod, M., "Interrelationships Among Mechanics, Numerical Analysis, Compensated Compactness, and Oscillation Theory", in *Oscillation Theory, Computation, and Methods of Compensated Compactness*, Dafermos, C., Ericksen, J., and D. Kinderlehrer, eds., *IMA Volume in Math and Applications*, vol. 2 (1986), Springer-Verlag, N.Y.

[165] Smoller, J., *Shock Waves and Reaction-Diffusion Equations*, (1982), Springer-Verlag, N.Y.

[166] Stratton, J. A., *Electromagnetic Theory*, (1941), McGraw-Hill Pub., N.Y.

[167] Szepessy, A., "Measure-Valued Solutions of Scalar Conservation Laws with Boundary Conditions", *Arch. Rat. Mech. Anal.*, vol. 107, (1989), 181-193.

[168] Tartar, L., "Compacité par Compensation: Résultats et Perspectives", in *Nonlinear P.D.E. and Their Applications: Collége de France Seminar*, vol. IV, (1983), Pitman Pub., Marshfield, Mass.

[169] Tartar, L., "Compensated Compactness and Applications to Partial Differential Equations", in *Heriot-Watt Symposium*, vol. IV, (1979), Pitman Publ., Marshfield, Mass.

[170] Tartar, L., "The Compensated Compactness Method Applied to Systems of Conservation Laws", in *Systems of Nonlinear P.D.E.*, J. M. Ball, ed., 263-285, (1983), D. Reidel Pub. Co., Dordrecht.

[171] Tartar, L., *Cours Peccot*, Collége de France (1977).

[172] Toupin, R. A., "A Dynamical Theory of Elastic Dielectrics", *Int. J. Eng. Sci.*, vol. 1, (1963), 101-126.

[173] Toupin, R. A., "The Elastic Dielectric", *J. Rational Mech. Anal.*, vol. 5, (1956), 849-915.

[174] Varley, E., Mortell, M. P., and A. Trowbridge, "Modulated Simple Waves", in *Wave Propagation in Solids*, 95-114, (1969), A.S.M.E., N.Y.

[175] Venkataraman, E. and R. S. Rivlin, "Propagation of First Order Electromagnetic Discontinuities in an Isotropic Medium", Tech. Rep. No. CAM-110-9, (1970), Lehigh University.

[176] Verhulst, F., *Nonlinear Differential Equations and Dynamical Systems*, (1990), Springer-Verlag, N.Y.

[177] Wang, J., "Wave Propagation Problem in a Nonlinear Transmission Line", preprint.

[178] Wang, J. and Li, C., "Globally Smooth Resolvability and Formation of Singularities of Solutions for Certain Quasilinear Hyperbolic Systems with a Dissipative Term", preprint.

[179] Wang, J., private communication.

[180] Weinan, E. and R. V. Kohn, "The Initial-Value Problem for Measure Valued Solutions of a Canonical 2×2 System with Linearly Degenerated Fields", *Comm. in Pure and Appl. Math*, vol. 44, (1991), 981-1000.

[181] Weinberger, H. F., *A First Course in Partial Differential Equations*, (1975), Xerox Publishing Co.

[182] Werner, K.-D., "The Evolution of Discontinuities in Solutions of Inhomogeneous Scalar Hyperbolic Conservation Laws in Several Space Dimensions", *Applic. Anal.*, vol. 31, (1988), 11-33.

[183] Wolfe, P., "Bifurcation Theory of an Elastic Conducting Rod in a Magnetic Field", *Quart. J. Mech. Appl. Math.*, vol. 41, (1988), 265-279.

[184] Wolfe, P., "Equilibrium States of an Elastic Conductor in a Magnetic Field; a Paradigm of Bifurcation Theory", *Trans. Amer. Math. Soc.*, vol. 278, (1983), 377-387.

[185] Wolfe, P. and T. Seidman, "Equilibrium States of an Elastic Conducting Rod in a Magnetic Field", *Arch. Rat. Mech. Anal.*, vol. 106, (1988), 307-329.

[186] Yariv, A., *Quantum Electronics*, (1967), John Wiley & Sons, N.Y.

[187] Yoshida, K., *Functional Analysis*, 6th ed., (1980), Springer-Verlag, N.Y.